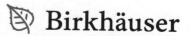

Trends in the History of Science

Trends in the History of Science is a series devoted to the publication of volumes arising from workshops and conferences in all areas of current research in the history of science, primarily with a focus on the history of mathematics, physics, and their applications. Its aim is to make current developments available to the community as rapidly as possible without compromising quality, and to archive those developments for reference purposes. Proposals for volumes can be submitted using the online book project submission form at our website www.birkhauser-sci ence.com.

More information about this series at https://link.springer.com/bookseries/11668

Michael Friedman · Karin Krauthausen
Editors

Model and Mathematics: From the 19th to the 21st Century

 Birkhäuser

Editors
Michael Friedman
The Cohn Institute for the History
and Philosophy of Science and Ideas
Humanities Faculty
Tel Aviv University
Ramat Aviv, Tel Aviv, Israel

Karin Krauthausen
Cluster of Excellence "Matters of Activity.
Image Space Material"
Humboldt-Universität zu Berlin
Berlin, Germany

The editors acknowledge the support of the Cluster of Excellence "Matters of Activity. Image Space Material" funded by the Deutsche Forschungsgemeinschaft (DFG, German Research Foundation) under Germany's Excellence Strategy – EXC 2025 – 390648296.

Matters of Activity · **Image Space Material**

ISSN 2297-2951 ISSN 2297-296X (electronic)
Trends in the History of Science
ISBN 978-3-030-97835-8 ISBN 978-3-030-97833-4 (eBook)
https://doi.org/10.1007/978-3-030-97833-4

This book is published under the imprint Birkhäuser, www.birkhauser-science.com by the registered company Springer Nature Switzerland AG
The registered company address is: Gewerbestrasse 11, 6330 Cham, Switzerland

Contents

Part II: Epistemological and Conceptual Perspectives

Part III: From Production Processes to Exhibition Practices

How to Grasp an Abstraction: Mathematical Models and Their Vicissitudes Between 1850 and 1950. Introduction

Michael Friedman and Karin Krauthausen

What is a model? Today, this question can only be answered either with a high degree of abstraction or generality, or with the most specific and precise contextualization, since the concept of the model and the practice of modeling are ubiquitous—in all the sciences and arts, in engineering and design. To underline this fact, already in 2003, the model theorist Bernd Mahr suggested that models could "become the semantic, combinatorial, and technical foundation of our culture, just as this was the case with numbers through mathematics and information technology [...]."[1] A brief look at the history of the model concept reminds us that the essential ambivalence attributed to models in the twentieth and twenty-first

---

[1] Bernd Mahr, "Modellieren: Beobachtungen und Gedanken zur Geschichte des Modellbegriffs," in *Bild, Schrift, Zahl*, ed. Sybille Krämer and Horst Bredekamp (Munich: Wilhelm Fink, 2003), 59–86, here 60: "in ähnlicher Weise zum semantischen, kombinatorischen und technischen Fundament unserer Kultur werden [könnten], wie dies die Zahlen durch die Mathematik und die Informationstechnik geworden sind [...]." Whereas in the present volume the object area is limited to mathematical models (mainly, if not entirely, between 1850 and 1950), and the authors have mostly chosen a historical approach, the investigations carried out by Mahr, a mathematician and computer scientist, are geared to an epistemology of the model. Mahr takes cultural history as a starting point to develop a generally valid logical-formal description of 'the nature of the model.' See also: Bernd Mahr, "On the Epistemology of Models," in *Rethinking Epistemology*, vol. 1, ed. Günter Abel and James Conant (Berlin and Boston: de Gruyter, 2011), 301–52.

M. Friedman (✉)
The Cohn Institute for the History and Philosophy of Science and Ideas, The Lester and Sally Entin Faculty of Humanities, Tel Aviv University, Ramat Aviv, 6997801 Tel Aviv, Israel
e-mail: friedmanm@tauex.tau.ac.il

K. Krauthausen (✉)
Cluster of Excellence "Matters of Activity. Image Space Material," Humboldt-Universität zu Berlin, Unter den Linden 6, 10099 Berlin, Germany
e-mail: karin.krauthausen@hu-berlin.de

© The Author(s) 2022
M. Friedman and K. Krauthausen (eds.), *Model and Mathematics: From the 19th to the 21st Century*, Trends in the History of Science,
https://doi.org/10.1007/978-3-030-97833-4_1

centuries—the ambivalence between concretion and abstraction (with the emphasis moving increasingly in the direction of abstraction)—can be traced much further back. Etymologically, the word 'model' is derived from the Latin *modulus*, the diminutive of *modus*. Whereas *modus* generally stands for 'measure' (also temporal measure), 'measuring stick,' and 'quantity,' as well as for 'aim,' 'rule,' or 'manner,' *modulus* (in Vitruvius, but also in the early Middle Ages) is essentially determined via the practice of architecture, where it stands in a technical sense for the dimensions of columns or the relations of their parts.[2] Both terms belong to the context of 'form giving' and design, but whereas *modus* "is a conceptual term designating something abstract that is posited and not given," *modulus* refers to "something concrete."[3] In the Italian architecture of the fifteenth and sixteenth centuries, the practical-concrete context of the model (or *modello*) becomes clearer with the growing importance of three-dimensional scale models of future (as well as finished) architectural projects, which, in the case of larger, more elaborate projects may have been used to win over a client—a famous example being the competition in 1418 for a model of the dome of Florence Cathedral (see Fig. 1).[4] While not providing a direct blueprint of the construction to be built, these models, which were mostly made from wood, acted as haptic-concrete elements in the design process. Despite their only moderate accuracy, they provided a convincing description of the construction's form—that is, they permitted a summary view of a future to be realized, but one that was sufficiently approximate in the detail to allow for adjustments and changes.

Later, three-dimensional models of this kind were also found in the natural sciences—an example is the series of crystal models constructed by Jean-Baptiste Louis Romé de l'Isle at the end of the eighteenth century (Fig. 2). Rather than representing a step in a design process for a building to be realized, however, these models are part of an epistemic process of 'form giving,' they visualize knowledge about crystals and allow both students and trained scientists to obtain (in combination with verbal explanations and visual representations in related treatises) an overview. In this way, they allow the student and scientist not only to acquire existing knowledge, but also to explore new lines of research. In the late 18th and early nineteenth centuries, crystallographers such as Jean-Baptiste Louis Romé de l'Isle and René Just Haüy manufactured numerous such models of

---

[2] Mahr, "On the Epistemology of Models," 255–59.

[3] Mahr, "Modellieren," 61: "ein Begriffswort ist, das etwas Abstraktes bezeichnet, das gesetzt ist und nicht vorgefunden wird;" "etwas Konkretes." For the English usage of the word, see the entry "Model" in *The Oxford English Dictionary*, vol. VI (Oxford: Clarendon Press, 1933), 568–69, as well as "model, n. and adj." in *The Oxford English Dictionary*, OED Online (Oxford University Press, December 2021): https://www.oed.com/view/Entry/120577?rskey=Hv9gMy&result=1&isAdvanced=false (accessed December 6, 2021).

[4] See: Andres Lepik, *Das Architekturmodell in Italien 1335–1550*, dissertation, series *Römische Studien der Bibliotheca Hertziana*, vol. 9 (Worms: Wernersche Verlagsgesellschaft, 1994)—on models in the process of designing and constructing Florence Cathedral, 59–89.

Fig. 1 Giorgio Vasari, "Filippo Brunelleschi and Lorenzo Ghiberti Presenting the Model of the Church of San Lorenzo (Florence)," ca. 1556–1558. Fresco in the Palazzo Vecchio, Florence. Photo: Peter Horree, 2017. © Alamy, all rights reserved

crystals made of paper, wood, or terra-cotta.[5] For Haüy, these models, which represented theoretical, idealized minerals, were "amenable to mathematical abstraction and geometrical analysis."[6] Yet scientific models also acted as prestigious objects intended for public and private collections. Moreover, in the nineteenth century, they played an important role in the self-promotion of university departments.

---

[5] On the history of crystallography, see: John G. Burke, *Origins of the Science of Crystals* (Berkeley: University of California Press, 1966); Henk Kubbinga, "Crystallography from Haüy to Laue: Controversies on the Molecular and Atomistic Nature of Solids," *Acta Crystallographica Section A: Foundations of Crystallography* 68, no. 1 (2012): 3–29. See also: Marjorie Senechal, "Brief History of Geometrical Crystallography," in *Historical Atlas of Crystallography*, ed. José Lima-de-Faria (Dordrecht, Boston, and London: Kluwer Academic Publishers, 1990), 43–59—on the paper models of crystals made by Nicolas Steno in 1669, 55.

[6] Lydie Touret, "Crystal Models: Milestone in the Birth of Crystallography and Mineralogy as Sciences," in *Dutch Pioneers of the Earth Sciences*, ed. Jacques L. R. Touret and Robert P. W. Visser (Amsterdam: Koninklijke Nederlandse Akademie van Wetenschappen, 2004), 43–58, here 57.

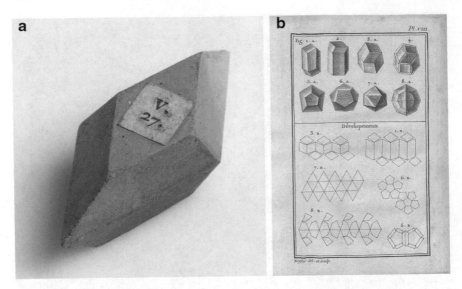

Fig. 2 **a** One of the 448 crystal models of unglazed porcelain made by Romé de L'Isle in Paris in ca. 1780. The complete collection was bought for Teylers Museum by Martinus van Marum in 1785 (Teylers Museum, Haarlem). CC BY-SA 3.0 NL (https://creativecommons.org/licenses/by-sa/3.0/nl/deed.en). **b** Planche VIII from Jean-Baptiste Louis Romé de L'Isle's *Essai de cristallographie*, 1772. (ETH Library Zürich, Rar 2708, https://doi.org/10.3931/e-rara-16480) Public Domain Mark)

Hence, alongside their analytic-epistemic function, these models had a strategic and political function within the university system: scientific collections became a way for institutes and universities to call attention to themselves—a task assumed in the twentieth and twenty-first centuries principally by the complex technologies of instruments and laboratory machines (such as the various large and small electron microscopes).[7]

But what is meant when one speaks of mathematical models in the nineteenth and twentieth centuries? That is the subject of the present volume, and this question is answered by the contributors to this volume in detailed historical studies of mathematical model practices in the long nineteenth century in France and Germany, and, going beyond this initial focus, in a series of interviews

---

[7] On the historical and current use of models in the sciences, see: Soraya Chadarevian and Nick Hopwood, eds., *Models: The Third Dimension of Science* (Stanford: Stanford University Press, 2004). On scientific university collections: David Ludwig, Cornelia Weber, and Oliver Zauzig, eds., *Das materielle Modell: Objektgeschichten aus der wissenschaftlichen Praxis* (Paderborn: Fink, 2014). On science policies in the nineteenth century, see the exemplary study on the mathematician and model builder Walther Dyck in: Ulf Hashagen, *Walther von Dyck (1856–1934): Mathematik, Technik und Wissenschaftsorganisation an der TH München* (Stuttgart: Steiner, 2003). On the history of material mathematical models in Germany at the end of the nineteenth century, see: Anja Sattelmacher, *Anschauen, Anfassen, Auffassen: Eine Wissensgeschichte Mathematischer Modelle* (Wiesbaden: Springer, 2021).

on model practices in the sciences of the nineteenth and twentieth centuries as well as on twenty-first century digital visualization techniques.[8] However, there can and will be no single answer that encompasses all mathematical models: mathematical models have been and continue to be so productive because their definition, function, and appearance was and is capable of change. When the talk is of mathematical models, then, from a historical perspective, these include both the haptic-concrete model constructions of the nineteenth century[9] and the abstract, formal model concepts of the twentieth century. The central concern of the present volume is to show the historical diversity and capacity for change as well as the pedagogical and epistemic importance of mathematical models.

In the following introduction we concentrate on the ambivalent meaning of mathematical models between concretion and abstraction. This is examined by means of a few paradigmatic protagonists, concepts, and concrete examples from the period between 1860 and 1950.[10] Here, it will be necessary, for the nineteenth

---

[8] Less space is given over in this volume to model practices in England. On these practices, see: June Barrow-Green, "'Knowledge Gained by Experience': Olaus Henrici—Engineer, Geometer and Maker of Mathematical Models," *Historia Mathematica* 54 (2021): 41–76—especially on the model tradition in Britain, 57–63. It is important to stress that James Clerk Maxwell, whose conception of models and modeling will be dealt with extensively in this introduction, was also interested in material mathematical models. As Barrow-Green notes, the Cambridge Modelling Club, founded in 1873, "[which] had its first meeting in February the following year, was founded 'to promote the making of models, machines and drawings, illustrative of geometry,' and its members included [Arthur] Cayley [...] and Maxwell (custodian of the models) [...]" (ibid., 59). However, she concludes that "there was considerable interest in models of surfaces in Britain in the half-century between 1860 and 1910. But equally evident is the fact that only a small number of mathematicians were involved in making such models, and those that did so, did so only occasionally." (Ibid., 63).

[9] On material mathematical models, see for example the following exemplary books: Gerd Fischer, ed., *Mathematical Models: From the Collections of Universities and Museums*, 2 vols. (Braunschweig: Friedrich Vieweg und Sohn, 1986); and more recently: Institut Henri Poincaré, ed., *Objets mathématiques* (Paris: CNRS Éditions, 2017); Ernst Seidl, Frank Loose, and Edgar Bierende, eds., *Mathematik mit Modellen: Alexander von Brill und die Tübinger Modellsammlung* (Tübingen: Museum der Universität Tübingen MUT, 2018).

[10] This can be done only in a very cursory way in an introduction—also with regard to the research literature. Apart from the research already mentioned in the footnotes or still to be discussed, we would like to refer here to the following anthologies and individual studies: Hans von Freudenthal, ed., *The Concept and Role of the Model in Mathematics and Natural and Social Sciences* (Dordrecht: Springer, 1961); Gert Schubring, "Searches for the Origins of the Epistemological Concept of Model in Mathematics," *Archive for History of Exact Sciences* 71 (2017): 245–78. On the theory and philosophy of the model, see in addition: Mary B. Hesse, *Models and Analogies in Science* (London: Sheed and Ward Ltd., 1963); Herbert Stachowiak, *Allgemeine Modelltheorie* (Vienna: Springer, 1973); Klaus Hentschel, "Die Funktion von Analogien in den Naturwissenschaften, auch in Abgrenzung zu Metaphern und Modellen," in *Analogien in Naturwissenschaften, Medizin und Technik*, series *Acta Historica Leopoldina*, vol. 56, ed. Klaus Hentschel (Stuttgart: Wissenschaftliche Verlagsgesellschaft, 2010), 13–66; Axel Gelfert, *How to Do Science with Models: A Philosophical Primer* (Vienna: Springer, 2016); Lorenzo Magnani and Tommaso Bertolotti, eds.,

century, to refer to models situated at the interface between physics and mathematics. In this way, it can be shown whether and which ontological questions are raised by the mathematical objects in each respective context (see section I.) and what meaning is attached to the models, taken by themselves and in relation to James Clerk Maxwell's notion of analogy, in research and theory (in section II.). From the question of *what* models represent, we then move on to considerations by prominent nineteenth century mathematicians and model builders (such as Felix Klein and Alexander Brill) about *how* models represent (section III.), before tracing the paradigm shift from '*Anschauung*' (intuition) and '*Bild*' (image) to 'word' or 'text' and 'formal logic' in the work of Felix Hausdorff and Alfred Tarski (section IV.(1)). In a final step we then briefly refer, with the example of mathematical biology, to the changed use of mathematical models in the natural sciences (section IV.(2)) and, with the example of structural anthropology (Claude Lévi-Strauss), to the deployment of mathematics and the model concept in the humanities of the first half of the twentieth century (section V.). The sequence of historical stations discussed in the introduction should show the development of mathematical models toward an increasing conceptual plurality. Whether and to what extent this plurality entailed an epistemic gain will be discussed at the end of the introduction (section VI.), before we close with an overview of the contributions to this volume.

I. Models at the End of the Nineteenth Century: Between Maxwell's 'Fictitious Substances' and Boltzmann's 'Tangible Representation'

When talking about mathematical models at the end of the nineteenth century, it is important to be clear which objects are being referred to. This concerns not only particular examples of models but also the history of the term 'model' within the field of mathematics. In nineteenth-century mathematics, 'model' clearly referred to material, three-dimensional objects, that is, to objects that one could pick up in one's hands—which is somewhat different from our current use of the term.

The nineteenth century understanding of the term 'model' is historically documented—for example, in Ludwig Boltzmann's article "Models" published in

---

Springer Handbook of Model-Based Science (Dordrecht et al.: Springer, 2017). For a comparison between models in the natural sciences and in the arts: Horst Bredekamp, "Modelle der Kunst und der Evolution," in *Debatte* 2 (2005), issue: *Modelle des Denkens*, ed. Präsident der Berlin-Brandenburgischen Akademie der Wissenschaften and Sonja Ginnow: 13–20, https://edoc.bbaw. de/opus4-bbaw/frontdoor/index/index/year/2007/docId/498 (accessed December 2, 2021); Reinhard Wendler, *Das Modell zwischen Kunst und Wissenschaft* (Paderborn: Fink, 2013). For the most recent and extensive review on the role and epistemic function of models in science, see: Roman Frigg and Stephan Hartmann, "Models in Science," in *The Stanford Encyclopedia of Philosophy* (Spring 2020 Edition), ed. Edward N. Zalta, https://plato.stanford.edu/archives/spr2020/ entries/models-science/ (accessed December 2, 2021).

1902 in the *Encyclopædia Britannica.*[11] In his article, Boltzmann, an Austrian physicist known for his discoveries in thermodynamics and the founding of statistical mechanics, provides a paradigmatic summary of the term 'model' as it was understood in his time. He starts with a general definition that attaches particular importance to the concrete, material execution: "The term model denotes a tangible representation, whether the size be equal, or greater, or smaller, of an object which is either in actual existence, or has to be constructed in fact or in thought."[12]

For Boltzmann, a model is a physical entity with definite spatial relations. Following on from this definition, he first distinguishes generally between stationary and moving models, before looking more closely at speculative 'kinematic' models (James Clerk Maxwell's models of the purely hypothetical particle motion of matter) and the more concrete categories of working models (showing the functioning of machines) and experimental models (for technical inventions). Finally, he looks briefly at the instrumental-mathematical and in no way trivial models found in the context of physics. These models facilitate complex calculations and (via a 'physical analogy'[13]) make such calculations possible in the first place.[14] In his examples and classifications, however, Boltzmann concentrates on the models in physics and mathematics, to which he attaches particular importance.[15] He describes mathematical models as follows: "In pure mathematics, especially geometry, models constructed of *papier-mâché* and plaster are chiefly employed to present to the senses the precise form of geometrical figures, surfaces, and curves."[16]

Here, it is important to note how restrictively Boltzmann describes the appearance of mathematical models: these are three-dimensional (i.e. haptic-concrete and material) models, which he understands as "precise" presentations of geometric figures. The function of these models was as an aid in teaching and research, whereby the modeling principally involved surfaces of second and third degree, and in the case of third order surfaces the model becomes a "complicated, not to say hazardous, construction" (see Fig. 3).[17]

---

[11] See: Ludwig Boltzmann, "Models," in *Encyclopædia Britannica*, 10th ed., vol. 30 (Edinburgh and London: Adam & Charles Black, 1902), 788–91. We (M.F./K.K.) presented the following remarks on Boltzmann's encyclopedia article at the conference "Models and Simulations 8" (March 15–18, 2018) at the University of Columbia in Columbia, South Carolina. For a more recent and systematic discussion of Boltzmann's text against the background of Maxwell's understanding of hypothesis, dynamical illustration, and model, see: Giora Hon and Bernard R. Goldstein, "Maxwell's Role in Turning the Concept of Model into the Methodology of Modeling," *Studies in the History and Philosophy of Science* 88 (2021): 321–33.

[12] Boltzmann, "Models," 788.

[13] We return to the concept of analogy with respect to models and modeling below.

[14] Boltzmann, "Models," 788–91.

[15] Ibid., 788: "Models in the mathematical, physical, and mechanical sciences are of the greatest importance."

[16] Ibid., 789.

[17] Ibid. Here, Boltzmann initially names stationary mathematical models, later adding moving models that "show the origin of geometrical figures from the motion of others," for instance several thread models that show the origin of surfaces in moving lines.

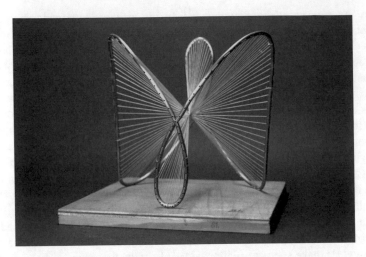

Fig. 3 Thread model by Hermann Wiener of a cone of third order and genus 0 (Georg-August-Universität Göttingen, Collection of Mathematical Models and Instruments, model 81). See: Martin Schilling, ed., *Catalog mathematischer Modelle*, 7th ed. (Leipzig: Verlag von Martin Schilling, 1911), 122, no. 79. © Collection of Mathematical Models and Instruments at Georg-August-University Göttingen, all rights reserved

In Germany beginning in the 1870s, three-dimensional mathematical models of this kind were produced in large series and sold through catalogues such as the *Catalog mathematischer Modelle* distributed from 1881 onward by the publishing house of Ludwig Brill.[18] These catalogues contained models designed by, among others, the mathematicians Alexander Brill (Ludwig Brill's brother) and Felix Klein (see Fig. 4). Nevertheless, such models could (and can) also be seen in the scientific collections in the mathematical departments of universities. Mathematical models were also presented outside of the universities, however, particularly in exhibitions of scientific instruments, models, and apparatuses, such as the Special Loan Exhibition of Scientific Apparatus in South Kensington, London, in 1876, and an exhibition organized by Walther Dyck in Munich in 1893, accompanied by the publication *Katalog mathematischer und mathematisch-physikalischer Modelle, Apparate und Instrumente*.

Boltzmann, however, speaks not only about three-dimensional models of surfaces in mathematics but also about models of surfaces in physics, above all in thermodynamics, where they mathematically represented the behavior of gases and fluids:

---

[18] This is not to imply that this tradition began in Germany during the 1870s; one of its beginnings may be located in France as early as the 1800s, with the rise of the *géométrie descriptive* of Gaspard Monge. Mathematicians in England and Germany followed this tradition; and from the 1870s at the latest, one sees a lively modeling practice and also an ongoing discourse on mathematical models, as this introduction and this volume intend to show.

44 V. Krümmung d. Flächen: C. Flächen von constanter mittlerer Krümmung, Minimalflächen:

einer Geraden, der Asymptote, constante Länge besitzen). Diese Fläche bildet den Uebergang zwischen den beiden vorgenannten Flächen und entspricht der Kugel bei den Flächen constanter positiver Krümmung. Die blau gezeichneten Curven auf ihr sind verschiedene geodätische Linien, die rothe ist eine Asymptotencurve. Von stud. math. Bacharach. Erläuterung beigegeben. (19—24 cm.) . . . ℳ. 9. —.

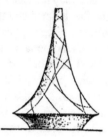

135. (V. xv.) Schraubenfläche von constantem negativen Krümmungsmass, deren Meridiancurve die Tractrix ist. Sie ist die einzige Schraubenfläche von der erwähnten Art, in deren Gleichung nicht elliptische Funktionen eintreten. (Bei Flächen constanter positiver Krümmung gibt es keine von dieser Eigenschaft.) Vergl. U. Dini, Comptes Rendus, Acad. Sc. Paris 1865, 1 Sem. pag. 340; Th. Kuen, Berichte der kgl. bayr. Acad. 1884. Von Dr. P. Vogel. Erläuterung beigegeben. (15—24 cm.) . . ℳ. 15. 50.

136. (VIII. xx.) Fläche von constantem negativen Krümmungsmass mit ebenen Krümmungslinien. Sie entsteht aus der Tractrixfläche dadurch, dass man auf den Tangenten an ein System von parallelen geodätischen Linien (Krümmung derselben $-\frac{1}{\rho^2}$) das Stück t in be-

stimmtem Sinn aufträgt. Die Fläche besitzt eine ebene und eine räumliche Rückkehrkante mit 2 Spitzen, sowie eine Doppelcurve. Das eine System von Krümmungslinien wird von Ebenen ausgeschnitten, welche durch eine (im Modell vertikal gestellte) Gerade hindurch gehen. Das andere System liegt auf Kugeln, deren Mittelpunkte in dieser Geraden liegen. (Vergl. Bianchi, Math. Annalen Bd. 16, sowie Enneper, Göttinger Nachrichten 1868; Th. Kuen, Sitzungsberichte der kgl. bayr. Acad. 1884, Heft II.) Modellirt von stud. math. Mack; Erläuterung hierzu von Assistent Th. Kuen. (16—25 cm.) . ℳ. 16. —.

137. (VIII. xxvi a.) Schraubenfläche, auf das Rotationsellipsoid Nr. 138 abwickelbar (nach E. Bour, Journal de l'Ecole Polyt. Bd. XXII.). Von Assistent Dr. P. Vogel. (12—26 cm.) ℳ. 10. 50.

138. (VIII. xxvi c.) Rotationsellipsoid, auf die vorige Fläche abwickelbar. (9—3 cm.) ℳ. 1. 50.

139. (VIII. xxvi b.) Rotationsellipsoid aus biegsamem Messingblech zur Demonstration der erwähnten Abwicklung. Die durch 2 gleich grosse Parallelkreise begrenzte Zone obigen Ellipsoids geht durch einen leichten Druck in die vorhin erwähnte Schraubenfläche über. ℳ. 2. 50.

In diesen Abschnitt gehört noch die in der folgenden Abtheilung aufgezählte windschiefe Schraubenfläche Nr. 145, welche auf das Catenoïd abwickelbar ist.

C. Flächen von constanter mittlerer Krümmung, Minimalflächen.

Die Flächen von constanter mittlerer Krümmung sind dadurch definirt, dass die Summe der reciproken Werthe ihrer 2 Hauptkrümmungsradien an jeder Stelle denselben Zahlenwerth

Fig. 4 Page from Ludwig Brill's *Catalog mathematischer Modelle* (Ludwig Brill, ed., *Catalog mathematischer Modelle für den höheren mathematischen Unterricht*, 3rd ed. (Darmstadt: L. Brill, 1885), 44). Note that models were usually depicted next to their descriptions, and that these depictions were only rarely photos

In thermodynamics, [...] models serve, among other purposes, for the representation of the surfaces which exhibits the relation between the three thermodynamic variables of a body, e.g., between its temperature, pressure and volume. A glance at the model of such a thermodynamic surface enables the behaviour of a particular substance under the most varied conditions to be immediately realized.[19]

Here, while Boltzmann does not explicitly mention which models and which model builders he is referring to, one may well be reminded of Maxwell's clay model of a thermodynamic surface from 1874.[20] Maxwell (1831–1879), one of the most important physicists of the nineteenth century, dedicated himself to the study of electromagnetism and the kinetic theory of gases. His clay model represents a substance with water-like properties by reproducing the latter's energy-entropy-volume coordinates, thus allowing the different possible states (gaseous, liquid, solid) to be read. This modeling of thermodynamic phenomena is based on a diagrammatic and geometric precedent—the 'graphical method' of the American mathematical physicist Josiah Willard Gibbs, who proposed this in 1873 for the combined representation of the properties of volume, pressure, temperature, energy, and entropy of a given body in any state (see Fig. 5a, b).[21] On his own three-dimensional realization of Gibb's geometric method of representation, Maxwell writes:

[Regarding] Prof. J. Willard Gibbs's [...] graphical methods in thermodynamics[:] [...] I made several attempts to model the surface, which he suggests, in which the three coordinates are volume, entropy and energy. The numerical data about entropy can only be obtained by integration from data which are for most bodies very insufficient, and besides it would require a very unwieldy model to get all the features, say of CO_2, well represented, *so I made no attempt at accuracy, but modelled a fictitious substance* [...].[22]

---

[19] Boltzmann, "Models," 789. Here is the complete quotation: "When the ordinate intersects the surface but once a single phase only of the body is conceivable, but where there is a multiple intersection various phases are possible, which may be liquid or gaseous. On the boundaries between these regions lie the critical phases, where transition occurs from one type of phase into the other. If for one of the elements a quantity which occurs in calorimetry be chosen—for example, entropy—information is also gained about the behaviour of the body when heat is taken in or abstracted." Here, one can claim that Boltzmann is suggesting making calculations on the model that would be too complex from a purely mathematical point of view.

[20] Moreover, Maxwell was not the first to construct such a model or to suggest this idea. During the 1860s James Thomson worked on data gathered by Thomas Andrews to make three dimensional graphs showing the relationship between the volume, temperature, and pressure of carbon dioxide, whether it was a gas or liquid. Maxwell was aware of their work. See: John Shipley Rowlinson, "The Work of Thomas Andrews and James Thomson on the Liquefaction of Gases," *Notes and Records of the Royal Society of London* 57, no. 2 (The Royal Society, 2003): 143–59.

[21] Josiah Willard Gibbs, "Graphical Methods in the Thermodynamics of Fluids," *Transactions of the Connecticut Academy* II (April–May 1873): 309–42; Gibbs, "A Method of Geometrical Representation of the Thermodynamic Properties of Substances by Means of Surfaces," *Transactions of the Connecticut Academy* II (December 1873): 382–404.

[22] James Clerk Maxwell, "Letter from Maxwell to Thomas Andrews, 15 July 1875," in *Maxwell on Heat and Statistical Mechanics: On "Avoiding All Personal Enquiries" of Molecules*, ed. Elisabeth

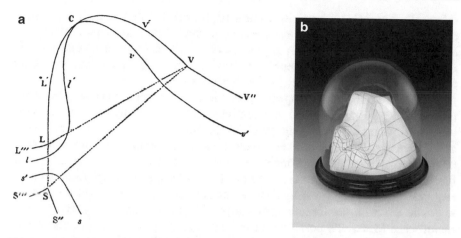

Fig. 5 **a** Josiah Willard Gibbs's presentation of the relationship between the various variables of a given body. From Josiah Willard Gibbs, "A Method of Geometrical Representation of the Thermodynamic Properties of Substances by Means of Surfaces," in Josiah Willard Gibbs, *The Scientific Papers of J. Willard Gibbs*, vol. 1 (New York and Bombay: Longmans, Green, And Co), 33–54, here 44, Fig. 2. **b** James Clerk Maxwell's thermodynamic surface Collections of the National Museums Scotland, T.1999.385 / PF4433). © National Museums Scotland, all rights reserved

From Maxwell's description one can infer the instrumental and productive character of his geometric model, which allows mathematical calculations, and thus also observations, about the development of the modeled object that are not possible with the real object. More important for our context, however, is Maxwell's statement about *what* he has modeled here: in part this is "a fictitious substance." The above quotation thus points to a fundamental question: how was the relation of a geometric-physical model of this kind to the reality investigated understood in the second half of the nineteenth century? Maxwell's formulation makes clear that, in the question of the model, he distances himself from an ontological commitment. The geometric-physical model refers to empirical relations—that is, it is not an arbitrary representation and not a product of the imagination. But since the theoretical knowledge about the physical objects is unreliable or insufficient, or the available data is incomplete or the calculation too complex, the model stands in for a well-reasoned but speculative, and in this sense 'fictitious,' content. Notable is also the difference that can be observed at this point between the models of physics and those of pure mathematics. At the time, both disciplines were facing a problem of representation (physics with respect to electromagnetism, and mathematics with respect to, for example, continuous but nowhere differentiable curves or complex curves and surfaces of increasing complexity and, among other

Garber, Stephen G. Brush, and C. W. Francis Everitt (Bethlehem and London: Lehigh University Press and Associated University Press, 1995), 247–48, here 248 (emphasis M.F./K.K.).

things, their singular points (see section III. below), but they reacted very differently: mathematics strove for exactness in the three-dimensional representation of curves and surfaces, whereas the physics of the 1870s suspended somewhat the faithfulness to reality of the representation in order to obtain the latitude needed to allow for further approaches to the object of study.

For Maxwell, the model (especially the kinematic model of the purely hypothetical particle motion of matter) is a continuation of thought by other means—this was Boltzmann's conclusion in 1902. In the nineteenth century, physical theory could no longer be understood as the clear and final determination of the structure of matter, but was "merely a mental construction of mechanical models."[23] And the functioning of these mechanical models must have just enough to do with the real phenomena to help the understanding of these phenomena, and thus have a heuristic effect on the formation of the physical theory. The radicality of Boltzmann's and Maxwell's position can be appreciated by recalling Pierre Duhem's conception of physics published in 1906 under the title *La théorie physique: son objet, sa structure*. For Duhem, physical theory was an achievement of abstraction, and indeed as a system of well-founded hypotheses and logical deductions—he explicitly excludes the material models of thermodynamics from knowledge.[24]

Already at the beginning of Boltzmann's article, such a broad function is attributed to all scientific and technical models. The mode of representation of models is compared here with the mode of functioning of thought. Thus, in order to form a mental representation of the world, thought has to link the things of the real world with concepts. For Boltzmann, therefore, the relation of similarity between mental representations and the things of the world necessarily remains incomplete and unverifiable—but without such a link between concept and thing, knowledge of the world becomes impossible. Boltzmann describes the act of mental representation as follows: "The essence of the process is the attachment of one concept having a definite content to each thing, but without implying complete

---

[23] Boltzmann, "Models," 790. The full quotation is: "Here again it is perfectly clear that these models of wood, metal, and cardboard are really a continuation and integration of our process of thought; for, according to the view in question, physical theory is merely a mental construction of mechanical models, the working of which we make to ourselves by the analogy of mechanisms we hold in our hands, and which have so much in common with natural phenomena as to help our comprehension of the latter."

[24] Pierre Duhem, *The Aim and Structure of Physical Theory* [French original 1906] (Princeton: Princeton University Press, 1954), 55: "[…] the mind contemplates a whole group of laws; for this group it substitutes a very small number of extremely general judgements, referring to some very abstract ideas; it chooses these primary properties and formulates these fundamental hypotheses in such a way that all the laws belonging to the group studied can be derived by deduction that is very lengthy perhaps, but very sure. This system of hypotheses and deducible consequences, a work of abstraction, generalization and deduction, constitutes a physical theory in our definition." On the devaluation of the model—compared with, for instance, an analogical method, which Duhem recognizes as a heuristic method in the preliminary stages of theorization—see the fourth chapter, "Abstract Theories and Mechanical Models," in ibid.

similarity between thing and thought."[25] For Boltzmann, the absence of a "complete similarity" between mental representation and real thing belongs to what he calls "symbolization"[26]—for which he cites language and writing as examples. If we transfer Boltzmann's considerations to the scientific model, then we can conclude that the model expands the space of thought and action, and this, one imagines, is because it can (and must) be a well-founded, but not a complete, representation of a physical or a geometric object.[27]

If Boltzmann ascribes such importance to the scientific model, then that is due not least to the model practice and theoretical considerations of a physicist whom Boltzmann mentions repeatedly in his article: Maxwell. In the next section, we must therefore turn our attention to Maxwell's understanding of model and analogy.

II. 1850s/1870s: 'Analogy' and 'Model' in Maxwell

How is 'analogy' defined in Boltzmann's article? 'Analogy' is first employed with respect to scientific quantities. If longitude, mileage, temperature, or other physical quantities are expressed by numbers, Boltzmann's term for these numbers is "arithmetical analogies," never 'models.'[28] According to Boltzmann, however, also the "tangible models" of mathematics and physics belong to the category of analogy: they create a "concrete spatial analogy in three dimensions."[29] The term 'analogy,' therefore, is broader and can include any representation of mathematical figures or scientific variables. The model, then, is a special class in the category of analogy insofar as it is extended in length, width, and height.

This power of analogy is founded in the changed claim of physics as represented by Maxwell. Maxwell is convinced that the structure of matter can be grasped neither by physical theory alone (i.e., a logical-deductive procedure) nor by a purely mathematical-analytic approach. And what is more, even the derivation of axioms from the experimental event—the procedure clearly favored by Maxwell—is not

---

[25] Boltzmann, "Models," 788.

[26] Ibid.

[27] In our reading of Boltzmann, we emphasize the heuristic function of models also for the scientific process of acquiring knowledge. Hon and Goldstein, on the other hand, understand Boltzmann's statements in the sense of a merely didactic function, also with reference to the term 'visualization' (*Veranschaulichung*), which Boltzmann uses in his article. We understand the terms *Veranschaulichung* and *Anschauung* in the nineteenth century in relation to the philosophy of Immanuel Kant, and hence, in relation to knowledge. Even if Kant's remarks on the role of *Anschauung* (intuition) for understanding lost their importance precisely as a result of new developments in mathematics in the nineteenth century, the German term *Anschauung* makes a broader claim in the sciences than the English 'visualization' or 'intuition.' See: Hon and Goldstein, "Maxwell's Role in Turning the Concept of Model into the Methodology of Modeling," 324.

[28] Boltzmann, "Models," 788–91, here 789.

[29] Ibid., 789.

possible in the case of the structure of matter.[30] For the investigation of elastic solids, the motion of gas molecules, or electromagnetic fields, and thus for domains for which no satisfactory theory was yet available and for which the conclusions from experiments were not sufficient, Maxwell recommended another procedure: the method of physical analogies.[31] Boltzmann is referring to this implicitly when he states: "The question no longer being one of ascertaining the actual internal structure of matter, many mechanical analogies or dynamical illustrations became available […]."[32]

Maxwell's method of analogy rests on the fact that different physical phenomena can have surprisingly similar mathematical formulas describing them. For Maxwell, this formal mathematical similarity becomes a motor for further calculations, new hypotheses, and theoretical inferences. Thus, in his 1856 article "On Faraday's Lines of Force," Maxwell uses the mechanical analogy of the motion of a fluid to investigate the geometry of electromagnetic 'lines of force' (for an example of these lines of force as they were drawn by Maxwell, see Fig. 6 from Maxwell's 1873 book *A Treatise on Electricity and Magnetism*).

The lines of force are a conjecture on Michael Faraday's part and not a certain, calculable finding. Maxwell assumes that these lines of force permeate space even when there are no objects on which such a force can act. Maxwell insists that electromagnetic phenomena are "not even a hypothetical fluid" and therefore appear in his considerations merely as a kind of imaginary substance.[33] The analogy to the motion of a fluid is nevertheless helpful and productive: it serves Maxwell to suggest to the understanding in a manageable form those mathematical ideas that

---

[30] Here, Maxwell is moving within a discussion that took place between the theoretically and hypothetically operating mechanico-molecular school (Pierre-Simon Laplace, Claude Louis Marie Henri Navier, Augustin Cauchy, Siméon Poisson) and the abstract, mathematical approach of an analytical mechanics (Joseph Louis Lagrange, Joseph Fourier). With the appeal to the facts of the experiment, Maxwell is taking up a third position in this conflict, thereby expanding the toolbox of physics. With the proclamation of a heuristic method of analogy as well as the geometric-graphic method, he expands physics once again. See on this: Robert Kargon, "Model and Analogy in Victorian Science: Maxwell's Critique of the French Physicists," *Journal of the History of Ideas* 30, no. 3 (July–September 1962): 423–36. See also: Mary Hesse, "Maxwell's Logic of Analogy," in *The Structure of Scientific Inference* (Berkeley: University of California Press, 2020), 259–82.

[31] James Clerk Maxwell, "On Faraday's Lines of Force" [1855/6], in *James Clerk Maxwell: The Scientific Papers of James Clerk Maxwell*, vol. I, ed. William Davidson Niven (Cambridge: Cambridge University Press, 1890), 156.

[32] Boltzmann, "Model," 790. The full quotation is as follows: "[…] Maxwell propounded certain physical theories which were purely mechanical so far as they proceeded from a conception of purely mechanical processes. But he explicitly stated that he did not believe in the existence in Nature of mechanical agents so constituted, and that he regarded them merely as means by which phenomena could be reproduced, bearing a certain similarity to those actually existing, and which also served to include larger groups of phenomena in a uniform manner and to determine the relations that hold in their case. The question no longer being one of ascertaining the actual internal structure of matter, many mechanical analogies or dynamical illustrations became available […]."

[33] Maxwell, "On Faraday's Lines of Force," 155–229, here 160.

Fig. 6 One of James Clerk
Maxwell's examples for
drawings of lines of force.
From James Clerk Maxwell,
*A Treatise on Electricity and
Magnetism,* vol. 1 (Oxford:
Clarendon Press, 1873),
Plates, fig. 4

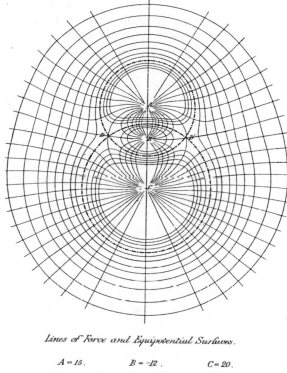

Lines of Force and Equipotential Surfaces.

$A = 15.$ $B = -12.$ $C = 20.$

are necessary for the study of the still largely unknown electricity.[34] In this case, according to Maxwell, neither a purely mathematical nor a purely hypothetical approach can help:

> In the first case [of a purely mathematical approach] we entirely lose sight of the phenomena to be explained; [...]. [In the second case of a theoretical approach], we see the phenomena only through a medium, and are liable to that blindness to facts and rashness in assumption which a partial explanation encourages.[35]

To compensate for the limitations of the mathematical formula and of the theoretical hypothesis, Maxwell employs the physical analogy (in this instance, the comparison between electromagnetic phenomena and the behavior of fluids) to allow mathematical descriptions and predictions even for the non-calculable and largely unknown domain. Maxwell defines this physical analogy as follows: "By a physical analogy I mean that partial similarity between the laws of one science and those of another which makes each of them illustrate the other."[36]

---

[34] Ibid., 157: "to bring before the mind, in a convenient and manageable form, those mathematical ideas which are necessary to the study of the phenomena of electricity."
[35] Ibid., 155–56.
[36] Ibid., 156.

Ernst Mach, who in 1902 declared Maxwell's method of analogy to be one of the most important methods of research, and considered physical analogy to be as central to scientific research as the experiment, aptly characterized physical analogy as an "abstract similarity."[37] Hence, analogy here is explicitly an abstract representation. It might be objected, however, that this abstractness still allows a proximity to the empiricism of the physical phenomenon.[38] According to Maxwell, this is precisely the advantage of analogy over mathematical description and physical theory. In Maxwell's work, analogy is thus distinguished by multiple and even partially contradictory properties: (1) it remains connected to empirical evidence and is therefore concrete; (2) it gives rise to a speculative but not at all arbitrary illustration, thus enabling the further examination of previously unknown domains; (3) it succeeds in this productive visualization by means of a mode of abstraction: precisely that abstract similarity of the actually different physical phenomena which is guaranteed in the identical mathematical formulas.

Yet Maxwell does not propagate analogy as the only method of research. In his research practice this method is combined with other representational practices, such as the diagram, the three-dimensional model, and mathematical calculation. The distinction found in Boltzmann's article between models as tangible representations and other forms of representation or symbolization (such as arithmetical analogies, and thus with numbers) appears to be less rigorous in Maxwell's case. When mentioning the drawings that help him to determine the lines of force of an electric field, Maxwell speaks explicitly of a "geometrical model of the physical phenomena."[39] For Maxwell, these geometric representations are valid representations of the physical laws at work in the electrical phenomenon—and the transition between the geometric drawing of a space traversed by lines of force and the three-dimensional model resulting from this seems fluid. Two-dimensional drawing and

---

[37] Ernst Mach, "Die Ähnlichkeit und die Analogie als Leitmotiv der Forschung," in *Annalen der Naturphilosophie* 1 (1902), ed. Wilhelm Oswald, 5–14, here 5: "Die Analogie ist jedoch ein besonderer Fall der Aehnlichkeit. Nicht ein einziges unmittelbar wahrnehmbares Merkmal des einen Objectes braucht mit einem Merkmal des anderen Objectes übereinzustimmen, und doch können zwischen den Merkmalen des einen Objectes Beziehungen bestehen, welche zwischen den Merkmalen des anderen Objectes in übereinstimmender, identischer Weise wiedergefunden werden. [...] [M]an könnte dieselbe auch eine abstrakte Ähnlichkeit nennen." ["Analogy is, however, a special case of similarity. Not a single immediately perceptible feature of one object need coincide with a feature of another object, and yet relations can exist between the features of one object in exactly the same way as those between the features of the other object. [...] [O]ne might also call this an abstract similarity."] (Unless otherwise noted, all translations of quotes by Benjamin Carter.) On Mach's understanding of analogy, see: Susan G. Sterrett, "Mach on Analogy in Science," in *Interpreting Mach. Critical Essays*, ed. John Preston, (Cambridge: Cambridge University Press, 2021), 67–83.

[38] See: James Clerk Maxwell, "On Physical Lines of Force," in *The Scientific Papers of James Clerk Maxwell*, vol. I, ed. William Davidson Niven (Cambridge: Cambridge University Press, 1890), 451–513, here 486: "I venture to say that anyone who understands *the provisional and temporary character of this hypothesis*, will find himself rather helped than hindered by it in his search after the true interpretation of the phenomena." (Emphasis M.F./K.K.).

[39] Maxwell, "On Faraday's Lines of Force," 158.

three-dimensional model allow an approach to the physical object that the object itself may not permit. In Maxwell's writings there is therefore a double extension of scientific representation: on the one hand, the model moves into the immediate vicinity of drawing (i.e., the difference between the two-dimensional and the three-dimensional visualization appears secondary); on the other, with the physical analogy, the methodological range of physics is expanded—this is done via a procedure that, as is expressly stated, is based on an incomplete, abstract similarity, and it is from this that it derives its heuristic force.[40] Nevertheless, for Maxwell as well as for Boltzmann, model and analogy do not coincide.[41] A partial overlapping (not a coincidence) of model and Maxwell's method of physical analogy is found only at the end of Boltzmann's encyclopedia article:

> It often happens that a series of natural processes [...] may be expressed by the same differential equations; and it is frequently possible to follow by means of measurements one of the processes in question [...]. If then there be shown in a *model* a particular case of [the first process] in which the same conditions at the boundary hold as in [the second process], we are able by measuring [...] in the *model* to determine at once the numerical data which [we may] obtain for the *analogous* case [...].[42]

As we have already seen, according to Maxwell, two different physical phenomena are formally analogous when both can be described mathematically by one and the same formula. Analogy, then, allows the transfer from the known (or better known) domain to the unknown domain. In Boltzmann's description the model now assumes a decisive position insofar as one domain of the analogy is not constituted by the physical object but already appears as a model of this object. This model allows the developments of the physical object to be read, and these findings can then be transferred to the other side of the analogy, which is to say, they can lead to a further analysis of that domain which, taken by itself, would not be accessible to measurement. In this case, via the access to the model, the physical analogy is made fruitful for further research. The model can therefore become

---

[40] Analogy can be understood as an aid in the development of a kind of provisional theory that can guide experimentation and mathematical treatment, and subsequently leads to a 'true' theory of electricity. Maxwell writes about the "laws" that he develops on the basis of Michael Faraday's research, the latter's idea of 'lines of force,' and with the help of the analogy of 'tubes,' ibid, 207: "In these six laws I have endeavoured to express the idea which I believe to be the mathematical foundation of the modes of thought indicated in the *Experimental Researches*. I do not think that it contains even the shadow of a true physical theory; in fact, its chief merit as a temporary instrument of research is that it does not, even in appearance, *account for* anything. [...] Besides, I do not think that we have any right at present to understand the action of electricity, and I hold that the chief merit of a temporary theory is, that it shall guide experiment, without impeding the progress of the true theory when it appears." The physical analogy thus promoted knowledge but did not found knowledge. See on this: Kargon, "Model and Analogy in Victorian Science," 433–34.

[41] On the difference between model and analogy in physics at the turn of the nineteenth to the twentieth century, see: David Hugh Mellor, "Models and Analogies in Science: Duhem versus Campbell?" *Isis* 59, no. 3 (Autumn 1968): 282–90.

[42] Boltzmann, "Models," 791 (emphasis M.F./K.K.).

part of the physical analogy and also support the latter's function as a motor of scientific knowledge.

At this point, however, the valorization of the model refers in Boltzmann's case to a hybrid: the geometric-physical model whose similarity to the real physical phenomenon need not be complete to be scientifically productive. In the course of the nineteenth century, one would invoke other concepts and descriptions to determine the use of these purely mathematical models in pedagogical and epistemic contexts. These were above all the concepts of *Anschauung* (intuition) and *Bild* (image), which are the subject of the next section.

III. 1880–1900: *'Anschauung'* and *'Bild'* (Klein and Brill)

Boltzmann's 1902 article "Models" gathers together the knowledge about models of the second half of the nineteenth century and in doing so discusses a variety of specific models: not only the traditional models of architecture and engineering, but also and in particular those of the natural sciences and mathematics. Here, not only the various functions of the models become visible, but also uncertainties about the definition and evaluation of in particular mathematical models. If one looks at the correspondence between Boltzmann and the man who requested the encyclopedia article from Boltzmann, Joseph Larmor (a professor of natural philosophy at Queen's College in Galway until 1885, then first a lecturer at St John's College, Cambridge, and from 1903 a professor of mathematics at Trinity College, Cambridge), then Boltzmann's discomfort becomes palpable. On January 7, 1900, in response to Larmor's request, Boltzmann writes:

> I have now hurriedly looked around for information about models. […] I also went to the Vienna Polytechnicum in order to look at models of house, roof, and bridge construction as well as an endless variety of machines but came away persuaded that what we have on hand in Vienna is obsolete and miserable. How much better the article could be written were one exactly familiar with models in London and America![43]

Boltzmann's first thought is about the use of models in architecture and engineering, which while being in use in technical universities were—this is the objection he raises—far more frequent in English-speaking countries than in Vienna, where Boltzmann was a professor of theoretical physics. These models had a clear instrumental or pedagogical function insofar as they were intended for the explanation of technical constructions—showing these constructions in a scaled-down form that often could be taken in the hands and occasionally even dismantled. Yet Boltzmann's misgivings were not only related to the dearth of good examples, but concerned, above all, the broad, hardly summarizable variety of models, whereby

---

[43] John T. Blackmore, ed., *Ludwig Boltzmann: His Later Life and Philosophy, 1900–1906. Book One: A Documentary History*, series *Boston Studies in the Philosophy of Science*, vol. 168 (Dordrecht: Springer, 1995), 57.

it was particularly with the mathematical models that he felt the most uncertainty. In the same letter he writes:

> I am also in the dark about how mathematical models should be handled. I don't even understand particularly much about algebraic surfaces of pure geometrical models. [...] If I dare to write about every kind of model, I am very much afraid that in spite of all of my efforts that I will fail to come up to the Encyclopaedia Britannica's standards. I simply don't know the material well enough. For this reason it would be my dearest preference if it would not cause any great difficulties to take the preparation of the article away from me [...].[44]

We should not be too quick to interpret Boltzmann's doubts about his competence as a personal shortcoming; these doubts were rather the expression of an objective overload with respect to the diversity of models found in polytechnic university collections in the second half of the nineteenth century. With respect to the mathematical models, another factor plays a role: their construction and active distribution began only in the nineteenth century (first in France at the beginning of the nineteenth century; then, in the second half of the century, also in England and Germany), and their function was less straightforward than that of the technical models. The importance of mathematical models was undoubtedly in teaching, since they were employed in university education for the teaching of mathematics, particularly at the technical universities. Their function was thus the visualization of complex mathematical objects. Yet they did act as a blueprint for the calculated construction of technical constructions—as was the case, for instance, with models of bridges. Rather, they were purely mathematical objects that the haptic model should make graspable—and Boltzmann doubted precisely whether his knowledge of these abstract mathematical objects (he explicitly mentions algebraic surfaces) was sufficient. He had no doubts, however, regarding the function of the mathematical models: their importance lay in their *Anschaulichkeit* (their capacity to make certain phenomena available to the senses), and this is based on their no longer being merely abstract objects of reason but concrete objects of sensory perception. This position can be gathered from Boltzmann's article "Über die Methoden der theoretischen Physik," which Dyck, the editor of the *Katalog mathematischer und mathematisch-physikalischer Modelle, Apparate und Instrumente*, requested in 1892. In this article Boltzmann writes that the material mathematical models serve "to make the results of a calculation intuitable [*anschaulich*], and indeed not merely for the imagination, but also visible to the eye, graspable by the hand, with plaster and card."[45] The material realization of the mathematical surfaces and curves translated into three dimensions addresses visual and haptic perception (eye and hand), and thus generates sensory evidence

---

[44] Ibid.

[45] Ludwig Boltzmann, "Über die Methoden der theoretischen Physik," in *Katalog mathematischer und mathematisch-physikalischer Modelle, Apparate und Instrumente*, ed. Walther Dyck (Munich: Wolf, 1892) 89–99, here: 90: "die Resultate des Calcüls anschaulich zu machen und zwar nicht blos für die Phantasie, sondern auch sichtbar für das Auge, greifbar für die Hand, mit Gips und Pappe."

for the mathematical contents. According to Boltzmann, the great advantage of mathematical models lies in the quicker understanding of the contents. That is why these models are more than mere illustrations of the results of the 'proper'— that is, the mathematical—procedure (the analysis); rather, they are equivalent to geometric construction:

> In mathematics and geometry it was at first undoubtedly the need to save labor that led from purely analytic methods back to the constructive methods as well as to the visualization by models. […] What an abundance of shapes, singularities, forms developing from one another the geometer of today has to commit to memory […].[46]

Mathematical models are suited to come to grips with the vastly increased set of objects of geometry. Their *Anschaulichkeit* aims at the economy of the acquisition of knowledge and the organization of knowledge. In this economy models represent, according to Boltzmann's hope, a way of saving of labor. And while he does not formulate this explicitly, it can be supposed that via models and geometric construction, the economy of the mathematically trained engineer can also benefit from such a saving of labor. For in the second half of the nineteenth century, the relation between mathematics and empiricism—and that means also the application of mathematics in other sciences and in engineering—was no longer self-evident. In the discussions of mathematicians on the foundation of their discipline, the reference to empiricism was now either asserted as a necessity and a virtue or dismissed in favor of formalization and logicalization. Boltzmann's formulation in his 1902 encyclopedia article that models "present to the senses" the form of geometric figures should therefore be taken seriously as a positioning.[47] Mathematical models stand for a mode of concretion that in the course of the development toward 'mathematical modernity' would be increasingly excluded. In this sense, these models were elements of a 'counter-modernity,' but this counter-modernity, as Herbert Mehrtens has pointed out, was an important accompaniment to the development toward modernity and not its reactionary adversary.[48] Mehrtens argument can perhaps be slightly reformulated: mathematical models played a key

---

[46] Ibid., 90–91: "In der Mathematik und Geometrie war es zunächst unzweifelhaft das Bedürfnis nach Arbeitersparnis, welches von den rein analytischen wieder zu den constructiven Methoden sowie zur Veranschaulichung durch Modelle führte. […] Welche Fülle von Gestalten, Singularitäten, sich aus einander entwickelnder Formen hat der Geometer von heute sich einzuprägen […]."

[47] Boltzmann, "Models," 789.

[48] Herbert Mehrtens has interpreted the so-called foundational crisis of mathematics at the turn of the nineteenth to the twentieth century as a complex interweaving of modernization tendencies and a counter-modern resistance, whereby elements of both directions can be found in part in one and the same mathematician. He considers the appeal to intuition and *Anschauung* as a sign of counter-modernity, whereas modernity can be described via the reference to language. See: Herbert Mehrtens, *Moderne–Sprache–Mathematik: Eine Geschichte des Streits um die Grundlagen der Disziplin und des Subjekts formaler Systeme* (Frankfurt am Main: Suhrkamp, 1990). Mehrtens's study is an important and original contribution to the history of nineteenth and twentieth century mathematics; however, it has also met with sharp criticism. The objects of this criticism

role insofar as they cushioned the development toward mathematical modernity. They could do this due to their dual nature, since while their contents was a purely mathematical object, the latter was lent *Anschaulichkeit* qua material modeling. Abstraction and concretion thus met in these models without this having to serve a purpose outside mathematics.[49]

It was indeed the case that arguments of modernization and arguments of counter-modernity went hand in hand in the writings of the proponents of model building. This applies in particular to Alexander Brill and Felix Klein, both of whom were advanced mathematicians. In nineteenth century Germany, Brill and Klein were among the driving forces behind the construction and acquisition of models. They advanced a reform in the education of future engineers and mathematicians, where the use of models was necessary to exemplify the new concepts and the abstract objects being used in class.[50] Klein's statements reflect the state of upheaval mathematics was undergoing at the time, and its search for its own foundations of knowledge. Thus, his influential Erlangen program, which he first set out in 1872, attempts a naming and unification of various geometries (parabolic, hyperbolic, and elliptic, as well as projective geometry). For this unification he suggested the use of an abstract, group-theoretical approach in order to investigate different spaces and manifolds, focusing on their groups of transformations. This new 'motion geometry' (*Bewegungsgeometrie*, as Hans Wussing termed it) positioned the concept of the group at the transition between geometry and analysis.[51]

---

are the almost exclusive limitation to the history of German mathematics and the strongly meta-mathematical perspective, which neglects the concrete mathematical developments of the period considered. See: Moritz Epple, "Styles of Argumentation in Late 19th Century Geometry and the Structure of Mathematical Modernity," in *Analysis and Synthesis in Mathematics: History and Philosophy*, ed. Michael Otte and Marco Panza (Dordrecht: Kluwer, 1997), 177–98; Moritz Epple, "Kulturen der Forschung: Mathematik und Modernität am Beginn des 20. Jahrhunderts," in *Wissenskulturen: Über die Erzeugung und Weitergabe von Wissen*, ed. Johannes Fried and Michael Stolleis (Campus: Frankfurt am Main, 2009), 125–58, Jeremy Gray, *Plato's Ghost: The Modernist Transformation of Mathematics* (Princeton: Princeton University Press, 2008), 9–12. In the English-speaking world, Gray's book *Plato's Ghost* describes, though from a broader point of view, also the 'modern' transitions in the late nineteenth century within mathematics and the mathematical sciences.

[49] While Mehrtens, in his book *Moderne–Sprache–Mathematik*, does not note the existence of material models, in a later article he does consider their multiple functions at the end of the nineteenth century. See: Herbert Mehrtens, "Mathematical Models," in *Models: The Third Dimension of Science*, ed. Soraya de Chadarevian and Nick Hopwood (Stanford: Stanford University Press, 2004), 276–306.

[50] For a general survey on Felix Klein's conception of visualization in mathematics and the role of three-dimensional models, see: Stefan Halverscheid and Oliver Labs, "Felix Klein's Mathematical Heritage Seen Through 3D Models," in *The Legacy of Felix Klein*, ed. Hans Georg Weigand, William McCallum, Marta Menghini, Michael Neubrand, and Gert Schubring, series *ICME-13 Monographs* (Cham: Springer, 2019), 131–52; Renate Tobies, *Felix Klein: Visions for Mathematics, Applications, and Education* (Cham: Birkhäuser, 2021), Sects. 2.4.3 and 4.1.1.

[51] See: Felix Klein, *Vergleichende Betrachtungen über neuere geometrische Forschungen* (Erlangen: A. Duchert, 1872), 6. On "*Bewegungsgeometrie*" see also: Hans Wussing, *The Genesis of*

In this way, Klein initially distances himself from a sensory-empirical approach to space and spatial figures, as can be inferred from the beginning of the text in which he set out his Erlangen program:

> We peel off the mathematically inessential sensory image [*sinnliche Bild*], and regard space only as a manifold of several dimensions, that is to say, if we hold to the usual idea of the point as spatial, three-dimensional element. By analogy with the transformations of space, we speak of transformations of the manifoldness; they also form groups.[52]

In order to talk about groups as a mathematical structure that characterizes manifolds, Klein had to remark that the sensory image was inessential, and hence the concreteness of the sensory impression has to be 'peeled off' in favor of the mathematical abstraction. Here, Klein undertakes a hard confrontation between abstraction (understood as a concept or as formal mathematics) and concretion (understood as a sensory impression or as a physical fact), one that will appear repeatedly in the development toward mathematical modernity.[53] On the other hand, however—and in a way similar to Maxwell's critique that one should avoid working only with a theoretical hypothesis or only with an abstract formula— Klein also distances himself from a complete, absolute abstraction of the research of space. Geometry aims at a "figurative reality" ("gestaltliche Wirklichkeit") of the spatial figures, and that means that the transformation groups stand for 'real' movements, and the groups are therefore attributed *Anschaulichkeit*.[54] For this faithfulness to reality of geometry claimed by Klein, mathematical models play a particular role:

---

the *Abstract Group Concept* [1969] (Cambridge: MIT Press, 1984), 144–45.; Mehrtens, *Moderne–Sprache–Mathematik*, 60–67.

[52] Klein, *Vergleichende Betrachtungen*, 7: "Streifen wir jetzt das mathematisch unwesentliche sinnliche Bild ab, und erblicken im Raume nur eine mehrfach ausgedehnte Mannigfaltigkeit, also, indem wir an der gewohnten Vorstellung des Punctes als Raumelement festhalten, eine dreifach ausgedehnte. Nach Analogie mit den räumlichen Transformationen reden wir von Transformationen der Mannigfaltigkeit; auch sie bilden Gruppen."

[53] Mehrtens has pointed out that the abstraction/concretion opposition was in fact posited by the representatives of counter-modernity, such as Klein, whereas the modernizer Riemann does not make a qualitative distinction between purely mathematical products and instrumental constructions tied to empiricism (i.e., between concept and fact). See: Mehrtens, *Moderne–Sprache–Mathematik*, 66–67.

[54] Klein, *Vergleichende Betrachtungen*, 42: "Es gibt eine eigentliche Geometrie, die nicht, wie die im Texte besprochene Untersuchungen, nur eine veranschaulichte Form abstrakter Untersuchungen sein will. In ihr gilt es, die räumlichen Figuren nach ihrer vollen gestaltlichen Wirklichkeit aufzufassen, und (was die mathematische Seite ist) die für sie geltenden Beziehungen als evidente Folgen der Grundsätze räumlicher Anschauung zu verstehen." ["There is an essential geometry that does not wish to be only a visualized form of abstract investigations—like the investigations discussed in the text. Its problem is to grasp the spatial figures in their full figurative reality, and (which is the mathematical aspect) to understand the relations valid for them as evident consequences of the principles of spatial intuition."].

For geometry, a model—be it realized and observed or only vividly imagined—is not a means to an end but the thing itself [*die Sache selbst*].[55]

Which is to say that Klein pointed out that three-dimensional models—despite being concrete, tangible objects, and therefore imposing concrete sensory images on the viewer—can serve not only as a means of visualization but also as "the thing itself." The reality of the mathematical objects can be grasped in the model, and indeed without this mathematical object being identified with a physical fact or a sensory impression. The plaster model of an algebraic surface does not merely illustrate the abstract mathematical object but presents a conceptually correct representation and, in this sense, the *Anschauung* (intuition) of a geometric reality.[56]

If one consults Klein's 1873 lecture "Über den allgemeinen Functionsbegriff und dessen Darstellung durch eine willkürliche Curve," with the distinction found there between a curve jotted arbitrarily on paper, a curve drawn on paper according to a specific law or formula, and a merely imagined curve as an aid, then three-dimensional mathematical models correspond to the curve drawn according to a specific law or formula on paper.[57] The latter is an empirical object of perception (unlike a mental image) and at the same time an idealization, since the conceptually conveyed law allows the representation to become mathematically precise in a way that is impossible for the merely approximate precision of the jotted curve. Even if the *Anschauung* of the built, drawn, or merely imagined models in Klein's sense does not yield a "*precision mathematics*," ["*Präzisionsmathematik*"] one can still infer that in a scientific respect the models operate in an orientating and even epistemic-heuristic way, and thus while not founding knowledge, are still able to guide it and give rise to new knowledge.[58] This is also suggested by Brill's statements in his lecture from November 7, 1886:

---

[55] Ibid.: "Ein Modell—mag es nun ausgefuhrt und angeschaut oder nur lebhaft vorgestellt sein—ist für diese Geometrie nicht ein Mittel zum Zwecke sondern die Sache selbst."

[56] Mehrtens, "Mathematical Models," 301: "[...] the models had, for Felix Klein and for a short time, been epistemic things; later he interpreted them as applied mathematics. But by the end of the century mathematicians took the models as imperfect representations of geometrical entities that could be used as an aid in communication about mathematics."

[57] See: Felix Klein, "Über den allgemeinen Fuctionsbegriff und dessen Darstellung durch eine willkürliche Curve," *Mathematische Annalen* 22 (1883): 249–59. For a discussion of these three methods, see: Klaus Thomas Volkert, *Die Krise der Anschauung: Eine Studie zu formalen und heuristischen Verfahren in der Mathematik seit 1850* (Göttingen: Vandenhoeck & Ruprecht, 1986), 228–31.

[58] For the notions "*precision mathematics*" ["*Präzisionsmathematik*"] and "*approximative mathematics*" ["*Approximationsmathematik*"] and the difference that Klein makes between a mere intellectual study on the one hand and applied mathematics on the other, see: Felix Klein, *Elementarmathematik vom höheren Standpunkte aus*, vol. I: *Arithmetik, Algebra, Analysis* [1908], reprint of the 4th slightly revised edition from 1933 (Berlin and Heidelberg: Springer, 1968), 39; and Felix Klein, *Elementarmathematik vom höheren Standpunkte aus*, vol. III: *Präzisions- und Approximationsmathematik* [1902], reprint of the 3rd slightly revised edition from 1928 (Berlin and Heidelberg: Springer, 1968).

The maker of a model was free to write a paper on this, the publication of which [...] played no small part in encouraging one to carry out the often-arduous calculations and drawings at the basis of the practical execution. Conversely, the model often prompted subsequent investigations into the specific features of the represented structure.[59]

Thus, according to Brill, mathematical models in the nineteenth century did not merely serve to visualize lengthy calculations; they were not simply a scientific tool. If one also considers the design stage, then they were also an object of research. The constructors of these models made use of solid theoretical knowledge and the methods connected with this (particularly calculation and geometric construction), but they needed to proceed innovatively and exploratively in order to understand a scarcely graspable—since abstract and possibly even speculative—object. The mathematical model is—at least in the design stage—an object of questioning, which qua construction should be transformed into an object that is both familiar and useful for the production of further knowledge. In this respect, the models served a process of intellectual appropriation and mathematical habitualization, however not as the passive learning of something already given, but as the active exploration of something at least partially unknown, and that means as the driving factor of present (linked to the object) and future scientific and application-oriented knowledge—what Klein calls "inventions and new mental connections."[60] It was this that gave the mathematical models of the nineteenth century their experimental and thus epistemic function.

If one wants to classify the role played by these models in the process of *Anschauung* more precisely, then, on the one hand, one has to recall what the history of science and the history of mathematics of the last decades has already

---

[59] Alexander Brill, "Über die Modellsammlung des mathematischen Seminars der Universität Tübingen (Vortrag vom 7. November 1886)," *Mathematisch naturwissenschaftliche Mitteilungen* 2 (1887), 69–80, here 77: "Dem Verfertiger eines Modells stand es frei, eine Abhandlung zu demselben zu schreiben, deren Veröffentlichung [...] nicht wenig dazu anreizte, die oft mühsamen Rechnungen und Zeichnungen, welche der praktischen Ausführung zu Grunde lagen, durchzuführen. Öfter veranlaßte umgekehrt das Modell nachträgliche Untersuchungen über Besonderheiten des dargestellten Gebildes."

[60] Klein, *Elementarmathematik vom höheren Standpunkt*, vol. III, 8: "*Da die Gegenstände der abstrakten Geometrie nicht als solche von der räumlichen Anschauung scharf erfaßt werden, kann man einen strengen Beweis in der abstrakten Geometrie nie auf bloße Anschauung gründen, sondern muß auf eine logische Ableitung aus dem als exakt vorausgesetzten Axiomensystem zurückgehen.* Trotzdem behält aber auf der anderen Seite *die Anschauung* auch in der Präzisionsgeometrie ihren großen und durch logische Überlegungen nicht zu ersetzenden Wert. Sie hilft uns die Beweisführung leiten und im Überblick verstehen, sie ist außerdem eine Quelle von Erfindungen und neuen Gedankenverbindungen." ["*As the objects of abstract geometry cannot be totally grasped by space intuition, a rigorous proof in abstract geometry can never be based only on intuition, but must be founded on logical deduction from valid and precise axioms.* Nevertheless *intuition* maintains, also in precision geometry, its irreplaceable value that cannot be substituted by logical considerations. Intuition helps us to construct a proof and to gain an overview; it is, moreover, a source of inventions and new mental connections."] On the heuristic, but not knowledge-founding function of *Anschauung* (intuition) in Klein, see also: Volkert, *Die Krise der Anschauung*, 238–42.

brought to light. Particularly in German-speaking countries, the power of *Anschauung* has been acknowledged as the prerequisite and achievement of knowledge since Immanuel Kant, but this ennoblement of the concept-led, so-called pure *Anschauung* (the nonempirical but sensory manifestation of a truth given qua reason) lost its persuasiveness in the course of the nineteenth century. Particularly the attempt to secure a pure *Anschauung* in an immutable, single 'a priori' of space and time was discredited with the rise of non-Euclidean geometries and could no longer be considered as the self-evident foundation of mathematical knowledge.[61] Moreover, the potential of sensory perception for deception was thoroughly explored by nineteenth century experimental physiology and psychology, for instance in relation to the blind spot of the eye or the misperception of continuous movement when looking at moving images (e.g., through a phenakistiscope). Visual perception was converted into an image production with its own dynamics.[62] Hence, the understanding of *Anschauung* underwent significant changes in the nineteenth century: on the one hand, the concept could no longer refer to a reliable a priori endowment; on the other, it was no longer related to visual perception and mental images alone, but was interpreted particularly by Charles Sander Peirce as an intuition of signs, and thus linked with the field of semiotics. For the mathematicians of the nineteenth century, however, this latter expansion was evidently not a well-known or convincing argument.[63]

In Klein's work, alongside *Anschauung*, one frequently finds also the concept of *Bild* (image), both in an empirical-concrete sense (visible drawing or haptic model) and, above all, in a sense related qua *Anschauung* to the imagination (mental image). Exemplary for this is an article from 1874, in which Klein discusses the *Bilder* (images) of mathematical functions and asks about their completeness. While explaining how to visualize a complex curve $y = f(x)$, he notes that the sketch of the real part of the function is incomplete. A second method concerns the representation of a surface whose coordinates are $(Re(x), Im(x), Re(f(x)))$, but although, according to Klein, this yields "a complete image,"[64] it does not allow

---

[61] Other factors were the increasing arithmetization of mathematics and the discovery of monster functions. See: Volkert, *Die Krise der Anschauung*, 251–59; Michael Friedman, *Kant and the Exact Sciences* (Cambridge: Harvard University Press, 1992). See also: Michael Friedman, "Kant on Geometry and Spatial Intuition," *Synthese* 186, no. 1 (2012): 231–55; Christophe Bouriau, Charles Bravermann, and Aude Mertens, eds., *Kant et ses grands lecteurs: L'intuition en question* (Nancy: Presses Universitaires de Nancy, 2016).

[62] On the experimentalization of visual perception and the consequences of this for the understanding and the practice of the image, see: Jonathan Crary, *Techniques of the Observer: On Vision and Modernity in the Nineteenth Century* (Cambridge, et al.: MIT Press, 1990).

[63] The link between *Anschauung* and semiotics is rigorously updated in: Volkert, *Die Krise der Anschauung*. For a similar argument, but with the additional assumption of a historical change in the conception of *Anschauung*, see: Johannes Lenhard, "Kants Philosophie der Mathematik und die umstrittene Rolle der Anschauung," *Kant-Studien* 97 (2006), 301–17.

[64] Felix Klein, "Ueber eine neue Art der Riemannschen Flächen (Erste Mitteilung)," *Mathematische Annalen* 7 (1874): 558–66, here 559: "ein vollständiges Bild."

a sufficient visualization of complex singularities. He formulates the problem as follows:

> In the investigation of the algebraic functions *y* of a variable *x*, one is accustomed to use two different *intuition-related aids* [*anschauungsmäßiger Hilfsmittel*]. One represents, namely, either *y* und *x* consistently as coordinates of a point of the plane—whereby the real values of these alone come into evidence and the image of the algebraic function becomes the algebraic curve—or one spreads the complex values of one variable *x* over a plane and designates the functional relation between *y* and *x* by the Riemann surface constructed over the plane. In many relations, it must be desirable to possess a transition between these two *intuition-based images* [*Anschauungsbildern*].[65]

Here, the meaning of the term 'image' (*Bild*) goes beyond an empirical-concrete visualization on paper (a drawn algebraic curve as an image of the algebraic function) and designates in addition a three-dimensional model (Riemann surface). Moreover, it should be noted that Klein invokes several "intuition-based images" ("Anschauungsbilder") (algebraic curve and Riemann surface) and searches for a "transition" between them. If *Anschauung* should serve the "investigation" of mathematical objects and productively advance this, then it must draw on several images. And to search for the transition between these images would then be—one must assume—the task of the understanding and the imagination. Klein's solution to the problem formulated in the above quotation is in any case to investigate, together with the given curve, a further curve: the dual curve in the projective complex plane.[66]

Most of Klein's drawings do not have the visualization of singularities for a theme but follow Julius Plücker's investigation on the relations between the invariants of algebraic curves and the corresponding invariants of their dual curves; specifically, Klein aims at a "Veranschaulichung" ("visualization")[67] of these relations. At the end of his contribution from 1874 there is a sketch of what a branch point of a singular curve of degree three looks like (see Fig. 7): "one [confers] to the surface an [...] outgoing branching [...], as it is visualized, for instance, in a symmetrical way by the included drawing."[68]

---

[65] Ibid., 558: "Bei der Untersuchung der algebraischen Funktionen *y* einer Veränderlichen *x* pflegt man sich zweier verschiedener *anschauungsmäßiger Hilfsmittel* zu bedienen. Man repräsentiert nämlich entweder *y* und *x* gleichmäßig als Koordinaten eines Punktes der Ebene, wo dann die reellen Werte derselben allein in Evidenz treten und das Bild der algebraischen Funktion die algebraische Kurve wird—oder man breitet die komplexen Werte der einen Variabeln *x* über eine Ebene aus und bezeichnet das Funktionsverhältnis zwischen *y* und *x* durch die über der Ebene konstruierte Riemannsche Fläche. Es muß in vielen Beziehungen wünschenswert sein, zwischen den beiden *Anschauungsbildern* einen Übergang zu besitzen." (Emphasis M.F./K.K.).

[66] The points of the dual curves correspond to the lines that are tangents to the original curves.

[67] Felix Klein, "Ueber eine neue Art der Riemannschen Flächen (Zweite Mitteilung)," *Mathematische Annalen* 10 (1876): 398–416, here 404.

[68] Felix Klein, "Ueber eine neue Art der Riemannschen Flächen (Erste Mitteilung)," 566: "man [erteilt] der Fläche eine [...] ausgehende Verzweigung [...], wie sie etwa, in symmetrischer Weise, durch die beigesetzte Zeichnung veranschaulicht ist."

Fig. 7 Felix Klein's visualization of a second order branch point. From Felix Klein, "Ueber eine neue Art der Riemannschen Flächen (Erste Mitteilung)," *Mathematische Annalen* 7 (1874): 558–566, here 566

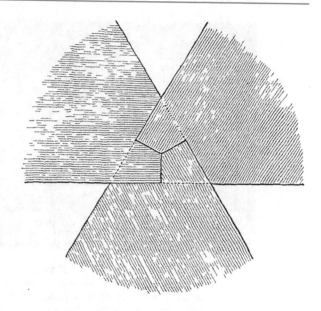

If one considers this example, and above all how Klein refers to the different 'images' of a complex curve, then one notices in the first quotation that while the first 'image' is a two-dimensional drawing (the algebraic curve as an image of the algebraic function), the second is a three-dimensional model (or even two models; see Fig. 8). Consequently, the term 'image' (*Bild*) functions in Klein as an umbrella term that, first, allows one to think together sensory-concrete three-dimensional models and two-dimensional representations, and, second, starting from this empirical level, points to a transition to be sought between the representations that has not yet been drawn or built, and in this respect should probably be conceived as an action of understanding and imagination. Klein's remarks suggest that he understands the two-dimensional drawing and the three-dimensional model as different but equally valid interpretations of the same mathematical object, whereby only their interaction does justice to the fact that there are several ways of exploring the mathematical properties of the object in question. That these two empirical methods of visualization—three-dimensional models and two-dimensional drawings—are thought together in order to revive and advance mathematical research is also clear in Ludwig Brill's introduction to the third edition of his catalogue of mathematical models in 1885: "[…] it will continue to be the publishing house's aim to serve those scientific circles that see the use of *models and drawings* as an aid and a strong support for the *promotion and stimulation* of mathematical studies."[69]

---

[69] Ludwig Brill, ed., *Catalog mathematischer Modelle*, 3rd ed. (Darmstadt: Brill, 1885), iv: "[…] so wird es auch künftig das Bestreben der Verlagshandlung sein, denjenigen wissenschaftlichen Kreisen zu dienen, welche in dem Gebrauch von *Modellen und Zeichnungen* ein Hülfsmittel und

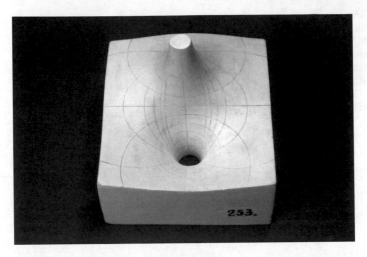

Fig. 8 Model by Adolf Wildbrett of the real part of the (complex) function $w = 1/z$. Similar models were made for the imaginary part of this function, as well as for other functions. (Georg-August-Universität Göttingen, Collection of Mathematical Models and Instruments, model 253). © Collection of Mathematical Models and Instruments at Georg-August-Universität Göttingen, all rights reserved

According to Klein, the ideal approach to mathematical objects lies in a 'transition' between different 'images.' This transition, however, appears to go beyond the visualization on paper and/or in plaster or with threads and is probably reserved for mental *Anschauung*, which qua combination of images can drive the investigation further. This combinatorics of the representations that aims, in the sum, to arrive at a better—since more precise and, in an epistemic respect, productive—image is stated more explicitly by Dyck in 1892. In connection with his construction of models for the real and imaginary part of complex functions, he notes:

> The present series of models was made following an introductory lecture on function theory. [...] In order to *visualize* the course of a function of a complex variable in the vicinity of certain singular points [...] by *a spatial representation*, both the real and the imaginary part of the function values are plotted in the familiar way over the plane of the complex argument as ordinates. Thus, each function of a complex argument is made sensible [*versinnlicht*] through two surfaces designated R and I, whose simultaneous observation provides an *image* of the function's course.[70]

---

eine kräftige Stütze zur *Förderung und Belebung* mathematischer Studien erblicken." (Emphasis M.F./K.K.).

[70] Walther Dyck, ed., *Katalog mathematischer und mathematisch-physikalischer Modelle, Apparate und Instrumente* (Munich: Wolf & Sohn, 1892), 176: "Die vorliegende Serie von Modellen ist entstanden im Anschluss an eine einleitende Vorlesung über Functionentheorie. [...] Um den Verlauf einer Function einer complexen Veränderlichen in der Umgebung gewisser singulärer Stellen

Only the simultaneity—and that means the combination of the sensory-concrete representations of mathematical objects—can deliver the 'image' (*Bild*) as such, which is the image of a process and basically a 'movement image' that arises in the imagination. As a result, the image concept is distanced in mathematics from a simple representational function (*Abbild*) and is to some extent abstracted as well as, and above all, made mental. The change in meaning of the image extends beyond mathematics and can be seen particularly clearly in the arts of the outgoing nineteenth and early twentieth centuries, for instance when in Impressionism the focus was on the (physiological) reality of the eye, or, in the programs of abstract painting, the autonomy of art became a theme, as in Wassily Kandinsky's book *Über das Geistige in der Kunst* (*Concerning the Spiritual in Art*), first published in 1912.[71] Yet while the fine arts programmatically expanded their understanding of the image, the image in mathematics was forced to take a back seat behind the word and language. When in a lecture given in 1886 Brill summarized the model movement as a successful one, he also noted a growing tendency toward or limitation to what he called 'the word,' that is, to a more formal language or to a more language-oriented conception of mathematics, embodied in synthetic geometry. This tendency led to an "underestimation" of the model in particular and of the image in general: "[...] this limitation to the word, which the synthetic geometers fondly favored for a time, could not fail to lead to an underestimation of the image and the technical skills required to produce it."[72] If one follows Brill here, one may think that it is the growing abstraction in mathematics at the turn of the nineteenth to the twentieth century that slowly brought the model tradition to its decline.

Indeed, for Brill and Klein, the word—that is, the abstract approach (represented here by synthetic geometry)—should and could not function as that which unifies two phenomena or two images as if to reconcile between them, since, to quote from Klein's 1893 lecture in Chicago, "mathematical models and courses in drawing are calculated to disarm [...] the hostility directed against the excessive abstractness of the university instruction [of mathematics]."[73] Less than ten years later, Boltzmann, in his article from 1902, suggested a rather different approach,

---

[...] durch eine *räumliche Darstellung zu veranschaulichen*, sind in der bekannten Weise sowohl der reelle als auch der imaginäre Teil der Functionswerte über der Ebene des complexen Argumentes als Ordinaten aufgetragen. So wird jede Function eines complexen Argumentes durch zwei mit *R* und *I* bezeichnete Flächen versinnlicht, deren gleichzeitige Betrachtung ein *Bild* des Functionsverlaufes liefert." (Emphasis M.F./K.K.).

[71] For a detailed study on this, see: W. J. Thomas Mitchell, "What Is an Image," in *Iconology* (Chicago: Chicago University Press, 1986), 7–46.

[72] Brill, "Über die Modellsammlung des mathematischen Seminars der Universität Tübingen," 71: "[...] diese von den Synthetikern eine Zeit lang mit Vorliebe gepflegte Beschränkung auf das Wort konnte nicht ermangeln, zu einer Unterschätzung des Bildes und der technischen Fertigkeiten, die zu dessen Herstellung erforderlich sind, zu führen."

[73] Felix Klein, *The Evanston Colloquium: Lecture on Mathematics* (New York: American Mathematical Society, 1911), 109.

pointing to a shift in how the term 'model' was considered: models in mathematics had been reduced to material models of surfaces and curves, only "elucidating [...] singularities,"[74] and the goal of mathematical modeling had shifted toward another scientific activity in which this modeling could be understood as 'reconciling' between two physical phenomena via a completely symbolic, abstract practice: the finding of an equation.[75]

IV. 1900s–1930s: From Material Analogies and 'Geometric Models' to Formal Analogies and Language-Oriented Models

Boltzmann's view that the aim of mathematical modeling was the finding of a unifying equation exemplifies how the understanding of the term 'model' changed in the first decades of the twentieth century from a material to a symbolic, language-based concept. We would like to examine two examples of this shift: to sketch the rise of the term 'model' as an instantiation of a mathematical system of axioms; and to examine the biological 'paper and pencil' models.

(1) 1891/1899/1936: Mathematics and the New Definition of 'Model'

One of the challenges for mathematicians in the second half of the nineteenth century consisted in finding a visualization for hyperbolic geometries. In the 1860s the Italian mathematician Eugenio Beltrami built material models that should do just that. He described this realization in the concrete model as an "interpretation" ("interpretazione") of hyperbolic geometry.[76] Later, one constructed also for other surfaces (or parts of surfaces) of non-Euclidean geometries equivalents in three-dimensional Euclidean space that can be understood as interpretations. In his paper "Les Géométries non Euclidiennes" from 1891, the French mathematician

---

[74] Boltzmann, "Models," 789.

[75] See: the quotation introduced above from the end of Boltzmann's article: "It often happens that a series of natural processes [...] may be expressed by the same differential equations." (Ibid., 791.) While, according to Hon and Goldstein, the physical understanding of modeling (i.e., the employment of models as a method of acquiring physical knowledge) had already been developed in Maxwell's late texts (esp. *A Treatise on Electricity and Magnetism*, 1873), this was only properly established by subsequent physicists, especially Oliver Lodge and George Francis Fitzgerrald. See: Hon and Goldstein, "Maxwell's Role in Turning the Concept of Model into the Methodology of Modeling," 327.

[76] Eugenio Beltrami, "Saggio di interpretazione della geometria non-Euclidea," in *Opere matematiche di Eugenio Beltrami*, vol. 1 (Milan: Mapli, 1868), 374–405.

Henri Poincaré discusses possible 'interpretations' of non-Euclidean geometries, and uses for this also the concept of the dictionary:[77]

> [...] let us construct a kind of dictionary by making a double series of terms written in two columns, and corresponding each to each, just as in ordinary dictionaries, the words in two languages which have the same signification correspond to one another [...]. [We can obtain theorems from one interpretation within the second one] as we would translate a German text with the aid of a German–French dictionary.[78]

The metaphor of the dictionary can be read as a sign that, already at the turn of the nineteenth to the twentieth century, mathematicians understood Beltrami's concept of interpretation with respect to mathematics as a language, although Beltrami himself probably hardly meant it that way. Poincaré's use of a metaphor borrowed from the realm of language did not remain without echo, however. In his book *Non-Euclidean Geometry*, published in 1906, Roberto Bonola points out that the model for non-Euclidean geometry was a material model, and he includes a photo showing one of Beltrami's paper models. Bonola emphasizes this analogical relation and describes it as a translation: "There is an analogy between the geometry on a surface of constant curvature [...] and that of a portion of a plane, both taken within suitable boundaries. We can make this analogy clear by translating the fundamental definitions and properties of the one into those of the other."[79] A few years later Hermann Weyl writes about "euclidian model" ("euklidisches Modell") of non-Euclidean geometry and understands this again as a translation with the help of a dictionary: "We now take up a dictionary with which the concepts of Euclidean geometry are translated into a foreign language, a 'non-Euclidean' one."[80] These examples already indicate that in the early twentieth century there was a shift in meaning that detached the mathematical model from its characterization qua visual *Anschaulichkeit* and concrete, material appearance and initiated a language-based

---

[77] Henri Poincaré, "Les Géométries non Euclidiennes," *Revue générale des sciences* 2 (1891), 769–74. The section dealing with models of non-Euclidean geometries is called "*Interprétation des géométries non-euclidiennes*" (ibid., 771).

[78] Ibid.: "[...] construisons une sorte de dictionnaire, en faisant correspondre chacun à chacun une double suite de termes écrits dans deux colonnes, de la même façon que se correspondent dans les dictionnaires ordinaires les mots de deux langues dont la signification est la même [...]. Prenons ensuite les théorèmes de Lowatchewski et traduisons-les à l'aide de ce dictionnaire comme nous traduirions un texte allemand à l'aide d'un dictionnaire allemand-français."

[79] Roberto Bonola, *Non-Euclidean Geometry: A Critical and Historical Study of Its Development* (Chicago: Open Court Publishing Company, 1912 [1906]), 134. See also: Jeremy Gray, "Anachronism: Bonola and non-Euclidean geometry," in *Anachronisms in the History of Mathematics: Essays on the Historical Interpretation of Mathematical Texts*, ed. Niccolò Guicciardini (Cambridge: Cambridge University Press, 2021), 281–306.

[80] Hermann Weyl, *Raum Zeit Materie—Vorlesungen über Allgemeine Relativitätstheorie* [1918], 3rd ed. (Berlin: Springer, 1919), 71–72: "Wir stellen jetzt ein Lexikon auf, durch das die Begriffe der Euklidischen Geometrie in eine fremde Sprache, die 'Nicht-Euklidische' übersetzt werden [...]." In this connection, Weyl also quotes Bonola's work, but fails to mention the models of Beltrami.

understanding of the model. This shift becomes explicit in the formal-logical concept of the model that became widespread following Alfred Tarski's writings of the 1930s. For Tarski, a Polish-American logician, the model should be defined as a correspondence between certain logical formulas and a certain mathematical structure. More precisely, the model was an instantiation—that is, a "realization"—of a formal axiom system, as in the following formulation from 1936:

> Let *L* be an arbitrary class of propositions. We replace all extra-logical constants occurring in the propositions of the class *L* by corresponding variables. [...] An arbitrary sequence of objects satisfying each propositional function of the class *L'* will be called a model or *realization of the propositional class L* (in just this sense one usually speaks about the model of the axiom system of a deductive theory).[81]

Tarski's approach does not come from nowhere, but is prepared for by Weyl, among others, who had already undertaken a dematerialization of the concept of model and understood this as the instantiation of an axiom system. Another influence was probably the writings of the mathematician David Hilbert. In his book *Grundlagen der Geometrie*, published in 1899, Hilbert discusses various axiom systems and presents a number of different geometries that satisfy these axioms. Here, Hilbert does not use the term 'model,' however, but speaks of a "system."[82] He is also familiar with Maxwell's research. In a letter to Gottlob Frege from December 29, 1899, shortly after the publication of *Grundlagen der Geometrie*, Hilbert mentions not only the mathematical theory of duality but also Maxwell's theory of electricity, and understands both as examples of a kind of abstract-analogical "transformation": "Any theory can always be applied to an infinite number of systems of basic elements. One only needs to apply [...] a transformation."[83] And this statement comes immediately after the famous sentence: "If among my points I think of some system of things, for example, the system love, law, chimney

---

[81] Alfred Tarski, "Über den Begriff der logischen Folgerung," in *Actes du Congrès International de Philosophie Scientifique, VII Logique, Actualités scientifiques et industrielles* 394 (Paris: Hermann & Cie, 1936), 1–11, here 8: "Es sei nun *L* eine beliebige Klasse von Aussagen. Wir ersetzen alle außerlogischen Konstanten, die in den Aussagen der Klasse *L* auftreten, durch entsprechende Variablen. [...] Eine beliebige Folge von Gegenständen, die jede Aussagefunktion der Klasse *L'* erfüllt, wollen wir als Modell oder *Realisierung der Aussageklasse L* bezeichnen (in eben diesem Sinne wird üblicherweise vom Modell des Axiomensystems einer deduktiven Theorie gesprochen)."

[82] David Hilbert, *Grundlagen der Geometrie* (Leipzig: Teubner 1899), 1, 4, and passim—and also: "System von Strecken" ("system of segments") (8, 37), "System von Dingen" ("system of things") (26), "System von Punkten, Geraden und Ebenen" ("system of points, lines, and planes") (39), etc.

[83] Hilbert to Frege in: Gottlob Frege, *Gottlob Freges Briefwechsel mit D. Hilbert, E. Husserl, B. Russell, sowie ausgewählte Einzelbriefe Freges* (Hamburg: Meiner, 1980), 6: "jede Theorie kann stets auf unendliche viele Systeme von Grundelementen angewandt werden. Man braucht nur eine [...] Transformation anzuwenden."

sweep…, and then assume all my axioms as relations between these things, then my propositions, for example, the Pythagoras, are also valid for these things."[84]

At this point it is essential to mention another protagonist who is important for the development of an abstract model concept, between Hilbert and Tarski: Hausdorff, in whose writings one can observe such a change in the use of the model concept already in 1903/4—thus even earlier than in Weyl. Hausdorff was among the early readers of *Grundlagen der Geometrie*.[85] Hilbert's axiomatic method attracted Hausdorff's attention and inspired him to undertake his own studies, that is, to find his own examples and counterexamples as well as single axioms and axiom groups of geometry. At first, however, he did not call these examples *models* of the systems of axioms. This happened only in 1903 in the context of his discussion on the consistency of non-Euclidean geometries—here, he already anticipates the later formal-abstract meaning of the term.

In a manuscript titled "Nichteuklidische Geometrie," probably written in 1901 or 1902,[86] Hausdorff declares *Abbilden* (the image of a mathematical mapping) to be a general strategy of a step-by-step proof of the consistency of an axiomatic system. Also the "Beltrami-Cayley'sche Bild" ("Beltrami–Cayley image") or generally a "euklidisches Bild" ("Euclidean image"),[87] as Hausdorff now formulates it, is given this new role as an instrument of a proof of consistency. Yet in Hausdorff's manuscript the term 'model' does not appear. This is no longer the case, however, in Hausdorff's later texts on the problem of space.[88] Thus, in "Das Raumproblem," which was his inaugural lecture at Leipzig University, he presents an axiomatic method that introduces a variety of different geometric systems that are limited solely by the criterion of consistency. Here, a terminological shift can be observed when Hausdorff writes: "The absence of a contradiction has been directly demonstrated by appropriate mappings [*Abbildungen*] of non-Euclidean

---

[84] Ibid.: "Wenn ich unter meinen Punkten irgendwelche Systeme von Dingen, z.B. das System: Liebe, Gesetz, Schornsteinfeger…, denke und dann nur meine sämtlichen Axiome als Beziehungen zwischen diesen Dingen annehme, so gelten meine Sätze, z.B. der Pythagoras auch von diesen Dingen."

[85] See: Moritz Epple, "Felix Hausdorffs Erkenntniskritik von Zeit und Raum," in Felix Hausdorff, *Gesammelte Werke*, vol. VI: *Geometrie, Raum und Zeit*, ed. Moritz Epple (Berlin: Springer 2021), 1–207, here 119–20.

[86] Felix Hausdorff, "Nichteuklidische Geometrie" in Felix Hausdorff, *Gesammelte Werke*, vol. VI: *Geometrie, Raum und Zahl*, ed. Moritz Epple (Berlin: Springer, 2021), 347–90.

[87] Ibid., 370.

[88] Hausdorff makes the problem of space the subject of his inaugural lecture "Das Raumproblem" in Leipzig on July 4, 1903, as well as of the lecture "Zeit und Raum," which he gives the following winter semester. See: Epple, "Felix Hausdorffs Erkenntniskritik von Zeit und Raum," 141–42, and Felix Hausdorff, "Das Raumproblem," in Felix Hausdorff, *Gesammelte Werke*, vol. VI: *Geometrie, Raum und Zahl*, ed. Moritz Epple (Berlin: Springer, 2021) 279–304; Felix Hausdorff, "Zeit und Raum," in Felix Hausdorff, *Gesammelte Werke*, vol. VI: *Geometrie, Raum und Zahl*, ed. Moritz Epple (Berlin: Springer, 2021), 391–450.

geometries onto Euclidean models and of Euclidean geometries onto pure arithmetic."[89] What Hausdorff had previously still designated as a "Euclidean image" now becomes a "Euclidean model," by which he does not mean material models but Euclidean geometry. As Moritz Epple has pointed out, this is one of the earliest occurrences of an abstract model concept.[90] At the same time, however, Hausdorff continues to use the term 'model' in its earlier mathematical sense (as a material model). This is the case, for example, in his inaugural lecture.[91]

One finds the new use of the term 'model' in an even more explicit form in the lecture "Zeit und Raum," which he gave during the winter semester 1903/04. Here, Hausdorff argues that, between two different "systems" of geometric objects that satisfy the same axioms, one can find a "translation" (thus drawing on the metaphor that had already been used by Poincaré in 1891): "One can visualize this principle of mapping [*Abbildungsprincip*] as a kind of translation from one language of geometry into another on the basis of a dictionary."[92] Here, it is a matter of the validity of Euclidean geometry, which is demonstrated by the mapping of geometric concepts onto their arithmetical counterparts.[93] For these images of mathematical mapping ((*Ab-*)*Bilder*), Hausdorff uses—as in his inaugural lecture—the term 'model,' but only with the image of a geometric system of objects of Euclidean geometry. For instance, he selects spherical geometry as the first example for the discussion of the consistency of a non-Euclidean geometry,

> because a Euclidean model can be immediately located for it. Instead of constructing a non-Euclidean geometry directly from the corresponding axioms (as Bolyai and Lobachevsky [...] have done), we want to search for Euclidean images of non-Euclidean relations, that is, to use our transformation principle or to conceive of a dictionary by means of which Euclidean propositions can be translated into non-Euclidean ones.[94]

---

[89] Felix Hausdorff, "Das Raumproblem," in *Ostwalds Annalen der Naturphilosophie* 3 (1903): 1–23, here 3: "Die Abwesenheit eines Widerspruchs ist durch geeignete Abbildungen der nichteuklidischen Geometrien auf euklidische Modelle und der euklidischen Geometrie auf die reine Arithmetik direkt bewiesen worden."

[90] Epple, "Felix Hausdorffs Erkenntniskritik von Zeit und Raum," 144.

[91] See, for example: Hausdorff, "Das Raumproblem" (1903), 8–9: "[...] man kann sagen: die sphärische Geometrie blieb unentdeckt, weil ihr euklidisches Modell, die Kugel, schon vorhanden war. Dass ein solches Modell auch für die pseudosphärische Ebene später gefunden wurde, in Gestalt der Flächen konstanter negativer Krümmung, erleichterte nicht ihre Erforschung [...]." ["[...] one can say: spherical geometry remained undiscovered because its Euclidean model, the sphere, was already available. That such a model was later found also for the pseudospherical plane, in the form of a surface of constant negative curvature, did not make their research any easier [...]."].

[92] Hausdorff, "Zeit und Raum," 417: "Man kann dies Abbildungsprincip als eine Art Übersetzung aus einer geometrischen Sprache in die andere veranschaulichen, unter Zugrundelegung eines Lexikons."

[93] Epple, "Felix Hausdorffs Erkenntniskritik von Zeit und Raum," 155.

[94] Hausdorff, "Zeit und Raum," 422: "weil sich für ihn unmittelbar ein euklidisches Modell auffinden lässt. Statt nämlich eine nichteuklidische Geometrie direct aus den zugehörigen Axiomen aufzubauen (wie es Bolyai und Lobatschefski [...] gethan haben), wollen wir euklidische Bilder

In his introduction to the abovementioned texts by Hausdorff, Epple argues convincingly that the above quotation as well as the corresponding brief passages in "Das Raumproblem" are among the earliest occurrences of the abstract model concept in Hausdorff's work, and in this respect they open up a path that leads beyond the author: "they simultaneously provide very early proof of a usage that points in the direction of later mathematical model theory."[95]

Thus, already at the beginning of the twentieth century, in Hausdorff's work, the model concept appears in an abstract usage. This does not occur consistently, however, but parallel to the earlier material meaning of the term. Hausdorff's significance for the development of abstract discourse is therefore difficult to determine. Rudolf Carnap and Kurt Gödel use the term 'model' explicitly in the 1920s in the sense of Hilbert's 'system.'[96] Carnap speaks of models of an axiomatic system in 1928,[97] and Gödel uses a similar terminology[98]—despite both coming from different logic traditions. These mathematicians are representative of the environment from which mathematical model theory emerged. This environment can be seen with Tarski's statement that a model is a realization of a class of propositions if one can verify that the theorems have been satisfied. The investigation of the differences between Carnap, Tarski, and Gödel, and the answer to the question of how the modern conception of the model emerged from these different approaches is beyond the scope of this introduction. The previous remarks will have to suffice to make clear that it was irrelevant for the mathematicians mentioned above whether a tangible model existed on which or with which the theorems could be located or demonstrated. What was relevant was now only the language-based reality of mathematics.

---

der nichteuklidischen Verhältnisse suchen, d. h. unser Transformationsprincip anwenden oder ein Lexikon ersinnen, vermöge dessen sich euklidische Sätze in nichteuklidische übertragen lassen."

[95] Epple, "Felix Hausdorffs Erkenntniskritik von Zeit und Raum," 155: "Sie bilden zugleich einen sehr frühen Beleg für eine Verwendung, die in Richtung der späteren mathematischen Modelltheorie weist." Epple points out, however, that Hausdorff also continued to use the earlier, material meaning of the term 'model' in his lecture "Zeit und Raum." Therefore, according to Epple, one cannot ultimately be certain "wie viel Gewicht Hausdorff zu diesem Zeitpunkt der abstrakten Rede von den (euklidischen) 'Modellen' beimaß." ["how much weight Hausdorff attached at this point to the abstract discourse of the (Euclidean) 'models.'"] Ibid., 156.

[96] See also: Paolo Mancosu, "Tarski on models and logical consequence," in The Architecture of Modern Mathematics: Essays in History and Philosophy, ed. Jose Ferreirós and Jeremy J. Gray (Oxford and New York: Oxford University Press, 2006), 209–37, esp. 210, footnote 2.

[97] Rudolf Carnap, Untersuchungen zur allgemeinen Axiomatik (Darmstadt: WBG, 2000), 94.

[98] Kurt Gödel, "Über die Vollständigkeit des Logikkalküls," in Kurt Gödel, Collected Works, vol. I: Publications 1929–1936 (Oxford: Oxford University Press 1986), 60–101, here 60.

(2) 1931/1925-6: The 'Pencil and Paper Models' of Biology and the Precursors of Modeling

When Tarski speaks of the realization of a propositional class, this underlines an aspect that was already implicit in Brill's lecture from 1886—the realization in language.[99] At roughly the same time as Tarski's statement about realization, Nicolas Rashevsky, one of the pioneers of mathematical biology, also introduced the term 'model' into his work. For Rashevsky, the mathematical formalization of biological phenomena was a central concern.

Rashevsky initially studied theoretical physics at the University of Kyiv in Ukraine, before fleeing to the USA in 1924 and eventually developing the idea of biomathematics at the University of Pittsburgh in the 1930s. His attempt to shed light on the complexity of biological phenomena led him to the study of the cell,[100] from whose actual appearance he nevertheless abstracted by equating its structure with a simpler geometric entity—a sphere or an ellipsoid. As Maya Shmailov has shown in her 2016 book *Intellectual Pursuits of Nicolas Rashevsky*,[101] Rashevsky was able to use this theoretical cell to develop mathematical equations that described the cell's growth and division. He referred to these equations as "pencil and paper models" and attached a higher value to them than to the "actual 'experimental' models" of physics. Already in 1931 he noted:

> The use of models is not unfamiliar to the physicist. However the physicist uses models in a somewhat different way. [...] [N]o one of them [the physicists] did actually 'build' any such models, nor experiment with them. These models were, if we may call them so, 'paper and pencil models.' [...] [A physicist] may satisfy himself by investigating mathematically, whether such a model is possible or not. The value of such 'paper and pencil' models is not only as great as that of actual 'experimental' models, but in certain respects it is even greater. [...] It is through the study of models that Maxwell finally arrived at his equations, which are and probably will forever remain one of the cornerstones of physics.[102]

In order to show whether and to what extent Rashevsky adopted and changed the understanding of the model in physics and mathematics, it will be necessary to refer to another example from the field of biology in which it is also a matter of mathematical representation, namely, the Lotka–Volterra model, as it is called today, or the predator–prey model.

---

[99] See: Brill, "Über die Modellsammlung des mathematischen Seminars der Universität Tübingen," 71: "Aber diese [...] Beschränkung auf das Wort konnte nicht ermangeln, zu einer Unterschätzung des *Bildes* [...] zu führen." ["[T]his limitation to the word [...] could not fail to lead to an underestimation of the *image* [...]."].

[100] ...and the fact that no two cells are exactly alike (here, we follow Maya Shmailov, *Intellectual Pursuits of Nicolas Rashevsky* (Basel: Birkhäuser, 2016), 66ff.). See: Nicolas Rashevsky, "Some Theoretical Aspects of the Biological Applications of Physics of Disperse Systems," in *Physics* 1, no. 3 (1931): 143–53.

[101] Shmailov, *Intellectual Pursuits of Nicolas Rashevsky*.

[102] Rashevsky, "Some Theoretical Aspects of the Biological Applications of Physics of Disperse Systems," 144.

Alfred James Lotka was an American mathematician, physicist, and statistician. Vito Volterra was an Italian mathematician and physicist known for his contributions to mathematical biology and integral equations. Both worked out the predator–prey equations independently and at approximately the same time, in 1925–26. Their equations describe the interaction of predator and prey populations, whereby the primary variable is the size of the predator or prey population. Because predator and prey populations interact with each other, their population dynamics are interconnected: whereas the predators reduce the population of the prey by eating them, the prey increase the population of the predators by providing them with food. In the current literature these equations are mostly designated as the Lotka–Volterra model—for example, in Michael Weisberg's 2013 book *Simulation and Similarity*.[103] Nevertheless, it is important to recall the actual terms used by Lotka and Volterra themselves. In Lotka's book *Elements of Physical Biology*, published 1925, for example, the word 'model' (and the corresponding verb 'to model') do not appear. In his preface, however, Lotka expressly notes that the spirit of his work is mathematical systematization:

> It is hoped that the mathematical character of certain pages will not deter biologists and others [...] from acquiring an interest in other portions of the book. [...] I may perhaps confess that I have striven to infuse the mathematical spirit also into those pages on which symbols do not present themselves to the eye.[104]

Similar formulations are found in Volterra's writings from 1926—although Volterra too does not describe his work as modeling or use the term 'model.' He does, however, set out the reasons for constructing a simplified representation, that is, a representation that reduces biological complexity and only provides an "approximate image" ("immagine approssimata") of this:

> In order to deal with the question [of predation] mathematically, it is better to start from hypotheses that, even if they distance themselves from reality, give an approximate image of it. Even if the representation will be, at least at first, coarse, [...] it will be possible to apply the calculation [...]. [Hence it is] advisable [...] to schematize the phenomenon by isolating the actions to be examined, assuming that they work alone and neglecting the others.[105]

Volterra addresses the problem of complexity by abstracting from it: the various elements have to be isolated in order to be represented mathematically. This is in

---

[103] Michael Weisberg, *Simulation and Similarity* (Oxford et al.: Oxford University Press, 2013), 3, 10, 11, 12, and passim.

[104] Alfred J. Lotka, *Elements of Physical Biology* (Baltimore: Williams and Wilkins, 1925), ix.

[105] Vito Volterra, "Variazioni e fluttuazioni del numero d'individui in specie animali conviventi," in Vito Volterra, *Opere Matematiche: Memorie e Note*, vol. 5: *1926–1940*, (Rome: Accademia Nazionale Dei Lincei, 1962), 1–111, here 1–2: "Per poter trattare la questione matematicamente conviene partire da ipotesi che, pure allontanandosi dalla realtà, ne diano una immagine approssimata. Anche se la rappresentazione sarà, almeno in un primo momento, grossolana, pure, [...], vi si potrà applicare il calcolo [...]. Quindi conviene, [...], schematizzare il fenomeno isolando le azioni che si vogliono esaminare, supponendole funzionare da sole e trascurando le altre."

keeping with Lotka's appeal to systematization and with Rashevsky's approach—although Rashevsky explicitly defines his activity as modeling.

Whereas, in the 1920s, the term 'model' was only rarely used explicitly for the procedures of the abstraction and mathematical representation of certain well-selected (biological) processes, this situation changed decisively in the 1950s and 1960s. The *Proceedings of the Fourteenth Symposium in Applied Mathematics of the American Mathematical Society* (1961) is devoted entirely to "Mathematical Problems in the Biological Sciences," and almost all the contributors describe their activity as 'modeling' or explicitly use the term 'model.' A few passages from the *Proceedings* should make this clear:

(1) "[...] [one can] show how a particular model is formulated mathematically."[106]
(2) "In all these respects our model idealizes reality [...]. [The drawn] conclusions [...] with respect to the model are of physical interest only insofar as they do not depend too grossly on these specific features of the model."[107]
(3) "One of the great advantages of setting up model biochemical systems is that they can be tested in new situations with little difficulty."[108]

Against the background of this frequent use of model vocabulary, it becomes apparent that the sense of the term 'model' differs considerably depending on whether it appears in a purely mathematical context or in the context of applied mathematics (in physics, biology, or in other disciplines). In the mathematics of the 1930s, the term 'model' was given a particular molding: here, it is defined as the 'theoretical realization' of a formal axiomatic system. Also in physics and biology the term underwent a clarification in the first half of the twentieth century; however, this differs considerably from the mathematical convention: in the natural sciences the mathematical model now designated a local, temporary mathematical abstraction of complex processes. Moreover, the term 'model' no longer appeared alone but as part of a terminological network: it was associated with other terms that already referred to the process of abstraction, such as 'theory,' 'analogy,' 'idealization,' and 'representation.'

---

[106] Herbert D. Landahl, "Mathematical models in the behavior of the central nervous system," in *Mathematical Problems in the Biological Sciences* (Rhode Island: American Mathematical Society, 1962), 1–16, here 1.
[107] Max Delbrück, "Knotting Problems in Biology," in *Mathematical Problems in the Biological Sciences* (Rhode Island: American Mathematical Society, 1962), 55–63, here 56.
[108] Arthur B. Pardee, "Biochemistry: Sterile or virgin for mathematicians?" in *Mathematical Problems in the Biological Sciences* (Rhode Island: American Mathematical Society, 1962), 69–82, here 73.

V. 1940s: Lévi-Strauss and Mathematical Models in Anthropology

In the twentieth century, this new conception of mathematical modeling—as a 'translation' of carefully selected 'specific features'—was not unique or limited to the discourse of the natural sciences. This is to be seen in the writings of Claude Lévi-Strauss (1908–2009), whose work was essential for the development of structural anthropology. His book *Les structures élémentaires de la parenté* (*The Elementary Structures of Kinship*), published in 1949, is considered one of the most important anthropological works on kinship of the period.[109] In this work Lévi-Strauss draws directly on the help of the mathematician André Weil to grasp the marriage customs of the Murngin of northern Australia as a coherent system of the exchange of women, and thus to decipher the various structures of kinship.[110] The empiricism of possible and impossible marriage ties seems too complex to be reduced to general laws and thus rules of combination. The ethnologist therefore had to limit himself to writing lists of commandments and prohibitions as well as descriptions of the observed relations, and could at best compile statistics. The mathematician Weil, however, kept himself at a distance as much from the qualitative descriptions of the ethnographer as from the quantitative assessment of social behavior. Instead, he enlisted the concept of the mathematical group[111]: first, he selected the characteristic permutations of the phenomenon at hand to then classify these in terms of their algebraic properties and deduce the group that enclosed them. Weil remarks:

> The most difficult thing for the mathematician, when it comes to applied mathematics, is often to understand what is at issue and to translate the data of the question into his

---

[109] On the mathematical approach of Lévi-Strauss and his understanding of model, see. Stephan Kammer and Karin Krauthausen, "Für einen strukturalen Realismus: Einleitung," in *Make it Real: Für einen strukturalen Realismus*, ed. Stephan Kammer and Karin Krauthausen (Diaphanes: Zürich, 2020), 7–79, here 56–64; Mauro W. Barbosa de Almeida, "Symmetry and Entropy: Mathematical Metaphors in the Work of Lévi-Strauss," in *Current Anthropology* 31, no. 4 (August–October 1990): 367–85; Jack Morava, "On the Canonical Formula of C. Lévi-Strauss," arXiv.org 2003, online: arXiv:math/0306174 (accessed December 2, 2021); Michael Bies, "Das Modell als Vermittler von Struktur und Ereignis: Mechanische, statistische und verkleinerte Modelle bei Claude Lévi-Strauss," in *Forum Interdisziplinäre Begriffsgeschichte* 5, no. 1 (2016), 43–54, online: https://www.zfl-berlin.org/files/zfl/downloads/publikationen/forum_begriffsgeschichte/ZfL_FIB_5_2016_1_Bies.pdf (accessed December 2, 2021).

[110] See: André Weil, "Appendix to Part One: On the Algebraic Study of Certain Types of Marriage Laws (Murngin System)," in Claude Lévi-Strauss, *The Elementary Structures of Kinship* (Boston: Beacon Press, 1969), 221–29.

[111] A group is a set of elements A together with an associated action, such that the four following axioms are fulfilled: closure, associativity, the existence of an identity element, and the existence of an inverse. An intuitive example is the set of integer numbers with addition. Weil uses the various characteristics of the permutation group. On the history of group theory in mathematics, see: Wussing: *The Genesis of the Abstract Group Concept*.

own language. Not without difficulty, I finally saw that it all came down to studying two permutations and the group they generate.[112]

Concretely, he begins with the observation that the Murngin are divided into marriage classes and that each man or each woman can seek his or her marriage partner only in certain classes that correspond to his or her own class. Membership of a class is determined by one's ancestry, so that the alliance of the parents dictates the possible alliances of the son and daughter. Weil assumes four classes and identifies first the four corresponding marriage types of the parents, then the marriage types of the sons and of the daughters derived from these. To these three series, in which the marriage type of the sons and of the daughters is already in a relation of permutation to the marriage type of the parents, he then adds specific conditions, for instance that each man may marry the daughter of his mother's brother. The types and the conditions, or the resulting possible constellations for son and daughter, are translated by Weil into an algebraic description, which in turn is subjected to certain hypotheses and in this way developed further. In the end he obtains two function formulas (one each for son and daughter), which together define the structures of the marriage system, and which should provide an answer with regard to both possible and impossible alliances.[113] The (utopian) long-term goal of this mathematically processed ethnology is to describe the marriage behavior and the resulting kinship systems of a large number of societies by means of algebraic formulas, and thus to guarantee a formal comparability. The precondition remains, however, that a manageable number of permutations can be identified, and that the added conditions have a broad validity (–neither could be guaranteed for the investigation of modern Western societies, for example).

When Weil describes applying mathematical structures to structural anthropology as a translation into one's own language, he is clearly borrowing or drawing on the common understanding of 'model' in the sciences of the time. This becomes evident in an explicit way when Lévi-Strauss, in the chapter "Social Structure" of his book *Structural Anthropology*,[114] complements the concept of structure with

---

[112] André Weil, *Œuvres Scientifiques—Collected Papers*, vol. 1, *1926–1951* (New York et al.: Springer, 1979), 563–64, here 563: "Le plus difficile pour le mathématicien, lorsqu'il s'agit de mathématique appliquée, est souvent de comprendre de quoi il s'agit et de traduire dans son propre langage les données de la question. Non sans mal, je finis par voir que tout se ramenait à étudier deux permutations et le groupe qu'elles engendrent." The selection of the decisive permutations from the manifold of the appearances is the first step of the analysis. The process of modeling is then developed further by the precise characterization of the permutations (are they associative or have an inverse?), which basically represent selected conditions of reality (e.g., that a man may marry the daughter of his mother's brother). Weil shows that in the case of the Murngin the two permutations found are interchangeable that is, commute and follow the mathematical definition of an abelian group. In this way, the system of kinship is formalized.

[113] See the confident conclusion in: Weil, "Appendix to Part One: On the Algebraic Study of Certain Types of Marriage Laws (Murngin System)," 226: "By means of these formulas, all the questions relating to this marriage law can easily be submitted to arithmetic examination." The formulas describe the various permutations.

[114] Claude Lévi-Strauss, *Structural Anthropology* (New York: Basic Books, 1963), 277–323.

the concept of model, whose definition he adopts in the form of a quotation from John von Neumann and Oskar Morgenstern's book *Theory of Games and Economic Behaviour* (1944):

> Such models [as games] are theoretical constructs with a precise, exhaustive and not too complicated definition; and they must be similar to reality in those respects which are essential in the investigation at hand. To recapitulate in detail: The definition must be precise and exhaustive in order to make a mathematical treatment possible. The construct must not be unduly complicated so that the mathematical treatment can be brought beyond the mere formalism to the point where it yields complete numerical results. Similarity to reality is needed to make the operation significant. And this similarity must usually be restricted to a few traits deemed 'essential' *pro tempore*—since otherwise the above requirements would conflict with each other.[115]

By taking up this model concept from the publication of the mathematician and economist,[116] Lévi-Strauss succeeds in strengthening his approach in three respects: First he reconciles the formalism and universalism of mathematics with the ongoing ethnographic observation of the richly detailed and changeable empirical phenomena by promoting the selective choice of significant aspects and thereby the transition to abstraction, while nevertheless insisting, with von Neumann and Morgenstern, on a "[s]imilarity to reality."

Second, he is able to do justice, via the legitimation of the game as a model, to the connection between necessity and arbitrariness, for example in his study of kinship systems. That is, he need relinquish neither the rational claim nor the contingency of the cultural phenomena. With the help of the model, the kinship relations become recognizable as a generalizable system made up of independent parts, precisely because this kind of modeling can describe the diverse and processual "form" of reality as an "action of laws which are general but implicit."[117]

Third and finally, the structuralist approach is conceived and legitimized as a strategic intervention of the scientist in the empirical data. In order to obtain an instructive modeling, both the ethnologist and the scientist must select wisely from the diversity of possible aspects. Hence, the object of ethnology is not a concrete family (or a concrete community of families), but the abstract relations between

---

[115] John von Neumann and Oskar Morgenstern, *Theory of Games and Economic Behaviour* (Princeton: Princeton University Press, 1944), 32–33, quoted in Claude Lévi-Strauss, "Social Structure," in Claude Lévi-Strauss, *Structural Anthropology* (New York: Basic Books, 1963), 277–323, here 316, footnote 3. The interpolation in brackets is by Lévi-Strauss.

[116] On the history and philosophy of the use of models in economics, see: Mary S. Morgan, *The World in the Model: How Economists Work and Think* (Cambridge: Cambridge University Press, 2012).

[117] Claude Lévi-Strauss, "Structural Analysis in Linguistics and in Anthropology," in Claude Lévi-Strauss, *Structural Anthropology* (New York: Basic Books, 1963), 31–54, here 34.

elements of what in the different cultures is recognized as a family.[118] The social structure is not defined inductively but must be read from the constructed model. While the hypothetical prototype in ethnology aims at an "empirical knowledge of the social phenomena," these are not objectively given but are only inferred with the help of modeling.[119] To achieve this it is important to limit the case examples and the aspects to be considered in order to obtain the striven for meaningful simplification and to transpose the observations into a systemic context—that is, to a model in which developments can be experimentally played through and universal structures deduced.

VI. Conclusion: The Model in the Twentieth Century: Fictitious, Fragmentary, Temporary

Lévi-Strauss's use of the term 'model' is remarkably similar to the way it was used at the same time in the natural sciences. Moreover, in the quotation from von Neumann and Morgenstern's book cited above, one can observe a terminological plurality that has already been noted: modeling demands "[s]imilarity to reality" and at the same time "formalism," although this similarity is usually "restricted" when choosing the traits to formalize.

To sum up, one can say that the range of meanings associated with the model concept had already been prepared by Maxwell's research and his use of the terms 'analogy' and 'model.' Maxwell emphasized the 'fictitiousness' of the object of his mathematical-physical model and the incomplete (or, for Mach, abstract) similarity that accompanied the physical analogy. In this way, scientific representation is provided with a crucial degree of latitude, which, rather than obstruct its heuristic function, strengthens it. Boltzmann, for his part, remarks that the mathematical models of the nineteenth century are 'tangible representations' that visualize abstract geometric objects—the ontological question is not present here.

For the period between 1850 and 1950, it can be tentatively stated that scientific representation moves away from the ideal of complete similarity in two respects: On the one hand, mathematical discourse distances itself from material mathematical models, and the term 'model' acquires a purely logical meaning that no longer depends on the similarity to a geometric or physical object. One can suppose that this specialization of the meaning was also prepared by Maxwell's understanding and use of the terms 'analogy' and 'model,' inasmuch as it was Maxwell's

---

[118] For the ethnologist, the family is not a biological but a social fact and should be described in terms of the functional relations that constitute it. See: ibid., 51: "Thus, it is not the families (isolated terms) which are truly 'elementary,' but, rather, the relations between those terms." Accordingly, kinship systems are "symbolic systems" (ibid.).

[119] Ibid., 31. See also: Lévi-Strauss, "Social Structure," 280: For Lévi-Strauss, it continues to hold that "there is a direct relationship between the detail and concreteness of ethnographical description and the validity and generality of the model which is constructed after it." Finally, the model should make the facts "immediately intelligible" (ibid.).

expansion that led Hilbert to discuss how, qua transformation, different systems could be derived from a theory. Nevertheless, starting in the 1930s, the concept of the model in mathematics becomes increasingly precise, since it was now grasped significantly more restrictively and independently of physical empiricism.

On the other hand, one can note that while in the discourse of the exact sciences the model operated as a shifting mediator between theory and reality, the model itself could be entirely temporary, as Maxwell observed. In the twentieth century this temporary character can be seen in the way the relation of the model to reality was to be understood as fragmentary and local (indeed, as a reduction of complexity). The scientific actors were well aware of this fragmentary, local relation of the model to reality. Here, one can detect a twist in the idea of the mathematical representation of 'reality': the models move away from the obligation to a complete representation of reality to become a productive tool of knowledge that can be readjusted according to the context. It is with this productivity of the model, with its local and fragmented character in mind that the contributions and interviews in this volume should be read.

* * *

The various contributions in this volume explore, from different perspectives, this productivity of the mathematical model and of modeling in mathematics. We start with a historical perspective, presenting both a *long durée* history and detailed case studies. We then continue in the second part with an examination of the epistemological and conceptual perspectives. Finally, the third part of the volume investigates not only how material mathematical models were produced, but also how they were exhibited.

Part 1: Historical Perspectives and Case Studies starts with the contribution by **Frédéric Brechenmacher**, which looks back at the early and parallel history of material mathematical models. At the center of Brechenmacher's investigation is the practice of the drawing and construction of mathematical models, a practice that was widespread in France starting in the late eighteenth century and still visible in the 1870s. Many studies on the golden age of material models have focused on the impressive models of higher mathematics from the period starting in 1860; however, in France the drawing of mathematical models of all kinds (i.e., beyond higher mathematics) goes back much further and became a well-established practice for the acquisition of geometric knowledge, particularly at the École polytechnique (which became so important at the end of the eighteenth century), and therefore in the teaching of technicians and engineers. The models of higher mathematics designed by prominent mathematicians partially break with this ideal of a general model-drawing practice, forcing it into the background. Brechenmacher's contribution corrects this by presenting model drawing as part of an important non-textual and practical tradition of (French) mathematics, while also identifying the central protagonists, model builders, and the communities in which the construction and the drawing of models took place. Finally, the author links the tradition of model drawing to the advance of the 'graphical method' and the related successful 'modelization' of empiricism at the end of the nineteenth century.

The following three contributions (by Klaus Volkert, David Rowe, and Tilman Sauer) provide detailed case studies of material mathematical models from the late nineteenth and early twentieth centuries. **Klaus Volkert**'s contribution examines the spectrum and function of mathematical models in polytechnic schools using the example of the ETH Zürich in the second half of the nineteenth century. At the center of this study is the model collection that Wilhelm Fiedler assembled there in the 1860s as a professor of descriptive geometry and geometry of position (*Geometrie der Lage*). This collection was so highly esteemed by Walther Dyck that he invited Fiedler to describe it for the catalogue of an exhibition planned to accompany the conference of the German Mathematical Society in Nuremberg in 1892 (an invitation that Fiedler turned down). The author examines Fiedler's importance for the teaching with and the research on models. Models were traditionally found in polytechnic schools, where they played an important role in the education of technicians and engineers, and in this case also of teachers. The teaching of descriptive geometry was not based on theorems and demonstrations, but presented problems that were solved by means of drawing or by graphic approximation. Fiedler, however, decisively expanded the existing model collection by acquiring additional models (e.g., a series of teaching models by Jakob Schröder) and by working on the production of models himself. As the author is able to show, this was the case also with regard to models for objects of higher mathematics. Already in 1865 Fiedler designed a model of the Clebsch surface (a cubic surface with 27 straight lines)—hence a few years before Christian Wiener's model was made, which today is considered the first. In doing so, Fiedler made use of relief perspectives: a geometric practice that was closer to the practice of artists and architects than to that of mathematicians. His rod model is only an approximation of the geometric Clebsch surface; nevertheless, it demonstrates the importance of polytechnic schools with respect not only to their use of simple models in teaching but also to their handling of models of higher mathematics.

David E. Rowe's contribution addresses the models of complex surfaces (specifically quartic surfaces) designed by Julius Plücker in the 1860s. At that time these models of higher mathematics represented "the more exotic geometrical knowledge" (Rowe), and were of some interest to mathematicians in England, but also to Plücker's student Felix Klein, as well as to Klein's students—before being forgotten at the beginning of the twentieth century. The author reconstructs the early history of these models and their nameable epistemic function in the mathematics of the late nineteenth century. To this end, he initially examines the interaction between mathematics and optics research using the example of the geometric representation and analysis of rays with the help of line congruences in three-dimensional space. Of central importance in this field is Fresnel's wave surface, a model that arose as a derivation from a quartic equation for a wave front of light passing through biaxial crystals, which led to further research in the 1820s, for example in the work of William R. Hamilton, who developed a general theory of ray systems (as infinitely thin pencils of rays), and, later, in the work of Plücker.

The path to Plücker's models passes via Ernst Kummer, who built on Hamilton's theory and developed it in a purely mathematical way for a large class of quartics, also designing thread models for these. The actual goal, however, and not a trivial one, was the general classification of quartic surfaces. This generalization was achieved by, among others, Sophus Liu, and as a result of the rethinking of models by Albert Wenker, Klein, and Klein's student Karl Rohn. Plücker's line geometry and his models of complex surfaces represent an important bridge for this progress in mathematical knowledge: they take up special quartics (that is, surfaces of degree four), which are enveloped by subsets of lines in a quadratic line complex. These central objects of line geometry are a degenerate type of the Kummer surface; but Plücker begins with canonical cases from which he systematically derives all other types. Klein and his student completed Plücker's models, and in this way were able to show the transition (by deformation) from Kummer surfaces to a number of central complex surfaces.

Tilman Sauer's contribution concentrates on a very specific type of model: the plaster models of curved surfaces by the Dutch geometer Jan Arnoldus Schouten. These were used by Schouten to illustrate the novel geometric concept of parallel transport, a concept developed in the framework of differential geometry for manifolds of an arbitrary number of dimensions. Sauer points out that, while this concept arose in a somewhat nonvisual mathematical setting (geometry of higher dimensions, n-dimensional manifolds and their associated curvature), Schouten, in order to assist the geometric *Anschauung* (intuition), turned to a material model to visualize geodesic transport. This material model was illustrative of a conceptual problem that was still being explored at the time; in that sense, the model's physical properties function epistemically—they not only help understand the abstract concept but may have played an important role for the conceptual development of Schouten's work.

The last three contributions in the first part deal with the transformation undergone by modeling during the later decades of the twentieth century in physics (Arianna Borrelli), 'applied' mathematics (Myfanwy E. Evans), and pure mathematics (Fernando Zalamea). **Arianna Borrelli**'s contribution deals with models and symmetry principles in early particle physics. Borrelli discusses the development of theoretical practices between the 1950s and the early 1960s and presents examples of the complex relationship between mathematics and the conceptualization of physical phenomena. Indeed, a closer look at theoretical practices in this period reveals a tension between the employment of advanced mathematical tools and the 'modeling' of observation, when 'model' is understood as a construction enabling the fitting and predicting of phenomena. Due to this tension, the question arises whether it is even possible to make a general claim about the relationship between mathematics and models. This contribution, then, examines a general tension expressed at the time: an opposition between local mathematical constructs that fit phenomena (such as models) and those expressing the more general (hidden) principles of the constructs' coming-to-be (such as theories or structures).

The interview with **Myfanwy E. Evans** explores the connections between pure and applied mathematics and the research on materials. More concretely, Evans shows how these fields intersect with material models and animation software. Through a consideration of the tradition of constructing material models of periodic minimal surfaces—a tradition initiated by Hermann Amandus Schwarz and his student Edvard Rudolf Neovius in the last third of the nineteenth century and continued by Alan H. Schoen during the 1960s—Evans shows how these models help to understand the structure of human skin, and how this applied research, in turn, prompts a mathematical theory of entanglement. Accordingly, these models and their twenty-first century digital visualizations can be considered as epistemic objects, both for applied and pure mathematics.

Last but certainly not least, **Fernando Zalamea** presents in his short but thorough contribution the work of Alexander Grothendieck, who, according to Zalamea, explored the pendulum movement between the abstract and the concrete, the universal and the particular in mathematics. Zalamea stresses two basic directions that Grothendieck examines in the space of transition between archetypes (universal categorical constructions) and types (concrete models): on the one hand, projecting archetypes to types in the 1950s (in his work on the *Tôhoku* paper and the Riemann–Roch theorem) and, on the other, embedding types into archetypes in the 1980s (in the works *Pursuing Stacks* and *Les Dérivateurs*). This pendulum movement between the abstract and the concrete is also seen in Grothendieck's general remarks about models in *Récoltes et semailles*. Zalamea points out, however, that it is both the abstract *and* the concrete, the universal *and* the particular (not the choice between one or the other) that are necessary to nurture the mathematical imagination.

The four contributions in *Part 2: Epistemological and Conceptual Perspectives* present epistemological perspectives on the model and modeling, while also expanding the discussion to include new terms and conceptions. The first contribution, by **Moritz Epple**, presents a panoramic view of the history of abstract representation in the sciences of the nineteenth and early twentieth centuries. Here, Epple concentrates on a number of terms that were employed during this period, such as 'analogy,' 'interpretation,' 'image,' 'system,' and, last but not least, 'model.' By focusing on actors' categories and reviewing various protagonists (physicists, mathematicians, and philosophers such as Ludwig Wittgenstein, Heinrich Hertz, James Clerk Maxwell, Hermann von Helmholtz, Eugenio Beltrami, Felix Klein, and Felix Hausdorff), the author delineates the epistemic ruptures, continuities, and symmetries between these various concepts. Hertz, for example, believed in a correspondence between causally structured reality, the activity of the mind, and scientific theory formation; however, this correspondence was based on an epistemic symmetry of the relations in his 'images' and 'models.' On the other hand, at the end of the nineteenth century and, more strongly, at the beginning of the twentieth century, geometry underwent an epistemological rupture, whereby its 'systems' and 'models' no longer necessarily reflected reality as such. As Epple points out, with the rise and development of the various interpretations, systems, and images of non-Euclidean geometry, one was forced to reshape and restructure the relations between inequivalent mathematical representations of physical space.

José Ferreirós's contribution explores a similar theme but from a different perspective by concentrating specifically on mappings and models and their affinity to acts of abstraction and imagination, and on how mathematics contributed to modern thinking at the turn of the nineteenth to the twentieth century. What was the role of the term '*Abbildung*' (mapping, representation) in the thought of Bernhard Riemann, Hermann von Helmholtz, or Richard Dedekind? What were the historical vicissitudes of the pictorial terms 'image' and 'representation,' '*Bild*' and '*Abbildung*' in the mathematics of this period? And how is this history reflected in theoretical physics—in Heinrich Hertz's reflections, for example, or in the conceptual and philosophical difficulties being encountered in physics and mathematics?

The contribution by **Axel Gelfert** deals with models from a more general perspective. For Gelfert, mathematical actions are mediated by symbol systems, notations, and formalisms, which actively shape mathematical practice. By stressing the role of mathematical practice and by reviewing the philosophies of this practice, this contribution examines the interweaving of notations, formalisms, and models in mathematics. This leads Gelfert to examine epistemic mathematical actions—that is, actions that can be considered as constitutive of mathematical practice. To answer in a concrete way what these epistemic actions in mathematics might be, Gelfert gives three main examples: the use of gestures and symbolic operations, the construction and use of material mathematical models, and the re-proving of mathematical theorems. Returning at the end of his paper to the issue of notation, the question arises whether notation can be considered as a long-term epistemic action. Moreover, while notations and formalisms, in order to function properly, may need to be aided by physical actions and material models, one can certainly ask how material models function in the long run with respect to formalism.

In the last contribution to the second part, **Gabriele Gramelsberger** describes the decline of *Anschauung* (intuition) in the nineteenth century and the subsequent declaration of *Anschaulichkeit* (palpable visuality) as a model in geometry. Gramelsberger surveys the debates on Immanuel Kant's concept of *Anschauung*, and follows these debates particularly in nineteenth century mathematics and twentieth century physics. Gramelsberger notes that the debate between Werner Heisenberg and Erwin Schrödinger, for example, can be reconstructed as a mismatching debate between Heisenberg's conception of *Anschauung* and Schrödinger's conception of *Anschaulichkeit*. While Heisenberg emphasized the loss of spatiotemporal *Anschauung* in the Kantian notion, Schrödinger ignored this loss in favor of the traditional spatiotemporal concept of (Newtonian) *Anschaulichkeit*. Moreover, while for a certain period—mainly in the mathematics of the nineteenth century—material models succeeded in replacing *Anschauung* by *Anschaulichkeit*, the loss of *Anschauung* in physics, especially in early quantum theory, was subsequently compensated for by formalism and—as in Gramelsberger's example—matrices. The author hence argues that in early quantum theory, in order to understand all possibilities of reality, one first needed to overcome certain notions inspired by geometric *Anschauung*.

The last part of this volume, *Part 3: From Production Processes to Exhibition Practices*, examines with three interviews the before and after of models—that is, how these models were produced (before they became finished models) and how they were presented (once they were ready). We start with an interview with **Anja Sattelmacher**, which partly continues the themes of the second part of the volume, dealing also with conceptions of *Anschauung* (intuition). The interview with Sattelmacher concentrates on the materiality of material mathematical models and on their processes of production. The interview has two interconnected foci: first, a consideration of how, during the nineteenth century, conceptions of *Anschauung* were to be seen as one of the goals of these models, that is, the visualizing of a mathematical formula; second, a detailed discussion of the models' materials and material techniques. At the center of this discussion is the question of how the choice of a material (such as paper, wire, or wax) influenced not only the production processes of the model, but also the epistemic values transmitted through them.

The interview with **Ulf Hashagen** examines the history of the numerous exhibitions of mathematical models, mainly in the second half of the nineteenth century and the first decades of the twentieth century. 'Exhibition' here refers not only to two exemplary exhibitions—the 1876 exhibition at the South Kensington Museum, London, and the 1893 exhibition at the Technische Hochschule München—but also to collections of models at various universities, such as the collection in Göttingen. How was the act of exhibiting mathematical models understood at the time? How did the manner of exhibiting mathematical material change during these decades? And how were the various agendas of the different disciplines at that time (engineering, mathematics, physics) reflected in and promoted by the use and presentation of models?

The last interview in this volume concerns not only future modes of presenting models, but also the future production of these material models. The interview with **Andreas Daniel Matt** examines the history of digital mathematical models at the end of the twentieth century. The IMAGINARY project developed by Matt and his team enables, one may claim, real-time mathematics—namely, digital models of surfaces presented within seconds in an interactive manner. The history of this computer program is more convoluted than one might think, however, and it is entangled with the history of the various exhibitions in which these digital models have been presented. If one considers the various exhibitions of material mathematical models during the nineteenth century—as discussed in the interview with Ulf Hashagen—the question arises whether IMAGINARY breaks with this former tradition? Or can one also detect certain continuities and common characteristics?

Translated by Benjamin Carter

Acknowledgements The authors acknowledge the support of the Cluster of Excellence "Matters of Activity. Image Space Material" funded by the Deutsche Forschungsgemeinschaft (DFG, German Research Foundation) under Germany's Excellence Strategy—EXC 2025-390648296.

We would like to thank our authors and interview partners for their wonderful contributions and for their patience. We thank also warmly Benjamin Carter for his careful copyediting and translations and Julian Kutsche and Elisabeth Rädler for their extremely helpful support during the last stages of preparing the manuscript.

Last but not least, we would like to remember a dear colleague and wonderful scholar who wrote for many years on the theory of models and who was the inspiration behind the idea for this volume: Bernd Mahr (1945–2015), a mathematician and professor for theoretical computer science at Technische Universität Berlin as well as (among many other things) the project leader of the research group "Models in Gestaltung" at the Cluster of Excellence "Image Knowledge Gestaltung. An Interdisciplinary Laboratory."

Part I
Historical Perspectives and Case Studies

Knowing by Drawing: Geometric Material Models in Nineteenth Century France

Frédéric Brechenmacher

Introduction

The university collections of mathematical models have aroused a growing interest since the turn of the 21th century. In Paris, Institut Henri Poincaré has recently enhanced the collection it had inherited when it was created in 1928 from the older cabinet of mathematics of the Sorbonne.[1] Several models have been restored through crowdfunding processes, both permanent and temporary exhibitions have been set up, the models that had fascinated the surrealists Man Ray and Max Ernst in 1934 have been loaned to several art museums,[2] the publication of a collective volume has been supported by the institute,[3] as well as the production of a documentary film (see Figs. 1, 2 and 3).[4]

---

[1] About this collection, see: Jean Brette, "La collection de modèles mathématiques de la bibliothèque de l'IHP," *Gazette des mathématiciens* 85 (July 2000): 4–8.

[2] See especially: the temporary exhibition "Le surréalisme et l'objet" set up at Centre Pompidou in Paris from 30 October 2013 to 3 March 2014. On Man Ray's photographs and paintings of several models displayed at IHP, see: Isabelle Fortuné, "Man Ray et les objets mathématiques," *Études photographiques* 6 (May 1999): 1–12; Edouard Sebline and Andrew Strauss, "Man Ray à l'Institut Henri Poincaré: des objets mathématiques aux équations shakespeariennes," in *Objets mathématiques*, ed. Institut Henri Poincaré (Paris: CNRS Éditions, 2017), 152–62.

[3] Institut Henri Poincaré, ed., *Objets mathématiques* (Paris: CNRS Éditions, 2017).

[4] *Man Ray et les équations shakespeariennes*, directed by Quentin Lazzarotto (Paris: Institut Henri Poincaré, 2019).

F. Brechenmacher (✉)
LinX, École polytechnique, IPParis, 91128 Palaiseau, IP, France
e-mail: frederic.brechenmacher@polytechnique.edu

© The Author(s) 2022
M. Friedman and K. Krauthausen (eds.), *Model and Mathematics: From the 19th to the 21st Century*, Trends in the History of Science,
https://doi.org/10.1007/978-3-030-97833-4_2

53

a   b

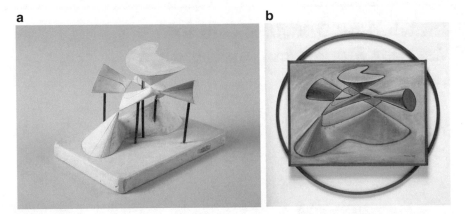

Fig. 1 **a** Model of a Kummer surface with eight double points, edited by Brill-Schilling in Halle. This object is one of the models of the collection of Institut Henri Poincaré which was photographed by Man Ray in 1934 and named after a play by Shakespeare: "King Lear" © Collections de l'Institut Henri Poincaré, all rights reserved. Photo: © Anne Chauvet. All rights reserved **b** Man Ray, "Shakespearean Equations: King Lear," 1948, oil on canvas, Hirshhorn Museum and Sculpture Garden, Washington, DC. © Man Ray2015 Trust/ VG Bild-Kunst, Bonn 2021, all rights reserved

Fig. 2 Model of a quartic surface with nine real double points, designed by Joseph Caron in Paris. This model was photographed by Man Ray in 1934 and named after Shakespeare's play "All's Well That Ends Well." © Collections de l'Institut Henri Poincaré, all rights reserved

Fig. 3 Model of an elliptic function, designed by Ludwig Brill in Darmstadt. This model was photographed by Man Ray in 1934 and named "The Merry Wives of Windsor." © Collections de l'Institut Henri Poincaré. Photo: Frédéric Brechenmacher, all rights reserved

In addition to the efforts of the mathematical community for preserving, enhancing, and publicizing collections of models, several publications and conferences have tackled various issues raised by the history of these collections. Most of these works have focused on what we shall designate in this paper as the 'models of higher mathematics' that were designed at the turn of the twentieth century and gave a material form to mathematical objects that were taught at the highest levels of mathematical education, such as the university lectures on the 'higher geometry' of cubic surfaces and their applications to mechanics.[5] These publications have usually identified two distinct periods during this golden age of mathematical models. The first, in the 1860 and 1870s, saw the emergence of models of higher geometry thanks to the growing individual commitment of various practitioners of mathematics, including several prominent mathematicians, especially in the United Kingdom, with James Joseph Sylvester, Arthur Cayley, Olaus Henrici, or Alicia Boole Stott, and in Germany, with Julius Plücker, Ernst

[5] See: Gerd Fischer, ed., *Mathematische Modelle* (Braunschweig: Vieweg + Teubner, 1986); Peggy Kidwell, "American Mathematics Viewed Objectively: The Case of Geometric Models," in *Vita Mathematica*, ed. Ronald Calinger (Washington: Mathematical Association of America, 1996), 197–208; Herbert Mehrtens, "Mathematical Models," in *Models: The Third Dimension of Science*, ed. Soraya de Chadarevian and Nick Hopwood (Stanford: Stanford University Press, 2004), 276–306; Irene Polo-Blanco, "Theory and History of Geometric Models" (Phd diss., University of Groningen, 2007); Jeremy Gray, Ulf Hashagen, Tinne Hoff Kjeldsen, and David E. Rowe, ed. "History of Mathematics: Models and Visualization in the Mathematical and Physical Sciences," *Oberwolfach Reports* 12, no. 4 (2015): 2767–858; Livia Giarcardi, "Models in Mathematical Teaching in Italy (1850–1950)," in *Mathematics and Art III. Visual Art and Diffusion of Mathematics*, ed. Claude P. Bruter (Paris: Cassini, 2015), 11–38; François Apéry, "Caron's Wooden Mathematical Models," in *Mathematics and Art III. Visual Art and Diffusion of Mathematics*, ed. Claude P. Bruter (Paris: Cassini, 2015), 39–48; Anja Sattelmacher, "Präsentieren: Zur Anschauungs- und Warenökonomie mathematischer Modelle," in *Sammlungsökonomien. Vom Wert wissenschaftlicher Dinge*, ed. Nils Güttler and Ina Heumann (Berlin: Kadmos, 2016), 131–55; Michael Friedman, *A History of Folding in Mathematics: Mathematizing the Margins* (Basel: Birkäuser, 2018), 127–205.

Eduard Kummer, Christian Wiener, Alfred Clebsch, Hermann Amadeus Schwarz, Felix Klein, and Alexander Brill. During the second period, the manufacturing of models of higher mathematics developed in Germany, starting with the publishing house of Ludwig Brill in Darmstadt in the 1880s, later merged with the editor Martin Schilling in Halle in 1899.[6] The production of models eventually culminated at the beginning of the twentieth century with semi-industrial manufacturers such as Brill/Schilling, Teubner, Mehrmittel anstalt, J. Ehrhard & Ci, and Polytechnisches Arbeits-Institut. These editors have disseminated mathematical models in universities and technological institutes all over Europe and the U.S.A. They boasted very large and diversified catalogues, which covered all branches of mathematics, such as geometry, mechanics, topology, and analysis, as well as of their applications to electricity, thermodynamics, shipbuilding, gearing, etc.

Several historical investigations have tackled the issue of the motivations that led to the development of such large and diversified collections. As a matter of fact, the heuristic value of models of higher mathematics for academic research seems to have been very limited, aside from a few, even though iconic, examples, such as the error of reasoning Felix Klein discovered by observing a model,[7] the counter-example Georges Brunel exhibited in a public demonstration in 1896 to a theorem of topology recently stated by Henri Poincaré,[8] or the key role played by the concrete folding manipulations of cardboard models in Henri Lebesgue's work on integration and developable surfaces.[9] Yet, most models of higher mathematics were actually designed only after algebraic research had been performed with pen and ink on the plane surfaces of papers or blackboards.[10]

---

[6] Martin Schilling, ed., *Catalog mathematischer Modelle für den höheren mathematischen Unterricht* (Halle: Martin Schilling, 1903).

[7] David Rowe, "On Franco-German Relations in Mathematics, 1870–1920," in *Proceedings of the International Congress of Mathematicians Rio de Janeiro 2018,* ed. Boyan Sirakov, Paulo Ney de Souza, and Marcelo Viana (Singapore: World Scientific, 2018), 21–36.

[8] In a public demonstration at the Bordeaux Society of physical sciences on 23 January 1896, Brunel gave a counter example to Poincaré's statement that any closed surface is a two-sided surface by displaying a model of a closed surface with just one side. See: Pierre Duhem, "Georges Brunel," *Association amicale des anciens élèves de l'Ecole normale supérieure* (1901): 103–16.

[9] Lebesgue was especially influenced by Darboux's lectures on higher geometry at the Sorbonne which, as shall be seen later, went along with sessions of practical works on mathematical models designed by Darboux's assistant, Joseph Caron. On Lebesgue's cardboard folding approach to integration, see: Sébastien Gandon and Yvette Perrin, "Le problème de la définition de l'aire d'une surface gauche: Peano et Lebesgue," *Archive for History of Exact Sciences* 63 (2009): 665–704.

[10] For a discussion of this issue in connection with the models of the 27 lines of a cubic surface, see: François Lê, "Around the History of the 27 Lines upon Cuvic Surfaces: Uses and Non-uses of Models," in *History of Mathematics: Models and Visualization in the Mathematical and Physical Sciences. Oberwolfach Reports* 14, no. 4, ed. Jeremy J. Gray, Ulf Hashagen, Tinne Hoff Kjeldsen, and David E. Rowe (2015): 2794–98, and Anja Sattelmacher, "Zwischen Ästhetisierung und Historisierung: Die Sammlung geometrischer Modelle des Göttinger mathematischen Instituts," *Mathematische Semesterberichte* 61, no. 2 (2014): 131–43.

In contrast to the limited heuristic value of models for mathematical research, most historical works have highlighted specific pedagogical values attributed to models, especially the ones of vizualization and manipulation. To be sure, the pedagogical values of models have been highly praised by several mathematicians,[11] especially in the international institutions that have been established at the turn of the twentieth century for promoting debates on mathematical education, such as the journal *L'Enseignement mathématique*, founded in 1899, and the International Commission on Mathematical Instruction established in 1908. Yet, historical sources about the actual pedagogical use of models of higher mathematics are scarce. Moreover, these sources are not as apologetic as public discourses. The use of models of higher geometry in classrooms or amphitheatres did not only raise practical difficulties—since models were often bulky, fragile, and costly—but the value of vizualization associated with them also conflicted with the important preliminary knowledge most models required from the students before they would be able to vizualize anything. Further, several teachers opposed the value of vizualization with the one of rigour associated with mathematical proofs performed on the blackboard.[12] This situation makes it difficult to assess the collective dimension of the pedagogical use of models of higher mathematics, beyond the individual commitment of iconic individuals, such as Klein, who had very strong and specific ideals about the roles of vizualization and experimentation, not only in mathematical research and education, but also in the very epistemological nature of mathematics.[13]

Mathematical models therefore call for further historical investigation on the social and cultural practices associated with models beyond the roles played by a few individuals at the turn of the twentieth century. This paper aims at shedding new light on such collective dynamics by investigating a period of time longer than the one of the golden age of models of higher mathematics. But this broader time scale forces us to restrict our corpus to a specific national setting. We shall, therefore, focus on the use of mathematical models in France from the late eighteenth century to the turn of the twentieth century.

---

[11] See in particular: Walther Dyck's opening speech in the first ever individual mathematics exhibition of models, apparatus, and instruments he organized in Munich on the occasion of the Annual Conference of the German Mathematician Society in 1893. In this speech, Dyck distinctly conceded the boundaries and restrictions of mathematical artifacts by explaining that, although many of the models did not have a practical use, they had an instructional purpose. Ulf Hashagen, *Walther von Dyck (1856–1934): Mathematik, Technik und Wissenschaftsorganisation an der TH München* (Stuttgart: Steiner, 2003), 431–36.

[12] See: Stanislas Meunier, "Reliefs à pièces mobiles destinés à l'enseignement de la géométrie descriptive," *La nature* 37 (Febrary 1874): 166–67; Paul Staeckel, "La préparation mathématique des ingénieurs dans les différents pays. Rapport général," *L'enseignement mathématique* 16 (1914): 307–28, here: 320.

[13] The importance attributed by Klein to geometric models as an *Anschauungsmittel* (illustration aid) in research and teaching of mathematics has especially been well studied. See David E. Rowe, "Klein, Hilbert, and the Göttingen Mathematical Tradition," *Osiris* 2nd Series 5 (1989): 186–213; David E. Rowe, "Mathematical Models as Artefacts for Research: Felix Klein and the Case of Kummer Surfaces," *Mathematische Semesterberichte* 60, no. 1 (2013): 1–24; Sattelmacher, "Zwischen Ästhetisierung und Historisierung;" Sattelmacher, "Präsentieren."

The renewed interest for mathematical models since the beginning of the 21th century has raised new issues about the specific situation of France. Several papers have highlighted the leading role of German mathematicians and manufacturers in the development of collections of models after the 1860s, while previous historical works had emphasised the legacy of Gaspard Monge's descriptive geometry, in the tradition of which Klein himself claimed he had been raised by his professor Plücker,[14] as well as the innovative models designed by Théodore Olivier in the 1840.[15] Both Monge and Olivier have therefore tended to be considered as French precursors of a movement that would eventually blossom in Germany, in a transition that may be understood in view of the larger historiographical perspective of a shift in the balance of power after the 1870 Franco-Prussian war. With regard to models, this idea of a transition between France and Germany seems to be underpinning the popularity acquired by the episode of Klein's trip to Paris in 1870,[16] during which the latter expressed his enthusiasm when discovering the collection of Olivier's string models displayed at the Conservatoire national des arts et métiers.[17] Even so, other collective dimensions have to be considered in between the very small scale of the individual experience and the very large one of the balance of European powers.

In this paper, we shall discuss the use of models in France in the framework of a broad mathematical practice, which was far from limited to the innovations of individuals such as Monge and Olivier. Over the course of the eighteenth and nineteenth centuries, the design and the use of mathematical models resulted from the strong belief that teaching geometry to engineers and technicians required to practice drawing, and more precisely model drawing, in contrast to other pedagogical methods such as plenary lectures: "it is in the drawing room that the master will judge the fruits of his teaching; it is there that he will recognize if the seeds he sowed from the pulpit chair has fallen on a good ground or on a stony soil."[18]

Teaching geometry, it was believed, required to "educate the hand and the eye," which was precisely the main pedagogical value attributed to the practice of drawing. On this issue, let us quote Jean-Jacques Rousseau's 1762 *Émile, or on*

---

[14] René Taton, *L'œuvre scientifique de Monge* (Paris: Presse Universitaires de France, 1951), 240.

[15] Joël Sakarovitch, "Théodore Olivier, Professeur de Géométrie descriptive," in *Les professeurs du Conservatoire national des arts et métiers, dictionnaire bibliographique 1794–1955*, ed. Claudine Fontanon and André Grelon (Paris: INRP/CNAM, 1994), 326–35.

[16] Felix Klein, *Gesammelte mathematische Abhandlungen,* vol. 2: *Anschauliche Geometrie. Substitutionsgruppen und Gleichungstheorie. Zur mathematischen Physik*, ed. Robert Fricke and Hermann Vermeil (Berlin: Springer, 1922), 3.

[17] Klein, *Gesammelte mathematische Abhandlungen*, vol. 2. The reviewer of this volume for *L'enseignement mathématique* especially focused on the episode of the discovery of the Olivier collection. See Grace Chisholm Young, "F. Klein–Gesammelte mathematische Abhandlungen," *L'enseignement mathématique* 24 (*1924–1925*): 167–69. See also: Felix Klein, *Vorlesungen über die Entwicklung der Mathematik im 19. Jahrhundert*, vol. 1, ed. Richard Courant and Otto Neugebauer (Berlin: Springer, 1979 [1926]), 63–93.

[18] Alphonse Bernoud, "Chr. Beyel. Ueber den Unterricht in der darstellenden Geometrie," *L'enseignement mathématique* 2 (1900): 63. All translations from French to English by the author, Frédéric Brechenmacher, if not stated otherwise.

Education, one of the main inspiration of the new national system of education set up during the French Revolution:

> One cannot learn to estimate the extent and size of bodies without at the same time learning to know and even to copy their shape; for at bottom this copying depends entirely on the laws of perspective, and one cannot estimate distance without some feeling for these laws. All children in the course of their endless imitation try to draw; and I would have Émile cultivate this art; not so much for art's sake, as to give him exactness of eye and flexibility of hand. Generally speaking, it matters little whether he is acquainted with this or that occupation, provided he gains clearness of sense-perception and the good bodily habits which belong to the exercise in question. So I would take good care not to provide him with a drawing master, who would only set him to copy copies and draw from drawings. Nature should be his only teacher, and things his only models [...]. I would even train him to draw only from objects actually before him and not from memory, so that, by repeated observation, their exact form may be impressed on his imagination, for fear that he should substitute absurd and fantastic forms for the real truth of things and lose his sense of proportion and his taste for the beauties of nature.[19]

The above quotation exemplifies a very strong pedagogical ideal associated with model drawing: the idea that the knowledge of forms and proportions required a direct contact with Nature with no mediation by any teacher, in contrast with the forms of knowledge transmitted by reading textbooks and listening to lectures. This ideal would have a lasting influence on the teaching of mathematics. From the eighteenth century to the turn of the twentieth century, mathematical models were usually considered as substitutes to natural forms and supported pedagogical methods that promoted action learning, relegated the role of the teachers to the one of supervisors, or even praised the mutual instruction of students by students: "Nature should be the only teacher." Geometric models, therefore, challenged the role of teachers in the teaching of mathematics.

Exactness of eye was a key issue in the philosophy of the Enlightenment: it was required for both the observation of sensible objects and the sense of proportion, which were the main instruments of knowledge for John Locke or Étienne

---

[19] Jean-Jacques Rousseau, "Émile, ou De l'éducation," in Jean-Jacques Rousseau, *Œuvres complètes de J.-J. Rousseau*, vol. 5, ed. Louis Barré (Paris: J. Bry, 1856), 97–98: "On ne saurait apprendre à bien juger de l'étendue et de la grandeur des corps, qu'on n'apprenne à connaître aussi leurs figures et même à les imiter; car au fond cette imitation ne tient absolument qu'aux lois de la perspective ; et l'on ne peut estimer l'étendue sur ses apparences, qu'on n'ait quelque sen timent de ses lois. Les enfants, grands imitateurs, essaient tous de dessiner : je voudrais que le mien cultivât cet art, non précisément pour l'art même, mais pour se rendre l'oeil juste et la main flexible; et, en général, il importe fort peu qu'il sache tel ou tel exercice, pourvu qu'il acquière la perspicacité du sens et la bonne habitude du corps qu'on gagne par cet exercice. Je me garderai donc bien de lui donner un maître à dessiner, qui ne lui donnerait à imiter que des imitations, et ne le ferait dessiner que sur des dessins : je veux qu'il n'ait d'autre maître que la nature, ni d'autre modèle que les objets. [...] Je le détournerai même de rien tracer de mémoire en l'absence des objects, jusqu'à ce que, par des observations fréquentes, leurs figure exactes s'impriment bien dans son imagination; de peur que, substituant à la vérité des choses des figures bizarres et fantastiques, il ne perde la connaissance des proportions et le goût des beautés de la nature.". English translation: Jean-Jacques Rousseau, "Rousseau's *Emile, or On Education*," trans. Barbara Foxley (Dover Publications: New York, 2013 [1911]), 128–29.

Condillac. Seeing was directly associated with intelligence: exactness of eye was a preliminary to the faculty of judgment because judging required comparison. The teaching of drawing was therefore intimately associated to that of geometry throughout the eighteenth and nineteenth century. As stated by the French report on the exhibition of geometric drawing at the 1862 world fair in London: "it is not sufficiently understood that drawing trains the eye, develops powers of observation, makes the finger more delicate [...]. Intimately linked with a few notions of geometry, drawing is useful in both the field and the workshop."[20] Further, as claimed by the mathematician and pedagogue Sylvestre François Lacroix, learning geometry through model drawing aimed at "training both judgment and the eye."[21]

Learning by drawing was associated with pedagogical issues very different from the values of visualization and manipulation that would be attributed to the models of higher geometry after the 1860s. Actually, the emergence of these models broke up with the long tradition of associating geometry with drawing, even though this rupture was less sudden in France where one of the main proponents of the models of higher geometry, Gaston Darboux, introduced the practice of drawing in the University of Paris in the 1870s. As we shall see in the first section of this paper, the specificity of the use of models in France was largely due to the distinction between universities and grandes écoles, and more precisely to the centrality of École polytechnique. This school indeed played a key role in the organization of the mathematical instruction in France because, on the one hand, of its centralized and national character as an institution, and, on the other hand, because its alumni dominated the mathematical sciences in France throughout much of the nineteenth century.

Several detailed investigations have already been devoted to the connections between the teaching of mathematics and drawing in France in the eighteenth and nineteenth centuries.[22] These works have especially shown the key role geometry played in the decline of the pedagogical approach promoted by the Académie des Beaux-Arts, which consisted in placing the model of the human figure at the core of the teaching of drawing. Yet, little attention has been paid to the specific models that were designed for the teaching of geometric drawing: it is on this specific issue that we shall focus on this paper. In order to investigate the specific French tradition of learning mathematics by model drawing, we shall pay a specific attention to the materiality of mathematical models. As we shall see in the second section of this paper, steel and string models were intertwined with industrialization, while plaster ones inherited from the arts of fortification, cardboard

---

[20] Jean Rapet, "Situation de l'enseignement chez les diverses nations représentées à l'exposition. Matériel scolaire," in *Exposition universelle de Londres de 1862. Rapports des membres de la section française du jury international sur l'ensemble de l'exposition,* vol. 6, ed. Michel Chevalier (Paris: N. Chaix, 1862), 16–79, here 69.

[21] Silvestre François Lacroix, *Essais sur l'enseignement en général et sur celui des mathématiques en particulier*, 4th ed. (Paris: Bachelier, 1838), 321: "On exerce ainsi le jugement en même temps que l'œil [...]".

[22] For a synthetic monograph on the teaching of drawing and of geometry in the eighteenth and nineteenth century, see Renaud d'Enfert, *L'enseignement du dessin en France. Figure humaine et dessin géométrique (1750–1850)* (Paris: Belin, 2003).

models supported the ideal of raising the mathematical instruction of the greatest number of children, while wooden models were, for a time, associated with the idea that the élite mathematicians of École normale supérieure had to be trained in handling saws and planes.

Specific attention to the materiality of models is also required by the very nature of the practices associated with models, such as drawing, fabricating, manipulating, observing, surveying, leveling, etc. Because of this practical nature, the use of models was usually not formalized by any textual knowledge. It is mainly for this reason that, as noted above, historical sources about the actual uses of models are scarce, especially when looking at the usual sources for the history of mathematics in the nineteenth century, such as books, periodical publications, and epistolary correspondence. To be sure, material models and épures, i.e. geometric drawings, form a rich pool of historical sources. But we shall nevertheless also look for textual historical sources by investigating the many reports that were devoted to technical innovations, crafts, and skills by the institutions involved in the industrialization of France, such as the Société d'encouragement pour l'industrie nationale, as well as local industrial exhibitions, and world fairs.

In the case of model drawing, in particular, the very epistemological essence of this activity was the transmission of a non-textual form of geometric knowledge, one that required practical work, apprenticeship and companionship. Reading texts or attending lectures could not subsume the knowledge associated with drawing: drawing was knowing. The investigation of skills, know-how, tacit knowledge, procedures, and scientific practices is a vivid field of research in the history of science. It has raised specific issues about historical sources as well as specific methodologies, such as that of reproducing experiments in order to access the material conditions, interpretations, and outcomes that emerge through investigations into matter. Such methodologies may be well adapted to the epistemological investigation of the type of mathematical knowledge associated with model drawing, in the interplay between the act of reading and attending lectures on descriptive geometry, and that of imitating, drawing and experimenting on models. This issue nevertheless goes beyond the scope of the present paper. But we shall emphasize a classical result in the history of non-textual knowledge: because of the absence of texts, and especially textbooks, non-textual knowledge cannot be dissociated from cultural practices and communities, it especially requires a direct transmission and may thus decline rapidly. In the third section of this paper, we shall therefore pay specific attention to identifying the communities associated with the practice of learning mathematics by model drawing, and whether this practice declined with the emergence of other forms of interplays between mathematics, models, and vizualization, such as those promoted by models of higher mathematics after the 1860s.

These discussions will eventually lead us in the fourth, and final, section of this paper: raising the issue of how the history of mathematical models may contribute to mathematical modelization. The etymology of the French 'épures,' which comes from 'épurer,' i.e. to refine, points to a typical activity in craftsmanship which consists in removing impurities or unwanted elements and which, when applied to drawing, involves a form of mathematisation that was theorized by Monge with the creation of descriptive geometry. We shall especially discuss how models were associated with a specific evolution of mathematisation in the view of the emergence of 'the graphical method,' which, at the turn of the twentieth century, would cover a very large range of graphical techniques, instruments, forms of vizualization, and knowledge.

Geometry and Model Drawing

The etymological origin of the word 'model' in the latin 'modello,' which derived from 'modulus,' i.e. measure or rhythm, highlights the ancient connections between mathematics and the arts in the uses of models for drawing, engraving, painting, sculpting, or constructing.[23] Renaissance humanism especially promoted such connections in the training of engineers. Construction drawing was developed in intimate connection with geometry and applied in various major concerns such as architecture, fortifications, cartography, wood and stonecutting, shading, shipbuilding, bridge building, and others in both civil and military engineering.[24] Model drawing thus came to be associated with a specific type of mathematical education through companionship and apprenticeship, which promoted practice and activity as opposed to, or as a complement to, reading books and attending lectures.

Drawing, Models, and Analysis

In the eighteenth century, drawing was considered as a critical factor for the progress of industry.[25] It participated in the promotion of manual work by the Encyclopedists who especially aimed at raising the value of the mechanical arts to the status of liberal arts. Drawing aimed not only at representing but also at explaining an operating process or a manufacturing process, it came to be considered as a kind of universal language and, until the nineteenth century, engineers had therefore to be "artist-engineers."[26]

The tradition of 'compagnonnage,' or fraternity, associated with 'corporations,' or guilds, was called into question during the age of Enlightenment. The creation of drawing schools and engineering schools played a key role in this interrogation: these schools indeed aimed at providing a vocational training for craftsmen and

---

[23] Peter Jeffrey Booker, *A History of Engineering Drawing* (London: Northgate, 1979); Antoine Picon, "Architecture, sciences et techniques. Problématiques et méthodes," *Les cahiers de la recherche architecturale et urbaine* 9–10 (January 2002): 151–60; Anne Coste and Joël Sakarovitch, "Construction History in France," in *Construction History. Research Perspectives in Europe*, ed. Antonio Becchi, Massimo Corradi, Federico Foce, and Orietta Pedemonte (Turin: Kim Williams Books, 2004).

[24] Joël Sakarovitch, *Epures d'architecture: de la coupe des pierres à la géométrie descriptive, XVIe-XIXe siècle* (Basel: Birkhäuser, 1998).

[25] See d'Enfert, *L'enseignement du dessin*, 32–35.

[26] Antoine Picon and Michel Yvon, *L'ingénieur artiste, dessins anciens de l'École des ponts et chaussées* (Paris: Presses des Ponts et Chaussées, 1989).

engineers as a complement to apprenticeship in workshops.[27] Yet the pedagogical method of fraternity would remain vivid in these schools even after the abolition of guilds and would especially play a key role in the mathematical training of engineers.

French military engineering schools, in particular, attributed a central role to mathematics in both the selection and the training of their students. The view that among all the sciences necessary to military engineers, mathematics have the most considerable rank became common in eighteenth century France.[28] The interest in mathematics arose not only because of its direct usefulness: mathematics, and especially instruction in mathematics, was seen to have valuable moral uses. It sharpened powers of reasoning and inculcated an orderly manner of thinking. Furthermore, the learning process of mathematics was considered to foster habits of work, self-control, and discipline. Mathematical education was also instrumental to the hierarchy between engineers and craftsmen while both were trained in model drawing.[29] The teaching of drawing actually aimed at both raising the qualification of the workforce and at disciplining it[30]: the practice of model drawing, in particular, was associated with the values of accuracy, heed, assiduity, and obedience.[31]

In the eighteenth century, the teaching of drawing was normalized as a progression from the simple to the complex. Models played a key role in a three-step progression: the students had first to copy drawings or prints, i.e. models of two dimensions, in order to acquire exactness of eye ("coup d'œil juste"), before passing to the "ronde-bosse," i.e. three dimensional models, and eventually to living and natural models. When their training was complete, students were supposed to be able to decompose a complex figure into a series of simple elements, corresponding to the models they had been trained with, and to recompose a complex drawing from its elementary parts (see Fig. 4).[32] This pedagogical method therefore followed the process of decomposition/recomposition that was formalized as the method of 'analysis' by Enlightenment philosophers such as Locke and Condillac, and which would be especially influential for the development of mathematical education.

In the tradition of the Académie des Beaux-Arts, the human figure played a key role in the teaching of drawing and most of the models that were used in drawing

---

[27] d'Enfert, *L'enseignement du dessin*, 43–44.

[28] Paris de Meyzieu, "Ecole royale militaire," in *Encyclopédie ou Dictionnaire raisonné des sciences, des arts et des métiers, par une Société de Gens de lettres*, vol. 5, ed. Denis Diderot and Jean Le Rond d'Alembert (Paris: Briasson, David, Le Bretton, Durand, 1755), 307–13.

[29] On the training of craftsmen in the eighteenth century, see: Antoine Léon, "Une forme typique de l'enseignement technique à la fin du XVIIIe siècle: Les écoles des dessin," *Bulletin du C.E.R.P.* 12, no. 1 (1963): 67–69; Arthur Birembaut, "Les écoles gratuites de dessin," in *Enseignement et diffusion des sciences en France au XVIIIe siècle*, 2nd ed., ed. René Taton (Paris: Hermann, 1986), 441–76; Yves Deforge, "Des écoles de dessin en faveur des arts et métiers," *Les Cahiers d'histoire du CNAM* 4 (July 1994): 14–30.

[30] Steven L. Kaplan, "L'apprentissage au XVIIIe siècle: le cas de Paris," *Revue d'histoire moderne et contemporaine* 40, no. 3 (1993): 436–79.

[31] d'Enfert, *L'enseignement du dessin*, 47.

[32] Ibid., 56.

Dessein,
Developemens du Mannequin.

Fig. 4 Robert Bénard, "Dessein, Dévelopemens du Mannequin," following the design by Louis Jacques Goussier. From "Recueil de planches sur les sciences, les art libéraux, et les arts méchaniques, avec leur explication [1763]," Plate VII; accompanying Denis Diderot and Jean-Baptiste le Rond d'Alembert, *Encyclopédie ou Dictionnaire raisonné des sciences, des arts et des métiers* (Paris: Briasson, 1762–1772).

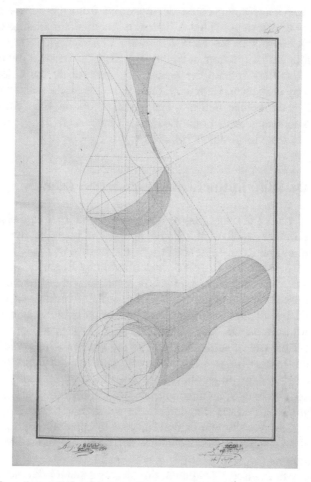

Fig. 5 Épure from the portfolio of Auguste Dupau, a student at École polytechnique in 1802. © Collections École polytechnique, Palaiseau, all rights reserved

schools were devoted to its representation and decomposition. Yet, other types of models were designed for special professions along the same analytical process of decomposition/recomposition, such as with the molds, capitals, and balustrades in architecture. At the turn of the nineteenth century, the model role of the human figure was challenged by geometric figures as well as by the mathematical models designed for the teaching of descriptive geometry (see Fig. 5).[33] For its creator Gaspard Monge, descriptive geometry embodied the "esprit d'analyse,"[34] not only

---

[33] Ibid., 98.
[34] Sakarovitch, *Epures d'architecture*, 214.

in the sense that its teaching could be organized from the simple to the complex, but also because it provided a heuristic "method for finding the truth."[35] As claimed by Lacroix in his 1805 *Essais sur l'enseignement*, in contrast with the slavish imitation of the human figure, models of geometric figure should be promoted because they formed the elementary parts of all the objects used, or manufactured, by craftsmen. Geometry, thus, provided models of a "more general usefulness" than the human figure.[36] For the same reason, the "dessin linéaire," created by the mathematician Louis-Benjamin Francœur in 1819, made only use of geometric models "for people's benefit."[37]

Geometric Drawing in the Royal Engineering Schools

The practice of model drawing played a key role in the first engineering schools established in France, such as the École royale des ponts et chaussées (Royal School of Bridges and Roadways; see Fig. 6), established in 1747,[38] and the École royale du génie de Mézières (Royal School of Military Engineering in Mézières), founded in 1748. The first was designed as a school without any professor. The students were mainly trained by drawing the models of various constructions that were deposited by visiting engineers, and second year students were supposed to advise first year students. More advanced students also trained their peers in mathematics by the use of textbooks such as the ones of Alexis Claude Clairaut, Charles Étienne Louis Camus, Charles Bossut, and Étienne Bézout. But most of the time was devoted to project-based learning in view of yearly competitions in mathematics, mechanics, architecture, stonecutting, planing and leveling, etc., which all required to perform drawings.[39] As we shall see later in greater detail, this important role given to companionship and to apprenticeship prefigured the method of mutual instruction that would develop in Europe in the beginning of the nineteenth century.

In contrast with the École des ponts, the Mézières school did include a professor of mathematics with the nomination of Charles Bossut in 1752. Yet the main

---

[35] Gaspard Monge, "Programme liminaire à la *Géométrie descriptive*, 20 janvier 1795," in *Leçons de mathématiques: Laplace, Lagrange, Monge*, ed. Jean Dhombres, vol. 1: *L'École normale de l'an III* (Paris: Dunod, 1992), 306: "un moyen de rechercher la vérité [...]."

[36] Lacroix, *Essais sur l'enseignement*, 359.

[37] Louis-Benjamin Francœur, *Le dessin linéaire d'après la méthode de l'enseignement mutuel* (Paris: Colas, 1819), 2: "[...] mais limité à la seule partie qui soit à l'usage du peuple, c'est-à-dire, l'enseignement du dessin linéaire."

[38] See: Antoine Picon, *L'invention de l'ingénieur moderne. L'École des Ponts et Chaussées 1747–1851* (Paris: Presses de l'École nationale des Ponts et Chaussées, 1992); André Grelon, Anousheh Karvar, and Irina Gouzevitch, *La formation des ingénieurs en perspective. Modèles de référence et réseaux de médiation, XVIIIe-XXe siècles* (Rennes: Presses Universitaires de Rennes, 2004); Joël Sakarovitch, "The Teaching of Stereotomy in Engineering Schools in France in the XVIIIth and

Fig. 6 Louis-Jean Desprez, "Vue imaginaire de l'École des Ponts et chaussées" (detail), circa 1750 (Musée Carnavalet, Paris). CC0 1.0 Universal (CC0 1.0) Public Domain Dedication (https://creativecommons.org/publicdomain/zero/1.0/)

role of this professor was not to lecture.[40] With regard to mathematical training, both schools relied mostly on the practice of model drawing and on mutual instruction.[41] At Mézières, the instruction was however less associated with project-based learning, and more and more organized in successive steps. It started with the construction of two épures of geometry. Next, this basic training in the elements of geometry was applied to the construction of épures in more special fields such as stonecutting, woodcutting, perspective, shadow drawing, and, in the second year of instruction, to fortification, survey work, buildings and machines. Because the

---

XIXth Centuries: an Application of Geometry, an 'Applied Geometry,' or a Construction Technique?" in *Entre Mécanique et Architecture. Between Mechanics and Architecture*, eds. Patricia Radelet-de Grave and Eduardo Benvenuto (Basel: Birkhäuser, 1995), 204–18.

[39] Bruno Belhoste, Antoine Picon, and Joël Sakarovitch, "Les exercices dans les écoles d'ingénieurs sous l'Ancien Régime et la Révolution," *Histoire de l'Éducation* 46 (1990): 56–73.

[40] Bossut only gave a series of short lectures on elementary mathematics, mechanics, and hydrostatics three days a week over a period of six months between 1752 and 1777, when the course of elementary mathematics was eventually cancelled. Further, Bossut's lectures on mathematics were hardly more than a repetition since they were based on the four volumes of Camus' textbook, which the students had already studied for passing the competitive entrance exam to the school. At the École des ponts, the students were, similarly, supposed to attend a series of lectures on chemistry and physics given by the professors of the Museum d'histoire naturelle. See René Taton, "L'École royale du Génie de Mézières," in *Enseignement et diffusion des sciences en France au XVIIIᵉ siècle*, ed. René Taton 559–615.

[41] Belhoste et al., "Les exercices," 53.

Fig. 7 Épure by Marchal, a student of Gaspard Monge at Mézières. From Gaspard Monge, "Petit traité des ombres à l'usage de l'école du genie." © Collections École polytechnique, Palaiseau, all rights reserved

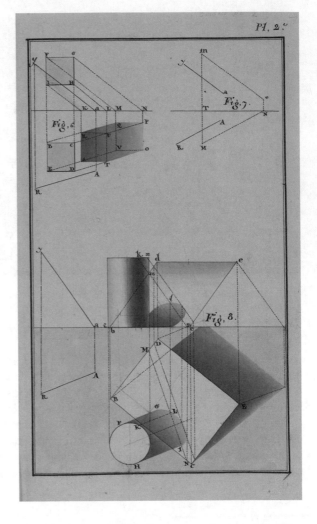

drawing of actual buildings or fortifications required time consuming outside activities, models played an important role at all stages of the education in the royal engineering schools.

Monge, who succeeded Bossut in Mézières, formalized the mathematical nature of model drawing with the creation of the 'dessin géométral,' which would later be renamed as descriptive geometry and would become one of the major branches of the mathematical sciences in the nineteenth century (see Fig. 7).[42] Descriptive geometry allows one too make "the intimate and systematic link between three-dimensional and planar figures:"[43] a three dimensional object is represented by two

---

[42] Taton, *L'œuvre scientifique de Monge.*

[43] Michel Chasles, *Aperçu historique sur l'origine et le développement des méthodes en géométrie* (Bruxelles: Hayez, 1837), 191: "l'alliance intime et systématique entre les figures à trois dimensions et les figures planes." See also: Victor Poncelet, *Traité des propriétés projectives des figures*

planar projections into mutually perpendicular directions; each of the two adjacent planar figures shares a full-scale view of one of the three dimensions of space. These figures may serve as the beginning point for a third projected view, such as of 'shadows' which facilitates the visualization of volumes. As Monge himself phrased it in his very first series of lectures on descriptive geometry at the École normale de l'an III in 1795:

> The purpose of this art [descriptive geometry] is two-fold. First it allows one to represent three-dimensional objects susceptible of being rigorously defined on a two-dimensional drawing [...]. Second [...], by taking the description of such objects to its logical conclusion, we can deduce something about their shape and relative positioning [...]. [It is] a language necessary for the engineer to conceive a project, for those who are to manage its execution, and finally for the artists who must create the different components.[44]

Monge had initially been hired at the Mézières school as a draughtsman in 1765 and assigned to the "atelier de la Gâche," a workshop devoted to the construction of models made of stucco. The construction of épures of fortifications provided Monge the opportunity to prove his mathematical abilities and he was elevated in 1766 to the position of *répétiteur* of mathematics, i.e. adjoint to Bossut, and eventually to the position of professor in 1769. Yet, as said before, Monge's role was not so much to lecture but to assist the students in their drawings. It was for the purpose of this companionship training that Monge established descriptive geometry as providing a mathematical formulation to the diversity of the graphical techniques of engineers.[45] On this issue, let us quote the historian Joël Sakarovitch:

> Descriptive geometry has been two-faceted from the time it was created. It is on the one hand an entirely new discipline [...] [which] offers an unprecedented manner of tackling three-dimensional geometry or, to be more exact, linking planar geometry with spatial geometry [...]. But it simultaneously appears as the last stage of a tradition that is losing momentum, as the ultimate perfecting of previous graphical techniques and, in that capacity, marks the endpoint of an evolutionary process as much as the birth of a new branch of geometry. As such, it can also be viewed as a transition discipline that allowed a gentle evolution to take place: from the 'artist engineer' of the Old Regime, whose training was based

---

(Paris: Bachelier, 1822); Jules de la Gournerie, *Discours sur l'art du trait et la géométrie descriptive* (Paris: Mallet-Bachelier, 1855).

[44] Gaspard Monge, quoted in Joël Sakarovitch, "Gaspard Monge: Géométrie Descriptive, First Edition (1795)," in *Landmark Writings in Western Mathematics, 1640–1940*, ed. Ivor Grattan-Guinness (Amsterdam: Elsevier Science, 2005), 225–241, here: 226. See: Gaspard Monge, "Programme liminaire à la Géométrie descriptive", 305–6: "Cet art a deux objets principaux. Le premier est de représenter avec exactitude, sur des dessins qui n'ont que deux dimensions, les objets qui en ont trois, et qui sont susceptibles de définition rigoureuse. Sous ce point de vue, c'est une langue nécessaire à l'homme de génie qui conçoit un projet, à ceux qui doivent en diriger l'exécution, et enfin aux artistes qui doivent eux-mêmes en exécuter les différentes parties. Le second objet de la géométrie descriptive est de déduire de la description exacte des corps tout ce qui suit nécessairement de leurs formes et de leurs positions respectives." See also: Gaspard Monge, "Stéréotomie," *Journal de l'École polytechnique* 1 (1794): 1–14.

[45] Sakarovitch, *Epures d'architecture*; Bruno Belhoste, "Du dessin d'ingénieur à la géométrie descriptive: l'enseignement de Chastillon à l'École royale du génie de Mézières," *In Extenso* 13, (June 1990): 103–35.

on the art of drawing rather than scientific learning, to the 'learned engineer' of the 19th century, for whom mathematics—and algebra in particular—is going to become the main pillar of his training.[46]

As we shall see in this paper, the growing importance of geometric models calls for reassessing the evolutions of descriptive geometry in the nineteenth century and its role in the interplay between textual knowledge and knowing by drawing.

In the royal schools of military engineering such as Mézières, the key role played by the practice of model drawing highlights a clear-cut distinction between the practical mathematical training provided within these schools and the more textual initial mathematical instruction of the students. One major feature of the royal military schools was indeed the distinction made between teaching and examining. Mathematics served as the dominant criterion in the entrance examinations, which took the form of an oral examination by a member of the Paris Academy of Science, such as Bossut, Bézout, or Pierre-Simon Laplace. These examinations were notoriously difficult and selective.[47] Lazare Carnot, for instance, succeeded to enter Mézières at his second attempt while Claude Rouget de Lisles did not succeed before his fifth attempt. Most candidates had received an elementary instruction in a Jesuit college, which included elements of arithmetic and of geometry in the tradition of Euclid. But the preparation for the entrance examinations of the royal military school was an individual affair based on the study of classical textbooks, such as Bezout's,[48] and required more advanced knowledge in arithmetic, algebra, geometry, and differential calculus.

The Foundation of École Polytechnique

In 1793, the schools of instruction and teaching were disorganized by the war that opposed revolutionary France to a coalition of European nations. In 1794, Jacques-Élie Lamblardie, director of the École des ponts, who lost a great number of his pupils, thought of creating a preparatory school for bridges and roads, and then for all engineers. Monge was enthusiastic about this idea and convinced several members of the Comité de Salut Public (French Public Welfare Committee) and the Convention. Under support of figures such as the chemist François Fourcroy, a decree of March 11, 1794 created the Central school of public works, which would be renamed École polytechnique one year later, on September 1, 1795.[49]

---

[46] Sakarovitch, "Gaspard Monge," 240.

[47] Roger Chartier, "Un recrutement scolaire au XVIIIe siècle: l'école royale de génie de Mézières," *Revue d'Histoire moderne et contemporaine* 20, no. 3 (1973): 369–75.

[48] Lilianne Alfonsi, *Étienne Bézout (1730–1783): mathématicien des Lumières* (Paris: L'Harmattan, 2011).

[49] On the history of the creation of Polytechnique, see: Ambroise Fourcy, *Histoire de l'École polytechnique* (Paris: Impr. de l'École polytechnique, 1828); Janis Langins, *La République avait besoin de savants: les débuts de l'Ecole polytechnique, l'Ecole centrale des travaux publics et les cours révolutionnaires de l'an III* (Paris: Belin, 1987); Jean Dhombres and Nicole Dhombres, *Naissance*

Its mission was to provide its students with a well-rounded scientific education with a strong emphasis on mathematics, physics, and chemistry. The Comité de Salut Public entrusted Monge, Lazare Carnot, and several other scholars with enlisting, by means of a competitive recruitment process, the best minds of their time, and teaching them science for the benefit of the French Republic. In 1795, all the other engineering schools were reorganized as special application schools for the students who had graduated from École polytechnique. The latter therefore acquired both a central and national role in the French educational system. It would spread its standards and pedagogical practices to other schools and would play a key role in imposing national standards of mathematical instruction in France and abroad through Napoléon's efforts to create a centralized, uniform system of education.

When the school was founded in 1794, its main features were the competitive entrance examination, the importance of mathematics, and the association of technical and mathematical education with military issues.[50] Monge elaborated the content of the first plan of instruction on two axes: the mathematics and the physics acquired by the experiment in the laboratories. The teaching of mathematics was divided between descriptive geometry, on the one hand, analysis and mechanics, on the other. Descriptive geometry had the most prominent role and was intimately associated to applications to the 'description of forms.' It started with stereotomy, i.e. the mathematical principles of descriptive geometry, and was then applied to architecture and fortifications. By contrast, the teaching of analysis was initially very limited, and focused on applications to the 'description of motions' in mechanics, hydrostatics and machines.

As was already the case in Mézières, the practice of model drawing in the teaching of geometry was given a prominent place. In the initial plan of instruction, about four fifths of the time (74 h) was devoted to practical activities (61 h) which consisted mainly in graphical activities in geometric drawing (30 h) and figure or landscape drawing (12 h). On a typical day, the short morning lecture mainly aimed at providing the students with the "knowledge, the instructions, and the methods" required for the graphical constructions of the day. As Monge phrased it in 1795: "the drawings constitute the ostensible work of the student [...] they require meditations, but there will not be any specific time devoted to these meditations, which will develop during the constructions, and the student who will

---

d'un nouveau pouvoir: sciences et savants en France, 1793–1824 (Paris: Payot, 1989); Ivor Grattan-Guinness, Convolutions in French Mathematics, 1800–1840: From the Calculus and Mechanics to Mathematical Analysis and Mathematical Physics (Basel: Birkhäuser, 1990); Charles C., Gillispie, Science and Polity in France: The Revolutionary and Napoleonic Years (Princeton: Princeton University Press, 2004); Bruno Belhoste, La formation d'une technocratie. L'École polytechnique et ses élèves de la Révolution au Second Empire (Paris: Belin, 2003).

[50] École polytechnique only became a military school in 1804 but military issues were already very strong in 1794. For a short synthesis on the role played by mathematics in the educational purpose of the school from 1794 to 1850, see: Ivor Grattan-Guinness, "The 'Ecole Polytechnique,' 1794–1850: Differences over Educational Purpose and Teaching Practice," The American Mathematical Monthly 112, no. 3 (2005): 233–50.

have trained simultaneously his intelligence and his skills using hands, will get, as the price of his double work, the exact description of the knowledge he will have acquired."[51]

The idea that mathematics established a hierarchy between engineers and craftsmen, while drawing was a 'common language' between them, shows continuity in the training of engineers before and after the French Revolution. As Antoine Lavoisier phrased it in his "reflections on public instruction:"

> Drawing is a sensitive language which speaks to the eye, gives shape to our thought and therefore expresses more than language; it is a mean of communication between the one who conceives or who commands and the one who executes. Considered as a language, drawing is an instrument for perfecting one's thoughts; drawing is therefore the primary education of those aiming at [a career in] the arts [i.e. the techniques].[52]

Yet, in contrast with the military schools in the Ancien Régime, the students of École polytechnique had to attend lectures of mathematics even though, according to Monge, the new school initially attached "much more importance to the works done by students with their own hands than to what they can learn by listening to professors or reading books. It is indeed the best method for fixing in the mind the knowledge that is acquired, for making it accurate, and for one to be certain that he fully possesses this knowledge."[53] The founding professors ('instituteurs') of analysis and mechanics were Joseph-Louis Lagrange and Gaspard Riche de Prony. Descriptive and differential geometry was in the hands of Monge, who also served as Director for two short periods. Each *instituteur* had an

---

[51] *Procès-verbaux du conseil d'administration de l'École centrale des Travaux publics*, séance du 20 pluviôse an III, Archives of École polytechnique, https://journals.openedition.org/sabix/703 (accessed April 15, 2022): "[…] ces constructions graphiques, c'est dans des dessins que consistera tout le travail ostensible des choses, ces Dessins, ces Constructions exigent de leur part des méditations ; mais il n'y aura aucun tems purement consacré à ces méditations ; elles auront eu lieu pendant toute la Durée des Constructions, et l'élève qui aura en même tems exercé son intelligence et l'adresse de ses mains, aura pour prix de ce double travail, la Descriptionexacte de la connoissance qu'il aura acquise."

[52] Antoine Lavoisier, *Réflexions sur l'instruction publique, présentées à la Convention nationale par le bureau de consultation des arts et métiers* (Paris: Du Pont, 1793), 15: "[…] le dessin est un langage sensible qui parle aux yeux, qui donne de l'existence aux pensées, et sous ce point de vue, il exprime plus que la parole ; c'est un moyen de communication entre celui qui conçoit ou qui ordonne un ouvrage et celui qui l'exécute, enfin considéré comme langue, c'est un instrument propre à perfectionner les idées : le dessin est donc la première étudie de ceux qui se destinent aux arts."

[53] Gaspard Monge, "Avant propos," *Journal de l'École polytechnique* 1 (1794): iii–viii, here iv: "Il faut dire encore que l'école est tellement montée, que l'on s'y attache bien plus au travail que l'élève exécute de ses propres mains, qu'à ce qu'il peut apprendre en écoutant les professeurs, ou en étudiant dans des livres. C'est en effet la meilleure méthode pour fixer dans l'esprit les connaissances que l'on acquiert s'assurer de leur justesse, et être certain qu'on le possède complètement."

assisting adjoint, who were named "répétiteurs" after 1798.[54] Among early notable adjoints or *répétiteurs* in mathematics, one may cite Jean Nicolas Pierre Hachette in geometry (another former member of the Mézières school who had been hired as a draughtsman and elevated to *répétiteur*) and Joseph Fourier in analysis. One major feature was the distinction made between teaching and examining so examiners were also appointed. For mechanics and analysis the initial examiners were Bossut and Laplace.

While the school had originally been conceived as the one and only institution to train engineers, the impracticability of the vision was soon recognized and the role of the school was thus changed in 1795 to that of a preparatory institution for the other schools, which were organized into a collection of 'écoles d'application,' such as École d'application de l'artillerie et du génie in Metz (School of Artillery and Engineering Applications), École des mines (School of Mining), and École nationale des ponts et chaussées (National School of Bridges and Roadways). This change would have important consequences on the roles attributed to mathematics at Polytechnique, especially through the influence of Laplace.[55] For six weeks in 1799 Laplace acted as Minister of the Interior. He proposed that the school have a governing council, the "Conseil de perfectionnement," to supplement the "Conseil d'Instruction" on teaching details, and a "Conseil d'Administration" for management.[56] Laplace was one of its founding members; and he exercised much influence there, in particular reducing the time given to descriptive geometry and transferring much of it to mechanics and analysis.[57] This opposition was led mainly by Laplace's desire to confine the programs at École polytechnique to teaching general theories, which would then be applied in the more specialized other schools. This kind of difference over curriculum policy in the school would continue for a long time: the archives of the reports of the school's councils highlight that the issue of the roles attributed to mathematics fuelled a never ending tension in the school curriculum, between the general and the special, and between the theoretical and the applied.[58] This tension would play an important role in the

[54] On the *répétiteurs* at Polytechnique in the nineteenth century, see: Yannick Vincent, "Les répétiteurs de mathématiques à l'École polytechnique de 1798 à 1900" (Phd diss., École polytechnique, 2019).

[55] Roger Hahn, "Le rôle de Laplace à l'École polytechnique," in *La formation polytechnicienne: 1794–1994*, ed. Bruno Belhoste, Amy Dahan Dalmedico, and Antoine Picon (Paris: Dunod, 1994), 54.

[56] Belhoste, *La formation d'une technocratie*, 50–51.

[57] Joël Sakarovitch, "La géométrie descriptive, une reine déchue," in *La formation polytechnicienne: 1794–1994*, ed. Bruno Belhoste, Amy Dahan Dalmedico, and Antoine Picon (Paris: Dunod, 1994), 77–93.

[58] See: Bruno Belhoste, "The École polytechnique and Mathematics in Nineteenth-Century France," in *Changing Images in Mathematics*, ed. Umberto Bottazzini and Amy Dahan (London: Routledge, 2001), 15–30. For the long shadow cast by this tension on the teaching of mathematics, see: Jean-Luc Chabert and Christian Gilain, "Debating the Place of Mathematics at the École polytechnique around World War I," in *The War of Guns and Mathematics: Mathematical Practices and Communities in France and Its Western Allies around World War I*, ed. David Aubin and Catherine Goldstein, vol. 42: *History of Mathematics* (Providence: American Mathematical Society, 2014), 275–306.

evolution of mathematical models in the nineteenth century, as shall be seen later in greater details.

From 1794 to 1800, École polytechnique thus passed from the "École de Monge" to the "École de Laplace."[59] Mathematics was given an increasing importance in the school's curriculum, from 50% in 1794 to 65% in 1800, while the teaching of applications was much reduced. Moreover, analysis came to play a more and more important role in the curriculum at the expense of descriptive geometry: while the respective proportions of descriptive geometry and analysis amounted to 50% and 8% of the curriculum in 1794, they amounted to 26% and 29% in 1800.[60]

But even though the first plan of instruction conceived by Monge was called into question one year after the creation of the school,[61] the intimate connection between the teaching of descriptive geometry and the practice of drawing would have a lasting influence at École polytechnique. Actually, geometric drawing was not much affected by the reduction of the teaching of descriptive geometry, and therefore of the lectures devoted to mathematical drawing: in the legacy of the pedagogical practices developed in the royal engineering school, drawing was indeed much more a practical activity at Polytechnique than a matter of plenary lecture. The industrialization of France in the century would even strengthen the importance of geometrical drawing with the creation of lessons on machinery distinct from the one of descriptive geometry.[62]

Mutual Instruction Versus Academic Pedantry

In 1794 École polytechnique was established under the label of the strong ideals that had been developed during the Enlightenment and which had called for the development of scientific education. Mathematics was especially valued as a way of emancipation because it was considered to provide results closer to the truth

---

[59] Théodore Olivier, "Monge et l'École polytechnique," *Revue scientifique et industrielle* 38 (1850): 64–68.

[60] To this proportion of 29%, one should actually add the 17% amounting to the teaching of mechanics in 1800 (in 1794, analysis and mechanics were not yet distinguished one from the other).

[61] On the evolutions of École polytechnique in the first half of the nineteenth century, see Bruno Belhoste, "Un modèle à l'épreuve. L'École polytechnique de 1794 au Second Empire," in *La formation polytechnicienne, 1794–1994*, ed. Bruno Belhoste, Amy Dahan Dalmedico, and Antoine Picon (Paris: Dunod, 1994), 9–30.

[62] Konstantinos Chatzis, "Mécanique rationnelle et mécanique des machines," in *La formation polytechnicienne: 1794–1994*, ed. Bruno Belhoste, Amy Dahan Dalmedico and Antoine Picon (Paris: Dunod, 1994), 95–108; Jean-Yves Dupont, "Le cours de machines de l'École polytechnique, de sa création jusqu'en 1850," *Bulletin de la Société des amis de l'École polytechnique* 25 (2000): 3–79.

than any other science.[63] For Nicolas de Condorcet in particular, mathematical education had a moral value: it aimed at ensuring the continuation of progress, not only in science and technology, but also in the morality of the younger generations. Further, the ideal of universality associated with mathematics was at the core of the evolution of the system of competitive recruitment process for the Ancient Regime royal military engineering school into a new system, which aimed at replacing hereditary privileges by individual merit, which was to be proved by solving mathematical problems.[64] Not only did the new system of competitive exams abolish any prerequisite of nobility but it also reduced the expectation of a prerequisite knowledge, acquired by studying textbooks, in order to favor intelligence over cramming.[65]

These ideals went along with the goal to create a new pedagogy that would promote both theoretical and practical knowledge. Inspired by the teaching of geometry by model drawing at Mézières as well by the pedagogical innovations made in mining schools such as the one of Schenitz in Hungary, the founders of École polytechnique aimed at promoting science activities and experiments.[66] This plan required both a library and a collection of scientific instruments. Both were initially constituted from the property seizures that had been taken under the exigencies of revolution in three waves from 1789 to 1793, especially in private library collections from the aristocracy and clergy. This collection quickly expanded with the publications and the new apparatus that were invented by the professors, often alumni of Polytechnique, for their teaching.

The important role attributed by Monge to action learning through the practice of drawing thus participated to a more general plan for articulating practice and theory. For the same purpose that laboratories were created for promoting experiments in chemistry, Monge introduced a distinction between the "grandes salles," where the *instituteurs* lectured, and the "petites salles," (see Fig. 8) in which the students were divided into several "brigades."[67] The students actually spent most of their time in the "petites salles," or the "salles d'études," at least five hours a

---

[63] Jean Antoine Nicolas de Caritat, marquis de Condorcet, "Discours de réception à l'Académie française," *Recueil des harangues prononcées par Messieurs de l'Académie française dans leurs réceptions et en d'autres occasions* 8 (1782): 413–49.

[64] On the continuities in the role devoted to mathematics in the entrance examinations of the royal military school and of École polytechnique, see: Janis Langins, "The Ecole polytechnique and the French Revolution: Merit, Militarization, and Mathematics," *Llull: Revista de la Sociedad Española de Historia de las Ciencias y de las Técnicas* 13, no. 24 (1990): 91–105; Janis Langins, "La préhistoire de l'Ecole polytechnique," *Revue d'histoire des sciences* 44, no. 1 (1991): 61–89.

[65] Circular to examiners in 1794 quoted in Fourcy, *Histoire de l'École polytechnique*, 34.

[66] Antoine-François Fourcroy, *Rapport sur les mesures prises pour l'établissement de l'école centrale des travaux publics, fait par Fourcroy au nom des comités réunis de salut public, d'instruction publique et des travaux publics. Du 3 vendémiaire an III* (Paris: Imprimerie du Comité de salut public, 1794).

[67] Monge, "Stéréotomie," 3–6; Belhoste, *La formation d'une technocratie*, 200.

Fig. 8 A student working in a 'petite salle.' From Gaston Claris, *Notre École polytechnique* (Paris: Librairies-imprimeries réunies, 1895), 140

day, and this time was mostly devoted to model drawing, studying daily lectures, and preparing for the regular oral examinations ("répétitions").[68]

---

[68] In the initial schedule established by Monge, the daily lectures of descriptive geometry took place from eight to nine in the morning and were followed by practical drawing exercices from 9 am to 2 pm. Four of the six weekly afternoon sessions, which took place from 5 pm to 8 pm, were devoted to drawing. See: Gaspard Monge, "Développement sur l'enseignement adopté pour l'école centrale des travaux publics", in Janis Langins, *La République avait besoin de savants: les débuts de l'Ecole polytechnique, l'Ecole centrale des travaux publics et les cours révolutionnaires de l'an III* (Paris: Belin, 1987), 227–69. On the evolution of this schedule in the beginning of the nineteenth century, see Fourcy, *Histoire de l'École polytechnique*, 376–79. Several testimonies of students on their schedule at the school are also available, such as the ones Jacques Louis-Rieu in

To fully understand the importance of the practice of drawing in the activities of the students, one has to recall the predominance of oral teaching and examinations at École polytechnique: until the 1830s the students were not expected to produce any mathematical writings other than the use of the blackboard during the examinations.

The promotion of the activity of the students was also motivated by the rejection of the model of university education—universities were abolished during the revolution—accompanied by a mistrust of professors and their "inevitable appetency for pedantry." As claimed by the chemist Antoine-François Fourcroy, one of the founders of Polytechnique, the new system of republican instruction should aim at "populating classrooms with students for avoiding the risk of populating them with professors."[69] The important role devoted to the practice of model drawing at Polytechnique therefore highlights, once again, the long-term legacy of companionship and mutual instruction in the transmission of the mathematical crafts of the engineers.

A form of mutual instruction was institutionalized at Polytechnique with the selection of a few 'chefs de brigades' among the best students who had passed the first entrance examination in 1794. Each of these 'chefs de brigades' was responsible for helping a group of students in their work on geometric drawing in one of the "petites salles."[70] A few years later, the 'chefs de brigades' were selected between the young graduates from the school rather than from the students themselves:

> [...] this disposition provides the opportunity to stay in Paris to the young men who may benefit the most from continuing their studies [...]. By making these positions temporary [...] we are protecting them against the pedantry, from which tenured professors so often fail to spare themselves.[71]

This experimentation of mutual instruction at Polytechnique would decline after 1798 when Laplace created the function of *répétiteurs*, i.e. adjunct professors who were in charge of the oral examinations of the students and of supervising the 'chefs de brigades' who, as a consequence, lost their autonomy and saw their role limited to the one of maintaining discipline in the "petites salles." Yet, several former 'chefs de brigades' and alumni of Polytechnique, such as Francœur and

---

1806 and of Auguste Comte in 1815. Auguste Comte, *Lettres d'Auguste Comte à M. Valat, professeur de mathématiques et ancien recteur de l'académie de Rhodez 1815–1844* (Paris: Dunod, 1870), 1–2; Jean-Louis Rieu, *Mémoires de Jean-Louis Rieu, ancien premier syndic de Genève* (Geneva: H. Georg, 1870), 16.

[69] Antoine-François Fourcroy, "Discours prononcé au corps législatif par Fourcroy, sur l'instruction publique, du 20 floréal an X," *Recueil des lois et règlements concernant l'instruction publique* 2 (May 10, 1802): 244.

[70] See Monge, "Stéréotomie," 4.

[71] Monge, "Développement," 241: "[...] Cette disposition fournit aux jeunes gens les plus en état d'en profiter, les occasions de continuer leurs études à Paris, et de perfectionner leur instruction. [...] En rendant ces places passagères, [...] on les met en garde contre le pédantisme, dont il est bien difficile que les instituteurs à poste fixe puissent se garantir."

Edme François Jomard, would get much involved in the movement for mutual instruction in France.[72] As will be seen later in greater detail, this movement would play a major role for the development of the mathematical instruction of the emerging working class through model drawing.

Monge's "Cabinet Des Modèles"

As mentioned above, models were instrumental in the distinction between the three main forms of pedagogical methods for teaching mathematics at École polytechnique: the plenary lectures of the *instituteurs*, the individual oral examinations of the *répétiteurs*, and the autonomous activities of groups of students in the petites salles. At the foundation of the school in 1794, Monge created a cabinet of models composed of the collection of all the drawing models that were to be used for the practical activities in the *petites salles*.[73] This collection consisted initially of models similar to the ones that had been previously used in the royal schools of engineering, such as épures, maps, models of architecture or fortifications, mechanical devices, etc.[74] As with the library and the scientific instruments of the laboratories, the cabinet of models was originally furnished with property seizures, especially in abolished royal institutions (such as the Mézières school),[75] but quickly expanded with the new publications and apparatus produced by students and professors. A drawing office with twenty-five draughtsmen was created for designing new models for the teaching of descriptive geometry and stereotomy.[76]

Monge's collection of models has unfortunately been lost, except a series of cardboard models of polyhedrons that were used for both the teaching of geometry and crystallography in physics and chemistry (see Fig. 9). The detailed inventory of the cabinet is known only for the year 1794, but historical sources document that Monge had commissioned a series of string models made of silk for the practice of geometric drawing. In 1814, the cabinet included two large-scale string models, one of the line generation of a revolution hyperboloid of one sheet, and the other of the line generation of a hyperbolic paraboloid (see Fig. 10).[77] But it is likely

---

[72] Renault d'Enfert, "Jomard, Francœur et les autres… Des polytechniciens engagés dans le développement de l'instruction élémentaire (1815–1850)," *Bulletin de la Sabix* 54 (2014): 81–94.

[73] Fourcy, *Histoire de l'École polytechnique*, 17.

[74] "Cabinet des modèles de l'École centrale. Citoyen Lesage, conservateur. Compte décadaire. 30 nivôse an 3, Lesage, conservateur" January 19, 1795; "Cabinet des modèles de l'École centrale. […] État de situation. Compte décadaire," Archives of École polytechnique. Fonds Prieur de la Côte-d'Or.

[75] Belhoste et al., "Les exercices," 98.

[76] Jean Nicolas Pierre Hachette, *Traité de géométrie descriptive*, 2nd ed. (Paris: Corby, 1828), xvii; Belhoste et al., "Les exercices," 101–107.

[77] Arthur Morin, *Conservatoire national des arts et métiers. Catalogue des collections publié par ordre de M. le Ministre de l'Agriculture et du Commerce*, 2nd ed. (Paris: de Guiraudet et Jouaust, 1855), 24.

Fig. 9 Cardboard model of a crystal. This set of cardboard models dates back to the creation of Monge's cabinet of models at École polytechnique. © Collections École polytechnique, Palaiseau, all rights reserved

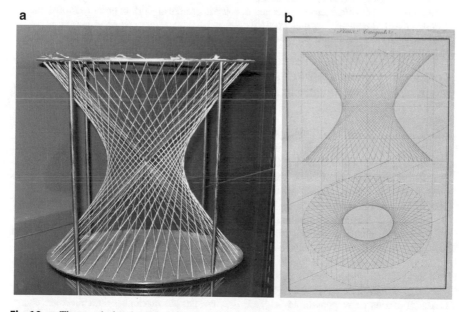

Fig. 10 **a** The revolution hyperboloid of one sheet in Gaspard Monge's cabinet may have been similar to the one above © Collections de l'Institut Henri Poincaré, all rights reserved. Photo: Frédéric Brechenmacher. **b** Drawing of the string model of the line generation of a revolution hyperboloid of one sheet in Monge's cabinet. Léon Duflos de Saint Armand, Épures 1823–1824. © Collections École polytechnique, Palaiseau, all rights reserved

that other types of models were designed since several craftsmen were attached to the *instituteur* of descriptive geometry: a fitter ('appareilleur'), a carpenter, a joiner, a locksmith, and a plaster modeler.

The development of the cabinet des modèles is also documented by the nomination in 1813 of Louis Brocchi as curator of the models ("conservateur des modèles"),[78] a position renamed in 1816, as "artist keeper of the cabinet of models," and again in 1820 as "artist curator of the cabinet of models."[79] Born in Veroli, Italy, Brocchi had arrived in Paris in 1799.[80] He had been at first hired temporarily by École polytechnique for restoring several models of Monge's cabinet that had been damaged by the students.[81] He was eventually offered a permanent position on June 4, 1813 with a larger scope of responsibilities and kept this position until his death in 1837.[82] His act of nomination provides rare information about the role of the models in the organization of the teaching:

> He is in charge of keeping the models of machines, architecture, woodwork, stonecutting, topography, &c, the brass models and the collections of épures that have to be distributed to the students.
> He receives instructions from various professors for maintaining his cabinet, for printing plates, for distributing épures and paper to students; for installing the models and drawings in the amphitheaters and the study rooms.
> He remains in his office during the hours devoted to graphical work in order to furnish the students with the paper they may need.
> He maintains in condition the plaster models of stonecutting; he restores the objects entrusted to him.[83]

In 1813, the students' graphical activities required Brocchi to remain in his office every morning, from Monday to Friday (between 8:30 to either 12:30 or 2:30) and

---

[78] "Le gouverneur de l'Ecole impériale Polytechnique nomme le sieur Brocchi…," 4 June 1813. Archives of École polytechnique, VI1d2.

[79] Letter from the director of École polytechnique to the minister of interior, 9 December 1820, Archives of École polytechnique, VI1d2.

[80] Act of naturalisation of Brocchi by the mayor of the 12th arrondissement of Paris, 9 January 1815, Archives of École polytechnique, VI1d2.

[81] "État supplémentaire de proposition de gratification en faveur d'agents attachés à la Direction des études," Archives of École polytechnique, 24 January 1814, VI1d2.

[82] "État pour servir à la liquidation de la pension revenant à Madame Antoinette Jeanne Darley, veuve de M. Louis Marie François Brocchi," 1837, Archives of École polytechnique, VI1d2.

[83] "Brocchi, artiste gardien du cabinet des modèles," 1816, Archives of École polytechnique, VI1d2: "Il a sous sa responsabilité la garde des modèles de machines, d'architecture, charpente, coupe des pierres, topographie &c, les cuivres et les collections d'épures qui doivent être distribuées aux élèves.

il prend les instructions des différens professeurs pour la tenue de son cabinet; pour le tirage des planches; les distributions d'épures et de papier à faire aux élèves; pour le placement des reliefs et dessins aux amphithéâtres et dans les salles d'étude.

il se tient à son cabinet pendant ses heures de travail graphique pour donner aux élèves le papier dont ils auraient besoin.

il maintien en bon état les modèles de coupe des pierres en plâtre; soigne les objets qui lui sont confiés; met de l'ordre dans son cabinet; provoque des mesures conservatrices et rend compte de l'emploi des objets de consommation qu'il distribue."

from 12: to 2:30 on Saturday.[84] Brocchi also enriched the cabinet by designing new models and instruments, such as a compass of stereotomy for measuring the dimension of a body in regard with three orthogonal planes. This instrument was based on the founding principles of descriptive geometry, i.e. orthogonal projections and the generation of a surface by the motion of a variable line along parallel planes corresponding to the sections of the surface by parallel planes.[85] It was especially used for designing geometric models of several types of mouldboards for modern plows. Brocchi also completed Monge's collection of string models, such as with a model of the line generation of a non-revolution hyperboloid of one sheet. He also designed plaster stonecutting models and molds of topographic landforms.[86] Several reports written by the administration of the school praise Brocchi's talent for designing new models of descriptive geometry "for the instruction of the students."[87] Brocchi's talent earned him an important reputation[88]: foreign engineering school such as the one of Saint-Petersbourg purchased several of his models.[89]

A Polytechnic Culture of Drawing

As seen before, the important role devoted to the practice of drawing in the teaching of mathematics at École polytechnique did not suffer from the relative decline of descriptive geometry in the curriculum.[90] The drawing skills of the students graduating from the school were actually a constant matter of preoccupation of the 'conseil de perfectionnement,' in which the schools of applications were represented. The school of artillery and engineering applications in Metz, in particular, often complained about the limited drawing skills of the students who entered Metz after having graduated from École polytechnique and pleaded for strengthening further the practice of model drawing. The épures of descriptive geometry were indeed crucial for both artillery and military engineering: they laid the ground

---

[84] "M. Brocchi, conservateur du cabinet des modèles," June 2, 1813, Archives of École polytechnique, VI1d2.

[85] Jean Nicolas Pierre Hachette, "Rapport fait par M. Hachette, au nom du Comité des arts mécaniques, sur plusieurs instrumens de mathématiques présentés à la société," *Bulletin de la Société d'Encouragement pour l'Industrie Nationale* 228 (November 1823): 145–60.

[86] The existence of these models is documented by a reclamation in which Brocchi's sons complained that École polytechnique made use of plaster models of topographies without having purchased the molds designed by their father for these models. "Auguste Brocchi à l'administrateur de l'École polytechnique," January 27, 1840, Archives of École polytechnique, VI1d2.

[87] Letter from the director of École polytechnique to the minister of interior for awarding a wage increase to Brocchi, december 20, 1818, Archives of École polytechnique, VI1d2.

[88] Brocchi was also charged by Louis Visconti, the architect of the Royal library, for restoring ancient vases and etruscan antiques.

[89] Letter from the count Johann von Sievers to Brocchi, October 4, 1818, Archives of École polytechnique, VI1d2.

[90] On the decline of descriptive geometry in the nineteenth century, see: Joël Sakarovitch, "La géométrie descriptive, une reine déchue," 77–93.

Fig. 11 Épure of a conical gearing by a student of École polytechnique in 1807. Epures 1794–1850. Cours de Jean-Nicolas-Pierre Hachette Lavis noir et blanc sur planche imprimée. © Archives de l'École polytechnique, all rights reserved

for the sciences of fortifications and topography and were necessary for designing various kinds of machinery and weaponry (see Fig. 11).

As a consequence, a drawing examination was added to the entrance examination of Polytechnique in 1804: it was the unique non-oral examination, and one of the two non-mathematical examinations along with a test in the French language.[91]

---

[91] Bruno Belhoste, "Anatomie d'un concours. L'organisation de l'examen d'admission à l'École polytechnique de la Révolution à nos jours." *Histoire de l'éducation* 94 (2002): 141–75.

Further, several épures were required for graduating from the school, which established a ranking of the students and therefore decided of the applications schools in which they would complete their training: four lavished épures of architectures, four lavished épures of machines, six épures of fortifications, represented by both descriptive geometry and perspectives, and six épures of maps. These requirements were even consolidated in 1812 with the addition of the elements of descriptive geometry in the program of entrance examinations, as well as of a special examination consisting in performing a geometric construction with only straightedge and compass.

Further, the teaching of drawing at Polytechnique was not limited to descriptive geometry and its various applications. As a matter of fact, engineers did not always have the adequate conditions to perform rigorous and precise geometrical drawings. They therefore also had to be trained in more classical forms of drawing, such as figure and landscape drawing, which involved professors trained at the École des Beaux-Arts (the Academy of Fine Arts) and carried humanistic values associated with the beaux arts, such as educating the "bon goût," i.e. the artistic sense of the students. These various forms of drawing and painting were commonly designated as "imitation drawing" (or "monkey drawing" in the student's slang) which, once again, highlights the key role models played. Imitation drawing was indeed learned by drawing various types of models,[92] including master drawings taken from revolutionary deposits, such as Jacques-Louis David's Bélisaire,[93] still life compositions, buildings, landscapes, and, after 1818, living models (see Fig. 12).[94]

The very large scope of the practice of model drawing at École polytechnique laid the ground for a common culture that exceeded the curriculum of the school. From 1818 to 1929, a yearly event was organized by the students, the "séances des Ombres," which consisted in a theater of Chinese shadows made of cardboard caricatures accompanied with songs written and performed by the students. Drawing caricatures was thus at the core of the social activities associated with the "Ombres," whose name also referred to the issue of shadow drawing in the teaching of both geometric and imitation drawings (see Fig. 13). This culture of drawing can also be seen in the richly illustrated student journal Le petit crapal, which was published from 1896 to 1932. Moreover, and more important for the topic of this paper, drawing played a central role in the professional careers of most students, whether they became engineers, military officers, scholars or, even, for a few of them, painters.

---

[92] François-Marie Neveu, "Dessin: Compte rendu par l'Instituteur de Dessin, relativement à cette partie de l'enseignement," Journal de l'École polytechnique 1 (1794): 78–80; François-Marie Neveu, "Cours préliminaire relatif aux Arts de Dessin," Journal de l'École polytechnique 1 (1794): 81–91.

[93] David's Bélisaire is a preparatory drawing for the 1794 painting, Bélisaire demandant l'aumône (Belisarius Begging for Alms).

[94] Jean-Baptiste Regnault, "Correspondance relative à l'introduction de l'étude du modèle vivant dans le cours de dessin, 22 avril 1818," Archives of École polytechnique, VIIb2.

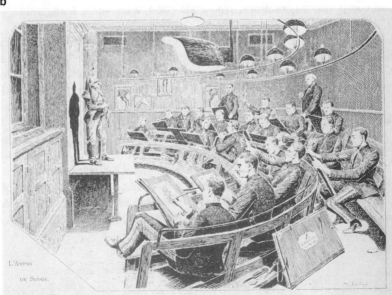

Fig. 12 **a** François Marie Neveu, "Suite du cours préliminaire relatif aux arts de dessin," *Journal de l'École polytechnique*, 4 (1796), n.p. © Bibliothèque nationale de France, all rights reserved. **b** An amphitheatre of life drawing, "L'amphi de singe." From Gaston Claris, *Notre École polytechnique* (Paris: Librairies-imprimeries réunies, 1795), 134

Fig. 13 Séance des ombres, 1882. © Collections École polytechnique, Palaiseau, all rights reserved

The Canons of Geometric Drawing: Models and the Artillery School

Monge's emphasis on descriptive geometry in the first plan of instruction of École polytechnique had laid the basis for a very coherent articulation between theory and practice as well as between the general and the special. The effectiveness of this plan was demonstrated during the French Campaign in Egypt and Syria (1798–1801), which included an important contingent of engineers and scholars assigned to the invading French force, 167 in total.[95] These scholars included several founding members of École polytechnique, such as Monge, as well as professors of the school, such as Fourier, and many alumni and students of the first promotions. They founded the Institut d'Égypte with the aim of propagating Enlightenment values in Egypt through interdisciplinary work. In this context, the young polytechnicians applied Monge's descriptive geometry for establishing fortifications in Cairo, surveying battle fields, mapping the cartography of Egypt, and for describing in minute details the monuments of Ancient Egypt, which gave rise to fascination with Ancient Egyptian culture in Europe and the birth of Egyptology:

> The most outstanding scholars were accompanied with engineers and architects of the highest merit in charge of surveying battlefields, cities, and the magnificent monuments of the pharaohs. We did not forget to arrange for them to have a staff of skillful draughtsmen working with them, and it even often happened that the skills of scholar, engineer, and draughtsman came together in the same operator [...]. Two of the youngest members of this Institut d'Égypte [...], Caristie and Jomard, who had just graduated from the new École polytechnique [...] told everyone that their colleagues and themselves had never separated the two fundamental elements of their task: the precision of measurements of all kinds of surveys and the artistic effect of the monuments, represented in perspective with the surrounding landscape as a frame.[96]

As we have seen before, Monge's plan was challenged as early as 1795 when the school was assigned the new role to provide a general instruction that would be specialized in the various application schools ('écoles d'applications'). The mathematical curriculum at the Polytechnique came to be conceived as fundamental

---

[95] Yves Laissus, *L'Égypte, une aventure savante* (Paris: Fayard, 1998); Patrice Bret, ed., *L'expédition d'Egypte, une entreprise des Lumières, 1798–1801* (Paris: Technique & Documentation, 1999).

[96] Aimé Laussédat, *Recherches sur les instruments, les méthodes et le dessin topographiques*, vol. 2. part 1 (Paris, Gauthier-Villars, 1901): 4-5: "Ainsi, en nous en tenant à l'expédition d'Égypte, où les plus illustres savants étaient accompagnés d'ingénieurs et d'architectes du plus grand mérite chargés de lever les plans des champs de bataille, des villes, des magnifiques monuments des pharaons, on n'avait pas oublié de leur adjoindre d'habiles dessinateurs, et il arrivait même souvent que les talents de savant, d'ingénieur et d'artiste se trouvaient réunis chez le même opérateur. [...] Deux des plus jeunes membres de cet Institut d'Égypte [...] Caristie et Jomard [...] déclaraient chacun à qui voulait l'entendre que leurs collègues comme eux-mêmes n'avaient jamais séparé les deux conditions essentielles de l'œuvre entreprise la précision des mesures et des relevés de toute sorte et l'effet artistique des monuments représentés en perspective ainsi que des paysages qui les encadraient le plus souvent."

instruction, which had to be theoretical and general in order to be applied, later on, in a great variety of special professions. This reformulation resulted in the growing importance of analysis at the expense of descriptive geometry and its applications. It therefore promoted an articulation between theory and application, as well as between the general and the special, very different from the one that had been designed by Monge. Yet, we shall see that Monge's legacy would remain vivid in the Metz school of artillery and engineering applications.

The Alliance Between Practice and Theory

Founded in 1794, the Metz artillery school would become the principal application school of École polytechnique after Napoléon merged it with the Mézières school of engineering in 1802.[97] In the first decades of the century, several alumni of the Metz school played a prominent role in the development of descriptive geometry and of its teaching in France, among whom the mathematician Jean-Victor Poncelet (who would become professor at Metz in 1825),[98] and two important promoters of geometric models, Théodore Olivier and Libre Bardin. As for the *Mémorial de* l'artillerie, a journal founded in 1824 and attached to both the school and the corps of artillery, it soon became a major periodical publication on descriptive geometry. Quite often, this journal published applications to engineering issues of more theoretical memoirs published in the *Journal de l'École polytechnique*, such as with the interplay of the publications of the colonel Lefevre on models of racks, pinions, and gearings, and of Olivier's mathematical research on space curves.

Bardin and Olivier graduated from Metz in 1816 and 1815 respectively, as lieutenants of artillery.[99] The first would quit the army in 1820 for experimenting with some business activities for a few years while the second would remain in the Metz school for a couple of years as adjunct to the *instituteur* of mathematical sciences and physics, before moving to Sweden between 1821 and 1826 for organizing the polytechnic instruction at the Royal School of Marienberg. They would meet again in the late 1820s at École polytechnique, where Bardin would be named professor of drawing and fortification (and would later be charged with managing all graphical works, in 1852), Olivier as *répétiteur* of descriptive geometry.

As most of the followers of Monge and supporters of geometry, Olivier and Bardin were active proponents of the industrialization of France, especially in

---

[97] Bruno Belhoste and Antoine Picon, eds., *L'École d'application de l'artillerie et du génie de Metz 1802–1870: Enseignement et recherches* (Paris: Musée des Plans-Reliefs, 1996).

[98] On Poncelet's teaching in Metz, see: Konstantinos Chatzis, "Les cours de mécanique appliquée de Jean-Victor Poncelet à l'École de l'Artillerie et du Génie et à la Sorbonne, 1825–1848," *Histoire de l'Éducation* 120 (2008): 113–38.

[99] On Olivier, see: Sakarovitch, "Théodore Olivier," 326–35; Eugène-Melchior, Peligot, Albert Perdonnet, Alexandre Dumas, "Funérailles de M. Théodore Olivier," *Bulletin de la Société d'Encouragement pour l'Industrie Nationale* 591 (September 1853): 502–10; Eugène Rouché, "La vie et les travaux d'Olivier," *Annales du Conservatoire des arts et métiers* 2, no. 8 (1896): 21–22.

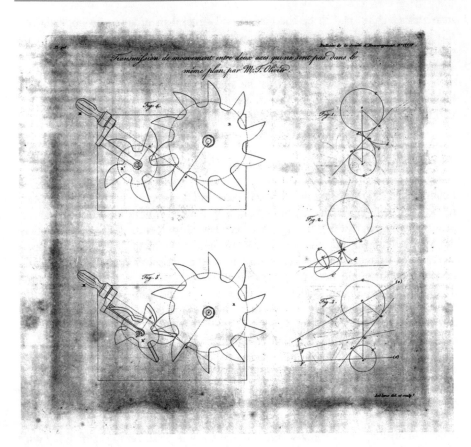

Fig. 14 Théodore Olivier, "Note sur un mode de transmission de mouvement entre deux axes qui ne sont pas dans un même plan," *Bulletin de la société d'encouragement nationale*, t. 304 (1829): 431

the Societé d'encouragement pour l'industrie nationale (Society for Encouraging National Industry), an organization established in 1801 to promote French industry.[100] The Société d'encouragement especially promoted innovations, by awarding prizes to inventors, and supported the development of technical education (see Fig. 14). Both Bardin and Olivier were especially active in the creation

---

[100] On the connection between the promoters of synthetic geometry and the industrialists, see: Lorraine Daston, "The Physicalist Tradition in Nineteenth Century French Geometry," *Studies in History and Philosophy of Science Part A* 17, no. 3 (1986): 269–95. On the technocratical role played by the engineers trained at Ecole polytechnique, see: Bruno Belhoste and Konstantinos Chatzis, "From Technical Corps to Technocratic Power: French State Engineers and their Professional and Cultural Universe in the First Half of the nineteenth century," *History and Technology* 23, no. 3 (September 2007): 209–25.

of new courses of descriptive geometry, which they conceived as the "writing of the engineer"("l'écriture de l'ingénieur"):

> [...] the one who knows how to read space can visit a factory or a manufactory without taking any note; after returning home, he can draw the tool and the machine that he has rightly seen and understood.[101]

Olivier, in particular, was a fierce opponent to the theoretical turn taken by École polytechnique under the influence of Laplace.[102] He blamed the "theoreticians," such as Laplace and Augustin Louis Cauchy, "who fashion themselves as pure scholars and consider that they form an aristocratic corp with the legitimacy to command and dominate practitioners."[103] Faithful to Monge, Olivier often used the Societé d'encouragement as a tribune for vindicating the articulation between practice and theory promoted in the first plan of instruction of École polytechnique. He never stopped insisting that "it is only through materialization that one can use the truths discovered by intelligence:"

> Without theory, practice is blind; theory is the torch that guides us. Without practice, the truths obtained by theoretical research are no more than idealities, which are useless to man's terrestrial condition, and which may only charm humans because they are intelligent beings [...], practice must precede theory. It is only through materialization that one can use the truths discovered by intelligence. Such is, always and everywhere, the law of useful labor [...]. Do not forget ever the principle, so powerful and fruitful, of the *alliance between practice and theory*.[104]

For Olivier, this alliance even involved political issues. He especially attributed the political turmoil of the 1848 revolution to utopian idealities and contrasted the love of vainglory and excessive freedom with the morality resulting from the love of work and of useful science. In 1849, he concluded a vibrant plea for developing a more important and diversified use of instruments in the teaching of geometry by claiming that:

> An education that would be limited to theoretical ideas, and in which science would only be studied from an abstract point of view, will produce a people of ideologues and dreamers; such an education will never train useful citizens. The most beautiful ideas are useless to

---

[101] Olivier, quoted in Sakarovitch, "Théodore Olivier," 328.

[102] Théodore Olivier, *Mémoires de géométrie descriptive, théorique et appliquée* (Paris: Carillan-Gœury & Dalmont, 1851), i-xxiii.

[103] Ibid. See also: Sakarovitch, "Théodore Olivier," 327.

[104] Théodore Olivier, "Notices industrielles," *Bulletin de la société d'encouragement pour l'industrie nationale* 580 (October 1852): 716-17, here: 717: "Sans la théorie la pratique est aveugle la théorie est le flambeau qui nous guide Sans la pratique les vérités dues aux recherches de la théorie ne sont que des idéalités inutiles à l'homme en sa condition terrestre et qui seulement peuvent le charmer parce qu'il est un être intelligent. [...] La pratique doit précéder la théorie Ce n'est qu'en les matérialisant que l'on peut utiliser les vérités découvertes par l'intelligence Telle est en tout et partout la loi du travail utile [...] N'oubliez jamais en toute chose le principe si puissant et si fécond de l'*alliance de la pratique et de la théorie*."

man until they are *materialized*. What is the use of moral truths to humanity, until they are put into practice, that is into traditions and into laws? It is only by making *use* of it that one can recognize whether a *thing* is good or bad; in order to know whether an *idea* is good or bad, one has therefore to *materialize* it, so that men can use it and appreciate its value. A materialized idea is the mind assuming a body, it *is the Word being made man*.[105]

As is exemplified by Olivier's discourses, the proponents of industrialization who supported the development of a technical education for the working class often associated popular education with moral issues.[106] The teaching of geometric drawing, in particular, was associated with the values of order, discipline and with the taste for work well done. The diffusion of geometric drawing in primary and technical education in the 1830 therefore participated to both the conservative political agenda of the constitutional monarchy and to the ideal of emancipation through education in mathematics.[107]

Because of his frustration with the evolution of his alma mater, which he blamed as having turned into an "École monotechnique" by focusing on analysis, Olivier participated in the foundation of the École centrale des arts et manufactures in 1826 (Central school for arts and manufactures),[108] in which he would be named professor of descriptive geometry and 'directeur des études' (dean of studies) in 1828, a position that would later be attributed to Bardin as well, from 1839 to 1841. Later on, Bardin and Olivier would both become professors at the Conservatoire national des arts et métiers (National Conservatory of Arts and Crafts), and Olivier would even be named director of the Conservatoire from 1852 to his sudden death in 1853. It was there that both Olivier and Bardin promoted the use of models for teaching descriptive geometry by designing innovative mathematical models. But before investigating these models further, we shall first discuss the specific educational model of the Conservatoire.

[105] Théodore Olivier, "Rapport fait par M. Théod. Olivier, au nom d'une commission spéciale, sur une nouvelle méthode de géométrie pratique, sans instruments, de M. Martin Chatelain, professeur à l'Athénée national," *Bulletin de la société d'encouragement pour l'industrie nationale* 544 (October 1849): 481–85, here: 485: "Un enseignement par lequel on ne donnera que des idées théoriques, dans lequel on n'étudiera les sciences que sous le point de vue abstrait, donnera un peuple d'idéologues et de rêveurs ; un pareil enseignement ne formera jamais de citoyens utiles. Les plus belles idées sont inutiles à l'homme tant qu'elles ne sont pas matérialisées. En quoi sont utiles à l'humanité les vérités morales, tant qu'elles ne sont pas passées dans la pratique, c'est-à-dire dans les meurs et dans les lois ? Ce n'est qu'à l'user qu'on peut reconnaitre si une chose est bonne ou mauvaise ; pour savoir si une idée est bonne ou mauvaise, il faut donc de, toute nécessité la matérialiser, pour que les hommes puissent d'abord s'en servir et ensuite en apprécier la valeur. Une idée matérialisée, c'est la pensée qui revêt un corps c'est le Verbe fait homme."
[106] d'Enfert, *L'enseignement du dessin*, 105.
[107] Pierre Rosanvallon, *Le moment Guizot* (Paris: Gallimard, 1985).
[108] Francis Potier, *Histoire de L'Ecole centrale des arts et manufactures* (Paris: Delamotte, 1887), 25–27.

Learning by Drawing at the Conservatoire and Beyond

The Conservatoire national des arts et métiers is one of the grandes écoles established by the National Convention during the French Revolution, along with École polytechnique and École normale.[109] In contrast with Polytechnique, the Conservatoire was not designed as a school but as a "depository for machines, models, tools, drawings, descriptions and books in all the areas of the arts and trades."[110] The Conservatoire was therefore charged with the collections of inventions, in which models and drawings played an important role (see Fig. 15). It did not originally provide lectures but was rather a place that could be visited, especially for the purpose of drawing the models of its collections. This activity was strictly regulated: the Conservatoire was opened to the public on Thursday and Sunday but the permission to practice drawing as well as to access the drawings in its archives required addressing a request to its director.

In 1798 though, Claude-Pierre Molard, the administrator of the Conservatoire, designed the project to create a "free school of drawing applied to the arts" (i.e. the techniques).[111] This school would eventually be created in 1806 with four professors, one of arithmetic and elementary geometry, one of descriptive geometry and its application to carpentry, stonecutting, etc., one of elementary architecture and drawing applied to mechanics, and one of figure drawing. In contrast with École polytechnique, the Conservatoire drawing school was designed for workers and not for engineers. It therefore did not include any lecture on higher mathematics, in accordance with the usual role played by mathematical knowledge in the hierarchy and the management of the French industrial and scientific institutions.[112] But the Conservatoire nevertheless appropriated the pedagogical method developed at Polytechnique for teaching descriptive geometry through model drawing.[113] Molard ordered several models of hyperbolic paraboloid designed by Hachette and fabricated by Brocchi for formalizing the moldboard plow attributed to Thomas Jefferson, a model of obtuse angle applied to a ship rudder designed after a drawing made by Poncelet and based on Hachette's *Traité des machines*, as well as Brocchi's stereotomy compass and it's applications to modern moldboards.

---

[109] Claudine Fontanon, "Les origines du Conservatoire national des arts et métiers et son fonctionnement à l'époque révolutionnaire (1750–1815)," *Les cahiers d'histoire du CNAM* 1 (1992): 17–44.

[110] Aimé Laussédat, "Le Centenaire du Conservatoire des Arts et Métiers," *Annales du Conservatoire des arts et métiers* 3, no. 1 (1899): 1.

[111] Alain Mercier, "Les débuts de la 'petite école.' Un apprentissage graphique au Conservatoire sous l'Empire," *Les cahiers d'histoire du CNAM,* 4 (July 1994): 27–55.

[112] On the similar role played by mathematics in the hierarchy and the management of observatory sciences in the nineteenth century, see: David Aubin, "Observatory Mathematics in the Nineteenth Century," *Oxford Handbook for the History of Mathematics*, ed. Eleanor Robson and Jacquelin Stedall (Oxford: Oxford University Press, 2009), 273–98.

[113] The first professor of mathematics at the Conservatoire, Louis Gautier, was an alumnus of polytechnique and a former student of Hachette. On the influence of Monge's followers on the CNAM, see: Émmanuel Grison, "L'École de Monge et les Arts et Métiers," *Bulletin de la Sabix* 21 (1999): 1–19.

Fig. 15 Émile Bourdelin (drawer) & Eugène Mouard (printer), "Salles rénovées du Conservatoire national des arts et métiers." From *Le Monde illustré*, (9 May 1863): 301

In turn, the Conservatoire was especially influential for the development of the teaching of geometric drawing in the other "écoles d'arts et métiers," which were created in the first half of the nineteenth century in Châlons, Angers, and Aix for providing medium level qualifications, especially to workshop foremen and machine-shop crew chiefs,[114] as well as in the practical mining schools of Alès and Saint-Étienne. Already in 1793, Monge had designed a project of establishing schools for workers and craftsmen,[115] but Monge's plan was not followed by the Convention and the teaching of descriptive geometry to workers would mainly be driven by the Conservatoire. After the 1820s, the education of the working class in geometry was especially supported by the idea that modern industry required exact drawing, and therefore the systematic use of orthogonal projection at the expense of perspective. Republican and Saint-Simonian theories about the intellectual improvement of the French working class renewed the interest in universal education that had achieved its first peak during the Revolution.

In 1819, the Conservatoire created three new public courses applied to the arts: in mechanics, chemistry, and industrial economy respectively. The course of applied mechanics was attributed to Charles Dupin, another former student of Monge at École polytechnique, who had graduated from the École d'application du Génie Maritime (naval engineering application school).[116] The academic work on geometry that Dupin had developed on the side of his activity as a naval engineer had earned him to be nominated to the body of the Paris Academy of Sciences one year before his nomination at the CNAM.

In 1825, Dupin promoted the creation of free public courses of applied geometry and mechanics in 57 cities.[117] These courses were inspired by the schools recently established in the United Kingdom for developing the instruction of workers in applied sciences. They targeted the various professions of the 'industrial class,' such as architects, carpenters, joiners, bricklayers, sculptors, painters, engravers, or even surgeons or anatomists. The charge of 20 of these 57 courses was attributed to former students of Ecole polytechnique, "true followers of the

---

[114] In Châlon, the "chef de travaux" was a former student of Polytechnique: François Emmanuel Molard. See: Charles R. Day, *Les écoles d'Arts et Métiers. L'enseignement technique en France, XIXe-XXe sicèle* (Paris: Belin, 1991).

[115] René Taton, "Un projet d'écoles secondaires pour artisans et ouvriers, préparé par Monge en septembre 1793," in *L'École normale de l'an III*, vol. 1: Leçons de mathématiques: Laplace, Lagrange, Monge, ed. Jean Dhombres (Paris: Dunod, 1992), 574–82.

[116] Konstantinos Chatzis, "Charles Dupin, Jean-Victor Poncelet et leurs mécaniques pour 'artistes' et ouvriers," in *Charles Dupin (1784–1873). Ingénieur, savant, économiste, pédagogue et parlementaire du Premier au Second Empire*, ed. Carole Christen, François Vatin (Rennes: Presses universitaires de Rennes, 2009), 99–113; Carole Christen, "Les cours pour les ouvriers adultes au Conservatoire des arts et métiers dans le premier XIXe siècle," *Cahiers de RECITS* 10 (2014): 33–56; Robert Fox, "Un enseignement pour une nouvelle ère: le Conservatoire des arts et métiers, 1815–1830,"*Cahiers d'histoire du CNAM* 1 (1992): 75–92.

[117] Charles Dupin, "Prospectus d'un cours de géométrie et de mécanique appliquées aux arts et métiers, à l'usage des chefs et sous-chefs d'ateliers et de manufactures," *Bulletin de la Société d'Encouragement pour l'Industrie Nationale* 255 (September 1825): 299–300.

illustrious Monge who will spread in the industrial class the enlightenment they have received from the genius of their master."[118] Even though the proportion of polytechnicians in these schools fell to about 25% in 1830,[119] many alumni of Polytechnique supported the development of free courses of geometry for workers. The Association Polytechnique, founded shortly after the 1830 Revolution, espoused the goal of raising the instruction of the working class. Its prototype was classes given by polytechnicians.[120]

The few actors, such as Dupin, Olivier, Bardin, or the colonel Arthur Morin,[121] who held teaching positions at the Conservatoire were therefore part of a much larger movement of engineers and artillery officers actively involved in the industrialization of France. Along the line of Saint-Simonian philosophy, they considered that industrial prosperity required putting the 'useful innovations' made by scholars in the service of the nation by increasing the instruction of the industrial class. In doing so they participated in spreading the pedagogical practices developed at Polytechnique, especially model drawing, as well as the ideals the school had inherited from the Enlightenment.

Geometry was indeed promoted by the polytechnicians as a mean of emancipation, by which workers would avoid the fate of being reduced to machines and thus the risk of proletarianization. Already in his 1793 project, Monge had promoted the project of teaching descriptive geometry to workers and craftsmen for developing not only rigor and exactness, but also the faculty of judgment, intelligence, and the "esprit d'analyse" ("analytical spirit"). As the mathematician Francœur phrased it when reporting on the textbook Dupin had designed for workers: "should the working people remain sunken in ignorance, they would badly serve the intelligence of the men who hire them, they could only be employed as a kind of machine, and would regress even further under the burden of a life much similar to the one of animals, that is limited to the exercise of physical strength [...]. The manufacturing

---

[118] Charles Dupin, "Exposé fait à la Société d'Encouragement sur les progrès du nouvel enseignement de la géométrie et de la mécanique, appliquées aux arts et métiers, en faveur de la classe industrielle," *Bulletin de la Société d'Encouragement pour l'Industrie Nationale* 257 (December 1825): 374–80.

[119] Renaud d'Enfert, "L'offre d'enseignement mathématique pour les ouvriers dans la première moitié du XIXᵉ siècle: concurrences et complémentarités," *Les Études Sociales* 159 (2014): 85–101.

[120] Gérard Bodé, "Les Associations polytechnique et philotechnique entre 1830 et 1869," in *Le Paris des Polytechniciens. Des ingénieurs dans la ville, 1794–1964*, ed. Bruno Belhoste, Francine Masson, and Antoine Picon (Paris: DAAVP, 1994), 63–68.

[121] An alumnus of the Polytechnique and of the Metz application school, Arthur Morin was nominated as professor of descriptive geometry at the Conservatoire in 1839 after having worked in Metz from 1829 to 1834 as adjunct to Poncelet. See: Claudine Fontanon, "Arthur Morin (1795–1880). Un ingénieur militaire au service de l'industrialisation," *Cahiers d'histoire et de philosophie des sciences* 29 (1990): 90–117.

and industrial prosperity of the realm will result from popular education [...]."[122] Dupin himself claimed that his textbook, which required no other prerequisites than the capacity to read and count, aimed not only at "accessing by simple steps to the intelligence of the methods of geometry and mechanics that are the most useful for the various branches of the industry," but also at "developing the most precious faculties of intelligence, comparison, reflection, judgment, and imagination as well as to allow the workers to execute their work more effectively and less painfully." In sum, it aimed at "preparing a new welfare for workers" and at "raising their morality by impressing in their mind the ideas and the habits of order and reason, which lay the most reliable ground for public peace and general happiness."[123]

Olivier's String Models

Théodore Olivier's scientific interests were mainly focused on the mechanical theory of gearing,[124] and more precisely on the mathematical determination of the shape of gear teeth, which involved investigations on space curves and therefore fundamental research in geometry, which he also applied to the tracing of railroads.

After 1825, he started publishing both on mathematics, with memoirs sent to the Paris Academy of sciences or to the *Journal de l'École polytechnique*, and on

---

[122] Louis Benjamin Francœur, "Rapport fait par M. Francœur sur la publication d'un ouvrage de M. Charles Dupin, de l'Academie des Sciences, destiné à répandre dans la classe des ouvriers l'enseignement des élémens de géométrie et de mécanique," *Bulletin de la Société d'encouragement pour l'industrie nationale* 257 (November 1825): 372–74, here 373: "Le peuple même des ouvriers, s'il demeure plongé dans l'ignorance, servira mal l'intelligence des hommes qui l'emploient, et ne pouvant même être employé que comme une espèce de machine, s'abrutira davantage sous le fardeau d'une existence trop semblable à celle des animaux dont sa force tient lieu. [...] La prospérité manufacturière et industrielle d'un royaume est donc la conséquence de l'instruction populaire."

[123] Dupin, "Exposé," 374-5: "[...] conduire [...] à l'intelligence des méthodes de géométrie et de mécanique les plus utiles aux différentes branches de l'industrie [...]. Un second but que le nouvel enseignement doit atteindre, c'est de développer dans les industriels de toute classe, et même dans les simples ouvriers, les facultés les plus précieuses de l'intelligence, la comparaison, la réflexion, le jugement et l'imagination ; c'est de leur offrir des moyens d'exécuter leurs travaux d'une manière moins pénible et plus fructueuse ; c'est de leur préparer un nouveau bien être ; c'est de rendre leur conduite plus morale en imprimant dans leurs esprits des idées et des habitudes d'ordre et de raison, qui sont les plus sûrs fondemens de la paix publique et du bonheur général."

[124] Hachette suggested Olivier start his research by providing a mathematization in descriptive geometry to a special kind of spur or level gear which had been presented in Paris in 1810 by White, a British engineer, as a frictionless gear. For a description of Olivier's works on gearings, see: Sakarovitch, "Théodore Olivier," 329 and Jacques M. Hervé, "Théodore Olivier (1793–1853)," in *Distinguished Figures in Mechanism and Machine Science*, ed. Marco Ceccarelli (Dordrecht: Springer, 2007), 296–319. See also: Théodore Olivier, *Théorie géométrique des engrenages: Destinés à transmettre le mouvement de rotation entre deux axes non situés dans un même plan* (Paris: Bachelier, 1842).

Fig. 16 Wooden models of gearings designed by Théodore Olivier for his teaching at the Conservatoire. © Musée des arts et métiers-Cnam, Paris/photos P. Faligot, all rights reserved

their technological applications to new types of gearings,[125] which he presented at the Société d'encouragement. Olivier would actually get very much involved in the Société d'encouragement. He wrote more than 15 reports for the "comité des arts mécaniques" of the society, on various types of innovative devices for railroads, gearings, riffles, machines, or even for rotating the biggest bell of the Metz cathedral, as well as on innovations in mathematical education, especially drawing instruments and geometric models.

After his nomination as a professor of descriptive geometry at the Conservatoire in 1839, Olivier designed a series of about 50 wooden models of gearings for teaching the applications of descriptive geometry by model drawing (see Fig. 16).[126] As for the more fundamental part of his teaching of descriptive geometry, Olivier developed an innovative approach by the methods of rotation, drawdown, and plane shift,[127] and adapted Monge's string models to his own mechanical concerns for ruled surfaces (see Fig. 17).[128]

The movable models he designed allow generating several ruled surfaces by changing the position of generating lines through the motion of an iron frame. These iron and string models were executed by Pixii and sons, a manufacturer of scientific instruments very close to École polytechnique. They were distributed

---

[125] About the interplay between geometry and technological applications, see, in particular, Francœur's report to a memoir presented by Olivier to the Société d'encouragement in 1829: Louis Benjamin Francoeur, "Rapport sur un Mémoire de M. *Olivier*, relatif à la vis sans fin," *Bulletin de la Société d'Encouragement pour l'Industrie Nationale* 295 (January 1829): 9–10.

[126] On Olivier's models of gearings, see: Hervé, "Théodore Olivier," 308–12.

[127] Sakarovitch, "Théodore Olivier," 331.

[128] Ruled surfaces play a key role in mechanics because any rigid-body motion (or displacement) is a screw motion. A screw has an axis and, therefore, when a rigid body moves with respect to another body, the locus of all the screw axes is a ruled surface. Olivier's interest in ruled surfaces can be seen as pioneering Plücker's 1869 geometry of straight lines.

Fig. 17 Théodore Olivier, movable string model of the intersection of two cylinders. © Musée des arts et métiers-Cnam, Paris/photo P. Faligot, all rights reserved

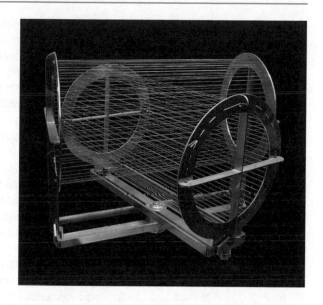

by the Société centrale de produits chimiques,[129] and had a large circulation in engineering schools in Europe and in the USA.[130] Olivier's models fall into two categories.[131] In the first, lines of a determined length generate surfaces and the strings are held taut on a quadrilateral metal frame in which sides are articulated by four parallel hinges. Such is especially the case of the models of hyperbolic paraboloids and of intersections of two cylinders. In a second category of models, the motion results from the variation of the length of the generating lines, which are made of silk strings passing through two metal wires and are attached to lead weights hidden in a wooden box. One of these models allows for turning the combination of a revolution cylinder and one of its tangent planes into the combination of a hyperboloid of one sheet and a hyperbolic hyperboloid, or to a cone and one of its tangent planes. In addition to designing dozens of new mathematical models, Olivier also very much extended the collection of mathematical models, which had been initiated when the Conservatoire had acquired the models designed by Brocchi at École polytechnique.[132]

At the Société d'encouragement, Olivier published several instructions for the organization of the teaching of descriptive geometry for workers.[133] He especially insisted on the differences between training workers and training engineers. While

---

[129] Société centrale de produits chimiques, *Catalogue général illustré* (Paris: Gauthier-Villars, 1891), 878–83.

[130] Sakarovitch, "Théodore Olivier," 333; Amy Shell-Gellasch, "The Olivier string models at West Point," *Rittenhouse* 17, no. 2 (December 2003): 71–84.

[131] Sakarovitch, "Théodore Olivier," 332; Hervé, "Théodore Olivier," 308–19.

[132] Morin, *Catalogue des collections*, 18–30.

[133] Théodore Olivier, "Instruction pour l'enseignement de la géométrie descriptive dans les écoles d'arts et métiers de Châlons, d'Angers et d'Aix," *Bulletin de la Société d'Encouragement pour l'Industrie Nationale* 546 (December 1849): 591–96.

descriptive geometry had to be taught as a science in engineering schools, it had
to be reduced to a tool when taught to workers. Descriptive geometry, thus, was to
be reduced to the "arts of projections" conceived as tools for "solving graphically"
practical problems in the workshops, especially surveying relief surfaces as well
as fabricating reliefs by the use of the drawings designed by engineers. Surveying required drawing the projections on two orthogonal planes of a relief model,
while fabricating reliefs consisted in the reciprocal operation. In contrast to a formal course of geometry, the goal was to learn geometry practically by drawing
and manipulating models of an increasing complexity: polyhedrons, plane sections of prisms and pyramids, plane sections of cylinders and cones, intersections
of prisms, pyramids, cylinders and cones, the generation of ruled surfaces by the
motion of a line, the flattening of developable surfaces on a plane, the construction
of tangents and of intersection curves between two surfaces, and helicoids (such as
screw-threads). Olivier therefore adapted to the training of workers in the *écoles
des arts et métiers* the usual pedagogical methods associated with the teaching of
drawing since the eighteenth century: the analytic decomposition/recomposition,
the importance of action-learning and practical work, and the central role of models as opposed to textual knowledge and to lectures. For Olivier, the lectures of the
professor had indeed to be limited to explaining the graphical methods required to
draw special épures, while "graphical work had to be considered as a *manipulation* which does not aim at having the students *copy* drawings but to teach them
to construct *exact épures* by using their knowledge and their intelligence."[134]

Olivier even designed a specific model and instrument for the training of workers in descriptive geometry. The 'omnibus' consisted in a box, whose top and
bottom were made of cork and could be articulated in order to represent the
two planes of projections in descriptive geometry. Four series of cards of various lengths and colors made possible a construction in space, by inserting red
cards in the bottom of the box, and representing both the projection of this construction and the projecting lines by cards of three other colors. This instrument,
Olivier claimed, "allows the students to touch by the *finger* and the *eye* all the
problems relative to points, lines and planes, as well as to *see*, before mobilizing
their intelligence by reasoning […] this instrument allows to teach students to *read
space* and to switch from projections to relief, and reciprocally."[135]

---

[134] Ibid., 595: "Le travail graphique doit être considéré comme une *manipulation* qui a pour but
non de faire *copier* des dessins aux élèves, mais de leur apprendre à construire des *épures exactes*
en se servant de leur *savoir* et de leur *intelligence*."
[135] Ibid., 596: "Cet instrument permet de faire toucher *du doigt et de l'œil*, aux élèves, tous les
problèmes relatifs au point, à la droite et au plan, et de leur faire *voir*, avant d'attaquer leur intelligence par le raisonnement, alors qu'il faut démontrer les solutions des problèmes ; cet instrument
a, de plus, l'avantage d'apprendre aux élèves *à lire dans l'espace*, et ainsi de passer des projections
au relief, et *vice versa*."

Bardin's Plaster Models

Libre Bardin's use of plaster for designing his own mathematical models highlights a practice of geometry and its applications very different from Olivier's concerns for mechanics and gearings. Bardin's scientific activities were mainly devoted to the applications of descriptive geometry to topography, which constituted the core of his teaching on fortifications at École polytechnique.

Since the sixteenth and seventeenth centuries, the art of fortifications had developed the tradition of using plan-reliefs, i.e. scale models made to visualize building projects or campaigns surrounding fortified locations. From the construction of wooden scale models of cities and fortifications, the practice of plan-reliefs evolved to the fabrication of plaster models of topographies in the eighteenth century.

From the 1830s to the 1860s, Bardin was considered one of the foremost specialists in the application of descriptive geometry to topography.[136] His plaster scale-models of notoriously difficult topographies,[137] such as islands and mountains, were exhibited in various industrial fairs, including the 1855 and 1867 world fairs in Paris,[138] and in London in 1862 (see Fig. 18).[139] His "plan relief stéréotomique" of the Mont Blanc was praised for providing "the geometrical form of the mountain."[140] In contrast with other plan-reliefs, which lacked geometrical precision, most commentators highlighted the interplay between Bardin's theoretical knowledge in geometry and his very practical manual skills for working with plaster:

> Even though relief representation is not new, its fecundity had remained buried and sterile because it lacked applications; until a man, who has been trained to both the exact sciences and to sensing by the use of his eyes and his hands, convinced himself of the usefulness of reliefs for instruction [...]. Thanks to the use of relief models, descriptive geometry has become much easier to teach [...]. In the hands of M. Bardin, wood, plaster, and carton-pierre are turned into true prodigies of precision and exactness.[141]

---

[136] On the evolutions of the methods and instruments of topographic drawing, see: Laussedat, "Recherches sur les instruments," 225–82.

[137] Several of Bardin's plan reliefs are preserved in the Musée des plans reliefs in Paris, such as the one of the island of Port Cros. Others are preserved in the collections of the Conservatoire, such as the ones of the Mont-Cenis or of the landscape surrounding the city of Metz. See: Morin, *Catalogue des collections*, 64–65.

[138] Félix Ferri-Pisani, "Cartes topographiques hydrographiques et géographiques," in *Exposition universelle de 1867 à Paris. Rapports du jury international*, vol. 2, ed. Michel Chevalier (Paris: Imprimerie administrative de Paul Dupont, 1868), 587–91.

[139] Auguste Daubrée, "Cartes et plans en relief," in *Exposition universelle de Londres de 1862. Rapport des membres de la section française du jury international sur l'ensemble de l'exposition*, vol. 6, ed. Michel Chevalier (Paris: Napoleon Chaix, 1862), 129–33.

[140] Jean-Gabriel-Victor de Moléon, ed., *Musée industriel et artistique, ou Description complète de l'Exposition des produits de l'industrie française faite en 1844* (Paris: Société polytechnique, 1844), 123.

[141] Ibid., 123–24: "Ainsi la représentation en relief n'est pas neuve, mais, faute d'application elle demeurait enfouie et stérile dans sa fécondité ; il a fallu qu'un homme exercé aux sciences exactes,

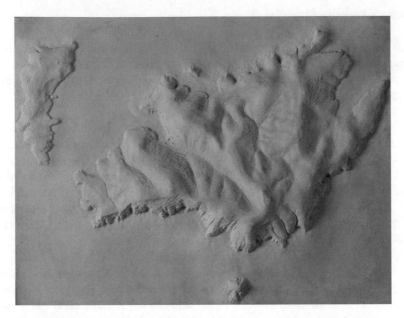

Fig. 18 Libre Bardin, plaster plan-relief of the Island of Port-Cros. © Musée des plans-reliefs (Paris)—Bruno Arrigoni, all rights reserved

Since Bardin's plan reliefs could be molded, and therefore reproduced industrially and quite cheaply, they became widely used for teaching topography.[142] When he became professor at the Conservatoire, Bardin used his skills for modeling plaster to fabricate plaster models of geometric solids that he used as drawing models for teaching descriptive geometry. One of his students, Charles Muret, who would himself become a surveying engineer and a professor at the Institut national agronomique,[143] continued and developed the use of plaster models for both topography and descriptive geometry.[144] Muret designed a collection of about

---

habitué à sentir par les lieux et par les mains se convainquit de l'utilité des reliefs dans l'instruction [...] Ainsi la géométrie descriptive à l'aide des modèles en relief est devenue bien plus facile à enseigner [...]. Le bois, le plâtre et le carton-pierre se transforment, sous la main de M. Bardin, en de vrais prodiges de précision et d'exactitude."

[142] Charles Combes and Eugène-Melchior Peligot, eds., "Séance du conseil d'administration du 9 août 1867," *Bulletin de la Société d'Encouragement pour l'Industrie Nationale* 2e série 14 (1867): 608.

[143] Charles Muret, *Topographie, levés ruraux, remembrement,* vol. 2, 3rd ed. (Paris: J.B. Baillière, 1934).

[144] Muret's plan reliefs were presented at the 1878 world fair in Paris. See: N.N., *Exposition universelle internationale de 1878. Section française. Deuxième groupe. Classe 16* (Paris: Imprimerie Delalain, 1878), 66. Several plan reliefs designed by Muret, such as of the city of Paris or of the Suez canal, are preserved in the collections of the Conservatoire. See: Conservatoire des arts et métiers, ed., *Catalogue du musée* (Paris: Conservatoire national des arts et métiers, 1953), 141–45.

Fig. 19 Charles Muret, plaster model of an icosaedron © Collections de l'Institut Henri Poincaré, all rights reserved. Photo: Frédéric Brechenmacher

600 models for the teaching of geometric drawing (see Fig. 19; see also Fig. 24). From 1865 to 1875, Delagrave edited this collection; a publishing house specialized in textbooks for both secondary and higher education. Muret's models had a very large circulation in both high schools and universities, all over Europe and the USA. The Paris faculty of science especially purchased the whole collection, which would form the seed of the Sorbonne cabinet of mathematics.

Model Drawing in Superior Primary Education

We have seen that a number of former students of École polytechnique had remained faithful to Monge's ideals about the role descriptive geometry and model drawing should play in the alliance between practice and theory. Several of them became involved in various experiments for developing the education of the working class, such as with Francœur and Jomard in the movement of mutual instruction at the beginning of the nineteenth century, and, after the 1820s, with Dupin, Bardin, Olivier and many of their fellow alumni of Polytechnique and Metz in the creation of free public courses of geometry all over the country.

In the 1830s, the French government eventually institutionalized a new system of public instruction for improving the education of children from modest households beyond primary education.[145] Established by the Guizot law of 1833, the "superior primary education" ("enseignement primaire supérieur") established a form of practical education parallel to the general secondary education provided by the lycées.[146] The curriculum of this new system of education especially included geometry and drawing, which were both usually taught by a professor of

---

[145] On the teaching of drawing in primary schools in France, see: Renaud d'Enfert, Daniel Lagoutte, and Myriam Boyer, *Un art pour tous. Le dessin à l'école de 1800 à nos jours* (Lyon: INRP, 2004). On mathematics in primary schools, see: Renaud d'Enfert, Hélène Gispert, and Josiane Hélayel, *L'enseignement mathématique à l'école primaire, de la Révolution à nos jours. Textes officiels*, vol. 1: *1791–1914* (Paris: INRP, 2003).

[146] Renaud d'Enfert, "Inventer une géométrie pour l'école primaire au XIXe siècle," *Tréma* 22 (2003): 41–49.

mathematics. The issue of challenging the industrial leadership of the British was instrumental in the development of technical education in France in the 1820s–1830s, and the promotion of geometric drawing was considered as a key issue for the construction machine industry.

The mathematician Louis-Benjamin Francœur, who had been both a student and a chef de brigade in the very first promotion of Polytechnique, conceived in 1819 a form of descriptive geometry adapted to mutual instruction: the "dessin linéaire."[147] Linear drawing was designed by Francœur as one of the four branches of primary education, along with reading, writing and arithmetic. It was organized by a progressive, analytic, method, from the drawing of straight lines and the simplest geometric figures to the complex patterns of architecture and eventually the human figure. The analytic method was more generally at the core of mutual instruction in which knowledge was decomposed into a series of "tableaux" ("tables") that were displayed in the classrooms where small groups of 8–10 children were supervised by a "moniteur" ("supervisor"). Francœur's dessin linéaire initially presented 5 tableaux of geometric models. In comparison, 125 tableaux were involved in Jomard's method for teaching reading, and 88 in the one for teaching arithmetic. But the importance of models was increased in the next editions of Francœur's dessin linéaire, with 10 tableaux in 1827, and 16 tableaux in 1832. The development of the diversity of models for the teaching of linear drawing was strongly supported by the Société pour l'instruction élémentaire in 1822 and a large number of textbooks were published after the 1830s along with plates of geometric models (see Fig. 20).[148]

With the Guizot law, the teaching of linear drawing was extended to superior primary education and to the écoles normales primaires established for training the teachers of primary schools. Several alumni of the Polytechnique were involved in designing and promoting the use of geometric models in superior primary instruction, such as with the patterns of cardboard models of polyhedrons published by Maximilien Marie in 1835,[149] and inspired by the models designed in London in 1758 by John-Lodge Cowley.[150]

At the Société d'Encouragement, Olivier strongly supported all pedagogical innovations based on the use of models and instruments. Their novelty was evaluated with the norms of technical and industrial innovations, especially manufacturing cost. For instance, in 1845, Olivier awarded the silver medal of the Société d'encouragement to the folded cardboard models of polyhedrons designed

---

[147] On linear drawing, see: d'Enfert, *L'enseignement du dessin*, 101–78.

[148] d'Enfert, *L'enseignement du dessin*, 128–42. See especially the exercices of model drawing in: Auguste Bouillon, *Exercices de dessin linéaire contenant un choix très varié de modèles pratiques d'architecture, de marbrerie, de charpente, de menuiserie, de serrurerie, et d'ameublement* (Paris: Hachette, 1847).

[149] François-Charles-Michel Marie, *Géométrie stéréographique, ou reliefs des polyèdres pour faciliter l'étude des corps*, (Paris: Hachette, 1835).

[150] On the history of folding in mathematics, and on Cowley in particular, see: Friedman, *A History of Folding*, 77–80.

Le maître peut, sans frais, exécuter ces modèles en relief, propres à montrer les contours apparents, les arêtes, etc. Ce n'est qu'après avoir acquis l'habitude de ces conceptions, en dessinant sur la planche noire, ayant d'ailleurs sous les yeux les tableaux et les modèles ; qu'on peut espérer que les enfants pourront copier de mémoire, sans trop altérer les perspectives.

Cette remarque s'applique aux pyramides, cônes, cylindres, sphères, qui sont dessinés dans les classes suivantes.

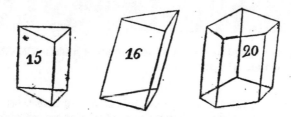

14 et 15. *Construisez un prisme triangulaire oblique,* fig. 16, ou *droit*, fig. 15.

Le *prisme* est un corps formé de deux polygones égaux et parallèles, dont les sommets semblables sont joints par des arêtes qui toutes sont parallèles et égales entre elles. Telles sont les figures 15, 16, etc., jusqu'à 22. La *hauteur* du prisme est la perpendiculaire aux deux bases qui en mesure la distance ; c'est une verticale qui se termine à sa rencontre avec les deux bases horizontales. On dit que le

Fig. 20 Francœur's 2nd tableau. From: Louis-Benjamin Francœur, *Dessin linéaire et arpentage*, 4th ed. (Paris: Bachelier, 1839), 41

by the civil engineer Louis Dupin, not so much because they innovated by integrating written text about the properties of each folded polyhedron on the folded cardboard itself (see Fig. 21),[151] but because they were much cheaper than the wooden models designed by several manufacturer in Paris:[152]

[151] On Dupin's models, see: ibid., 126–30.

[152] See, for instance: Albert Marloye, *Catalogue des principaux appareils d'acoustique et autres objets qui se fabriquent chez Marloye, à Paris, rue de la Harpe* (Paris: Ducessois, 1840), 15. See also the issue of the manufacturing cost as discussed by Olivier when awarding prices to the topographical relief models presented at the national industrial exhibition in 1844: Théodore Olivier, "Cartes

Fig. 21 Louis Dupin's folded models. © Musée des arts et métiers—CNAM/Photo: Aurélien Mole/Mudam Luxembourg, all rights reserved

> Relief models are so useful for teaching geometry that we must promote both the introduction and the continuation of their use in the primary school of art and crafts. Wooden models are pricey [...], M. Louis Dupin had the fine idea of constructing a series of cardboard polyhedrons, which can be juxtaposed in order to form a cube.[153]

From 1844 to 1847, Dupin augmented his collection and entrusted their execution to the manufacturer of scientific instruments Molteni and Sigler for "delivering these solids at prices that should facilitate the introduction in schools."[154] The mathematical models and instruments designed for the teaching of geometry were displayed on a regular basis in industrial exhibitions,[155] world fairs, and eventually

---

géographiques. Globes terrestres et célestes et terrestres. Cartes en relief. Modèles topographiques en relief. Planétaires," in *Exposition des produits de l'industrie française en 1844. Rapport du jury central*, vol. 2 (Paris: Fain & Thunot, 1844), 521–28. See also Olivier's focus on the issue of the manufacturing cost in regard with innovations in geometrical drawing instruments: Théodore Olivier, "Rapport fait par M. Théod. Olivier, au nom du Comité des arts mécaniques, sur un étui de mathématiques de M. Legey," *Bulletin de la Société d'Encouragement pour l'Industrie Nationale* 404 (February 1838): 46–48.

[153] Théodore Olivier, "Rapport sur les solides en carton de M. Louis Dupin," *Bulletin de la Société d'Encouragement pour l'Industrie Nationale* 487 (January 1845): 10–11: "Les modèles en relief sont si utiles pour l'enseignement de la géométrie, que l'on doit s'efforcer d'en introduire l'usage dans les écoles primaires et d'arts et métiers et de l'y conserver. Les modèles en bois sont coûteux [...] M. Louis Dupin a eu l'heureuse idée de construire une série de polyedres en carton, qui, par leur juxtaposition forment un cube."

[154] Théodore Olivier, "Extrait des procès-verbaux des séances du conseil d'administration," *Bulletin de la Société d'Encouragement pour l'Industrie Nationale*, 518 (August 1847): 443.

[155] For an early example, see the report of the 1849 national exhibition in Paris and its discussion of the various innovations developed in Europe for realizing cheap relief maps that may be used for teaching geography in elementary school as well of models for teaching elementary

in special exhibitions such as the gigantic Loan Exhibition of Scientific Apparatus in London in 1876,[156] and the exhibitions of mathematical devices which accompanied the first national and international congresses of mathematicians at the turn of the century. The world fairs, in particular, provided an international space of discussion, comparison, and competition for mathematical tools as for any other industrial innovations.[157] In the 1860s, the reports written by the French delegates to the world fairs highlight a growing sense of the increasing superiority of the models manufactured in Germany, especially in Darmstadt where Schroeder's manufacturing processes allowed to produce wooden models of descriptive geometry for a competitive price thanks to special machine tools and no less than 50 workers.[158]

The Models of Higher Geometry

In the 1860s–1870s, the exhibitions of models and instruments, that had been traditionally associated with technical and primary mathematical education, met with the new models of higher geometry that were designed by prominent mathematicians in the main centers of mathematical academic activity in Europe such as Göttingen, Munich, Cambridge, and Paris.[159]

Even though the models of higher geometry carried strong pedagogical ideals, these ideals were usually very different from the ones associated with the traditional use of models in technical and primary education. To be sure, both the traditional models and the new models of higher geometry were associated with the pedagogical values of visualization and manipulation, i.e. the issue of making use of the eye and the hand in the teaching of mathematics. But while the traditional use of models could not be dissociated from the idea that the eye and the hand had to be trained by the practice of model drawing, models of higher geometry were often designed for universities in which drawing was usually not associated with mathematical education.

geometry: Mathieu, "Globes célestes et terrestres; Machines planétaires; Cartes en relief; Modèles géométriques," in *Rapport du jury central sur les produits de l'agriculture et de l'industrie exposés en 1849*, vol. 2 (Paris: Imprimerie nationale, 1850), 562–66.

[156] *Handbook to the Special Loan Collection of Scientific Apparatus 1876*, Piccadilly: Chapman and Hall, 1876.

[157] See: Ed. Grateau, "Instruments de mathématiques et modèles pour l'enseignement des sciences," in Chevalier, ed., *Exposition universelle de 1867 à Paris*, 521–47.

[158] Ibid, 540.

[159] Ibid. 541.

Naturalistic Mathematics

Pedagogical ideals were not the only force driving several prominent mathematicians to design models of higher geometry in the 1860s and 1870s. The issue of the classification of cubics and quartics required the careful investigation of the singularities and special configurations that allowed classifying species of surfaces in a naturalistic approach to mathematics.[160] In the 1820s, the classification of conic sections and quadrics had already highlighted the limits of algebraic methods: while analysis had allowed to classify all the surfaces of the second order by the algebraic character of the roots of their characteristic equations, this same method had failed to classify the types of intersections of two quadric surfaces. This issue had fuelled a criticism of the genericity of algebraic methods, which resulted in both the promotion of the geometrical investigation of singularities and the attention to algebraic singularities with the development of specific forms of representations such as invariants, determinants, and matrices.[161]

The geometrical characterization of singularities and special incidence configurations, as well as their combinatoric enumeration, played a key role in the classification of surfaces of order higher than two, especially cubic and quadric surfaces, like Hesse's inflection point configuration for cubic curves, or Schläfli's double six in connection with the 27 lines of a cubic surface (see Figs. 22 and 23). The wooden models designed by Plücker for displaying select features of a certain class of quartic surfaces linked to quadratic line complexes, or the ones constructed

Fig. 22 Model of the 27 lines on a cubic surface. © Collections de l'Institut Henri Poincaré, all rights reserved

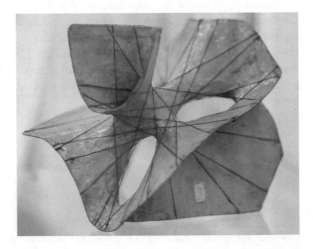

[160] Rowe, "Mathematical Models," 5.
[161] On the role of the intersection of conics and quadrics in Cauchy's criticism of the genericity of algebra and ideals of rigor, see: Thomas Hawkins, "Cauchy and the Spectral Theory of Matrices," *Historia Mathematica* 2, no. 1 (1975): 1–20. On the evolution of this issue in algebra in the nineteenth century, see: Frédéric Brechenmacher, "La controverse de 1874 entre Camille Jordan et Leopold Kronecker," *Revue d'histoire des mathématiques* 13 (2007): 187–257.

Fig. 23 Alfred Clebsch diagonal surface. ©

by Christian Wiener for representing Clebsch's classification of surfaces, played a role similar to the collections of species or minerals in natural history.[162]

The idea of working with models was especially derived from the traditional use of models and instruments in experimental physics. Plücker himself experimented with rarefied gases, built one of the first gas discharge tubes, and carried out his mathematical models with the assistance of Heinrich Geissler, the inventor of the eponym glass tubes.[163] Klein, who had assisted Plücker in designing his mathematical models, emphasized the connection between these models and Plücker's earlier research in physics as well as the influence of Michael Faraday who, according to Klein, had given Plücker the initial impetus to build models illustrating different types of the surfaces he unveiled as the centerpiece of his new line geometry.[164] It was also thanks to Plücker's reputation as an experimental physicist that his mathematical models had an almost immediate circulation in England.[165] As for the special quartics studied by Kummer in the mid 1860, they were directly associated with the caustics of geometrical optics.[166]

---

[162] The issue of the classification of conics and cubics was especially brought to the fore in Henri Fehr's reviews on geometric models in the journal *L'enseignement mathématique*. About the several types of classifications materialized by Wiener's models, see: Henri Fehr, "Herm. Wiener. Die Einteilung der ebenen Kurven und Kegel dritter Ordnung in 13 Gattungen," *L'enseignement mathématique* 3 (1901): 310.

[163] Rowe, "Mathematical Models," 5.

[164] Klein, *Gesammelte mathematische Abhandlungen*, vol. 2, 3–7.

[165] Rowe, "Mathematical Models," 6.

[166] Ibid., 8.

These academic ideals contrasted with the ones associated with the traditional use of models and instruments in engineering schools or in industry. Klein, in particular, even though he had been much impressed by Olivier's mathematical models during his trip to Paris in 1870, was chiefly influenced by his two masters, Plücker and Clebsch, whose premature deaths left him with the responsibility of their legacies.[167] A few years later, Klein promoted the use of models in the mathematical laboratory he had founded with Alexander Brill—another former student of Clebsch—at the Munich Technische Hochschule.[168] Even though he carried on with the tradition of action-learning associated with models, as opposed to reading textbooks or assisting plenary lectures, and as providing a direct contact with natural forms, he did not aim at carrying on the tradition of teaching by drawing but was rather inspired by the role of the laboratory as a place of experimentation in physics, which had been especially promoted in Munich by his colleague Carl Linde. Rather than model drawing, it was the construction and the manipulation of mathematical models that Klein promoted as a way to deepen the mathematical training of doctoral students such as Walther Dyck and Hermann Wiener (the son of Christian Wiener).[169]

The naturalistic ideal to "render a great service to geometrical science by calling attention to the concrete shapes of objects, which are too apt, even in the mind of the serious student, to exist only as conceptions very imperfectly realized,"[170] was also the motto of the Cambridge modeling club founded by Arthur Cayley in 1873, and which especially benefited from the contribution of Olaus Henrici,[171] another former student of Clebsch who held a position at University College London where he aimed at developing a 'modern' pedagogical approach to geometry by breaking with both the logico-deductive tradition of Euclid and the algebraic formalism of analytical geometry. Both the classification of geometric surfaces and the material representation of specific mathematical properties aimed at promoting 'observation' in 'pure mathematics,' i.e. a value which had developed in the natural sciences, more precisely in observational sciences, and very much associated

---

[167] On this issue, and on the first series of Zinc model designed by Klein and his friend Wenker, see: Klein, *Gesammelte mathematische Abhandlungen,* vol. 2, 3.

[168] See: Rowe, "Göttingen Mathematical Tradition," 191; Rowe, "Mathematical Models," 15; Renate Tobies, "Felix Klein in Erlangen und München: ein Beitrag zur Biographie," in *Amphora. Festschrift für Hans Wussing zu seinem 65. Geburtstag,* ed. Sergei S. Demidov, David Rowe, Menso Folkerts, and Christoph J. Scriba (Basel: Birkhäuser, 1992), 751–72.

[169] On Hermann Wiener's use of models in his teaching in Darmstadt, see: Anja Sattelmacher, "Geordnete Verhältnisse. Mathematische Anschauungsmodelle im frühen 20. Jahrhundert," in *Berichte zur Wissenschaftsgeschichte* 36, no. 4: "Bildtatsachen," special issue, vol. 2, ed. Ina Heumann and Axel Hüntelmann (2013): 294–312.

[170] Henry John Stephen Smith, "Geometrical instruments and models," in *The Collected mathematical papers of James Joseph Sylvester*, vol. 2, ed. Henry John Stephen Smith and James Whitbread Lee (Cambridge: Cambridge University Press, 1894), 698–710, here: 699.

[171] June Barrow-Green, "Clebsch took notice of me: Olaus Henrici and surface models," *Oberwolfach Reports* 14, no. "History of Mathematics: Models and Visualization in the Mathematical and Physical Sciences," ed. Jeremy J. Gray, Ulf Hashagen, Tinne Hoff Kjeldsen, and David E. Rowe (2015): 2788–90.

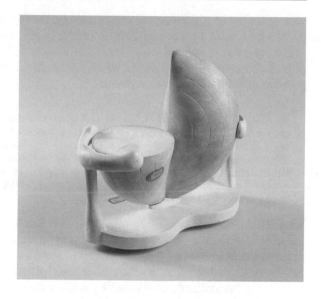

with the emergence of the ideal of objectivity.[172] In topology in particular, and in contrast to the classification of cubics in algebraic geometry:

> […] no complete *corps de doctrine* has yet been formed of the properties of situation of figures […]. We cannot therefore expect to find this part of the science of geometry extensively illustrated by models, or by drawings expressly prepared for the purpose. But any great collection of geometrical objects cannot fail to supply examples of such properties; and what is of more importance, may be expected to suggest entirely new points of view in a branch of inquiry, which, more than almost any other within the range of pure mathematics, is dependent on direct observation.[173]

These naturalistic academic ideals were to be disseminated all over Europe and the U.S.A. with the emergence of semi-industrial manufactures of mathematical models, starting with the editor Ludwig Brill, Alexander Brill's brother, who had inherited the familial printing house in Darmstadt. Klein's own specific philosophy of 'anschauliche Geometrie' has already been the subject of several historiographical studies. Let us simply recall the role played by Klein's discovery in 1870 of an error in a statement on the singularities of the asymptotic curves that lie on a fixed Kummer surface by the observation of a physical model of such a surface made by his friend Albert Wenker.[174] Klein developed the conception that, even though the realization of models usually comes only after algebraic studies with pen and paper, only the material representation of a model can demonstrate the

---

[172] Lorraine Daston and Peter Galison, *Objectivity* (New York: Zone Books, 2007).

[173] Smith, "Geometrical instruments and models," 700.

[174] See the letter from Klein to Sophus Lie cited in: David E. Rowe, "Klein, Lie, and their early Work on Quartic Surfaces," in S*erva di due padroni: Saggi di Storia della Matematica in onore di* Umberto *Bottazzini,* ed. Alberto Cogliati (Milano: Egea, 2019), 189–198. See also: Rowe, "On Franco-German Relations," 21–36.

very existence of a geometrical object and impress its 'true character' in the mind. As we have seen in the introduction of this paper, this idea of 'impressing the mind' by a direct contact with nature was already crucial to Rousseau's philosophy of education, but, in contrast to Rousseau, Klein did not associate it with the practice of drawing but with observation:

> There is an essential [*eigentliche*] geometry, which does not only mean to be, as the investigations discussed in the text are, a visualized [*veranschaulichte*] form of abstract investigations. Here it is the task to grasp the spatial figures in their full figurative reality [*gestaltliche Wirklichkeit*], and (which is the mathematical side) to understand the relations valid for them as evident consequences of the principles of spatial intuition [*Anschauung*]. For this geometry, a model—be it realized and observed or only vividly imagined—is not a means to an end but the thing itself.[175]

Visualization thus played a key role in the 'anschauliche Geometrie,' which Klein associated with a naturalistic philosophy of mathematical objects as both real objects and witnesses of the very nature of the human mind. Accordingly, Klein developed a pedagogical practice of models disconnected from drawing, even though the "visualization" allowed by models did require a preliminary training of the eye and the hand. Significantly, the creation of Klein's special seminar and laboratory of mathematics in Munich was made possible because, unlike most other such institutions, the Munich Technical Hochschule trained not only engineers and architects but also teaching candidates: this situation provided Klein with students who had already been trained in geometric drawing, as well as with the opportunity to disconnect his *anschauliche* approach to geometry with the practice of drawing, since his seminar was limited to the students pursuing the teaching candidates program.[176]

The Darboux-Caron Wooden Models

In contrast with the use of models promoted by Klein, the development of models of higher geometry in France carried on the tradition of model drawing. It actually resulted from an importation in the University of Paris of pedagogical practices developed in technical and primary education. Gaston Darboux, who held the chair of higher geometry at the Sorbonne, played a central role in this evolution. In the

---

[175] Felix Klein, *Vergleichende Betrachtungen über neuere geometrische Forschungen* (Erlangen: Andreas Deichert, 1872), 42: "Es gibt eine eigentliche Geometrie, die nicht, wie die im Texte besprochenen Untersuchungen, nur eine veranschaulichte Form abstracterer Untersuchungen sein will. In ihr gilt es, die räumlichen Figuren nach ihrer vollen gestaltlichen Wirklichkeit aufzufassen und (was die mathematische Seite ist) die für sie geltenden Beziehungen als evidente Folgen der Grundsätze räumlicher Anschauung zu verstehen. Ein Modell – mag es nun ausgeführt und angeschaut oder nur lebhaft vorgestellt sein – ist für diese Geometrie nicht ein Mittel zum Zwecke sondern die Sache selbst." English translation from Mehrtens, "Mathematical models," 289. See also: Felix Klein, "A comparative review of recent research in geometry," *Bulletin of the New York Mathematical Society* 2, no. 10 (1893): 244.
[176] Tobies, "Felix Klein," 751–72.

Fig. 25 The cabinet of mathematics of the Sorbonne (before 1914). © Collections de l'Institut Henri Poincaré, all rights reserved. Photo Ch. Barenne, Paris. ca. 1914

early 1870s, he promoted the development of the mathematical cabinet of the Paris faculty of science, whose collection of models had been initiated with the acquisition of the Muret collection (see Fig. 25).

Darboux also attentively followed the innovations developed abroad in the design of new models. In the *Bulletin des sciences mathématiques et astronomiques*, a review journal he managed since 1868,[177] he especially reviewed the models designed by Brill and Klein in Munich (see: Fig. 26):

> One has often wondered whether drawings and models are useful for mathematical education. [...]. Whatever the opinion one may have on this issue [...], everyone will agree that models provide a lively and striking ingredient for both students and professors, models allow displaying the results obtained after painful computations, or arduous discussions, in a real, concrete, and elegant form.[178]

---

[177] On Darboux's editorial work in the *Bulletin*, see: Barnabé Croizat, "Gaston Darboux: naissance d'un mathématicien, genèse d'un professeur, chronique d'un rédacteur," (Phd diss., University of Lille, 2016).

[178] Gaston Darboux, "Comptes rendus et analyses," *Bulletin des sciences mathématiques et astronomiques* 2, no. 6 (1882): 5–14, here: 5-6: "Souvent on s'est demandé s'il était utile

Fig. 26 Photos of models at
the inner cover of Ludwig
Brill, ed., *Catalog
mathematischer Modelle für
den höheren mathematischen
Unterricht* (Darmstadt: L.
Brill, 1881)

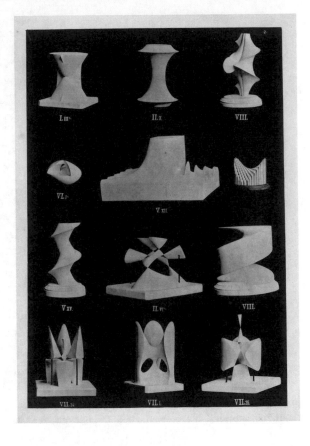

Yet, in contrast to most other prominent European mathematicians, Darboux did
not only promote the value of vizualization of models but also carried on the traditional practice of model drawing. His lectures at the Paris faculty of science were
accompanied with practical drawing activities in the tradition of the pedagogical
methods developed by Monge at the Polytechnique and which, as we have seen,
had broadly circulated in both technical and primary education in the nineteenth

---

d'employer des dessins et des modèles dans l'enseignement mathématique [...] Et pourtant toute
personne, quelle que soit son opinion sur la question posée précédemment, voudra bien convenir
que le modèle fournit non seulement à l'élève, mais aussi au professeur, un élément plein de
vie, saisissant, alors qu'après un calcul pénible ou après une discussion ardue le résultat peut
être présenté sous une forme réelle, concrète et élégante." See also: Gaston Darboux, "Revue
bibliographique," *Bulletin des sciences mathématiques et astronomiques* 8 (1875): 7–17.

century, but not yet in the university system.[179] The drawing activities were supervised by Joseph Caron, who had been appointed director of graphical works at École Normale Supérieure in 1872, a position identical to the one Bardin used to hold at École polytechnique.

Darboux has often been presented as the personification of the shift that occurred in the figure of the mathematician in France at the turn of the twentieth century, from the 'ingénieurs savants' trained at École polytechnique to the university professors trained at École normale. Darboux was indeed one of the first prominent mathematicians who favored École normale over the Polytechnique after having been ranked first in the competitive exams of both schools in 1861. While such a choice was very uncommon in the 1860s, it would become almost obvious for aspiring mathematicians in the 1880s. But Darboux's association with Caron for importing in the university the pedagogical methods developed at the Polytechnique also highlights a form of continuity in the evolutions of mathematics in France in the late nineteenth century.

In the tradition of Brocchi, Bardin, Olivier, and many others, Caron designed material models for the practical drawing activities associated with Darboux's lectures on curves and surfaces (see Fig. 27).[180] From 1872 to 1915, he supplied the Cabinet de mathématiques of the Sorbonne with about a hundred models, mainly made of wood.[181] He also published several textbooks of descriptive geometry for the candidates preparing for the competitive exams of the grandes écoles, such as École polytechnique and École normale supérieure.[182]

But while the use of models by Monge's followers had focused on either the basic elements of descriptive geometry or on its applications to the engineering sciences, Caron's wooden models aimed at representing the much more theoretical configurations presented in Darboux's lectures, such as a Kummer surface with twelve real double points, a rational algebraic surface of degree eight generated by the plane section of a cylinder rolling on another cylinder, the envelope of the normals for a Plücker conoid, minimal surfaces, etc. Even though they carried on the tradition of model drawing, Caron's series of wooden models of higher geometry thus broke up with the issue of the alliance between theory and applications, which, as we have seen, had been very much associated with the use of models at Polytechnique and, more generally, in technical education (see Fig. 28).

---

[179] At the time when Poncelet held the chair of applied mechanics at the Sorbonne, the latter could do nothing more than display a collection of models during his plenary lectures, since the university did not have any laboratory or other facility for practical work and model drawing. See: Paul Appel, "L'enseignement scientifique à l'Université de Paris," *L'enseignement mathématique*, 8 (1906): 337–42. The practical orientation of Poncelet's course at the Sorbonne was scarcely tolerated. See: Chatzis, "Les cours de mécanique appliquée," 128.

[180] Gaston Darboux, *Leçons sur la théorie générale des surfaces et les applications géométriques du calcul infinitésimal*, 1st ed. (Paris: Gauthier-Villars, 1887–1896).

[181] Apéry, "Caron's Wooden Mathematical Models," 38–48.

[182] See Joseph Caron, *Cours de géométrie descriptive. À l'usage des classes de mathématiques spéciales* (Paris: E. Foucart, 1883).

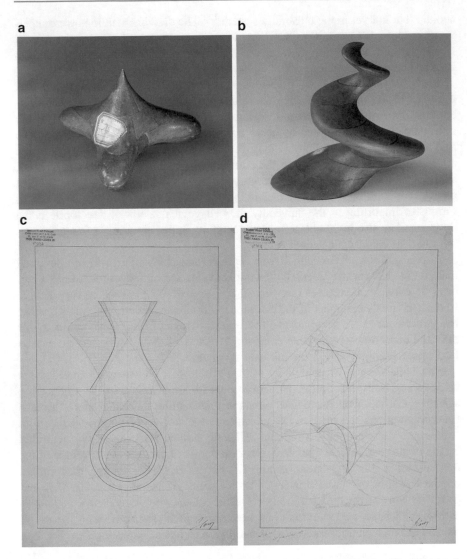

Fig. 27 **a** Quartic surface with a peak by Joseph Caron. © Collections de l'Institut Henri Poincaré, all rights reserved. Photo: François Apéry. **b** Spiral surface by Joseph Caron. © Collections de l'Institut Henri Poincaré, all rights reserved. Photo: François Apéry. **c** Épure of a hyperboloid, with a work on shadows, by Joseph Caron. © Collections de l'Institut Henri Poincaré, all rights reserved, épure n°032. **d** Épure by a student of Joseph Caron in 1907: two cones with a common tangent plane. © Collections de l'Institut Henri Poincaré, all rights reserved, épure n°014

a   b

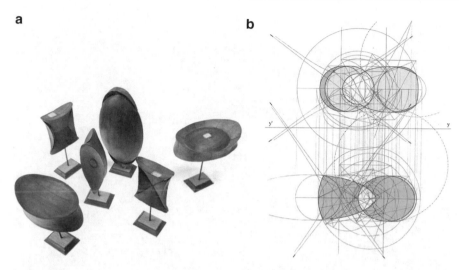

As with the models of higher geometry promoted by Klein in Germany, the ones of Darboux and Caron proceeded from an autonomization of mathematics as a discipline in the context of the development of higher scientific education. Even though Klein and Darboux were both active advocates of the interplay between theory and applications, this interplay did not take the same meaning in universities, even German technical universities, as the one which had been promoted in École polytechnique or in the Conservatoire des arts et métiers. Its focus was on the interplay between mathematics and other academic scientific disciplines rather than aiming at a direct usefulness for engineering sciences or the industry.

Darboux's course of higher geometry at the Sorbonne was very much oriented to fundamental applications to mechanics and optics. Several models designed by Caron display configurations of cinematic geometry and mechanics, such as space curves generated by the motion of a cylinder on a plane. Another series of eight models made between 1912 and 1914 were devoted to the caustics generated by a wave front in optics.[183] In the early 1890s, Darboux himself got involved in the conception of a drawing instrument based on geometrical notions but designed for experimental activities in mechanics. The herpolhodographer, designed with

---

[183] Apéry, "Caron's Wooden Mathematical Models," 44–47.

Fig. 29 Herpolodographer of Gaston Darboux and Gabriel Koenigs. © Collections École polytechnique, Palaiseau, all rights reserved

Gabriel Koenigs, made the tracing of herpolodies possible, i.e. space curves generated by the rotation of a rigid body around its center of gravity (see Fig. 29). It was fabricated by the manufacturer Château Père et fils in 1900 and presented to the world fair in Paris that hosted the second International Congress of Mathematicians.

While the traditional mathematical models and instruments had been displayed in industrial exhibitions and world fairs since the 1840s, exhibitions of models of higher geometry participated in exhibitions of scientific instruments and equipment, such as the major international exhibition held at the South Kensington Museum in 1876,[184] as well as to the emergence of congresses devoted specifically to mathematics, such as the conferences of German mathematicians in Munich in 1893 and in Hamburg in 1902, the international conference organized after the Chicago World fair in 1893, and the international congress of mathematicians in

---

[184] Henry John Stephen Smith, "Geometrical Instruments and Models," in *The Collected Mathematical Papers of Henry John Stephen Smith,* vol. 2, ed. James Whitbread Lee Glaisher (Oxford: Clarendon Press, 1894), 698–710.

Heidelberg in 1904.[185] Collections of models therefore participated in the emergence of both national and international communities of mathematicians and in the shaping of their public image.[186] Significantly, the founding congress of an association of professors of mathematics, organized by David Eugene Smith in New York in 1904, displayed both a collection of models and a collection of photographs and portraits of famous mathematicians.[187] By contrast, École polytechnique never purchased any of the models of higher mathematics manufactured at the turn of the century: since mathematical models were associated with model drawing at the Polytechnique, the more elementary models of descriptive geometry were undoubtedly more relevant than the ones of higher mathematics, especially since geometric drawing was a part of the elementary training of the students and was not anymore an issue in the more advanced courses of analysis, geometry, or mechanics.[188]

Models and the 1902 Educational Reform in France

We have seen that the development of collections of models of higher geometry participated in the much larger phenomenon of the autonomization of mathematics as an academic discipline, in contrast to the broad spectrum covered by the 'mathematical sciences' of the first part of the nineteenth century. The emergence of a market for model manufacturers was, in particular, a consequence of both the development of higher education in Europe and of the increasing role mathematics played in both general and technical education. At the turn of the twentieth century, several European nations initiated large educational reforms that aimed at promoting the links between pure and applied sciences in a context of fierce industrial, economical and military competition. Darboux and Klein played a key

---

[185] Henri Fehr, "Le 3ᵉ Congrès international des Mathématiciens; Heidelberg, 1904. Les expositions de bibliographie et de modèles et instruments," *L'enseignement mathématique* 6 (1904): 476–81. On models in the ICMEs and ICMIS, see: Gert Schubring, "Historical comments on the use of technology and devices in the ICMEs and ICMI," *ZDM Mathematics Education* 42 (2010): 5–9.

[186] Ulf Hashagen, "Mathematics on Display: Mathematical Models in Fin de Siècle Scientific Culture," in *Oberwolfach Reports* 14, no. 4: "History of Mathematics: Models and Visualization in the Mathematical and Physical Sciences," ed. Jeremy J. Gray, Ulf Hashagen, Tinne Hoff Kjeldsen, and David E. Rowe (2015): 2838–41.

[187] N.N., "Association des professeurs de mathématiques des Etats Moyens et du Maryland," *L'enseignement mathématique* 6 (1904): 63–64.

[188] At the turn of the twentieth century, descriptive geometry was taught in the preparatory classes to the grandes écoles in the Lycées, i.e. before entering the Polytechnique. See: C. Roubaudy, *Traité de géométrie descriptive à l'usage des élèves des classes de mathématiques spéciales et des candidats à l'École polytechnique* (Paris: Masson, 1916).

role in these reforms.[189] Both aimed at developing the connections between general and technical education and both promoted action learning in the teaching of mathematics, especially by the use of models, in contrast to the tradition of Euclid's elements.[190]

In France, the 1902 reform played a crucial role in the development of scientific education in the lycées, i.e. the system of general secondary education, which, until then, had been dominated by the humanities. In contrast to primary and technical education, secondary education had maintained the teaching of imitation drawing in the tradition of the Beaux-Arts, with a focus on the model of the human figure. In 1852, the distinction between a scientific section and a literary section in the lycées had allowed the introduction of linear drawing,[191] and the role of linear drawing in the scientific sections of the lycées had been strengthened in the 1880, but it had nevertheless remained within the scope of the teaching of imitation drawing until 1902 when it was incorporated into the teaching of mathematics.[192]

University professors who aimed at adapting secondary education to the 'modern world' conducted the 1902 reform. But, in contrast with what had happened decades before in superior primary education and technical education, the development of a general scientific education was not legitimated solely by the usefulness of the applications of sciences. The reformers aimed not only at training "practical and useful men" but also at founding a "new humanism" in which "scientific humanities" would be no less involved than the literary humanities in the "formation de l'esprit" ("formation of the human spirit").[193]

As the president of the commission for the revision of the programs of mathematics, Darboux promoted the activity of the students, the "experimental method,"

---

[189] In Germany, the "Meraner Lehrplan" presented in 1905 at the meeting of the Deutschen Mathematiker-Vereinigung in Meran was a new syllabus adopting some of the basic points of Klein's reform movement. It proposed approaching geometry via intuitive geometry and especially promoted the use of models for strengthening spatial intuition, the consideration of geometrical configurations as dynamic objects, making room for applications, and in connection with practical activities of drawing and measuring with the use of straightedge and compass, aiming especially at a coordination of planimetry and stereometry. See: August Gutzmer, "Bericht betreffend den Unterricht in der Mathematik an den neunklassigen höheren Lehranstalten," *Zeitschrift für mathematischen und naturwissenschaftlichen Unterricht* 36 (1905): 543–53; Felix Klein, *Vorträge über den mathematischen Unterricht*, vol. 1 (Leipzig: Teubner, 1906), 208–19.

[190] This evolution was promoted at an international level by a great many contributors to *L'enseignement mathématique*. See, for example: "Conférence sur l'enseignement scientifique en Allemagne," *L'enseignement mathématique* 12 (1910): 387–93.

[191] d'Enfert, *L'enseignement du dessin*, 172.

[192] Renaud d'Enfert, "Entre mathématiques et technologie: l'enseignement du dessin géométrique dans le primaire et le secondaire (France,1880-début XXe siècle)," *Revista de História da Educação Matemática* 2, no. 2: "HISTEMAT," special issue (2016): 39–55.

[193] Bruno Belhoste, "L'enseignement secondaire français et les sciences au début du XXe siècle. La réforme de 1902 des plans d'études et des programmes," *Revue d'histoire des sciences* 43, no. 4 (1990): 371–400; Bruno Belhoste, Hélène Gispert, and Nicole Hulin, eds., *Les sciences au lycée: Un siècle de réformes des mathématiques et de la physique en France et à l'étranger* (Paris: Vuibert-INRP, 1996).

and a focus on "concrete problems" as opposed to the abstract and logical reasoning of the traditional framework of Euclid's geometry.[194] In addition to transferring to mathematical education several ideals of the natural sciences, such as "observations," "experimentation," and "classification," the reform promoted the adaptation to general education of the pedagogical methods of primary and technical education, which aimed at rendering mathematics more accessible to more students. Learning by drawing was especially considered as one of the best methods to promote the activity of the students, "with the use of collections of models and of elementary instruments."[195] It aimed at developing the student's intuition of the geometrical space by experimenting with its "reality," and thereby developing a "lively perception" of the theorems of geometry as well as of their applications and industrial potentialities.[196] The use of plaster and string models was, in particular, associated with several pedagogical values such as the one of motivation, by "making geometry more lively and interesting," the valorization of manual skills, as a counterweight to "purely verbal definitions," as well as the capacity to "judge the usefulness" of theorems by experimental activities.[197]

The goal of the 1902 reform in transferring to general education the pedagogy of model drawing developed in technical education is especially highlighted by the professional trajectory of Célestin Roubaudi at the turn of the twentieth century (see Figs. 30, 31 and 32). Roubaudi had been trained at the École normale spéciale of Cluny, a school established in 1865 for training the teachers of technical schools and which stressed the important role of geometric drawing. After having passed the special competitive exam for teaching mathematics in technical schools, Roubaudi became professor of descriptive geometry at Cluny between 1880 and 1891, when this special school was cancelled. After the 1902 reform, Roubaudi moved from technical to general education by teaching descriptive geometry to students who prepared for the competitive exams of École polytechnique and École normale supérieure at the lycée Saint Louis and the lycée Louis-Le-Grand in Paris. He published several books on the teaching of geometric drawing in both general secondary education and in the grandes écoles and came to be considered as the one of the best specialists in descriptive geometry. After 1909, Roubaudi succeeded

---

[194] N.N., "Modifications apportées au plan d'études des lycées et collèges de garçons," *L'enseignement mathématique* 7 (1905): 491–97.

[195] Ibid., 495: "une collections de modèles et d'appareils simples". For an example of a textbook published in line with the 1902 reform and promoting the connection between geometry, drawing, and motion, see: Carlo Bourlet, *Cours abrégé de géométrie* (Paris: Hachette & Cie, 1907).

[196] N.N., "Modifications apportées", 495. Among the many discourses promoting this evolution of mathematical education, see, in particular: Louis Crelier, "Le dessin de projection dans l'enseignement secondaire," *L'enseignement mathématique* 6 (1904): 300–4. See also: Christian Beyel, "L'enseignement de la géométrie descriptive dans les écoles moyennes," *L'enseignement mathématique* 3 (1901): 431–36.

[197] N.N., "Modifications apportées", 493–4. See also: "Circulaire, adressée par M. Le Vice-Recteur de l'Académie de Paris à MM. les Inspecteurs d'Académie, Proviseurs, Principaux et Professeurs de Mathématiques et de Physique du ressort," *L'enseignement mathématique* 9 (1907): 231–34.

a   b

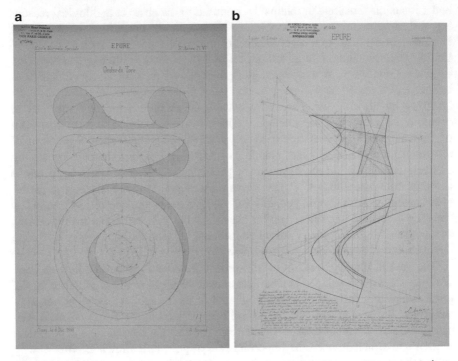

Fig. 30 **a** Épure of the shadow of a torus by Jouvent, a student of Célestin Roubaudi at the École normale spéciale de Cluny in 1888. © Collections de l'Institut Henri Poincaré, all rights reserved, épure n°074. **b** Épure by a student of Roubaudi at the lycée Saint Louis in Paris in 1911: intersection of a paraboloid and a cone. © Collections de l'Institut Henri Poincaré, all rights reserved, épure n°025

Joseph Caron as the director of graphical works at École normale supérieure and thus participated to the training of French elite mathematicians.

Even though the 1902 reform focused on secondary education, it also resulted in setting new goals for teacher training and thus impacted the universities. The agrégation of mathematics, i.e. the selective competitive exam one had to pass to become a professor in the lycées, was adapted to the reform in 1904. Faculties of science were encouraged to promote experimental activities by creating "laboratories of mathematics," "furnished with models and instruments as numerous and as diverse as possible."[198] Mathematical models thus participated to the hybridation

---

[198] N.N., "Les études générales des Mathématiques pures et appliquées et de la Physique," *L'enseignement mathématique* 10 (1908): 11–18, here 13 and 16: "laboratoires de mathématiques," and "Nous recommendons aussi des collections de modèles mathématiques […]. L'étendue de ces installations devrait être comprise à peu près […]."

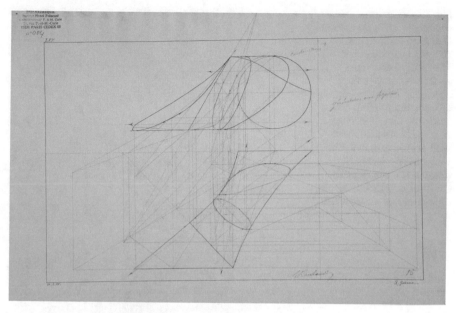

Fig. 31 Épure of the intersection of a hyberboloid and a cylinder by René Gateaux, a student of Célestin Roubaudi at École normale supérieure in 1908. Considered as one of the most promising mathematician of his generation, René Gateaux was killed in action during the first months of World War I. © Collections de l'Institut Henri Poincaré, all rights reserved, épure n°084

of two places of knowledge, the university library, a traditional place of mathematical practice, and the laboratory, which had become one the most emblematic places of both scientific and industrial activity in the century.[199]

In Paris, the mathematicians Jules Tannery and Émile Borel, who had both been strong supporters of the 1902 reform,[200] created the "laboratory of mathematics education" of the École normale supérieure with financial support of the faculty of science, secured by its dean, the mathematician Paul Appel.[201] It is likely that

---

[199] For the historiography of space in science studies see e.g.: Michel de Certeau, *The practice of everyday life*, trans. Steven Rendall (Berkeley: University of California Press, 1984); Owen Hannaway, "Laboratory Design and the Aim of Science: Andreas Libavius versus Tycho Brahe," *Isis* 77 (1986): 585–610; Adi Ophir and Steven Shapin, "The place of knowledge: a methodological survey," *Science in Context* 4, no. 1 (Spring 1991): 3–21; David N. Livingstone, "The spaces of knowledge: contributions towards a historical geography of science," *Environment and Planning D: Society and Space* 13, no. 1 (1995): 5–34; David Aubin, Charlotte Bigg, and Otto H. Sibum, eds., *The Heavens on Earth: Observatories and Astronomy in Nineteenth-Century Science and Culture* (Durham: Duke University Press, 2010).

[200] Émile Borel, "Les exercices pratiques de mathématiques dans l'enseignement secondaire," *Revue générale des sciences pures et appliquées* 15 (1904): 431–40.

[201] On laboratory of mathematics, especially in Italy, see: Livia Giacardi, "The Emergence of the Idea of the Mathematics Laboratory in the Early Twentieth Century," in *"Dig where You Stand"*

Fig. 32 Épure of a surface of
constant slope by a student of
Roubaudi at the École
normale supérieure in 1910.
The issue of drawing a
surface of constant slope
highlights the autonomization
of mathematics as an
academic discipline in regard
with the issues associated
with geometric drawing in
elementary or technical
schools. © Collections de
l'Institut Henri Poincaré, all
rights reserved, épure n°088

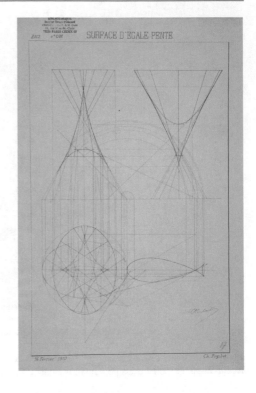

this laboratory was inspired by the evolution of the preparation to the German certificate of capacity for teaching in superior secondary education which, after 1901, included seminars and, on the model of the laboratory established by Klein in Munich in the 1870s, aimed at promoting both the activity of trainee teachers and the relationships between pure and applied mathematics, especially by the fabrication of models and the manipulation of instruments.[202]

The École normale laboratory aimed at training future teachers: models in wood, cardboard, or wire and cork were conceived and built for teaching geometry and mechanics. The didactic uses of other instruments such as mechanical linkages, pantographs, inversors, calculating machines, and instruments for geodesy and land surveying were also taught. The establishment of this laboratory thus participated to the expansion of the collection of the mathematical cabinet of the

---

2: *Proceedings of the Second International Conference on the History of Mathematics Education*, ed. Kristín Bjarnadóttir, Fulvia Furinghetti, José Matos, and Gert Schubring (Lisbon: UIED, 2011), 203–25; Livia Giacardi, "The School as a 'Laboratory.' Giovanni Vailati and the Project for the Reform of the Teaching of Mathematics in Italy," *International Journal for the History of Mathematics Education* 4, no. 1 (2009): 5–280.

[202] Friedrich Pietzker, "L'enseignement universitaire et l'instruction de maîtres des Gymnases," *L'enseignement mathématique* 3 (1901): 13–25. See also: Henri Fehr, "L'enseignement des mathématiques supérieures à Iéna," *L'enseignement mathématique* 3 (1901): 126.

Sorbonne, through both the local production of new models and regular acquisitions from the catalogs of German manufacturers such as Brill/Schilling. The local production of models was nourished by the creation of a woodcraft workshop at the École normale, where the students who prepared for the agrégation of mathematics had to practice woodcraft on a weekly basis under the supervision of a craftsman. This training in the handling of "the saw, the plane and the jointer plane" provided the students with the opportunity to design new models for the teaching of mathematics. As one of them, the later mathematician Albert Châtelet, phrased it in 1909:

> [...] it is very useful [for teachers] to be aware of the skills required to master the design of small wooden models as well as to be able, when needed, to conduct the work of a craftsman for the reproduction of a model [...]. A few collections of mathematical models are available in the market in France but they are mostly intended for primary education. Teachers, therefore, have to either design themselves the devices they would like to use in the classroom or to conduct their fabrication.[203]

The students were especially inspired by the new pedagogical practices developed by the French society of physics for promoting experimental education, as well as by the collections of instruments of the laboratory of physical mechanics. This laboratory had been created at the Sorbonne at the end of the nineteenth century for extending from geometry to applied mechanics the pedagogical method promoted by the tandem Darboux/Caron.[204] But the students also developed new practices specific to mathematical education, by designing "actual duplicates of proofs, by the use of figures in space instead of drawings on the blackboard."[205] Such models aimed at visualizing both traditional methods, such as the computation of the volume of a polyhedron by its decomposition into elementary polyhedrons, or the new concepts recently introduced in the lycées such as isometries, displacements, dilatations, and inversions: "motion cannot be represented on the blackboard."[206]

> Finally, we have a much smaller number of models associated with the geometry of the 5th book [of Euclid]. One of the most serious difficulties for beginners is to 'see' what is represented by the more or less rough figures in perspective used for illustrating the main proofs of the 5th book. This difficulty would be radically diminished if, before drawing a figure

---

[203] Albert Châtelet, "Le laboratoire d'enseignement mathématique de l'École Normale Supérieure de Paris," *L'Enseignement mathématique* 11 (1909): 206–10, here: 207-8: "D'autre part, il lui serait également très utile de se rendre compte des difficultés à surmonter pour la confection d un petit modèle en bois, et, de pouvoir au besoin diriger le travail d'un ouvrier pour la reproduction du modèle. [...] On trouve actuellement dans le commerce, en France, quelques collections de modèles mathématiques mais destinées surtout à l'Enseignement Primaire. Les Professeurs sont donc encore obligés de faire confectionner sur place ou de confectionner eux-mêmes les appareils qu'ils désireraient utiliser dans leur classe."
[204] See Paul Appel, "L'enseignement scientifique à l'université de Paris," *L'Enseignement mathématique* 8 (1906): 337–43.
[205] Châtelet, "Laboratoire d'enseignement," 208.
[206] Ibid., 209.

on the blackboard, the teacher showed the true figure in space to his students—a figure on the blackboard being basically nothing more than a diagram whose nature is more algebraic than geometric. For doing so, no more is required than a few cork slides, some wire and a little ingenuity. [...] We do not have any model for descriptive geometry [...].[207]

The Golden Age of Mathematical Models in View of the Decline of Model Drawing

As is illustrated by the absence of models of descriptive geometry at the École normale laboratory, the use of models in teacher training participated to the autonomization of mathematics as a specific teaching discipline. This practice of models broke with both the traditional association of mathematics with Euclid geometry in general education and with the intimate relationship between descriptive geometry and applications in technical education. It is in this context that the use of mathematical models in the teaching of mathematics began to be truly disconnected from the practice of drawing in France and that models came to be considered as a tool of vizualization, complementary to the figures drawn on the blackboard, rather than as a way to educate the hand and the eye.

It is also in this context that collections of models of both elementary and higher mathematics were established and developed in a great number of faculties of science and in the lycées. But it is difficult to assess whether these collections were actually used by teachers and mathematicians beyond specific areas such as the École normale laboratory, the Darboux/Caron lectures at the Sorbonne, as well as primary and technical education where the use of models had been established decades before,[208] and where the links between the teaching of mathematics and drawing were still very vivid. Historical sources on the actual use of models of higher mathematics are scarce. The overviews of both local and national pedagogical methods published in the journal *L'Enseignement mathématiques* usually promote the use of models and instruments in accordance with the progressive editorial line of this journal. But they often come with tempered criticisms about the practical difficulties associated with models, described as cumbersome, fragile,

---

[207] Ibid., 209: "Enfin nous avons en beaucoup plus petit nombre des modèles relatifs à la géométrie du Ve livre. Une des difficultés les plus sérieuses pour les commençants est en effet de 'voir' ce que représentent les figures de perspective plus ou moins grossières qui servent à illustrer les principales démonstrations du Ve livre. Cette difficulté serait bien diminuée si, avant de faire une figure au tableau—figure qui n'est au fond qu'un schéma plus algébrique que géométrique—le professeur montrait aux élèves la figure elle-même de l'espace. Pour cela il suffit de quelques plaques de liège, quelques fils de fer et d'un peu d'ingéniosité. [...] Nous n'avons encore aucun modèle pour la géométrie descriptive [...]."

[208] On the continuation of the practice of model drawing in technical education in France, see: "Conférences sur l'enseignement technique moyen en France," *L'enseignement mathématique* 12 (1910): 393–412.

costly or even locked out in a closet...[209] A few contributors developed more in depth criticisms about whether models of higher mathematics were actually helpful for visualizing mathematical properties: since these models often display singular configurations rather than a global point of view, they usually required from the observer an important preliminary knowledge in mathematics.

More importantly, "in education, innovations do not get a foothold overnight" as a contributor to *L'Enseignement mathématique* stressed it in 1914, "and thus the use of models and instruments has not yet become very common."[210] To be sure, evolutions of official national programs of instruction do not guarantee the local evolution of the actual practices of teachers. Especially in the case of a reform such as the one of 1902 in which the official national programs of instruction were designed by a few university professors with little consultation of secondary school teachers. The traditional opposition in France between primary and secondary education, technical and general education, grandes écoles and universities was another obstacle for the adaptation in the secondary and general education of pedagogical methods developed in the primary and technical education.

The 1902 reform had withdrawn geometric drawing from the scope of the teaching of drawing and attributed it to the teaching of mathematics. As a result, the traditional technical dimension of linear drawing was marginalized in the lycées, while it remained at the core of primary and technical education. Thus, the reform eventually increased the opposition between the geometric drawing taught by draughtsmen, often architects or engineers, with a focus on its applications and through a large and diversified use of models, and the geometric drawing taught by professors of mathematics as an auxiliary to geometry promoted as an "instrument of culture" in general education.[211] While the alliance between theory and application was at the core of the teaching of geometric drawing promoted by Monge and his followers, the turn of the century saw the growing autonomization of technical, or industrial, drawing from geometric drawing,[212] and especially from descriptive geometry.[213]

In *L'Enseignement mathématique*, several secondary school teachers opposed the value of visualization and manipulation associated with models to the traditional ideal of rigor associated with mathematics in general education.[214] While

---

[209] See: Meunier, "Reliefs à pièces mobiles," 167; Henri Vuibert, *Les Anaglyphes géométriques* (Paris: Vuibert, 1912), 8.

[210] Staeckel, "La préparation mathématique," 320: "[...] l'usage des modèles et des appareils n'est encore que fort peu répandu."

[211] d'Enfert, "Entre mathématiques et technologie," 46.

[212] Crelier, "Le dessin de projection," 300.

[213] S. May, "De la concordance entre le dessin technique et la géométrie descriptive," *L'enseignement mathématique* 14 (1912): 53–57; Louis Kollros, "Le dessin linéaire et la géométrie descriptive dans les écoles réales," *L'enseignement mathématique* 14 (1912): 63–64. Staeckel, "La préparation mathématique," 321.

[214] On a comparison between the new and the ancient pedagogical methods, see: J.-P. Dumur, "Les mathématiques pratiques dans les 'Public Schools,'" *L'enseignement mathématique* 16 (1914): 148–49.

they did not reject entirely models and instruments, they pleaded for limiting their use to the primary and elementary schools since no manipulation or visual demonstration should challenge the rigor of a mathematical proof on the blackboard.[215] Even though collections of models participated in shaping the place of mathematics, by the hybridation of libraries and laboratories, as well as the persona of mathematicians, by public exhibitions, a tension arose with a more ancient symbolic attribute of the professor of mathematics: the blackboard.

Even in the reformist camp, new devices of visualization such as projection devices, cinema, photogrammetry, and stereoscopy quickly challenged the value of modernity associated with models. Stereoscopy, in particular, made the visualization of relief with plane pictures possible. The mathematical principles of stereoscopic photography had been laid in the 1850s and photographs had been used since then as a form of visualization complementary to the use of models, as is exemplified by the two stereoscopic photographs of the first model of the 27 lines on a cubic surfaces that were shot very soon after the model had been designed by Wiener in 1868.[216] Stereoscopy would become more and more popular after 1905 and, because stereoscopic plates were cheaper and less cumbersome than actual models, they tended to be seen as a 'modern' alternative to collections of models.[217] Challenged by new techniques of visualization, models tended to be reduced to manipulation. But manipulation in mathematics was often more effectively performed by the actual construction of models by students,[218] than by the use of preexisting collections, which soon collected dust in forsaken closets.[219]

Open Questions: Models, Mathematical Modelization, and the Graphical Method

The golden age of mathematical models at the turn of the twentieth century coincided with a decline of the traditional pedagogical practice of model drawing in the teaching of mathematics. The advent of large collections of models of higher mathematics all over Europe and the U.S.A. therefore carried with it the onset of

---

[215] Meunier, "Reliefs à pièces mobiles," 167; Staeckel, "La préparation mathématique," 320–21.

[216] Christian Wiener, *Stereoscopische Photographien des Modelles einer Fläche dritter Ordnung mit 27 reellen Geraden: Mit erläuterndem Texte* (Leipzig: Teubner, 1869).

[217] Henri Fehr and G. Stiner, "Vues stéréoscopiques pour l'enseignement de la Géométrie," *L'enseignement mathématique* 8 (1906): 385–90; Henri Fehr, "Le stéréoscope et ses applications scientifiques," *L'enseignement mathématique* 9 (1907): 142–46; Charles Perregaux and Adolphe Weber, *Le relief en Géométrie par les couleurs complémentaires: 50 planches de stéréométrie et de géométrie descriptive* (Bienne: E. Magron, 1916).

[218] Vuibert, *Les Anaglyphes*, 8.

[219] When he discovered it in 1934, Man Ray referred to the collection at the Institut Henri Poincaré as collecting dust. See: Man Ray, "Notes sur les Équations shakespeariennes, To be continued unnoticed, Some Papers by Man Ray in connection with his exposition," in *Catalog of Man Ray's exhibition at the Copley gallery* (Beverly Hills: Copley gallery, 1948), 7–9. Man Ray, *Autoportrait* (Paris: Laffont, 1964): 314.

obsolescence, the function of models reduced to visualization and manipulation. Both the grandeur and the decadence of models have therefore to be analyzed in view of the long-term relationship between mathematics and drawing.

This relationship especially raises open historical questions about the role that may have been played by models in the emergence of mathematical modelization.[220] The history of modelization has tended to focus on theoretical developments in mathematics and neighboring sciences such as mechanics and physics. Even though historians have investigated several practices of visualization, of writing, and of computations, research in the history of mathematics has often laid the emphasis on practices of visualizations associated with academic publications, while the palette of model drawing techniques and devices encapsulated in the mathematics of the engineers in the nineteenth century have rather been associated with the history of technology.[221] Significantly, drawing has often been considered as a burden in the training of prominent mathematicians such as Camille Jordan and Henri Poincaré, who failed to rank first when they graduated from École polytechnique because of their bad grades in drawing. Yet, geometric drawing may be considered retrospectively as one of the roots of mathematical modelization, because of both its ubiquitous use in technology and its intimate relationship with mathematical education and academic publications.

We shall especially argue that descriptive geometry played an exemplary role for innovative graphical methods and visualization devices throughout the nineteenth century. This role calls for reassessing the usual historiographical description of Monge's descriptive geometry as a transitional discipline, understood as both the ultimate perfecting of previous graphical techniques and the "last stage of a tradition that is losing momentum,"[222] while algebra and analysis would become increasingly important in the training of engineers. It especially raises new questions about the history of descriptive geometry in the nineteenth century from the perspective of the evolution of the graphical methods associated with it.

An important issue that calls for further investigation is the role played by model drawing in the development of the very techniques of visualization that would eventually render both models and mathematical drawing obsolete. In the nineteenth century, several forms of mathematical visualization were developed without being subjected to any reflexive discourses or theoretical developments. Quite often, these forms of representations were not considered as mathematical objects, or methods, for decades, and could not be dissociated from specific, and

---

[220] Moritz Epple, "A plea for Actor's Categories: On Mathematical Models, Analogies, Interpretations, and Images in the 19th Century," *Oberwolfach Reports* 14, no. 4: "History of Mathematics: Models and Visualization in the Mathematical and Physical Sciences," ed. Jeremy J. Gray, Ulf Hashagen, Tinne Hoff Kjeldsen, and David E. Rowe (2015): 2773–79.

[221] Yves Deforge, *Le graphisme technique: son histoire et son enseignement* (Seyssel: Champ Vallon, 1981).

[222] Sakarovitch, "Gaspard Monge," 240; Belhoste et al., "Les exercices," 109.

often tacit, cultural practices.[223] By contrast, we have seen that model drawing had been formalized early on in the eighteenth century, with the interplay of a mathematical theory, descriptive geometry, and its applications. We have seen also that drawing was at the core of the mathematical training of the polytechnicians who, in the nineteenth century, were active in all the branches of the mathematical sciences and involved in both academic and engineering activities.

Geometric drawing provided these polytechnicians a model for designing various new forms of visualization, which would eventually fall under the designation of 'graphical method' at the turn of the twentieth century.[224] Several alumni of Polytechnique especially supported the emergence of photography, which they considered as an improvement of the épures of descriptive geometry. When he committed himself to convince the French government to fund the daguerreotype, François Arago, an illustrious alumnus and professor of École polytechnique, contrasted the precision and fastness of Louis Daguerre's innovation with the épures drawn by polytechnicians during the campaign of Egypt:

> When looking at the first tableaux that M. Daguerre exhibited to the public, everyone thought about the immense advantage that such an exact and swift means of reproduction would have provided during the campaign of Egypt; everyone was struck by the reflection that, should photography had been known in 1798, the faithful picture of so many iconic tableaux would not have been lost for the scholarly world [...]. Had the Institute of Egypt been furnished with two or three of M. Daguerre's devices, [...] vast areas of the fictional or conventional hieroglyphs that are represented in several plates of its celebrated masterpiece [i.e. *L'expédition d'Égypte*] would have been replaced by real hieroglyphs; and their design would have surpassed in accuracy, and local color the works of the most skilled painters; and photographic images, the formation of which is submitted to the rules of Geometry, would have allowed to reassemble, with only a small set of data, the exact dimensions of the highest and most inaccessible parts of the ancient monuments.[225]

---

[223] For a case study on the algebraic cultures associated with the use of the tabular representation of matrices, see: Frédéric Brechenmacher, "Une histoire de l'universalité des matrices mathématiques," *Revue de Synthèse* 131, no. 4 (2010): 569–603. On the analytical representation of substitutions, see: Frédéric Brechenmacher, "Self-portraits with Évariste Galois (and the shadow of Camille Jordan)," *Revue d'histoire des mathématiques* 17, no. 2 (2011): 271–369.

[224] Dominique Tournès, "Mathematics of Nomography," in *Mathematik und Anwendungen*, ed. Michael Fothe, Michael Schmitz, Birgit Skorsetz, and Renate Tobies (Bad Berka: Thillm, 2014), 26–32; Dominique Tournès, "Une discipline à la croisée de savoirs et d'intérêts multiples: la nomographie," in *Circulation Transmission Héritage, Actes du XVIIIe colloque inter-IREM*, ed. Pierre Ageron and Évelyne Barbin (Caen: Université de Caen Basse-Normandie, 2011), 415–48; Dominique Tournès, "Pour une histoire du calcul graphique," *Revue d'histoire des mathématiques* 6, no. 1 (2000): 127–61.

[225] François Arago, *Rapport de M. Arago sur le Daguerréotype, lu à la séance de la Chambre des Députés, le 3 juillet 1839* (Paris: Bachelier, 1839), 25–31: "A l'inspection de plusieurs des tableaux qui ont passé sous vos yeux, chacun songera à l'immense parti qu'on aurait tiré pendant l'expédition d'Égypte, d'un moyen de reproduction si exact et si prompt ; chacun sera frappé de cette réflexion, que si la photographie avait été connue en 1798, nous aurions aujourd'hui des images fidèles d'un bon nombre de tableaux emblématiques [...] Munissez l'institut d'Égypte de deux ou trois appareils de M. Daguerre [...] [...] de vastes étendues d hiéroglyphes réels iront

As is illustrated by Arago's early use of the daguerreotype to shoot pictures of the moon, the issue of providing a precise mathematical visualization of inaccessible areas was an early and important application of photography, which fuelled several innovations, such as photogrammetry and metrophotography. These innovations raised new mathematical problems, such as of the rectification of the photographs shot from aerostats.[226] These problems were associated with important issues in both civil and military topography, as is illustrated by the siege of Sévastopol in 1854–1855, when British and French photographers made use of aerostats for scouting the fortifications of the Russians. After Sévastopol had fallen, the colonel Langlois was put in charge of painting a panorama of the siege. A former student of the Polytechnique who had become a painter and had specialized in the painting of military scenes, Jean Charles Langlois surveyed the topography of the scene by making use of both the drawing techniques of descriptive geometry and photography: "he surveyed the map of the scene and the positions of the armies from the top of the Malakoff tower [...] by the use of photographic devices and thus applied, for the first time, photography to surveying panoramic maps."[227]

Panoramas were a specific form of geometric visualization based on conic, spherical, or cylindrical perspectives. The issue of surveying panoramic maps of both the topography and of the geological nature of mountains gave rise to the development of the field of topophotography in the late 1850s, in which several former students of the Polytechnique and of the Metz application school where involved, such as the geologist Aimé Civiale. Again, the rectification of photographs, as well as their use for measurement in topography, raised difficult mathematical issues, which were tackled in academic publications in the *Comptes rendus de l'Académie des sciences*. It is in this context that Libre Bardin designed the plans-reliefs that would eventually lead him to fabricate plaster mathematical models. As a matter of fact, Bardin made use of photography for surveying the Mont-Blanc,[228] as well as for exploring innovating forms of mathematical visualization, such as radiant panoramas, which consisted in the mathematical anamorphosis of a whole panoramic view on a plane surface (see Fig. 33).

Radiant panoramas allow a direct visualization of all the angles between any vertical plane and any point of the panorama: concentric circles are drawn around

remplacer des hiéroglyphes fictifs ou de pure convention; et les dessins surpasseront partout en fidélité, en couleur locale, les couvres des plus habiles peintres; et les images photographiques étant soumises dans leur formation aux règles de la géométrie, permettront, à l'aide d'un petit nombre de données de remonter aux dimensions exactes des parties les plus, élevées les plus inaccessibles des édifices."

[226] Laussédat, "Recherches sur les instruments," 241.

[227] Germain Bapst, *Essai sur l'histoire des Panoramas et des Dioramas: Extrait des Rapports du Jury international de l'Exposition universelle de 1889* (Paris: G. Masson, 1889), 25: "[...] le colonel Langlois [...] leva du haut de la tour Malakoff, au moyen d'appareils photographiques, les plans des positions occupées par les armées et appliqua ainsi pour la première fois, comme nous l'avons déjà vu, la photographie à la levée des plans panoramiques."

[228] Laussédat, "Recherches sur les instruments," 251.

Fig. 33 Libre Bardin's radiant panoramas of the environs of Metz. From Aimé Laussédat, *Recherches sur les instruments, les méthodes et le dessin topographiques* (Paris: Gauthier-Villars, 1901), plate IV

the center of perspective, each circle representing the points of a same angular height.[229] This direct and simple visualization of angles was considered as especially helpful for surveying and leveling. Thus radiant panoramas were considered as providing a solution to the mathematical problem of photography. Further, Bardin's radiant panoramas highlight, once again the role played by model drawing in graphical innovations: the panorama of the environs of Metz was constructed by the mathematical transformation of a preexisting developed cylinder panorama designed by one of the draughtsmen involved in the teaching of mathematical drawing at the Metz school.

Because mathematical drawing played a key role in the engineering sciences, innovative graphical techniques of visualization were rather evaluated with the

---

[229] The development of radiant panoramas had especially been initiated by the panographer designed in 1827 by the geographer and mathematician Louis Puissant.

criteria of industry than of the academy. The criteria of precision, effectiveness, and production cost were especially favored over the one of conceptual novelty. These criteria, which, as we have seen, Olivier applied when evaluating mathematical models, were actually applied to all graphical techniques. In his report on a new drawing machine for reproducing, enlarging, or reducing any épure, Olivier claimed that, even though there was nothing new in the design of this camera obscura, its realization was nevertheless innovative since it allowed to "save time" with no loss of "mathematical exactness":

> [...] inventions rarely show new principles, most of the time a truly new invention is based on a new way to materialize known principles; it is often a new modality that provides the effective simplification of a mechanism that used to be too complicated and costly; it is more importantly the achievement of a simpler machine, a machine that can be used with more speed and more security, and which can be delivered to the industry for a cheaper price.[230]

In turn, these criteria of precision, effectiveness and simplicity gave rise to a new approach to geometric constructions, such as with Émile Lemoine's geometrography.[231] In many ways, mathematical models can be considered as falling in the more general category of graphical methods for a large part of the nineteenth century. Investigating further this more general context would allow us to understand more precisely the emergence of collections of models of higher mathematics after the 1860s, which broke with the tradition of model drawing in a process of autonomization of the forms of visualization specific to academic mathematics. This prowess was not limited to material models and went along with theoretical developments on the mathematical properties of forms of visualization. It therefore played a role in the emergence of the concept of mathematical modelization.

For this reason, the evolution of mathematical models in the nineteenth century should not be reduced to a unique path, from applied, or engineering mathematics, to academic mathematics. The collections of models of higher mathematics are only one of the many forms of evolution of the variety of graphical techniques designed in the nineteenth century. A striking example is provided by Etienne Jules Marey's "méthode graphique."[232] While Marey is best remembered today for his

---

[230] Théodore Olivier, "Rapport fait par M. Théodore Olivier, au nom du comité des arts mécaniques, sur une machine à dessiner présentée par M. Grillet," *Bulletin de la société d'encouragement pour l'industrie nationale* 488 (February 1845): 49–52, here: 49–50: "[...] dans les inventions, les principes nouveaux, sont rares, et presque toujours ce qui constitue une invention qui doit être considérée comme réellement nouvelle, c est une nouvelle manière de matérialiser un principe connu; c'est souvent un mode nouveau, qui simplifie d'une manière heureuse un mécanisnie trop compliqué et trop coûteux; c'est surtout arriver à une machine plus simple que celles connues, et que l on puisse faire fonctionner avec plus de rapidité et plus de sûreté, et dont les produits puissent être livrés à l'industrie à un prix moins élevé."

[231] Émile Lemoine, "La géométrographie ou l'art des constructions géométriques," *Association française pour l'avancement des sciences* 21 (1892): 36–100.

[232] Marey developed the idea of the superiority of the graphical method through its application to the investigation of blood circulation. See: Étienne-Jules Marey, *Du mouvement dans les fonctions de la vie, Leçons faites au Collège de France* (Paris: Germer Baillière, 1868).

motion pictures of chronophotography, and often celebrated as a forerunner of cinema, his main aim was to develop a mathematical description of motion by the use of what he designated as "photographic épures." He eventually named his approach "the graphical method," subsuming all graphical techniques and instruments, from drawings to photographs or even the visualization of timelines in textbooks of history:

> The graphical method has driven progresses in almost all the branches of science and, for this reason, has benefited considerable development recently. Arduous statistics have given way to tables in which the inflexions of a curve throw light on all the phases of a phenomenon. Moreover, tracing devices can draw automatically the curve of either physical or physiological phenomena which could not be observed directly because of their speediness, slowness, or weakness. Yet, the inscription of phenomena in the form of curves may sometimes prove defective; a more powerful method has been created: Chronophotography.[233]

Photography, Marey claimed, "is increasingly replacing drawings, maps, and relief figures" (i.e. models).[234] Aiming at developing a mathematical representation of motion in space through photography, he started with the investigation of the mechanical motions of the basic elements of geometry, i.e. the point and the line. His first chronophotographical épures were devoted to generating ruled surfaces, such as a cylinder, a hyperboloïd and a cone, by the motion a single string, with an explicit reference to Olivier's mechanical string models and their use for the teaching of descriptive geometry at the Conservatoire (see Fig. 34).

For the investigation of more complicated motions, such as the one of a runner, the surface was then reduced 'geometrically' to a series of points and lines that allowed to superpose several photographs in what constituted an 'épure' of 'geometric chronophotography.' (see Fig. 35) Recall that the concept of 'épure' is underpinned by a process of 'reduction' in the mathematization, or the modelization, of a phenomenon.

The emergence of the graphical method can be seen as an evolution of mathematical models different from the emergence of models of higher mathematics. Both broke with the tradition of model drawing but not for the same reason. On the one hand, the practice of drawing had never been a legitimate activity for teaching mathematics in universities and, as we have seen, the use of mathematical models traditionally aimed at developing action-learning pedagogical methods in opposition with reading texts or attending lectures. Often presented as substitutes for direct contact with nature, the knowledge associated with mathematical models was opposed to the one of professors, and had even fuelled criticisms about the pedantry of academic knowledge. On the other hand, drawing was considered as obsolete because slower and less precise than new graphical devices.

As aforementioned, the graphical method subsumed a great variety of visualization techniques and instruments, including the ones developed in statistics, such

---

[233] Étienne-Jules Marey, *Le mouvement* (Paris: G. Masson, 1894), Avant-propos. English translation: Étienne-Jules Marey, *Movement* (London: Heinemann, 1895), Preface (translation modified).
[234] *Marey, Le mouvement,* 18.

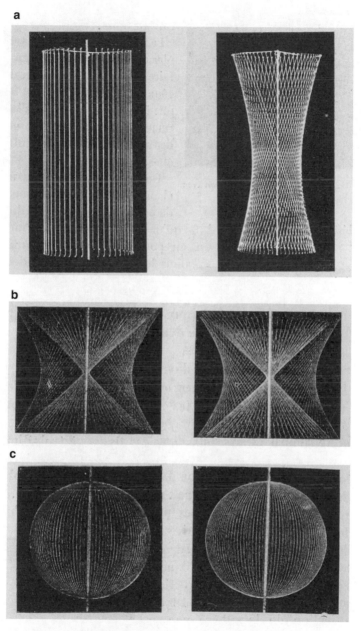

Fig. 34 **a** Cylinder and hyperboloid generated by the rotation of a white string, Étienne-Jules Marey, *Le mouvement* (Paris: G. Masson, 1894), 25, Figs. 14, 15. © Bibliothèque nationale de France. **b** Sphere generated by the rotation of a half-ring of white string. From ibid., 28, Fig. 19. © Bibliothèque nationale de France, all rights reserved. **c** Hyperboloid and its asymptotic cone. From ibid., 28, Fig. 20 © Bibliothèque nationale de France, all rights reserved

Fig. 35 **a** Pictures of a runner reduced to shiny lines (geometric chronophotography). From Étienne-Jules Marey, *Le mouvement* (Paris: G. Masson, 1894), 61. © Bibliothèque nationale de France, all rights reserved. **b** Photographic épure of a jumper. From ibid., 138

a

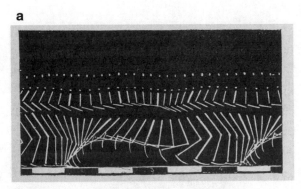

b

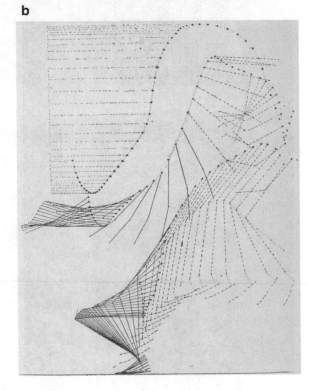

as with the choropleth map designed by Dupin in 1826 for representing the distribution of illiteracy in France,[235] or in the field of graphical statics,[236] geodesy,[237] the promotion of function graphs and their use for approximating the roots of algebraic equations,[238] and the methods of graphical calculation that would give rise to a specific mathematical theory: nomography.[239] When Olivier reviewed Léon Lalanne's pioneering "abaques" of graphical computation at the Société d'encouragement, it was plain to him that the contour lines used in topographic maps for representing reliefs were a major source of inspiration for the graphical layout designed by Lalanne (see Fig. 36).[240] The specificity of Lalanne's method for graphical calculation was that it displayed only straight lines, and was therefore based on the transformation of the curves of several functions. For Olivier, this approach was inspired by a fundamental principle of descriptive geometry, i.e. the transformation of a surface into a simpler one, and of systems of lines in space into systems of lines on a plane.

As with the models of higher mathematics, the graphical method aimed at enhancing visualization, and therefore precision, but it rather focused on an ideal of 'clarity' in the representation than on the ideal of objectivity associated with academic sciences. Clarity was a necessary preliminary to effectiveness; it required not only simplicity but more importantly to make the 'choice' of what should be simplified, and therefore carried on the main value traditionally associated with the teaching of drawing: training the eye for improving the capacity of judgment. This traditional value was still emphasized by Carlo Bourlet, in his inaugural lecture as the new professor of descriptive geometry at the Conservatoire in 1906:

> One should never forget that the unique purpose of descriptive geometry is to represent pieces of stones, of woods, of machines, and architectural details with the precision and clarity required for any effective achievement. The artisan to whom the drawing will be transmitted needs to recognize at first glance the form and the details of the piece he has

---

[235] Gilles Palsky, "Connections and Exchanges in European Thematic Cartography. The Case of XIXth Century Choropleth Maps," *Belgeo* 3, no. 4 (2008): 413–26.

[236] Konstantinos Chatzis, "La réception de la statique graphique en France durant le dernier tiers du XIXe siècle," *Revue d'histoire des mathématiques* 10, no. 1 (2004): 7–43.

[237] Martina Schiavon, *Itinéraires de la précision. Géodésiens, artilleurs, savants et fabricants d'instruments de précision en France, 1870–1930* (Nancy: Presses universitaires de Nancy, 2014).

[238] See in particular the report devoted to the graphical method of Carl de Ott, a professor of descriptive geometry in Prague, by: Jules Morin, "De l'utilité de l'application de la géométrie aux calculs algébriques," *Bulletin de la société d'encouragement pour l'industrie nationale*, 502 (April 1846): 447–84; 714–15.

[239] As with model drawing, the main actors of the development of nomography were engineers trained at the Polytechnique and aimed at providing a mathematical formulation to key issues in engineering science such as cuttings and embankments. See: Dominique Tournès, "Mathematics of Engineers: Elements for a New History of Numerical Analysis," *Proceedings of the International Congress of Mathematicians* 4 (2014): 1255–73.

[240] Théodore Olivier, "Rapport fait par M. Théodore Olivier, au nom du comité des arts mécaniques, sur un abaque ou compteur universel de M. Léon Lalanne," *Bulletin de la société d'encouragement pour l'industrie nationale*, 502 (April 1846): 161–62.

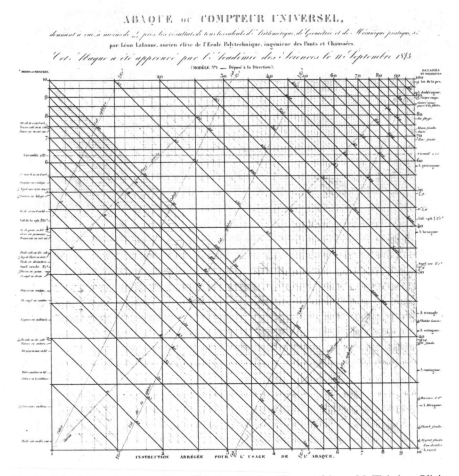

Fig. 36 Léon Lalanne's abaque. From Théodore Olivier, "Rapport fait par M. Théodore Olivier, au nom du comité des arts mécaniques, sur un abaque ou compteur universel de M. Léon Lalanne," *Bulletin de la société d'encouragement pour l'industrie nationale* 502 (April 1846): 161

to fabricate. The role—and I shall even say the *duty*—of the draughtsman is to represent objects in a simple manner. He cannot choose randomly between projection planes, or even modes of representation, but he has to be very judicious in his choices. He has to *make see* and therefore being able to *see by himself*.[241]

---

[241] Carlo Bourlet, "La géométrie descriptive au conservatoire des arts et métiers de Paris," *L'enseignement mathématique* 9 (1907): 89–93, 91–92: "Il ne faut pas, en effet, oublier que le but unique de la Géométrie descriptive et de la Stéréotomie est de représenter des morceaux de pierre, des pièces de bois, des organes de machines, des détails et ensembles architecturaux d'une façon claire et précise qui en permette l'exécution. L'artisan, auquel on transmettra le dessin, doit pouvoir, d'un premier coup d'œil, connaître la forme et les détails de la pièce qu'il est chargé d'exécuter. Le rôle—je dirai plus—le devoir du dessinateur est donc de présenter ces objets d'une

Reassessing the history of mathematical models in view of the development of the graphical method calls especially for further investigation on the interplay between models and instruments.[242] The origin of the graphical method was indeed usually attributed to the device designed by Poncelet and Morin for tracing automatically the altitude of a body in free fall.[243] As most proponents of descriptive geometry and model drawing, Poncelet did not separate the use of models from the one of instruments.[244] Recall that already at the creation of École polytechnique, Monge's cabinet of models as well as the practice of drawing in the 'petites salles' was considered as an adaptation to mathematical education of the practice of experimenting with instruments in chemistry laboratories. Moreover, Monge and his pupils often studied geometrical problems closely connected with experimental physics, especially geometrical optics.

Investigating further the role models and instruments played in the interactions between mathematics and experimental sciences would allow us to shed new light on the emergence of models of higher mathematics. As we have seen, these models were often associated with a naturalistic approach to the sciences of surfaces, which involved not only geometry but also mechanics and optics. As a matter of fact, several of the earliest models of Monge's cabinet were used not only in the teaching of mathematics, but also of physics, and chemistry, such as with cardboard crystallographical models. This versatility of models is also exemplified by the plaster models designed by Augustin Fresnel for his work on the theory of light.[245] Ampère's 1815 theory of the internal organization of molecules is another typical example of the interplay between crystallography, chemical combinations,

manière simple. Il doit, non pas s'imposer au hasard des plans de projection, ni même un mode de représentation, mais les choisir judicieusement pour atteindre le maximum d'effet utile. Il doit faire voir; et pour cela, il faut d'abord qu'il voie lui-même."

[242] The development of nomography provides an example of graphical method in which representations cannot be dissociated from instruments: Lalanne's "abaques" were both graphical tables and computing devices. See: Dominique Tournès, "Construire pour calculer," in Les constructions mathématiques avec des instruments et des gestes, ed. Évelyne Barbin (Paris: Ellipses, 2014), 265–96.

[243] Marey, Du mouvement dans les fonctions de la vie, 107; Gustave Le Bon, "La méthode graphique et les appareils enregistreurs. Leurs applications aux sciences physiques, mathématiques et biologiques," in Études sur l'exposition universelle de 1878, ed. Eugène Lacroix (Paris: Librairie scientifique, industrielle et agricole, 1878), 7:329–432.

[244] Olivier was an especially strong proponent of the use of instruments in the teaching of geometry and particularly promoted the inventors of new tracing devices. Among the several reports he devoted to this issue at the Société d'encouragement, see the powerful plea for instruments by which he concluded a report on folding procedures, Théodore Olivier, "Rapport sur une nouvelle méthode de géométrie pratique, sans instruments, de M. Martin Chatelain," Bulletin de la société d'encouragement pour l'industrie nationale, 544 (October 1849): 481–85.

[245] Letter from Roquet to Barré de Saint Venant, "Le remercie pour le don au lycée de modèles en plâtre (recherches de Monge et de Fresnel)," July 3, 1863, Archives of École polytechnique. Fonds Barré de Saint Venant.

and geometry,[246] while Louis Poinsot's theory of order highlights another kind of interplay between polyhedrons, mechanics, algebra, and number theory, which turned to be instrumental to the development of both group theory and topology.[247]

Both the design and the use of crystallographical models required specific instruments, i.e. goniometers, for measuring angles between crystal faces. Goniometers quickly evolved from mechanical devices into optical devices, and turned out to be instrumental for the intimate connections between crystallography, geometry, mechanics, and optics in Fresnel's wave theory of light in the 1820s, which, in turn, resulted in the design of what may be considered as the first models of higher mathematics, i.e. Fresnel wave surfaces made of plaster, to which Kummer would provide a generalization with his research on quartics in the 1860s. In the experimental sciences the design of models involved a close collaboration of scholars and manufacturers of instruments, such as with Fresnel and the optician Jean-Baptiste Soleil in Paris, for designing both instruments and models of wave surfaces,[248] and with the physicist Gustave Magnus and the draughtsman Ferdinand Engel in Germany, whose models of Fresnel wave surface of crystals were exhibited at the world fairs of London and Paris in 1851 and 1855, before being commercialized in Germany and in the USA. Plaster models of Fresnel wave surfaces can already be found in the catalogues of the Parisian manufacturers Soleil/Duboscq and Hoffman in the 1840s, accompanied with specimens of crystals, wooden models of crystallographic polyhedrons, goniometers as well as artifacts for visualizing surfaces by luminous projections.'

Conclusions

Over the course of the eighteenth and nineteenth centuries, the history of mathematical models in France cannot be dissociated from the one of model drawing in mathematical education. The specificity of the French educational system was mainly due to the centuries-long trend towards centralization, which culminated during the French Revolution with the creation of several central and national institutions such as École polytechnique, École normale and the Conservatoire national des arts et métiers. These 'grandes écoles' were created in opposition to

---

[246] André-Marie Ampère, "Lettre de M. Ampère à M. le Comte Berthollet, sur la détermination des proportions dans lesquelles les corps se combinent, d'après le nombre et la disposition respective des molécules dont leurs parties intégrantes sont composées," *Journal des mines* 37, no. 217 (January 1815): 5–40.

[247] Jenny Boucard, "Louis Poinsot et la théorie de l'ordre: un chaînon manquant entre Gauss et Galois?" *Revue d'histoire des mathématiques* 17 (2011): 41–138; Frédéric Brechenmacher, "The theory of order, a specific *nineteenth* century model of scientificity," *Oberwolfach Reports* 14, no. 4: "History of Mathematics:Models and Visualization in the Mathematical and Physical Sciences," ed. Jeremy J. Gray, Ulf Hashagen, Tinne Hoff Kjeldsen, and David E. Rowe (2015): 2808–11.

[248] Letter to Barré de Saint–Venant, "Remerciements pour le modèle (surface d'onde de Fresnel) que Saint Venant a offert à la société philomatique," 1863, Archives of École polytechnique. Fonds Barré de Saint–Venant.

the traditional universities. In contrast to the pedagogical method of plenary lectures, they aimed at promoting the activity of the students through the practice of science experiments and geometric drawing. École polytechnique, in particular, continued the long tradition of apprenticeship and companionship in the training of engineers. But in contrast with the royal engineering schools of the eighteenth century, Polytechnique articulated the practice of geometric drawing with theoretical lectures. It is in this specific contest that Monge's descriptive geometry fully blossomed as a new branch of the mathematical sciences.

On the one hand, this new science carried on the traditional idea that teaching geometry to engineers required to 'educate the hand and the eye' through model drawing. Models were thus considered as substitutes for natural forms and supported pedagogical methods that promoted action learning, relegated the role of the teachers to the one of supervisors, or even praised the mutual instruction of students by students. The very epistemological essence of this activity was the transmission of a non-textual form of geometric knowledge, one which required practical work and could not be subsumed to reading texts or attending lectures: drawing was knowing. But on the other hand, the practice of drawing was articulated to theoretical lectures in the most advanced sciences of the time, especially analysis. The teaching of Monge's descriptive geometry was organized by the process of decomposition/recomposition at the core of the "esprit d'analyse." It followed a progression from the simple to the complex: the students had first to copy geometric figures and their intersections, i.e. models of two dimensions, in order to acquire exactness of eye, before passing to three dimensional geometric models, and eventually to the natural models of topographical landscapes, buildings, or technological devices. When their training was completed, students were supposed to be able to decompose a complex figure into a series of simple elements, corresponding to the models they had been trained with, and to recompose a complex drawing from its elementary parts.

Even though the instruction plan devised initially by Monge in 1794 was quickly challenged by the increasing role attributed to analysis at the expense of descriptive geometry, the practice of model drawing continued to play an important role over the course of the nineteenth century. Because of the central role played by the École polytechnique in the emergence of a national educational system, this pedagogical approach to the teaching of geometry spread to the other institutions of technical education that were created in the first decades of the nineteenth century, starting with the drawing school created by the Conservatoire national des arts et métiers, the movement for mutual instruction, and eventually with the institutionalization of the national system of superior primary education in 1833.

For Gaspard Monge, descriptive geometry embodied the "esprit d'analyse," not only in the sense that its teaching could be organized from the simple to the complex, but also because it provided a heuristic 'method for finding the truth.' The important role played by model drawing at École polytechnique participated in a more general plan for articulating practice and theory. In the first half of the nineteenth century, the legacy of Monge's ideals about the role descriptive geometry and model drawing should play in the alliance between practice and theory remained especially vivid in the school for artillery and military engineering applications at Metz. The followers of Monge who graduated from the Metz school were strong proponents of industrialization. Along the line of the Saint-Simonian philosophy, they considered that industrial prosperity required putting the 'useful innovations' made by scholars in the service of the nation by increasing the instruction of the industrial class. In doing so, they participated in spreading the pedagogical practices developed at the Polytechnique, especially model drawing, as well as the ideals the school had inherited from the Enlightenment. These engineers and artillery officers often associated the teaching of geometry by model drawing with moral issues, especially the value of discipline and the taste for work well done. The diffusion of geometric drawing in primary and technical education in the 1830s therefore participated in both the conservative political agenda of the constitutional monarchy and in an ideal of emancipation through education in mathematics.

Several important innovations in the design of geometric models were made in this context in the 1840s and 1850s. The diversity of these innovations highlights the variety of the public and issues associated with the teaching of geometric model drawing. Olivier's movable string models were designed for his teaching of descriptive geometry at the Conservatoire national des arts et métiers, with a view to applications in the drawing of gearings. By contrast, Olivier designed a simple and cheaper cardboard model, the 'omnibus,' for raising the elementary mathematical instruction of the greatest number of children. Bardin's plaster models emerged from his research in topography and inherited from the practice of designing plans-reliefs in the arts of fortification and topography. The novelty of these innovations was evaluated with the norms of technical and industrial innovation, especially manufacturing cost, in various local industrial fairs, on the national scene of the Société d'encouragement pour l'industrie nationale, and on the international setting of the world fairs.

The emergence of models in higher mathematics in the 1860s broke with the long tradition of model drawing. Even though the models of higher geometry carried strong pedagogical ideals, these ideals were usually very different from the ones associated with the models designed for technical and primary education. To be sure, both the traditional drawing models and the new models of higher geometry were associated with the pedagogical values of visualization and manipulation, i.e. the issue of making use of the eye and the hand in the teaching of mathematics. But while the traditional use of models could not be dissociated from the idea that the teaching of geometry required first to train the eye and the hand by the practice of model drawing, the models of higher geometry were often designed for universities in which drawing was usually not associated with mathematical education. The idea of working with models in the universities was rather derived from the use of models and instruments in experimental physics. Both the classification of geometric surfaces and the material representation of specific mathematical properties and singularities aimed at promoting observation in pure mathematics, i.e. a value that had developed in observational sciences. These academic ideals contrasted with the ones associated with the traditional use of models and instruments in engineering schools or in industry. The key role devoted to models in Klein's approach to geometry was based on the traditional idea, already much valued in Rousseau's philosophy of education, that only the material representation of a model can impress the 'true character' of a geometric object in the mind. Even so, Klein did not associate this ideal with the practice of drawing but with the one of observation and with a naturalistic philosophy of mathematical objects as both real objects and witnesses to the very nature of the human mind.

To be sure, the rupture between the new models of higher geometry and model drawing did not happen overnight. On the contrary, Gaston Darboux attempted to introduce in the general education of the lycées and universities the pedagogical practices of model drawing that had developed in technical and primary education. Yet, the series of wooden models designed by Joseph Caron for Darboux's lectures on higher geometry nevertheless participated in the shaping of mathematics as an academic discipline in the context of the development of higher scientific education. Even though Klein and Darboux were both active advocates of the interplay between theory and application, this interplay did not take on the same meaning in universities as had been promoted by Monge and his followers. Its focus was on the interplay between mathematics and other academic scientific disciplines, rather than at aiming at a direct usefulness for engineering sciences or the industry. The development of collections of models of higher geometry participated in the much larger phenomenon of the autonomization of mathematics as an academic discipline, in contrast to the broad spectrum covered by the mathematical sciences in the first part of the nineteenth century. The emergence of a market for model manufacturers was, in particular, a consequence of both the development of higher education in Europe and of the increasing role mathematics played in both general and technical education.

In addition to transferring to mathematical education several ideals from the natural sciences, such as observation, experimentation, and classification, the French educational reform of 1902 attempted to promote the adaptation to general education of the pedagogical methods of primary and technical education, such as the use of models. But the use of models in the lycées aimed mostly at rendering mathematics more accessible to more students. It broke with both the traditional association of mathematics with Euclidian geometry in general education and with the intimate relationship between geometry and application in technical education. It is in this context that the collections of models of both elementary and higher mathematics were established and developed in a great number of faculties of science and the lycées. But it is also in this context that the use of mathematical models in the teaching of mathematics began to be truly disconnected from the practice of drawing in France and that models came to be considered as a tool of visualization, complementary to the figures drawn on the blackboard, rather than as a way to educate the hand and the eye. This evolution is especially exemplified by the absence of models of descriptive geometry in the collection that was set up for teacher training at École normale supérieure. The use of models in general education participated to the autonomization of both mathematics as a specific teaching discipline and of geometric drawing as specific to technical education.

The decline of the production and the use of mathematical models after World War I have often been seen as a consequence of the evolution of mathematics, such as with Herbert Mehrtens' claim that models had a place neither in modernism nor in the traditions of counter-modernism within mathematics.[249] But the discussions on the fading golden age of models have usually focused on the collections of models of higher geometry and even more precisely on the issue of the influence of Klein's *anschauliche* approach to mathematics, especially with regard to both formalism and intuitionism. Yet, in view on the more ancient tradition of model drawing for teaching mathematics, the increasing autonomization of mathematics with regard to drawing at the turn of the twentieth century was a major cause for the decline of geometric model in the following decades. Another important aspect is that the golden age of models of higher mathematics rose and fell during the time of the emergence of the figure of the mathematician as a university professor. Models of higher mathematics were designed or ordered by professors, while mathematical models had been traditionally challenging the role of professors in the teaching of mathematics and promoting pedagogical approaches to mathematics such as action learning and companionship. While models had usually been considered as substitutes to natural forms, their decline coincided with the increasing role of textual knowledge and lectures in the teaching of mathematics.

The golden age of mathematical models at the turn of the twentieth century coincided with a decline of the traditional pedagogical practice of model drawing in the teaching of mathematics. The advent of large collections of models of higher mathematics all over Europe and the U.S.A. therefore carried within it the forthcoming obsolescence of models, the function of which was reduced to the one of visualization and manipulation. Both the grandeur and the decadence of models have therefore to be analyzed in view of the long-term relationship between mathematics and drawing. This relationship especially raises open historical questions about the role that may have been played by models in the emergence of mathematical modelization. Geometric drawing may indeed be considered retrospectively as one of the roots of mathematical modelization, because of the model role it played for the development of the graphical methods and visualization devices that would eventually render both models and mathematical drawing obsolete. Model drawing especially carried on an ideal of clarity in the representation, in contrast to the ideal of objectivity associated with academic sciences and the models of higher mathematics. This ideal was especially consubstantial to the concept of épure, which implied the making of choices, and to the main pedagogical value associated with drawing: training the eye for improving the capacity of judgment. Its transfer to other graphical methods implied new issues in the formalization of the choice, or judgment, of what should be simplified in a representation, and these issues could not be dissociated from the instruments used for observation, experiments, as well as for tracing and representing. Reassessing the history of mathematical models in view of the development of the graphical method therefore calls for further investigations on the manifold links of mathematical models to the history of instruments, experiments, the natural sciences and the variety of graphical devices developed in the late nineteenth century between pedagogical, instrumental and research goals. The emergence of models of higher mathematics in the 1860s is only of the many lines of development of the traditional association between geometry, drawing, and models.

---

249 ...

Wilhelm Fiedler and His Models—The Polytechnic Side

Klaus Volkert

In this article I will present some information on the history of mathematical models (in the sense of concrete objects referring to something mathematical and abstract) in the world of polytechnic schools, in particular at what is known today as the *Eidgenössische polytechnische Schule* (ETH) in Zurich. My story is focused on the person and the work of Wilhelm Fiedler (1832–1912), professor of descriptive geometry and geometry of position at that polytechnic from 1867 until 1907. I will provide some unique information on the role of models in everyday life of a polytechnic school by looking at Fiedler's large correspondence and at his publications as well as at some other documents. What I want to emphasize is the importance of the tradition of polytechnic schools concerning the production and the use of models. This tradition differed profoundly from that of universities; in particular, it paved the way for the establishment of models as scientific and pedagogical objects. A threefold continuity characterized Fiedler's use of models: It was a continuation of the teaching he probably had as a pupil, it was in accordance with the traditions of the polytechnic teaching in general, and it was in continuity with the conception and practice of the Polytechnikum in Zurich.

Wilhelm Fiedler

Otto Wilhelm Fiedler was born April 3, 1832 in Chemnitz. In those days, Chemnitz was an important industrial center called the Manchester of the east. Because

K. Volkert (✉)
AG Didaktik und Geschichte der Mathematik, Universität Wuppertal, Gaußstr. 20, Raum F 12.08, 42119 Wuppertal, Germany
e-mail: volkert@uni-wuppertal.de

© The Author(s) 2022
M. Friedman and K. Krauthausen (eds.), *Model and Mathematics:
From the 19th to the 21st Century*, Trends in the History of Science,
https://doi.org/10.1007/978-3-030-97833-4_3

his parents lacked the financial means to pay the fees of a *Gymnasium*, Wilhelm went first to the *Bürgerschule*, then to the *Gewerbeschule* in Chemnitz. Afterwards, he studied at the *Bergakademie* at Freiberg—not at a university because he had no *Abitur* and no money—where the founder of scientific mine surveying J. Weisbach influenced him. After finishing his studies, he worked as a teacher for a short time in Freiberg, then at his former school in Chemnitz. In 1859, he presented his thesis on descriptive geometry ("Die Zentralprojektion als geometrische Wissenschaft," published 1860 in Chemnitz in the *Wissenschaftliche Abhandlung* of his school) at the university of Leipzig; its referee was August Ferdinand Möbius. In 1864, Fiedler was called to the Polytechnic in Prague where he was deeply involved in the struggles between 'German' and 'Czech' professors. He left Prague in 1867 in order to go to Zurich. Here he was appointed professor of descriptive geometry and geometry of position (darstellende Geometrie und Geometrie der Lage) or what is known today as projective geometry. Fiedler was a member and for many years also head of the VI. *Abtheilung* (VI. department), also called *Fachlehrerabtheilung*. This department was responsible for the training of future teachers and for providing courses on basic subjects like descriptive geometry, mathematics and physics for the students of the other departments—typically future engineers. Besides Fiedler there were two chairs for German speaking professors of (pure) mathematics in the department. Here we find illustrious names like Bruno Elwin Christoffel, Georg Frobenius, Adolf Hurwitz, Hermann Minkowski, Friedrich Prym, Friedrich Schottky, Hermann Amandus Schwarz, Heinrich Weber—to cite only a few. Fiedler taught for forty years in Zurich. He was responsible for descriptive geometry, but he also delivered courses on projective geometry and other geometrical subjects for future teachers. He had two assistants who helped him with his teaching. In particular, they were occupied with the practical work done by the students in the classrooms for drawing.

These assistants were not supposed to do research; they were hired for teaching tasks only. It is important to note that the Polytechnikum in Zurich—like many other polytechnics—offered an atmosphere that was very favorable to the use, the production and the presentation of material models of all kinds. Models were an integral and highly estimated part of the polytechnic tradition.[1] In Gottfried Semper's building at Zurich, they were ubiquitous and displayed in prominent places.[2] When entering the building through its main entrance, one ran into the

---

[1] See: Uta Hassler and Torsten Meyer, "Die Sammlung als Archiv paradigmatischer Fälle," in *Kategorien des Wissens: Die Sammlung als epistemisches Objekt*, ed. Uta Hassler and Torsten Meyer (Zurich: vdf Hochschulverlag AG, 2014), 7–74, as well as Uta Hassler and Christine Wilkening-Aumann, "'Den Unterricht durch Anschauung fördern': Das Polytechnikum als Sammlungshaus," in *Kategorien des Wissens: Die Sammlung als epistemisches Objekt*, ed. Ute Hassler and Torsten Meyer (Zurich: vdf Hochschulverlag AG, 2014), 75–98, and Klaus Volkert, "Mathematische Modelle und die polytechnische Tradition," *Siegener Beiträge zur Geschichte und Philosophie der Mathematik* 10 (2018): 161–202.

[2] See: Hassler and Meyer, "Die Sammlung als Archiv," 13, for a list of collections presented at the Polytechnikum in Zurich.

collection of sculptures. The role of the collections is explained in a report on the polytechnic written on the occasion of the world-exposition at Paris (1889):

> These [the collections of the Polytechnikum] are primarily intended for teaching and study in the relevant fields, for which purpose they are made available to students at all times for private study. In addition, they serve to further the research of the teaching staff, advanced students, and scholars outside the institute. Finally, they are used for the instruction of the general public, and it is to this end that at least the more important collections are made available at certain hours to all visitors, and their exhibits displayed and labelled in such a way that the layperson without expert guidance is still able to benefit from them.[3]

Nowadays, Fiedler is remembered—if he is remembered at all—as the German editor of several books written by George Salmon: the combination Salmon–Fiedler was well known during the second half of the nineteenth century. It should be noted that Fiedler did much more than simply translate Salmon's book; he inserted a lot of new texts, made corrections, added problems, notes, and so on. But he published also a lot of papers and three books of his own:

> *Die Elemente der neueren Geometrie und der Algebra der binären Formen. Ein Beitrag zur Einführung in die Algebra der linearen Transformationen* (Leipzig: Teubner, 1862). [The elements of the newer geometry and algebra of binary forms. A contribution to the introduction to the algebra of linear transformations.]
> *Die darstellende Geometrie in organischer Verbindung mit der Geometrie der Lage. Für Vorlesungen an technischen Hochschulen und zum Selbststudium* (Leipzig: Teubner, 1871); 2nd ed. (Leipzig: Teubner, 1875); 3rd ed. in 3 vols.: vol. 1 (Leipzig, 1883), vol. 2 (Leipzig, 1885), vol. 3 (Leipzig, 1888); 4th ed. of the first vol. (Leipzig, 1904).[4] [Descriptive geometry in organic relation with geometry of position. For lectures at technical universities and for self-study]
> *Cyclographie oder Construction der Aufgaben über Kreise und Kugeln und elementare Geometrie der Kreis- und Kugelsysteme* (Leipzig: Teubner, 1882). [Cyclography or construction of tasks on circles and spheres and elementary geometry of circle and sphere systems]

---

[3] "Dieselben [die Sammlungen des Polytechnikums] sind in erster Linie für den Unterricht und das Studium in den betreffenden Fächern bestimmt, wofür sie den Studirenden jederzeit zum Privatstudium offen stehen; daneben dienen sie auch den wissenschaftlichen Fortschritten der Lehrkräfte, vorgerückter Studirender und von Gelehrten ausserhalb der Anstalt; endlich sind sie noch der Belehrung des Publikums gewidmet, indem wenigstens die wichtigeren Sammlungen zu bestimmten Stunden jedermann geöffnet und deren Gegenstände so aufgestellt und etiquettirt sind, dass auch der Laie, ohne fachmännische Führung, daraus Belehrung schöpfen kann." Wilhelm Fiedler, *Die Eidgenössische Polytechnische Schule in Zürich: Herausgegeben im Auftrage des Schweizerischen Bundesrathes bei Anlass der Weltausstellung in Paris 1889* (Zürich: Zürcher & Furrer, 1889), 59.

[4] Translation into Italian: Wilhelm Fiedler, *Trattato di geometria descrittiva, tradotto da Antonio Sayno e Ernesto Padova. Versione migliorata coi consigli e le osservazioni dell'Autore e liberamente eseguita per meglio adattarla all'insegnamento negli istituti tecnici del Regno d'Italia* (Firenze: Successori Le Monnier, 1874).

The teaching of geometry to future architects and civil engineers at the Polytechnikum in Zurich was minimized around 1885—in particular, projective geometry[5] was eliminated in those courses. But Fiedler always provided a substantial teaching of geometry to future teachers. To the students of the VI. department, geometry was taught by a series of four courses:

> Descriptive geometry. Part one: parallel projection (3 hours a week, winter term 1879/80);
> Descriptive geometry. Part two: central projection (2 hours a week, summer term 1880);
> Geometry of position (4 hours a week, winter term 1880/81);
> Projective coordinates (2 hours a week, winter term 1880/81 and summer term 1881).[6]

By means of this teaching, Fiedler could survive—so to speak—as a researching geometer combining his research with instruction.[7]

Some Remarks on Teaching and Early Models

In order to understand Fiedler's occupation with models better, it is helpful to look closer at the teaching delivered at *Gewerbeschulen*[8] and polytechnics, in particular of descriptive geometry. The latter was considered a hybrid theme: on one hand mathematics, on the other hand practice. Often, it was not understood as a genuine part of mathematics itself. This is illustrated by the fact that the Polytechnikum in Zurich—and most other polytechnics in the German speaking area—had both professors for mathematics and descriptive geometry. Descriptive geometry did not have a lot of theorems with proofs but many, many problems to solve by drawing. Often, these solutions are only graphic approximations, which can be checked by drawings.

It should be remembered that the teaching of drawing had a long-standing tradition going back to the eighteenth century with drawing schools for craftsmen. In that teaching, material models were used to present the objects to be depicted (a

---

[5] In the 1860s, Karl Culmann, a famous civil engineer and creator of graphical statics, established courses on projective geometry ("Geometrie der Lage") for his students. His idea was that these courses should provide the basis for understanding the methods of graphic statics developed by himself. After a short period when Theodor Reye provided this teaching, it was taken over by Fiedler after his arrival at Zurich.

[6] This cycle was held 1879 to 1881; Fiedler's eldest son, Ernst, was among the students. He took notes and prepared excellent booklets with his notes written in shorthand (ETH-Bibliothek, Hochschularchiv, Hs 110 Mappe 1).

[7] Fiedler himself explained this in a letter to Friedrich Kick (ETH-Bibliothek, Hochschularchiv, Hs 87: 498a).

[8] *Gewerbeschulen* (or *Gewerbsschulen*) were secondary schools preparing pupils for technical and business professions or for entering a polytechnic. They provided no *Abitur*, so their graduates could not enter a university. During the nineteenth century these two orientations were separated step by step. For the sake of simplicity, I will speak simply of technical schools.

cube, for example) to the pupils—or to control the drawings made by the students.[9] Of course, these models were rather simple and crude, but the tradition existed. Traces of this tradition are visible in some *Schulprogramme*; these were reports on the academic year, often supplemented by a paper written by a teacher.[10] Concerning models, I want to cite three examples where they were discussed: explicitly three books by J. Gierer (1847), Franz Harter (1847) and Paul Wiecke (1873).[11] This is to give but a few examples. Obviously, this strand of the history of models deserves further investigation. The Polytechnikum in Zurich owned several models of elementary character like cubes made out of glass or metal,[12] even as late as 1930, a dodecahedron made out of marble.

At the Gewerbeschule, which gradually replaced the drawing schools during the nineteenth century, descriptive geometry was taught at a basic level. Fiedler grew up in this tradition. As a professor of the Gewerbeschule, he himself taught descriptive geometry. This was not his first occupation when he came to Chemnitz, but in 1857 he was obliged to do so because his colleague, who was responsible for teaching descriptive geometry, fell ill.[13] In order to understand the teaching of descriptive geometry, it is important to remember that there were other more concrete courses in drawing—for example, in technical drawing or on perspective. This means that descriptive geometry was not understood as cultivating drawing

---

[9] See: Antonius Lipsmeier, *Technik und Schule: Die Ausformung des Berufsschulcurriculums unter dem Einfluß der Technik als Geschichte des Unterrichts im technischen Zeichnen* (Wiesbaden: Steiner, 1971), Wolfgang König, *Künstler und Strichezieher: Konstruktions- und Technikkulturen im deutschen, britischen, amerikanischen und französischen Maschinenbau zwischen 1850 und 1930* (Frankfurt: am Main Suhrkamp, 1999), and Kerrin Klinger, *Zwischen Gelehrtenwissen und handwerklicher Praxis. Zum mathematischen Unterricht in Weimar um 1800* (München: Fink, 2014). It should be remarked that Pestalozzi was a great advocate of intuition and its role in learning mathematics. Later, in the second half of the nineteenth century, Kant's ideas on pure intuition were broadly discussed in connection with geometry.

[10] There are thousands of such papers devoted to mathematics and its teaching, see: Gert Schubring, *Bibliographie der Schulprogramme in Mathematik und Naturwissenschaften (wissenschaftliche Abhandlungen) 1800–1875* (Bad Salzdetfurth: Franzbecker, 1986). A more recent and more complete source on *Schulprogramme* (on all themes) is provided by Franz Kössler (see: Franz Kössler, "Personenlexikon von Lehrern des 19. Jahrhunderts," preprint, published 26th Sept, 2008, http://geb.uni-giessen.de/geb/volltexte/2008/6106/ (accessed November 23, 2021)).

[11] See: Paul Wiecke, "Ueber die abwickelbaren Normalflächen an Hyperboloiden," (attachment to *Programm der Königlichen Höheren Gewerbeschule zu Kassel*, Kassel: Hof- und Waisenhausdruckerei, 1873), J. Gierer, "Der Zeichenunterricht nach Körpern," (*Schulprogramm Landwirthschafts- und Gewerbschule Fürth [Bayern]*, 1847), and Franz Harter, "Ueber die Ausführung der Körperschnitte in Modellen als Beitrag zum Unterrichte in der descriptiven Geometrie," (*Programm der Landwirthschafts- und Gewerbschule Amberg*, 1840). Wiecke is referring in his paper to models presented by his school at the world-exhibition taking place in Vienna in 1873. I thank Hellmuth Stachel (Vienna) for providing me this information.

[12] See: ETH-Bibliothek, Hochschularchiv, Hs 1196: 30 and Hs 1196: 50.

[13] See: Wilhelm Fiedler, "Meine Mitarbeit an der Reform der darstellenden Geometrie in neuerer Zeit. Schreiben gerichtet an den Herausgeber dieses Jahresberichts," *Jahresbericht der Deutschen Mathematiker Vereinigung* 14 (1905): 493–503, here 493.

itself but as a theory basic for different types of drawing used by engineers, architects, etc. Nevertheless, the students had to do a lot of drawing for their courses on descriptive geometry, but normally of a more abstract nature (no machines or buildings, etc.) Descriptive geometry is more than drawing as is indicated by its name.[14] This statement was very important to Fiedler.

As noted above, at polytechnic schools there were a lot of models in use and in production; hence, to use models in mathematical training was in complete conformity with the practice of these schools. Let me note that elementary mathematical models of the type normally used at Gewerbeschulen were produced and sold by a company named *Polytechnisches Arbeits-Institut J. Schröder* in Darmstadt (see Fig. 1); they were explicitly called *Unterrichtsmodelle* ('models for teaching'). Around 1840, Jakob Schröder,[15] a former teacher, invented a model to illustrate plane and elevation (see Fig. 2). Composed of three planes meeting orthogonally, the planes were mobile, such that they could be rotated into one single plane.[16]

Shortly after his arrival in Zurich, Fiedler bought a collection of forty models in four series produced by J. Schröder at a price of 40 sfr.[17] These models were made out of wood (two series), by rods (one series) or by strings (one series). Some of them still exist in Zurich (in the *Sammlungen* of the ETH and at the Kantonsschule Rämibühl).

The focus of Schröder's company was on producing models of machines and instruments, but the company also sold models of solids (see Fig. 3) and 'models of descriptive geometry' like the plane and elevation of a straight line, two cylinders penetrating each other, and shadows. There were also wooden cones that could be decomposed into two parts in order to illustrate the resulting conic section and wooden ellipsoids that could be decomposed into two parts showing the range of circles on the surface. Schröder's models are rather robust[18]—really usable in everyday teaching. Despite the fact that they were widespread in the nineteenth century,[19] seemingly most of these models have disappeared since then, perhaps because they were considered to be neither of high monetary (compared to other

---

[14] It is neither called 'constructive geometry' nor 'linear drawing.'

[15] On Jakob Schröder's work and life see: Lipsmeier, *Technik*, 216–17.

[16] See: Lipsmeier, *Technik*, 290. Some years before Schröder, Théodore Olivier at Paris constructed a similar model called "Omnibus." Concerning this model see: João P. Xavier and Eliana M. Pinho, "Olivier string models and the teaching of descriptive geometry," in *"Dig where you stand" 4*, Proceedings of the Fourth International Conference on the History of Mathematics Education, ed. Kristín Bjarnadóttir, Fulvia Furinghetti, Marta Menghini, Johan Prytz, and Gert Schubring, (Roma: Edizioni Nuova Cultura, 2017), 399–413.

[17] See: ETH-Bibliothek, Hochschularchiv, Hs 1196: 30, 1 and Hs 1196: 50, 1. At that time, Fiedler earned 4,300 Swiss francs a year.

[18] See the more detailed discussion of the nature of Schröder's models and their dispersion by Lipsmeier, *Technik*, 217 n. 600. The great importance of Schröder's company in the field of technical education is also underlined by Hermann Ludewig, *Das technische Unterrichtswesen auf der Weltausstellung in Wien 1873, mit besonderer Berücksichtigung des maschinentechnischen Unterrichts* (München: Ackermann, 1875), 16, 25–26, 32–33, 38, 41.

[19] See: Lipsmeier, *Technik*, 217, for more information on the dissemination of Schröder's models.

GEGRÜNDET 1837.

ILLUSTRIRTER CATALOG

FÜR

UNTERRICHTS-MODELLE

UND

APPARATE

Medaillen.

1837. Darmstadt.
1839. Darmstadt
1844. Berlin.
1850. Leipzig.
1851. London.
1854. München.
1861. Darmstadt.
1862. London.
1867. Paris.
1869. Carlsruhe
1873. Wien.
1876. Darmstadt.
1876. Philadelphia.
1879. Sidney.
1880. Melbourne.
1883. Amsterdam
(Ehrendiplom).

Orden.

RITTERKREUZ I. CL.
des Grosshzgl. Hessischen
Philippsordens.

RITTERKREUZ
des Grosshzgl. Hessischen
Ludwigsordens.

Goldene Ordensmedaille
für
Wissenschaft, Kunst &
Industrie.

POLYT. ARBEITS-INSTITUT

J. SCHRÖDER

DARMSTADT.

1885.

Preisliste wird gratis und franco übersandt.
Illustrirter Katalog à Mark 5 (enthaltend 96 Seiten) — kann direct oder durch den Buchhandel
bezogen werden.

Fig. 1 Inner cover of Jakob Schröder, ed., *Illustrirter Catalog für Unterrichtsmodelle und Apparate* (Darmstadt: Schröder, 1885)

Fig. 2 Jakob Schröder's model for plane and elevation. From Antonius Lipsmeier, *Technik und Schule: Die Ausformung des Berufsschulcurriculums unter dem Einfluß der Technik als Geschichte des Unterrichts im technischen Zeichnen* (Wiesbaden: Steiner, 1971), 291, Fig. 28.

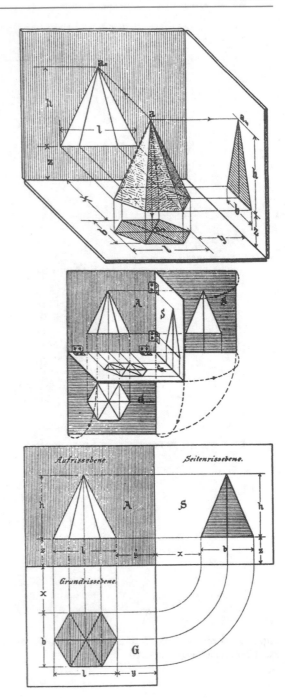

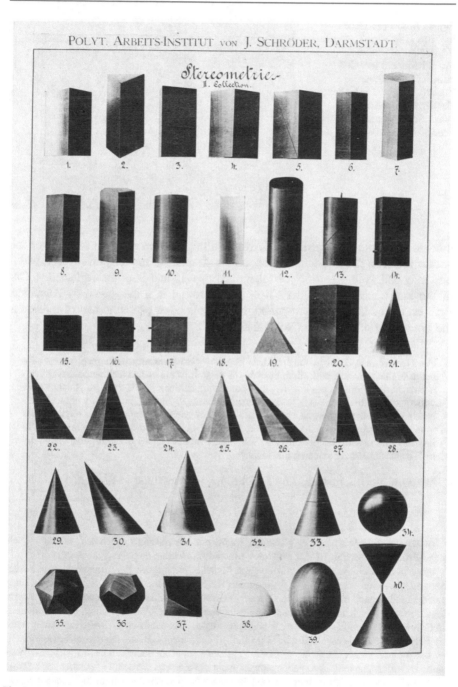

Fig. 3 Models of solids produced and sold by J. Schröder. From Jakob Schröder, ed., *Illustrirter Catalog für Unterrichtsmodelle und Apparate* (Darmstadt: Schröder, 1885)

Fig. 4 Model produced by
the firm J. Schröder: a torus.
(Collection of scientific
instruments and teaching
aids, ETH-Bibliothek
Zurich). © ETH-Bibliothek
Zurich, CC BY-SA 4.0
(https://creativecommons.org/
licenses/by-sa/4.0/)

models later sold by L. Brill or Martin Schilling they were rather cheap) nor scientific value. In the recent discussions on mathematical models, this aspect, which one may call 'sub-scientific' or 'elementary', was and is often neglected.[20] One of the rare traces of Schröder's firm can be found in a talk given by Alexander Brill on the occasion of the opening of the collection of mathematical models at the University of Tübingen (November 7, 1886):

> For a long time in Germany too there have been models whose purpose has been to serve the needs of mathematical education. Probably in most universities and technical schools there is a cupboard or dusty corner somewhere containing cardboard models of polyhedra, cones with planar sections, curves of intersection of cones and cylinders, etc., mostly of indeterminate provenance, which in the current state of higher education are no longer required, and thus no longer replaced by the newer more beautiful forms that numerous suppliers of teaching aids (the oldest being the estimable Schröder model factory in Darmstadt) habitually give these elementary teaching materials.[21]

Wooden models produced by J. Schröder are presented in Figs. 3, 4, and 8.

---

[20] By the way, there was also a long-standing tradition to use materials which one may call models in the elementary teaching of mathematics—like rods presenting numbers, and so on.

[21] "Auch in Deutschland giebt es seit Langem Modelle, welche den Bedürfnissen des mathematischen Unterrichts dienen sollen. Wohl an den meisten Universitäten und technischen Schulen finden sich in irgend einem Schrank oder einem staubigen Winkel Pappmodelle von Polyedern, Kegeln mit ebenen Schnitten, Durchdringungskurven von Kegeln mit Cylindern u.s.w. von meist unbestimmter Herkunft, die bei dem heutigen Stande des Hochschul-Unterrichts nicht mehr gebraucht werden und deshalb durch die neueren schöneren Formen, welche zahlreiche Lehrmittel-Anstalten (darunter als älteste die verdiente Schröder'sche Modellfabrik in Darmstadt) diesen elementaren Unterrichtsmitteln zu geben pflegen, nicht mehr ersetzt werden." Alexander Brill, "Über die Modellsammlung des mathematischen Seminars der Universität Tübingen," *Mathematisch-naturwissenschaftliche Mitteilungen* 2, no. 1 (1889): 69–80. See also: Walther Dyck, "Einleitender Bericht über die mathematische Ausstellung in München," *Jahresbericht der Deutschen Mathematiker-Vereinigung* 3 (1892–93): 39–56, here 51, who cites the "altbekannte Werkstätte von Schröder in Darmstadt" [well-known workshop of Schröder in Darmstadt].

There was another usage of the term 'model' which was important (and current) in the second half of the nineteenth century—at least for German speaking scholars working in descriptive geometry. This usage is related to the problem of representing a part of space[22] by another part of space such that the impression is that of seeing the scene by one eye. In the language of descriptive geometry, this is the problem of projecting a part of space from a point to another part of space—that is, a central projection. If the spaces were imbedded in a four-dimensional space, everything could be done in strong analogy to the case of the plane.[23] If not, things become complicated. A practical situation in which this problem arises is the construction of a stage design in a theater. Jean-Victor Poncelet, the first who studied the problem from an abstract point of view, coined the term 'relief-perspective.'[24] In German, it was common to speak of *Reliefperspektive* or *Modellierung* (see Fig. 5), its product being a *Modell*. Note that this type of model is an abstract entity. It is an image of a mapping (in modern terms), not a material object.

Here is a passage in which Fiedler explains the significance of this construction[25] and the motivation behind it:

> The significance of this construction is that it includes the geometric basis of *artistic modelling*, that is, one based on pleasing deception, and that it [the construction] encompasses all other technically feasible *modeling methods*, and finally the *methods of presentation through drawings on planes* as special cases. [...] When, in addition, the *models* constructed in this way are lit by light radiating from a suitably chosen point of the opposing plane *Q*, the eye *C* believes, instead of its collinear transformation, to see the object itself endowed with the correct sun shadows. (*Stage scenery*—the space between the planes *S* and *Q* is the space of the stage, *S* the plane of the curtain.)[26]

---

[22] Often, this was the half-space behind a plane—the front of the stage, so to speak. In that case, the other plane is the vanishing plane—the back plane of the stage.

[23] Thus, here the motivation can be found to introduce four-dimensional space or to create models for that situation. See: Klaus Volkert, *In höheren Räumen: Der Weg der Geometrie in die vierte Dimension* (Berlin: Springer, 2018). Henrik de Vries, a former student of Fiedler, discussed that possibility at length in his book *Die Lehre von der Zentralprojektion im vierdimensionalen Raum* (Leipzig: Göschen, 1905).

[24] *Perspective en relief*. But note that the name was used before Poncelet; see below on Breysig's book: Johann Adam Breysig, *Versuch einer Erläuterung der Reliefsperspective, zugleich für Mahler eingerichtet* (Magdeburg: Keil, 1798).

[25] The construction itself is indicated in Fig. 5. The principal idea behind it is to reduce the spatial problem to the plane by using plane cuts passing through the center of projection.

[26] "Die Bedeutung dieser Construction liegt darin, dass sie die geometrische Grundlage *künstlerischer* d. h. auf angenehme Täuschung ausgehende *Modellirung* mit enthält, und dass sie alle anderen technisch brauchbaren *Modellirungsmethoden* u. schliesslich die *Darstellungsmethoden durch Zeichnungen auf Ebenen* als spezielle Fälle umfasst. [...] Wenn dem entsprechend hergestellte *Modelle* noch beleuchtet werden durch Licht, das von einem geeignet gewählten Punkte der Gegenebene *Q* ausgeht, so glaubt das Auge *C* statt seiner collinearen Umformung das mit dem richtigen Sonnenschatten ausgestattete Object selbst zu sehen. (*Decoration der Schaubühne*—der Raum zwischen den Ebenen *S* und *Q* ist Raum der Bühne, *S* die Vorhangebene.)" Wilhelm Fiedler, *Darstellende Geometrie: Autographierte Ausarbeitung einer Vorlesung* (Zürich: E. Zimmer, 1894), 48. An exemplar of Fiedler's lecture notes can be found at the library of the ETH

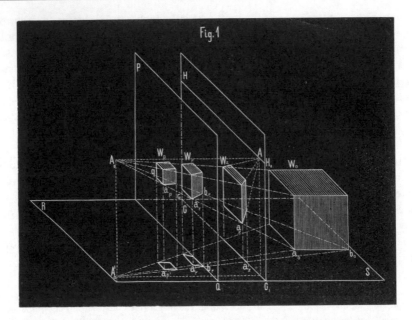

Fig. 5 The construction of several models of a cube. From Rudolf Staudigl, *Grundzüge der Reliefperspective* (Wien: Seidel & Sohn, 1868), 3

This text is taken from an autographed[27] version of Fiedler's course on descriptive geometry. Here, he is more explicit than in his book on descriptive geometry and he provides more background information than in his textbook. One of the first models produced in Fiedler's surroundings was that of a relief-perspective constructed by his assistant Rafael Morstadt in Prague.[28] Seemingly, this model has disappeared, but models of this type, constructed by Burmester, still exist (see Fig. 6).[29]

(Rar 9099). I thank Urs Stammbach (Zurich) for providing me with an exemplar of these lecture notes.

[27] Autography or lithography was a technique of printing invented by Alois Senefelder around 1800, which was used to produce lecture notes. Well-known examples are due to Felix Klein, but he was not the only person to see the utility of that technique for teaching. As early as 1874, Fiedler ordered 400 copies of an autographed page on the method of least squares in order to distribute them to the students (see: ETH-Bibliothek, Hochschularchiv, Hs 1796: 50, 2). Later, he bought also autographies by Klein.

[28] See: Rafael Morstadt, "Ueber die räumliche Projection (Reliefperspective), insbesondere diejenige der Kugel," *Zeitschrift für Mathematik und Physik* 12 (1867): 326–39. We will see below that Morstadt constructed also another model for Fiedler. Note that Morstadt (and Burmester) produced a material model of the abstract model constructed by a *Reliefperspektive*.

[29] I thank Ulf Hashagen (Munich) for providing me this information.

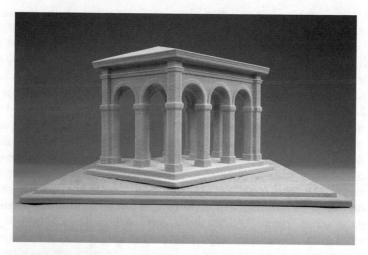

Fig. 6 Daniel Lordick, 2004: Reconstruction of a relief perspective originally created 1883 by Ludwig Burmester. 3D print based on starch and plaster. Photo: © Lutz Liebert, TU Dresden, 20055, all rights reserved

The interest in the construction of models by using the *Reliefperspektive* is underlined by the fact that R. Staudigl published a book on that theme in 1868.[30] In its preface, he emphasizes the fact that the *Reliefperspektive* is of a great practical value and that it was neglected by the geometers. In a letter dated Riga, 27 March / 8 April, 1877,[31] Fiedler's former assistant, Alexander Beck, reported:

> Some time ago I had a plaster model made as an example of a relief-perspective, for which I provided the sculptor with a drawing. [...] The model has turned out very well and has become a centerpiece of my collection. I also intend to acquire the new Munich models.[32]

The *Reliefperspektive* is a recurrent theme in Fiedler's writing—not the least because it is so near to the practice of painters and architects. In a paper published in 1882, Fiedler presented the history of this subject, in particular the forgotten book *Versuch einer Erläuterung der Reliefperspective zugleich für Maler eingerichtet von J. A. Breysig, Professor der schönen Künste und erstem Lehrer an der k. Kunstschule zu Magdeburg, 1798*, written by Johann Adam Breysig.[33] Fiedler

---

[30] Staudigl, *Grundzüge*.

[31] The two dates are noted because the Julian calendar was still in use at Riga.

[32] "Ich habe mir vor einiger Zeit ein Gypsmodell als Beispiel einer Reliefperspective anfertigen lassen, zu welchem ich dem Bildhauer die Zeichnung lieferte. […] Das Modell ist sehr gut gelungen und bildet ein Prunkstück meiner Sammlung. Ich gedenke auch, die neuen Münchner Modelle anzuschaffen." ETH-Bibliothek, Hochschularchiv, Hs 87: 40.

[33] It is specific to the early history of descriptive geometry in German speaking countries that it was closely linked to the arts of painting and drawing. See: Nadine Benstein, "A German Interpreting of Descriptive Geometry and Polytechnic," in *Descriptive geometry, The Spread of a Polytechnic Art: The Legacy of Gaspard Monge*, ed. Évelyne Barbin, Marta Menghini, and Klaus Volkert (Cham: Springer 2019), 139–66.

remarks that careful models of a *Reliefperspektive* were built by his former assistant Morstadt in Prague, by Burmester in Dresden and by his assistant Johannes Keller.[34] The *Reliefperspektive* was also discussed in a *Schulprogramm*.[35]

In his paper "Über die niedere Sphärik" [On Elementary Spherics] (1832), Christoph Gudermann described another early model (or apparatus) for teaching purposes.[36] In this paper, Gudermann announces his treatise on spherical geometry, published in 1835, and discusses the theory of area for spherical triangles based on—in modern terms—equidecomposability. At the end of the paper, he deplores the state of the art of spherical geometry, that is, the fact that it had fallen into neglect for some time. One reason for this—following Gudermann— were the difficulties of representing the situation on the sphere by plane drawings. So Gudermann constructed a model of a sphere including tools to draw great circles and to transport distances such that constructions can be performed. He called it a "sphärographischer Apparat."[37] This model was produced and sold by J. V. Albert, a firm in Frankfurt. It contained a sphere, a ruler (an instrument to draw great circles on the sphere), and a transporter (an instrument to move segments or lengths from one place to another, a compass, so to speak). Gudermann underlined the fact that his model is also beneficial for the teaching of geography and astronomy. In sum, we can state that Gudermann's model was a paradigmatic case for the construction of models motivated by pedagogical needs.

In concluding this section, I want to add that Fiedler also used wall charts in his teaching in order to illustrate themes (like the common perpendicular of two skew straight lines in space).[38] Seemingly, the use of wall charts was a widespread custom in (elementary) teaching in that period. They were produced and sold by different companies.

Models in Fiedler's Correspondence

In this paragraph, I will present some traces in Fiedler's work left by his pre-occupation with models, in particular, in his large correspondence. When Fiedler came to Zurich in 1867, he replaced Joseph Wolfgang von Deschwanden, the first professor of descriptive geometry at the recently founded Zurich Polytechnikum (1855). Deschwanden was a man with highly developed abilities as a painter, but

---

[34] See: Wilhelm Fiedler, "Zur Geschichte und Theorie der elementaren Abbildungs-Methoden," *Vierteljahrsschrift der Naturforschenden Gesellschaft Zürich* 27 (1882): 125–75, here 135–36.

[35] Wilhelm Burghardt, "II. Beitrag für den Unterricht in der Reliefperspective," (*Programm Nordhausen Realschule*, 1861) [Published together with: "I. Anleitung zur Analyse vermittelst des Lötrohres"].

[36] That is synthetic geometry of the sphere (in contrast to spherical trigonometry). At the time this paper was written, Gudermann was a teacher at the *Gymnasium* in Cleve (now: Kleve [Niederrhein]).

[37] Christoph Gudermann, "Ueber die niedere Sphärik," *Journal für die reine und angewandte Mathematik* 8 (1832): 363–69, here 369.

[38] See: ETH-Bibliothek, Hochschularchiv, Hs 1196: 30 and Hs 1196: 50 for more details.

Fig. 7 Cover of Wilhelm Fiedler's inventory of models (ETH-Bibliothek, Hochschularchiv, Hs 1196: 50). © ETH Bibliothek, all rights reserved

he was also an expert in the construction of machines.[39] It is not known whether Deschwanden used models or not. But, at least in other places in the school, Fiedler encountered many models.

There are two main types of sources concerning Fiedler's relation to models: on one hand there are documents by the administration and the inventories cited above (see Fig. 7),[40] on the other hand there are letters to Fiedler. Let us start with the first. The highest level of administration of the Polytechnikum was the *Schulrath* (board of the school) with five members—among them the president of the board, the *Schulrathspräsident*.[41] In the records of the Schulrath, we find remarks that Fiedler received money in order to buy models: January 19, 1871 and May 8, 1877 he got 100 sfr. In 1871, the sum was dedicated to buy models in gypsum, 1877 to purchase models to be used in teaching. Thus, it is clear that the officials approved the acquisition of models.

Luckily, we can rely on a more detailed document. The Polytechnikum, that is, the Schulrath, reported once a year to the *Bundesrath* in Bern on the state

---

[39] He served also as director of the Polytechnikum and as professor at the university.

[40] One of them (ETH-Bibliothek, Hochschularchiv, Hs 1196: 30) was an official document approved by the authorities, the other (Hs 1196:50) was a private document in the possession first of Fiedler, then of Grossmann. The first was established in 1907 when Fiedler was replaced by his successor Grossmann; it was controlled annually until the 1930s by the authorities. It seems plausible that the inventories cite only models which were bought or which resulted in other costs (e.g. for the material). Home-made models produced for free are not mentioned at all. I thank Wibke Kolbmann for providing me information on the inventory of: ETH-Bibliothek, Hochschularchiv, Hs 1196: 50.

[41] For a long time this was Karl Kappeler.

of the affairs of the school.[42] In the annual report for the year 1878, we find a long passage on the collection of mathematical models[43]: it is stated first that the collection of mathematical models is presented in the report for the first time,[44] and that it is not intended to present all objects in the collection in the report.

> Furthermore, however, for a number of years wire or rod models of algebraic surfaces etc. have been made by the various assistants of descriptive geometry under the direction of the professor [that is, Fiedler]—which should be mentioned here as a speciality. It is due to the great and regular workload of the assistants that work on these models has been slow to advance, and that the number of models has grown only slowly.[45]

Then, some special models (or types of models) are mentioned in more detail[46]:

1. "A wire model of an orthogonal system in the bundle."
2. "A model of the main points and main planes of the central collineation of spaces in connection with the central projection of planar systems."
3. "A string model of the developable surfaces of a cylindrical helix."
4. "Wire models of cubic surfaces with 27 real straight lines with systems of parallel cross sections"—three types of this model were present in the collection.
   "Models of quartic surfaces"—two types of this model were present in the collection.
   "A model of a curve of intersection of two second-degree surfaces and their developable surfaces with representation of their double curves is in progress."[47]

This list is interesting for several reasons. First, it is clear that the cited models in the collection were not elementary (like a cube, etc.) In particular, the models of the fourth type, the famous cubic surface with 27 real straight lines (among them Alfred Clebsch's *Diagonalfläche*), were rather sophisticated. They were the result

---

[42] It should be remembered that the Polytechnikum in Zurich was a foundation by the Swiss Federation whereas all other Swiss schools and universities were (and are) run by the cantons—except, of course, private institutions and the French-speaking counterpart of the ETH in Lausanne.

[43] *Bericht des Eidgenössischen Polytechnikums an den Bundesrath 1876* (Bibliothek ETH-Archiv), 10–11.

[44] It was also the last time.

[45] "Ueberdies aber sind seit einer Reihe von Jahren von den jeweiligen Assistenten der darstellenden Geometrie unter Leitung des Professors Stab- oder Drahtmodelle algebraischer Flächen usw. gefertigt worden, welche als Specialität hier einmal Erwähnung finden mögen. Es ist in der großen regelmäßigen Arbeitsbelastung der Assistenten begründet, daß diese Modellarbeiten nur langsam gefördert werden können und daß die Zahl der Modelle nur langsam wächst." *Bericht des Eidgenössischen Polytechnikums*, 10.

[46] *Bericht des Eidgenössischen Polytechnikums*, 10–11.

[47] 1. "Ein Drahtmodell des Orthogonalsystems im Bündel." / 2. "Ein Modell der Hauptpunkte und Hauptebenen der centrischen Collineation der Räume im Zusammenhang mit der Centralprojection ebener Systeme." / 3. "Ein Fadenmodell der entwickelbaren Flächen einer cylindrischen Schraubenlinie." / "Drahtmodelle von Flächen dritter Ordnung mit 27 reellen Graden mit Systemen paralleler Querschnitte" / "Modelle von Flächen vierter Ordnung" / 4. "Ein Modell der Durchdringungscurve von zwei Flächen zweiten Grades und ihrer entwickelbaren Fläche mit Darstellung ihrer Doppelcurven ist in Arbeit." *Bericht des Eidgenössischen Polytechnikums*, 11.

of contemporaneous advanced research.[48] Their existence presupposed a certain level reached in the theory of surfaces. After the discovery by Arthur Cayley and Salmon (1849) that a cubic surface has 27 straight lines, several mathematicians worked on that surface—among them Clebsch. In 1868, Christian Wiener (Karlsruhe) constructed a material model of such a surface.[49] Adolf Weiler's example (1872) was a substantial improvement (see below).[50] It is clear that the progress of geometric theory played a decisive role here: only after the progresses made by algebraic (in those times often called analytic) geometry[51] did it become possible to think of models of sophisticated surfaces. A discovery of a certain importance was Schläfli's double-six; it enabled Wiener to construct his model.[52] Therefore, it is surprising that the story of these models began only in the 1840s.

It is commonly accepted that Wiener's model was the first model of this type of surface. But Fiedler reported the story in a different way:

> The editor [Fiedler] has owned a rod model of the generic cubic surface with 27 real straight lines since 1865 [sic!]; he has provided information about its simple construction in vol. 14 of the *Zeitschrift für Math.* (*Literaturzeitung*). A plaster model resulting from a suggestion by Clebsch and constructed by Wiener has been in wide circulation since 1869. Calculations based on this have been communicated by Cayley in *Transactions of Cambridge*, vol. 12 (1873), pp. 366–383. Recently, the cubic surfaces with four nodes and the diagonal surface (also as a rod model) have also been modeled. The principle of the rod model can be applied without difficulty in numerous other cases.[53]

In his paper mentioned in the citation above, Fiedler gave a detailed description of stereoscopic photographs of Wiener's model published by Wiener himself. Then,

---

[48] See: David E. Rowe, "Mathematical Models as Artefacts for Research: Felix Klein and the Case of Kummer Surfaces," *Mathematische Semesterberichte* 60 (2013): 1–24, for more details on that research.

[49] An exemplar of it was bought by Fiedler in 1870 at a price of 115 Swiss francs (see: ETH-Bibliothek, Hochschularchiv, Hs 1796: 30, 9).

[50] Concerning the history of those models which we may call scientific ("artefacts for research" [David Rowe], that are not primarily constructed for teaching purposes), see: Oliver Labs, "Straight Lines *on* Models of Curved Surfaces," *The Mathematical Intelligencer* 39, no. 2 (2017): 15–26; Rowe, "Mathematical Models" and David E. Rowe, "On Building and Interpreting Models: Four Historical Case Studies," *The Mathematical Intelligencer* 39, no. 2 (2017): 6–14. The story of Clebsch's models and other surfaces is reported by Klein himself several times, see: Felix Klein, *Gesammelte mathematische Abhandlungen*, vol. 2, ed. Robert Fricke and Hermann Vermeil (Berlin: Springer, 1922), 3–5. Concerning the history of the surface itself, see: François Lê, "Entre géométrie et théorie des substitutions: une étude de cas autour des vingt-sept droites d'une surface cubique," *Confluentes Mathematici* 5, no. 1 (2013): 23–71, and Labs, "Straight Lines."

[51] We should also think of differential geometry with its central notion of curvature as an important source for new and interesting examples of modeling—a fact often underlined by Brill in his autobiography.

[52] See: Labs, "Straight Lines," 19–21.

[53] "Von der allgemeinen Fläche dritter Ordnung mit 27 reellen Geraden besitzt der Herausgeber [Fiedler] ein Stabmodell seit 1865 [sic!], über dessen bequeme Herstellung er Auskunft gegeben hat in Bd. 14 der 'Zeitschrift für Math.' (Literaturzeitung). Seit 1869 ist ein von Wiener construirtes Gypsmodell mehrfach verbreitet, das auf Anregung von Clebsch entstanden ist. Berechnungen aus

Fiedler presents the story of his own model, beginning in 1861. After failing with his attempt to construct the model using calculations, which turned out to be too difficult, Fiedler decided to construct it with the help of the *Reliefperspektive*. Of course, the result is only an approximation of the geometric surface. The model was brazed together by Fiedler's assistant Rafael Morstadt in Prague.[54] Fiedler called the result a *Stabmodell* (model made by metal wiring). Note that in the report cited above it was called a "speciality" of Fiedler's approach.[55] Fiedler concluded:

> In this way the model in the form of a rod model provides a complete view of the surface with its very curious openings. It has a width, length, and height of approximately 0.8 m. I have always found it extremely serviceable as a means to clarify general conceptions regarding the theory of algebraic surfaces; its accuracy is entirely adequate. I have occasionally indicated a pair of conjugate Steiner's tritangent planes [*Trieder*]; I have determined the double points of the involutions on the lines of a tritangent plane and thereby shown [*anschaulich gemacht*] how the six parabolic points of the surface lie on the same plane four times in threes on a line, etc.
>
> The relative ease of construction makes a repetition for the purpose of the formation of varieties and special cases possible; perhaps this communication will encourage this. May it at least awaken a broad interest in the models and photographs of Herr Professor Wiener. Fluntern, near Zürich, W. Fiedler.[56]

Anlass derselben hat Cayley mitgetheilt in 'Transactions of Cambridge' Bd. 12 (1873), p. 366–383. Neuerdings sind auch die Flächen dritter Ordnung mit vier Knotenpunkten und die Diagonalfläche (auch als Stabmodell) modelliert worden. Das Prinzip der Stabmodelle ist auf sehr viele andere Fälle bequem anwendbar." George Salmon and Wilhelm Fiedler, *Analytische Geometrie des Raumes*, vol. 2: *Analytische Geometrie der Curven im Raume und der algebraischen Flächen*, 2nd ed. (Leipzig: Teubner, 1874), 662–63 [Fiedler is referring to the *Zeitschrift für Mathematik und Physik* which included a "Literaturzeitung" with a separate pagination]. This book contains also a very detailed description of another model made out of paper; see: Salmon and Fiedler, *Analytische Geometrie*, 624–25. Both texts are in the notes, written by Fiedler. They are not in Salmon's original version of his book.

[54] Felix Klein knew of the existence of this model and other models in Zurich; this is shown by his letter, dated Erlangen, February 2, 1873, cited below.

[55] Fiedler's model does not exist anymore. There are no pictures of it, only Fiedler's report and some traces in his correspondence are left.

[56] "So giebt das Modell als Stabmodell doch die vollständige Anschauung der Fläche mit ihren höchst merkwürdigen Oeffnungen. Es ist circa 0,8 m breit, lang und hoch. Ich habe es immer ganz trefflich brauchbar gefunden, um die allgemeinen Anschauungen der Theorie algebraischer Flächen daran zu verdeutlichen; seine Genauigkeit reicht dafür völlig aus. Ich habe gelegentlich ein Paar der Steiner'schen conjugirten Trieder markirt, ich habe die Doppelpunkte der Involutionen auf den Geraden einer dreifach berührenden Ebene bestimmt und dadurch anschaulich gemacht, wie solche sechs parabolische Punkte der Fläche in derselben Ebene viermal an dreien in einer Geraden liegen etc. / Die relative Leichtigkeit der Herstellung macht eine Wiederholung derselben behufs Bildung von Varietäten und Specialfällen möglich, vielleicht regt diese Mittheilung dazu an. Möge sie wenigstens nicht verfehlen, für die Modelle und Photographien des Herrn Professor Wiener vielseitig Interesse zu erwecken. / Fluntern, bei Zürich, W. Fiedler." Wilhelm Fiedler, "Besprechung von Stereographische Photographien des Modelles einer Fläche dritter Ordnung mit 27

Second, some of the examples cited in the report are directly linked to Fiedler's specific interests: the first one to his rather original way of constructing duality in a (projective) plane,[57] the second is related to the *Reliefperspektive* discussed above.

Who were Fiedler's assistants in the period from 1867 (Fiedler's arrival at Zurich) to 1876 (publication of the report just cited)? With the help of the course catalogue,[58] we find the following names: Albert Fliegner (1842–1928),[59] J. J. Hemmig, Alexander Beck (he became professor of descriptive geometry in Riga), Adolf Weiler[60] and Johann Keller. Marcel Grossmann, besides Martin Disteli and Giuseppe Veronese perhaps the best-known pupil of Fiedler, was his assistant after 1900 and later (1907) his successor to the chair for descriptive geometry at the Polytechnikum.

As mentioned above, Adolf Weiler (1851–1916) is well known for having constructed a model of Clebsch's surface during his stay in Göttingen.[61] After studying at the Polytechnikum in Zurich (1867–1871), where he got a diploma of the *Fachlehrerabteilung*,[62] Weiler left Zurich for Göttingen. It seems plausible that Fiedler proposed this to Weiler.[63] Weiler remained in contact with his "Sehr geehrter Herr Professor" via letters. In his second letter from Göttingen, dated June 15, 1872, Weiler reported:

> Last week I spoke at the colloquium on ruled surfaces & made models for this. These were rather small & and made of small cigar boxes with sheets of paper stuck in holes & strings threaded between.
> That led me to execute a similar model on a slightly larger scale & to send it to you & it is the first of its kind. Through trial and error I have obtained rather favorable results. On the

reellen Geraden. Mit erläuternden Texten von Dr. Chr. Wiener, Professor am Polytechnikum zu Carlsruhe," *Zeitschrift für Mathematik und Physik* 14 (1869), *Literaturzeitung*: 34.

[57] It is described in Fiedler, *Die darstellende Geometrie*, 2nd ed., 113–14.

[58] Officially, these catalogues were "Anhänge zu den Protokollen der Sitzungen des Schulrathes," usually they are called "Polyprogramme."

[59] Fliegner became professor of mechanics at the Polytechnikum in 1872.

[60] Not to be confused with August Weiler, a math teacher at Mannheim and author of textbooks. On Adolf Weiler, see: *Universität Zürich. Rektoratsreden und Jahresberichte. April 1916 bis Ende März 1917* (Zürich: Orell Füssli, 1917), 54–55.

[61] This model was presented to the Göttingen Academy by Clebsch in 1872. See: Alfred Clebsch, "Ueber Modelle von Flächen dritter Ordnung," in *Nachrichten von der K. Gesellschaft der Wissenschaft und der Georg-August-Universität aus dem Jahre 1872* (Göttingen: Dieterich, 1872), 402–4. Weiler is cited by Klein in his letter to Fiedler dated Erlangen, February 2, 1873, mentioned above. In his *Gesammelte mathematische Abhandlungen* (vol. 2), Klein explains that Weiler brought with him the "tradition of Fiedler" (Klein, *Gesammelte mathematische Abhandlungen*, 3).

[62] The department concerned with future teacher training.

[63] Fiedler maintained an important correspondence with Clebsch, whom he considered as an authority; see: ETH-Bibliothek, Hochschularchiv, Hs 87: 160–70.

surface the double curves with two cuspidal points as well as two fixed tangent planes are clearly visible.[64]

Weiler spoke of common work with Felix Klein some months later (January 14, 1873):

> I'm currently working on an assignment in line geometry & if there is enough time I will produce a few models of cubic surfaces with Klein. Klein is also very busy still, so the plan will perhaps be put into action later.[65]

On the basis of the correspondence with Fiedler, we may state that Weiler was an important pioneer in the construction of models, inspired in Zurich by Fiedler. After his return to Switzerland, Weiler served for a short time as a math teacher at Stäfa, 1874 he returned to Zurich becoming one of Fiedler's assistants. He received his habilitation at the Polytechnikum in 1875 and at the university in 1891. In 1899, he was promoted to professor there. Note that there is a difference here: the well-known model by Weiler is one in gypsum (or plaster),[66] whereas Fiedler's models of type 4 were made by metal wiring.[67]

For every semester, the Polytechnikum published the catalogue of courses offered to the students. In those catalogues, some institutions and collections are described under the heading "Sammlungen und Institute" [collections and departments]. Here, we find hints at the collection of models for mathematics and descriptive geometry in the catalogues for the winter terms 1892/93, 1894/95, and 1896/97. It is more or less obvious that this collection was not considered to be very important because it is mentioned only on few occasions. The most important collection, getting a lot of money from the Schulrath, was the *Maschinensammlung* (collection of models of machines). The collection of mathematical models

---

[64] "Letzte Woche sprach ich im Colloquium über Regelflächen & verfertigte dazu Modelle. Diese waren ziemlich klein & bestanden aus Zigarrenkistchen, in denen Blätter in Löchern steckten & dazwischen Fäden durchgezogen.

Das führte mich darauf, ein solches Modell in etwas größeren Dimensionen auszuführen & Ihnen zu übersenden & es stellt dasselbe die erste Gattung dar, durch Ueberlegen und Probieren habe ich ziemlich günstige Verhältnisse erhalten. Auf der Fläche sind die Doppelcurven mit zwei Cuspidalstellen sowie zwei stationären Tangentialebenen deutlich sichtbar." Weiler to Fiedler, June 15, 1872, ETH-Bibliothek, Hochschularchiv, Hs 87: 1491, https://doi.org/10.7891/e-manuscripta-85659 (accessed November 23, 2021). This model is noted in Fiedler's inventory as no. 7, p. 2, together with the commentary: "(von Herrn Weiler ausgeführt in Begleit. seiner Preisbewerbungsschrift 1872)."

[65] "Gegenwärtig arbeite ich an einer Aufgabe aus der Liniengeometrie & wenn Zeit dafür bleibt, werde ich mit Klein einige Modelle über Flächen dritter Ordnung ausführen. Noch ist auch Klein sehr beschäftigt, so daß der Plan vielleicht später angegangen wird." Weiler to Klein, January 14, 1873, ETH-Bibliothek, Hochschularchiv, Hs 87: 1494, https://doi.org/10.7891/e-manuscripta-85682 (accessed November 23, 2021). In the meantime, Weiler followed Klein to Erlangen, where he received his doctorate in 1873. Its theme was a problem from line geometry.

[66] See: *Göttinger Sammlung mathematischer Modelle*, Modell 135. The firm of L. Brill later sold a plaster model of Clebsch's surface designed by Rodenberg based on Weiler's model. It was bought later also by Fiedler for his collection.

[67] Not to be confused with rods—a material often used by Fiedler.

was probably stored at the second floor of Semper's building in the rooms reserved for descriptive geometry.[68]

It should be mentioned that most polytechnical schools run *Modellirwerkstätten* (workshops for modeling), where students could construct models in wood, clay, or metal—mainly of machines, buildings, and so on. Qualified persons with a great degree of practical competence headed these workshops. Thus, the *Modellirwerkstätten* were also a good place to construct mathematical models. But we have no traces for such an activity. It seems plausible that products of the *Werkstätten* were occasionally presented to the public—e.g. in the context of exhibitions.[69] Fiedler hired and paid sculptors to also produce models for him in plaster (Lobde [Berlin], Spiess).[70]

Let us look at Fiedler's letters. Among his correspondents there are many geometers—often working at polytechnic schools—like Christian and Hermann Wiener, Alexander Beck, Ludwig Burmester, Oskar Schlömilch. A very interesting correspondence is that with Klein. It started in 1872 when Klein sent his now famous 'Erlanger Programm' to Fiedler. In his letter, dated Erlangen, February 2, 1873, Klein speaks of models in the context of an intended meeting of mathematicians in Göttingen.[71] In particular, he invites Fiedler to participate in the planned exhibition of mathematical models:

> We want to combine this with a model exhibition. The committee will pay for the cost of postage and packing and guarantee safe transport. Could you send us models from Zürich, perhaps also from the models of third-degree surfaces you've had made? These should be sent to Dr. Riecke in Göttingen, assistant at the Physics Institute. We would be greatly in your debt.[72]

---

[68] See the plan of the Polytechnikum reproduced in: Hassler and Meyer, "Die Sammlung als Archiv," 18. The area reserved for descriptive geometry in Semper's plan was situated at the south-east corner of the building on the second floor. The way in which mathematical models were presented is discussed in Anja Sattelmacher, "Zwischen Ästhetisierung und Historisierung: Die Sammlung geometrischer Modelle des Göttinger mathematischen Instituts," *Mathematische Semesterberichte* 61 (2014): 131–43. Here one finds also some photographs. The order in which the models were presented is an interesting problem that is wide open to my knowledge.

At the Polytechnikum in Zurich, there was also a room with a library for the assistants of descriptive geometry.

[69] See: Hassler and Meyer, "Die Sammlung als Archiv" and Hassler and Wilkening-Aumann, "Das Polytechnikum" for more information on the *Maschinensammlung* and the *Modellierwerkstatt*.

[70] ETH-Bibliothek, Hochschularchiv, Hs 1796: 50.

[71] This meeting took place in Göttingen from April 16 to April 18, 1873. Partly, it was held to commemorate Clebsch who died just before. It is also considered as a step towards the foundation of the *Deutsche Mathematiker-Vereinigung* (1890).

[72] "Wir möchten mit derselben eine Modell-Ausstellung verbinden. Das Comité übernimmt die Versendungs- als auch die Verpackungskosten, so wie die Garantie für ungefährdeten Transport. Könnten Sie von Zuerich aus uns Modelle zukommen lassen, vielleicht auch von den Modellen betr. Flächen dritten Grades, die sie haben anfertigen lassen? Diesselben wären an Dr. Riecke in Goettingen, Assistent am physikalischen Institute, zu adressieren. Sie würden uns dadurch sehr verpflichten." Klein to Fiedler, February 2, 1873, ETH-Bibliothek, Hochschularchiv, Hs

Alexander Brill wrote about this meeting:

> These endeavors [the construction of models] were further stimulated by a gathering of mathematicians in Göttingen (1873), where an exhibition of mathematical models with broad participation aroused interest. Alongside the collections of Plücker, Klein, and Muret, there were several models by Schwarz and Wiener, as well as a cubic surface ('diagonal surface') constructed by Weiler (then at Göttingen) at the instigation of Clebsch. In addition, there were several models representing singular points of a cubic surface (one with 4 nodes) constructed in collaboration with Klein, and an elliptic paraboloid represented by assembled pieces of card in the form of semicircles from its circular sections—and much else besides.[73]

Hence, no trace of Fiedler.

In a second letter, dated Erlangen, November 11, 1873, Klein mentions "nice models of minimal surfaces," which were constructed following instructions by Schwarz[74] and which were presented at the Göttingen meeting. He asks Fiedler for the address of the craftsman who produced his models. In a third letter, dated Erlangen, November 21, 1873, Klein thanks Fiedler for his explications of Schwarz's models. He is pleased by the fact that Adolf Weiler—who had just got his doctorate with Klein in Erlangen—returned to Zurich in order to become Fiedler's assistant. Some years later, after Klein was appointed at the polytechnic in Munich (1875), he wrote to Fiedler (dated Munich, October 22, 1876) about the importance of improving the teaching of mathematics for future engineers: "When

---

87: 576, https://doi.org/10.7891/e-manuscripta-84491 (accessed November 23, 2021). Thanks to Peter-Maximilian Schmidt (Zurich) for transcribing Klein's letters to Fiedler.

[73] "Eine weitere Anregung wurden diesen Bestrebungen [zur Konstruktion von Modellen] zu Teil auf einer Versammlung von Mathematikern in Göttingen (1873), wo eine vielseitig beschickte Ausstellung von mathematischen Modellen Interesse erregte. Es waren dort ausser den Collectionen von Plücker, Klein, Muret mehrere Modelle von Schwarz, Wiener u. A. auch eine gewisse Fläche 3. Ordnung ('Diagonalfläche'), welche auf Clebschs Veranlassung von Weiler (damals Göttingen) hergestellt worden war, ferner einige Modelle zur Repräsentation singulärer Punkte einer Fläche 3. Ordnung (eine solche namentlich mit 4 Knoten), unter Mitwirkung von Klein konstruiert, dann ein elliptisches Paraboloid, das durch zusammen gesteckte Kartonscheiben in der Form von Halbkreisen aus seinen Kreisschnitten dargestellt war, u. A. m." Brill, "Über die Modellsammlung," 75–76. In a letter (ETH-Bibliothek, Hochschularchiv, Hs 87: 1495, https://doi.org/10.7891/e-manuscripta-85665 (accessed November 23, 2021)), dated Erlangen, April 20, 1873, Weiler informed Fiedler on this meeting, in particular on the models presented there. He told Fiedler that three of his models were presented at the meeting, one model by Neesen (Bonn), some models by Johann Eigel Sohn (Cologne [Weiler says Bonn, which seems to be false]) and by Schwarz. He mentioned also "French" models made of strings. Eigel produced models designed by Plücker and Klein; see: Rowe, "Mathematical models," 6.

[74] Schwarz was professor of mathematics at the Polytechnikum in Zurich from 1869 to 1875. So, Fiedler and Schwarz were colleagues—seemingly, on good terms: Schwarz allowed Fiedler to use a manuscript on minimal surfaces written by him as an appendix to the second edition of Salmon and Fiedler, *Analytische Geometrie des Raumes*, vol. 2, 683–90. Fiedler was also one of the godfathers of the first daughter of the Schwarz couple, Kronecker was the other (see: letter to Cremona, October 30, 1869).

I was in Zurich a year ago, I saw the thread models of F_3 and of F_n, and its associated. I would like to have these for our institute as well." ["Ich sah, als ich vor einem Jahr in Zürich war, die Fadenmodelle der F_3 und der F_n und dazugeh. $\bar{C}_2$. Ich möchte für unser Institut sehr gerne diese ebenfalls haben."][75] It is well known that Klein, in collaboration with Alexander Brill, created an important collection of models at the polytechnic in Munich.[76]

L. Brill,[77] a firm that specialized in producing and selling mathematical models later sold Weiler's sophisticated models. In 1899, L. Brill was sold to the firm Martin Schilling and moved from Darmstadt to Halle and later to Leipzig. Many of the models still existing are models sold by those firms. Fiedler himself bought models from Schilling, as is proved by letters from Schilling to Fiedler conserved in Zurich.[78] In a letter to Schilling, dated Zurich, November 25, 1901, Fiedler remarked that he himself had constructed models of the type of nr. 234[79] in Schilling's catalogue long ago. He also remarked that he had bought some other models from L. Brill's firm.[80] In his letter, Fiedler also remarks that the collection of mathematical models in Zurich was reopened recently.

A letter written by Walther Dyck to Fiedler (dated Munich, December 28, 1891) expressed the interest of wanting to organize an exposition of models in Nuremberg in 1892 in the context of the planned meeting of the *Deutsche Mathematiker-Vereinigung*.[81] He asked Fiedler to participate: "Now I know that your institute in particular contains a large number of specifically interesting models and apparatuses that are not available elsewhere."[82]

---

[75] Klein to Fiedler, October 22, 1876, ETH-Bibliothek, Hochschularchiv, Hs 87: 583, https://doi.org/10.7891/e-manuscripta-84497 (accessed November 23, 2021). F_3 is a cubic surface, F_n a surface of degree n. C means here a conic surface, 2 indicates that the class of the surface is reduced by 2 (because of the presence of knot-points). The creation of an adequate nomenclature including symbolic representations is a typical feature linked to exhibiting models; see: Hassler and Meyer, "Die Sammlung als Archiv," 10.

[76] Some information on Brill, Klein and their models can be found in Brill's unpublished autobiography "Aus meinem Leben" (conserved at the library of the Technische Universität Munich). I thank this library for providing me a copy of Brill's text.

[77] Ludwig Brill was a brother of the mathematician Alexander Brill just cited. Alexander Brill proposed to his brother the idea to sell mathematical models in order to save his firm which was in financial difficulties. The firm was the heritage of their father; in its beginnings it was a printing house.

[78] Fiedler to Schilling and Schilling to Fiedler, December 28, 1900 to January 15, 1904, ETH-Bibliothek, Hochschularchiv Hs 87: 1082a–1084. The acquisition took place in 1900 and 1901. See also the inventories Hs 1196: 30 and Hs 1736: 50.

[79] These models are string models of curves in space (Series XXI in: Martin Schilling, ed., *Catalog mathematischer Modelle für den höheren mathematischen Unterricht*, 6th ed. (Halle an der Saale: Martin Schilling, 1903), 145–151).

[80] This is confirmed by the inventories ETH-Bibliothek, Hochschularchiv, Hs 1796: 30 and Hs 1796: 50.

[81] The Deutsche Mathematiker-Vereinigung (DMV) was founded in 1890. The planned exhibition was perhaps intended to promote the newly founded association in public.

[82] Dyck to Fiedler, December 28, 1891, ETH-Bibliothek, Hochschularchiv Hs 87: 249, https://doi.org/10.7891/e-manuscripta-84098: "Nun weiß ich, daß besonders Ihr Institut eine große Reihe spezifisch interessanter und auswärts nicht vorhandener Modelle und Apparate enthält."

Dyck asked Fiedler to write a paper on his collection in order to describe it for the public. The meeting at Nuremberg was canceled because of an epidemic there; the exposition was displaced and presented in Munich (September 1 to September 20, 1893).[83] In a second letter, Dyck asked Fiedler to write a paper on his collection for the planned exposition in Chicago—once again in vain. In his letter, Dyck praised Fiedler's importance for the development of the collection in Zurich by his "excellent teaching activities."

In sum, we get the impression that Fiedler kept a certain distance to the mainstream movement in Germany promoting models during the 1890s. This might have been due to his age (he was in his sixties) or maybe it was the result of a certain distrust that he felt against the activities of university professors taking up ideas from the world of polytechnics. In a letter by Klein, dated Munich, July 30, 1880, speaking of his ideas to introduce courses on descriptive geometry at the university (of Leipzig), Fiedler noted (August 13, 1880):[84]

> Method of execution—not exclusively graphical-technical—not modern geometry adrift from the Mother of Geometry—only body, only spirit! He won't fall short, but let the effort be rewarded. I didn't expect the decision [to take up geometry again] after his recent papers.[85]

Fiedler was deep-rooted in the tradition of polytechnic education. He was not—like the great majority of the other mathematicians at the polytechnics, including his own colleagues in Zurich, Klein being a typical example—a mathematician trained at a university who took up a job at a polytechnic for a certain time exporting ideas from there to here.[86]

In conclusion, I want to emphasize that the correspondence was an important instrument for Fiedler to get and to provide information (of all kinds). Partly, this was due to his geographic position, but partly it was also a consequence of his style. In 1905, looking back at his career, Fiedler wrote[87]: "Durch die briefliche Verbindung mit vielen der Besten unter den Mathematikern der Zeit, wie

---

[83] Dyck published a long paper describing the exposition (Walther Dyck, "Einleitender Bericht über die mathematische Ausstellung in München," *Jahresbericht der Deutschen Mathematiker-Vereinigung* 3 (1894): 39–56) and a catalogue of the exposition (Walther Dyck, ed., *Katalog mathematischer und mathematisch-physikalischer Modelle, Apparate und Instrumente* (Munich: Wolf & Sohn, 1892)). Fiedler is not mentioned in the catalogue (but Schwarz). In describing the history of mathematical models, Dyck cites long passages from Brill, "Über die Modellsammlung."

[84] Klein to Fiedler, July 30, 1880, ETH-Bibliothek, Hochschularchiv Hs 87: 584, https://doi.org/10.7891/e-manuscripta-84531 (accessed November 23, 2021).

[85] "Art u. Weise der Ausf.—nicht ausschl. graph.-techn.—nicht mod. Geom. losgelöst von der Mutter d. G.—bloß Körper, bloß Geist! Er wird's nicht verfehlen, aber die Mühe möge ihren Lohn finden. Erwartet habe ich s. Entschl. nicht nach seinen neueren Abh."

[86] This aspect is discussed in detail in the dissertation by Nadine Benstein (Wuppertal): *Zwischen Zeichenkunst und Mathematik: Die darstellende Geometrie und ihre Lehrer an den Technischen Hochschulen und deren Vorgängern in ausgewählten deutschen Ländern im 19. Jahrhundert* (2019; available at UB Wuppertal: http://elpub.bib.uni-wuppertal.de/edocs/dokumente/fbc/mathematik/diss2019/benstein).

[87] Fiedler, "Meine Mitarbeit," 503.

Möbius, Plücker, Hesse, Aronhold, Clebsch, Kronecker, Cayley, Salmon, Brioschi, Beltrami, Cremona, um nur bereits Abgerufene zu nennen, hat mich aber meine einsame Arbeit immer beglückt." Fiedler spoke on the topic of isolation on several occasions, e.g. in his correspondence.[88]

Models in Fiedler's Teaching and Publishing

Of course, there are other places in Fiedler's large correspondence that hints at models, and, of course, we do not know all this correspondence in detail. Nevertheless, I stop here in order to look at Fiedler's published books and papers, including some materials found in the ETH archive.

There are three types of publications in Fiedler's large body of work: first, there are the books by Salmon and Fiedler, Fiedler's greatest success. Whereas Fiedler reworked the books by Salmon with additions, changes, and so on, they remained the books of their author. We find only few personal notes by Fiedler in them. Second, there are Fiedler's own books. Among them, only his treatise on descriptive geometry had a certain success with different editions. In it, we find a lot of personal remarks underlining the fact that for Fiedler research and teaching were strongly linked. Thirdly, there are his papers. During his life in Zurich, Fiedler published many papers in the *Vierteljahrsschrift der Naturforschenden Gesellschaft Zürich*. He was an active member of the society, serving even as its president for a certain time. These papers often have the character of reports on work in progress. In one of these papers, models play a notable role: "Herr Prof. Fiedler hält einen Vortrag über Geometrisches mit Vorweisungen" (Prof. Fiedler gives a talk on geometric themes with demonstrations).[89] Fiedler brought two models with him and commented on them. The two models illustrated how two cones of second degree penetrate each other—in particular, the curves of intersection.[90] Each has a different size—one large, one small—and they were made of different materials—strings and rods, or rods only. The discussion of the nature of the curves of penetration is very detailed; in particular, it becomes clear how the research on the abstract nature of the curve and the construction of the concrete object interacted. Fiedler does not mention the name of the person who constructed the models; perhaps he did it himself.

---

[88] See the letters to Cremona, April 25, 1864, August 5, 1866, June 22, 1867, in *Correspondence of Luigi Cremona (1830–1903)*, ed. Giorgio Israel, 2 vols. (Turnhout: Brepolis, 2017), vol. I, 642–43, 647–49, 653–55).

[89] Wilhelm Fiedler, "Vortrag über Geometrisches mit Vorweisungen," *Vierteljahresschrift der Naturforschenden Gesellschaft Zürich* 28, no. 2–3 (1883): 289–90. It was the secretary of the society, Robert Billwiller, who wrote the published report—not Fiedler himself. In most cases, the *Vierteljahrsschrift* published papers written by Fiedler himself. The talk was delivered July 23, 1883.

[90] A typical problem of descriptive geometry, going back at least to the days of Monge, was to construct such curves of intersection. A model of that type was announced in the report for the Bundesrath in 1876.

There was a second report on a talk delivered by Fiedler (February 19, 1883): "Herr Prof. Fiedler spricht unter Vorweisung von Modellen über eine Singularität algebraischer Oberflächen" (Prof Fiedler speaks about a singularity of algebraic surfaces and shows some models).[91] Also, in this talk, Fiedler presented some models. First, he showed a model of Clebsch's surface, then he referred to a *Diplomarbeit* (master's thesis) written 1875 in the VI. department[92] and to some investigations of a former pupil of his at the school in Chemnitz, Eckart, now professor at Chemnitz. Here again, we see that the construction of models was closely linked to Fiedler's research and that of his students. Fiedler presented also a second model of a surface—the first of a surface of third degree, the second of fourth, both made by strings—referring once again to a *Diplomarbeit*. It was a Hungarian student, Béla Tötössy, who presented this thesis in 1880. Fiedler was very content with it[93] and, due to his contacts with Klein, he arranged its publication in the *Mathematische Annalen*. Tötössy had constructed a model of a ruled surface of fourth degree made out of strings; it was given to the collection of the Polytechnikum. In his talk, Fiedler explained this model in some detail. At the end of his talk, he announced some results presented by Tötössy at the "mathematische Seminar des Polytechnikums" (June 14, 1882).[94] By all these remarks, one gets the impression that Fiedler worked in a busy and productive situation at the Polytechnikum. It seems that the years around 1882 were a sort of peak in this respect.

In 1891, Fiedler presented a summary of his studies on curves of penetration to the *Naturforschende Gesellschaft*[95]—without any hint at models. He provides many, many details and, in sum, one gets the impression that Fiedler's mathematics was rather old fashioned in the 1890s. Contemporary research in mathematics with its tendency for abstraction, axiomatics, etc.[96] no longer had such an interest

[91] Wilhelm Fiedler, "Vortrag über eine Singularität algebraischer Oberflächen unter Vorweisung von Modellen," *Vierteljahrsschrift der Naturforschenden Gesellschaft Zürich* 28, no. 2–3 (1883): 290–93.

[92] Because the information provided by Fiedler is meager, it is not absolutely clear to which thesis he is referring here. There is a list of candidates in Fiedler's papers (ETH-Bibliothek, Hochschularchiv, Hs 87a: 71); in it we find two entries which can fit here: Johannes Keller "Über reciprokc Fundamentalgebilde der zweiten und dritten Stufe" and E. Mosiman "Über die Eigenschaften von Flächen dritter Ordnung." Keller was Fiedler's assistant from 1875 to 1887. He received his PhD in 1879 from the University of Zurich (in collaboration with Fiedler): "Die einander doppelt conjugirten Elemente in allgemeinen reciproken Systemen"; later he passed his habilitation at the Polytechnikum. As mentioned by Fiedler on another occasion, Keller produced a model of a *Reliefperspektive*.

[93] This is the only model cited by Fiedler in his correspondence with Cremona; see: letter to Cremona dated February 12, 1882.

[94] Wilhelm Fiedler, "Vortrag," 292. Tötössy later became a professor at the polytechnic at Budapest.

[95] Wilhelm Fiedler, "Ueber die Durchdringungen projectivischer Kegel," *Vierteljahrsschrift der Naturforschenden Gesellschaft Zürich* 36 (1891): 87–113.

[96] This trend was often related to the Berlin School (Kronecker, Weierstrass), Hilbert's *Grundlagen der Geometrie* (1899) later reinforced it.

in details; it was increasingly oriented towards general aspects. In other words, Fiedler had lost the connection to the new trends.

Fiedler was not only a researcher and an engaged teacher, he also contributed to the didactical discussion (to use a modern term, Fiedler himself preferred the term "pedagogical"). In 1877, he published a paper "Zur Reform des geometrischen Unterrichts."[97] This was a plea to integrate descriptive and projective geometry into the teaching of geometry at school in order to provide a clear structure ("system"). Fiedler discusses the contents of the teaching of geometry, not its methods. But there is one place where we get a glimpse of methods too—that is, the use of models. Here, he speaks of drawings made on the base of models made of rods ("Stabmodell")[98]; he concludes:

> Actual descriptive geometry is not needed to deal with such problems; they form a simple connection between the practice of drawing after rod models and the fundamental definitions of geometry, and automatically lead to the correction of failings in perception, for instance, and to the realization of the indispensability of such corrections in all cases in which it is a matter of mathematically defined forms.[99]

It should be mentioned that there are some objects in Fiedler's estate, preserved at the ETH archive, which are directly linked to his teaching: there are photographs of star polyhedra (see Fig. 8) and of Clebsch's surface and drawings of little rods which perhaps served to build up polyhedra.[100] There are also many notes, written by Fiedler himself or taken by his students, reproducing his lectures or parts of them, but there are no hints as to the use of models (Fig. 9).

Conclusions

At the end of this article, I want to conclude by making some remarks. First, it seems important to note that the use of mathematical models was not new in the second half of the nineteenth century. At least in the context of technical education, in the drawing schools, the Gewerbeschulen, and so on, models were known and used for a long time; models were an integral and important aspect of polytechnic tradition. For Fiedler, coming from that tradition, it was quite natural to use models in his teaching. And, of course, the Polytechnikum in Zurich with all

---

[97] Wilhelm Fiedler, "Zur Reform des geometrischen Unterrichts," *Vierteljahrsschrift der Naturforschenden Gesellschaft Zürich* 22 (1877): 82–97. This paper was also translated into Italian. Three letters written by Fiedler to the editor, Torelli, were included in the Italian edition. In the first of these letters, Fiedler speaks of teaching drawing with the help of models.

[98] Fiedler, "Zur Reform," 88.

[99] "Eigentliche descripitive Geometrie ist zur Ausführung solcher Probleme nicht erforderlich; sie bilden eine einfache Verbindung der Uebung im Zeichnen nach Stabmodellen mit den fundamentalen Definitionen der Geometrie und führen sofort zur Correctur des etwa der Wahrnehmung nicht treu genug Abgesehenen und zur Einsicht von der Unentbehrlichkeit einer solchen Correctur in allen Fällen, wo es sich um mathematisch bestimmte Formen handelt." Fiedler, "Zur Reform," 89.

[100] ETH-Bibliothek, Hochschularchiv, Hs 87a: 29.

Fig. 8 A cylinder penetrating a cone. Model produced by J. Schröder, 1869 (Collection of scientific instruments and teaching aids, ETH-Bibliothek Zurich). © ETH Bibliothek Zurich, CC BY-SA 4.0 (https://creativecommons.org/licenses/by-sa/4.0/)

Fig. 9 In Wilhelm Fiedler's estate there are several photographs of this subject: a twin tetrahedron. Of course, those photographs are not in color and they are of poor quality. Thus, we reproduce here a recent picture of an object from the ETH-collection (Collection of scientific instruments and teaching aids, ETH-Bibliothek Zurich). © ETH Bibliothek Zurich, CC BY-SA 4.0 (https://creativec ommons.org/licenses/by-sa/ 4.0/)

its collections and *Werkstätten* was favorable to the use of models of all kinds; skilled experts were present in the *Werkstätten*. In particular, students were familiar with the production and handling of models, they knew how to understand things through them. As we have seen, Fiedler produced and used models rather early in his career. Afterwards, the construction of models was an important part of his teaching and of his research. In his eyes, the construction of models offered a possibility to control and correct the results that were deduced in a theoretical way; the integration of students and assistants into the process of research became possible by models. And, of course, they were a means to train spatial intuition.

Fig. 10 A 'home-made' model of a hyperboloid with one sheet (Collection of scientific instruments and teaching aids, ETH-Bibliothek Zurich). The author and the date of construction are unknown. © ETH-Bibliothek Zurich, CC BY-SA 4.0 (https://creativecommons.org/licenses/by-sa/4.0/)

During his first years, Fiedler took part in a network of researchers interested in the construction of models. It must be added, however, that later on, he did not participate in the public activities of this network—such as the expositions in Göttingen and Munich; nor did he engage himself in the 'mass production' (to follow David E. Rowe) of models. Fiedler felt more or less isolated, also at his Polytechnikum. The only colleague who worked on mathematical models there was Schwarz, who left Zurich in 1875.[101] Fiedler was interested mainly in 'home-made' models;[102] apparently, he did not want to see them produced by a firm.[103]

He installed an important collection of mathematical models at the Polytechnikum in Zurich, also by buying models from firms like L. Brill, Martin Schilling,

---

[101] After arriving at Göttingen, Schwarz engaged himself in building up a collection of models there; see: Sattelmacher, "Zwischen Ästhetisierung und Historisierung," 131. Models in the possession of Schwarz are also mentioned in the inventory ETH-Bibliothek, Hochschularchiv, Hs 1796: 50.

[102] See Figure 10 for an example.

[103] In Dyck's catalogue (1892), we find a firm in Zurich named Coradi, a *Mathematisch-mechanische Werkstatt*, which produced mathematical instruments like integraphs (see: Dyck, ed., *Catalog*, 197–201). Hence, it would not have been difficult for Fiedler to find a partner for the production of models in his own town—if he would have wanted to. Coradi presented its instruments at the International Congress in Heidelberg (1904). The exhibition of models and instruments at that congress was organized and presented by Martin Disteli (then professor in Strasbourg i.e., before Karlsruhe, later he became professor in Dresden and, once again, in Karlsruhe and finally at the University of Zurich), a former student and assistant of Fiedler.

and J. Schröder. The models were presented to a restricted public[104]; thus, one may state that they served also as publicity for his chair and his field, that is, descriptive geometry. The way that the models were presented also allowed for their use (at least in a passive way—by looking at them) and to understand things through them in this way.[105] The presentation of models was a common practice in other collections of the Polytechnikum; the collections were places of learning. We may suppose that the students of the Polytechnikum were accustomed to learn in this way ("lehrerlos"[106]). Fiedler used his models also to address a larger public including non-mathematicians by presenting them to the *Naturforschende Gesellschaft*.

In sum, we may state that the triad research–teaching–representation was realized by Fiedler's way of working with models—for him, the most important aspect was teaching.[107]

---

[104] That means students, professors and external experts. Other collections at the Polytechnikum, such as the famous entomological collection, were open to a general public. Therefore, they had a warden.

[105] Fiedler considered himself as an autodidact. See his letter to Klein, Hottingen–Zürich XII 1885, Staats- und Landesbibliothek Göttingen, Cod. Ms. F. Klein 19. Hence, learning on his own was familiar to him.

[106] Hassler and Meyer, "Die Sammlung als Archiv," 12. Of course, some indications concerning the models could be made by the teachers during their courses.

[107] Note added in proofs: some letters to and by Fiedler are now accessible in Sara Confalonieri, Peter-Maximilian Schmidt, and Klaus Volkert, eds., *Der Briefwechsel von Wilhelm Fiedler mit Alfred Clebsch, Felix Klein und italienischen Mathematikern*, series *Siegener Beiträge zur Geschichte und Philosophie der Mathematik*, vol. 12 Siegen: universi, 2019.

Models from the Nineteenth Century Used for Visualizing Optical Phenomena and Line Geometry

David E. Rowe

Introduction

In 1891, one year after the founding of the Deutsche Mathematiker-Vereinigung (DMV), the Munich mathematician Walther Dyck faced a daunting challenge. Dyck was eager to play an active role in DMV affairs, just as he had since 1884 as professor at Munich's Technische Hochschule.[1] Thus, he agreed, at first happily, to organize an exhibition of mathematical models and instruments for the forthcoming annual meeting of the DMV to be held in Nuremberg. Dyck's plan was exceptionally ambitious, though he encountered many problems along the way. He hoped to obtain support for a truly international exhibition, drawing on work by model-makers in France, Russia, Great Britain, Switzerland, and of course throughout Germany. Most of these countries, however, were only modestly represented in the end, with the notable exception of the British. Dyck had to arrange for finding adequate space in Nuremberg for the exhibition, paying for the transportation costs, and no doubt most time consuming of all, he had to edit single handedly the extensive exhibition catalog in time for the opening. At the last moment, however, a cholera epidemic broke out in Germany that threatened to nullify all of Dyck's plans. When he learned that the DMV meeting had been canceled, he immediately proposed that the next annual gathering take place in 1893 in Munich. There he had both helpful hands as well as ready access to the

[1] On his career, see: Ulf Hashagen, *Walther von Dyck (1856–1934). Mathematik, Technik und Wissenschaftsorganisation an der TH München* (Stuttgart: Steiner, 2003).

D. E. Rowe (✉)
Institute of Mathematics, Johannes Gutenberg-University Mainz, Staudingerweg 9, 5. OG, 55128 Mainz, Germany
e-mail: rowe@mathematik.uni-mainz.de

M. Friedman and K. Krauthausen (eds.), *Model and Mathematics: From the 19th to the 21st Century*, Trends in the History of Science,
https://doi.org/10.1007/978-3-030-97833-4_4

necessary infrastructure. Moreover, he had already completed the most onerous part of his task: the exhibition catalog.[2]

This paper is concerned with events that long preceded Dyck's exhibition. It does, however, deal with some of the models found in one fairly small section of his catalog, entitled 'line geometry.'[3] Under this heading, he subsumed five topics: caustic curves, ray systems, caustic surfaces, evolute surfaces, and line complexes. All except the fourth topic will be discussed below, although with attention to only a few of the models in Dyck's exhibition. My main focus in the final sections of this paper will be on models of so-called complex surfaces. These are special fourth-degree surfaces that Julius Plücker introduced in the 1860s for visualizing the local structure of a quadratic line complex. Plücker's complex surfaces turned out to be closely related to Kummer surfaces,[4] and both of these types of quartics are examples of caustic surfaces, which arise in geometrical optics. Indeed, Kummer surfaces represent a natural generalization of the wave surface, first introduced by Augustin Fresnel (1788–1827) to explain double refraction in biaxial crystals.

Plücker designed his models to illustrate the various shapes of very special quartic surfaces with a double line (Fig. 1 as one example). These models, made using a heavy metal such as zinc or lead, were displayed for a brief time both in Germany and abroad. In 1866, Plücker took some with him to Nottingham, where he spoke at a conference about these complex surfaces. Arthur Cayley (1821–1895), Thomas Archer Hirst (1830–1892), Olaus Henrici (1840–1918), and other leading British mathematicians quickly took an interest in them, as they were certainly among the more exotic geometrical objects of their day.[5] At Hirst's request, Plücker later sent a set of boxwood models to the London Mathematical Society (see Fig. 2). These have recently been put on display at De Morgan House, the headquarters of the London Mathematical Society,[6] though without any information about the mathematics behind them. More extensive collections of Plücker models can be found in Göttingen, Munich, and Karlsruhe. Those in Göttingen are on display in the collection of models at the Mathematics Institute, which can be viewed online, whereas those at technical universities in Munich and Karlsruhe are held in storage. These Plücker models are generally recognizable not only due to their exotic shapes but also because they are far heavier than the more familiar geometric models made in plaster.

Until very recently, Plücker's models had languished in dusty display cases or in cob-webbed cellar rooms. One has to look quite diligently to find any mention of

---

[2] Walther Dyck, ed., *Katalog Mathematischer und Mathematisch-Physikalischer Modelle, Apparate und Instrumente* (München: Wolf & Sohn, 1892).

[3] "Modelle zur Liniengeometrie," see: Dyck, ed., *Katalog*, 279–85.

[4] Felix Klein, "Ueber die Plückersche Komplexfläche," *Mathematische Annalen* 7 (1874): 208–11, and Ronald W. H. T. Hudson, *Kummer's Quartic Surface* (Cambridge: Cambridge University Press, 1905).

[5] Arthur Cayley, "On Plücker's Models of certain Quartic Surfaces," *Proceedings of the London Mathematical Society* 3 (1871): 281–85.

[6] See: https://www.lms.ac.uk/archive/plucker-collection (accessed April 18, 2022).

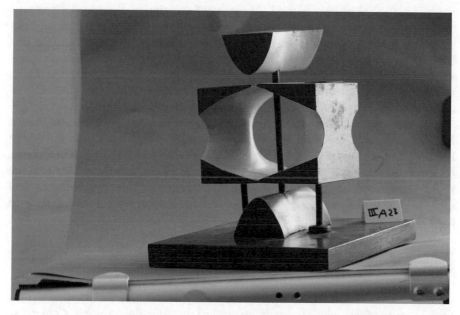

Fig. 1 Julius A Plücker model from the collection at Technical University Munich. Photo: courtesy of Gerd Fischer, all rights reserved

Fig. 2 Julius Plücker's
model of a complex surface
with eight real nodes. ©
Science Museum Group, all
rights reserved

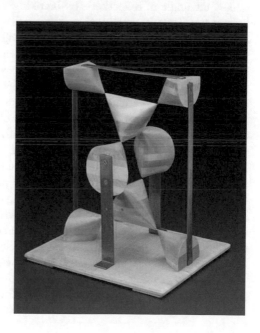

them in historical sources as well. A very brief description, though, can be found in the final entry on line geometry in Dyck's catalog. This was written by Felix Klein, who during the mid 1860s worked closely with Plücker at the time he first designed twenty-seven models of complex surfaces. Klein's work on line geometry showed how the Plücker quartics can be treated as degenerate Kummer surfaces, which afterward were regarded as the surfaces of principle interest for the theory of quadratic line complexes. Since Klein promoted models and visualization of geometric structures nearly all his life, many mathematicians must have known about Plücker surfaces in the 1890s. Yet, despite the attention his models attracted during the final decades of the nineteenth century, particularly among British geometers, by the early twentieth century they had nearly disappeared without a trace. What follows can be understood at least in part as an attempt to bring these long lost objects back to life. We begin, though, with some background setting relating to the longstanding interplay between optical and mathematical knowledge.

Optics Stimulating Mathematics Simulating Optics

Geometrical optics, a field of inquiry nearly as old as mathematics itself, was often closely intertwined with purely mathematical investigations. In his *Catoptrica*, Heron of Alexandria derived the law of reflection—that the angle of incidence equals the angle of reflection—by invoking a minimal principle, namely, that light beams travel along paths that minimize distance. Heron's main interest, though, seems to have been directed toward achieving a variety of surprising optical effects by means of carefully arranged mirrors. He presumably never reflected on the phenomenon of refraction, however, whereas Claudius Ptolemy performed experiments to determine the angles of refraction for light passing from air to other media, such as water and glass. In the tenth century, the Baghdad mathematician Ibn Sahl wrote a commentary on Ptolemy's *Optics* as well as an independent study in which he obtained a law of refraction equivalent to Snell's law.[7] The latter result was found experimentally in Leyden at the beginning of the seventeenth century. Fermat then formulated the principle of least time to account for this refraction law, assuming that the speed of light differs in passing from one medium to another. Leibniz then re-demonstrated Fermat's argument in his famous paper of 1684, in which he introduced the rules for his differential calculus. During the decade that followed, Johann Bernoulli made ingenious use of Fermat's principle to solve the famous brachistochrone problem by treating the earth's gravitational field as a medium with a continuously varying 'refraction index.'

Already in antiquity, it was known that the conic sections had numerous optical properties. Thus, rays emanating from one focal point of an ellipse will be reflected so that they gather in the second focal point. Light rays can also be directed so

---

[7] Roshdi Rashed, "A Pioneer in Anaclastics: Ibn Sahl on Burning Mirrors and Lenses," *Isis* 81 (1990): 464–91.

as to gather in a focal point by means of a paraboloid, which reflects light rays parallel to its axis into the focus of the surface. The partially extant work of Diocles, *On burning mirrors*, shows this genre of interest. This work exerted a strong influence on optics in the Arabic world, particularly on the most important authority on geometrical optics during the medieval period, Ibn al-Haytham. Fermat's contemporary and rival, Descartes undertook numerous investigations of optical phenomena. His study of the focal properties of lenses led to his discovery of the special quartic curves known today as Cartesian ovals.

In general, however, reflected and refracted rays will form caustic curves rather than converging to a point. Thus, a continuous family of plane light rays will be reflected by a smooth curve so as to envelope a second curve, a catacaustic; if refracted, the corresponding envelope is called a diacaustic. Similarly, one speaks of caustic surfaces formed by envelopes of rays that are reflected or refracted by a smooth surface. A glance at the table of contents in L'Hôpital's famous *Analyse des infiniment petits pour l'intelligence des lignes courbes* from 1696, the first textbook for the differential calculus, shows that the determination of caustics was a major topic of interest already in the seventeenth century. By the nineteenth century, these optical phenomena were being studied as systems of rays, or congruences of lines, in 3-dimensional space. Ray systems are 2-parameter families of lines that undergo reflections and refraction. In 1808 Etienne Malus introduced the notion of normal congruences, which are families of lines orthogonal to a surface.[8] The theorem of Malus-Dupin then states that a normal congruence that undergoes any number of reflections and refractions on given surfaces will remain a normal congruence afterward.

Constructing Fresnel's Wave Surface

The Fresnel wave surface was an object of considerable interest in the 1820s and long afterward; Plücker studied it in some detail in "Discussion de la forme générale des ondes lumineuses."[9] Fresnel derived it from a quartic equation for a wave front of light passing through biaxial crystals, in which the light rays undergo double refraction. This turned out to be a two-leaved surface, though its deeper geometrical properties eluded him. During the 1830s, the Fresnel surface attracted the attention of two natural philosophers at Trinity College Dublin, James MacCullagh (1809–1847) and W.R. Hamilton (1805–1865), who revealed several new properties. MacCullagh constructed the wave surface directly from Fresnel's index ellipsoid, which represents the strength and orientation of refraction in the

---

[8] Eisso Atzema, "The Structure of Systems of Lines in 19th Century Geometrical Optics: Malus' Theorem and the Description of the Infinitely Thin Pencil," (PhD diss., Utrecht University, 1993), 19–37.

[9] Julius Plücker, "Discussion de la forme générale des ondes lumineuses," *Journal für die reine und angewandte Mathematik* 19 (1938): 1–44.

Fig. 3 This picture suggests a simple continuity argument that proves why exactly two planes through the center of an ellipsoid will cut the surface in circles. From David Hilbert and Stephan Cohn-Vossen, *Anschauliche Geometrie* (Berlin: Springer, 1932), 16 © Springer, all rights reserved

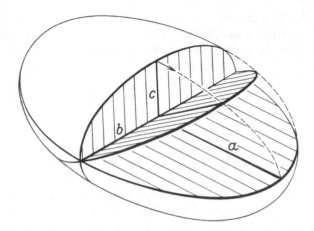

crystal.[10] This will be a surface of revolution in the case of uniaxial crystals, but for biaxial crystals the lengths of the ellipsoid's three axes are unequal. From the classic text *Traité des surfaces du second degré* by Monge and Hachette (1st ed., 1802), geometers knew the following two fundamental properties of central quadrics: (1) Each plane section of the surface yields a conic and the conic sections that lie in parallel planes are all similar (i.e., their shape is identical, so they only differ in size). (2) There are two distinguished directions in which the conics cut by parallel planes will be circles. A simple continuity argument shows why property (2) holds (see Fig. 3).

These two distinguished directions correspond to the four singular points of the wave surface where its two leaves fall together, as illustrated in Fig. 4. Here the space between the leaves has been filled in, so this model is actually a quarter section of a solid figure whose outer and inner shells belong to the surface. Using laser-in-glass technology, Oliver Labs has produced models of the Fresnel surface that provide a far clearer picture of its geometrical structure. An idea of what such a model looks like can be gathered from Fig. 5. This model was designed so as to enhance the regions surrounding the singular points. The transparent glass makes it far easier to visualize the conical structure formed in passing from a singular point to the outer surface.

MacCullagh's colleague in Dublin, the mathematician-astronomer W.R. Hamilton, was the first to recognize the significance of this geometrical structure for optics. In "Third supplement to an essay on the theory of systems of rays,"[11] he predicted that light rays propagating in two special directions should split into a

---

[10] James MacCullagh, "On the Double Refraction of Light in a Crystallized Medium, According to the Principles of Fresnel," *Transactions of the Royal Irish Academy* 16, no. 2 (1830): 65–78.

[11] William Rowan Hamilton, "Third supplement to an essay on the theory of systems of rays," *Transactions of the Royal Irish Academy of Sciences* 17, no. 1 (1837): 1–144. The publication date is misleading, as Hamilton wrote this "Third Supplement" some years earlier; he discovered conical refraction already in 1832.

Fig. 4 Model of the interior of a Fresnel wave surface showing two singular points. Model made by Otto Bökler in 1880: Fresnel's wave surface, a quartic surface discovered by Augustin-Jean Fresnel in 1822. Material: Plaster (gypsum). (Collection of Technische Universität Dresden). Photo: Lutz Liebert, TU Dresden, 2003, all rights reserved

Fig. 5 Fresnel's Wave Surface as a 3D-laser-in-glass object, produced by MO Labs Dr. Oliver Labs. This recent technique has the advantage over historical models of exposing the surface alone, so that its structure, including the four singular points, can be visualized quite adequately. This semi-transparency also makes it possible to see both the outer and inner shells simultaneously; older historical models required two parts in order to be able to look inside. Photo: © Oliver Labs, all rights reserved

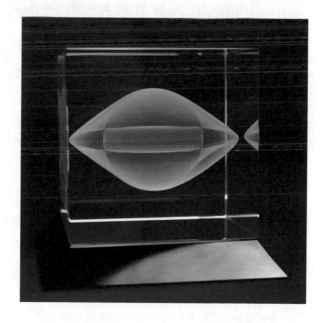

whole circle of light, a phenomenon that came to be known as conical refraction. Hamilton discussed this theoretical result in November 1832 with a local
expert on experimental optics, Humphrey Lloyd (1800–1881), who confirmed this
prediction less than a month later.[12] This shows how a purely mathematical feature of the Fresnel surface led to the discovery of a new physical phenomenon.
Hamilton's successful prediction of conical refraction was celebrated by several
contemporary investigators. Plücker later wrote: "No experiment of physics ever
made such an impression on me [...] it was a thing unheard of and completely
without analogy."[13]

MacCullagh, Hamilton, and George Salmon all helped to make the Fresnel
wave surface the most prominent example of a quartic surface in the mathematical literature. By introducing complex numbers, Salmon showed that the Fresnel
surface actually has sixteen singular points, only four of which are real.[14] It thus
belongs to the special class of quartics that would come into prominence with
Ernst Eduard Kummer's (1810–1893) work in the 1860s. Salmon showed further
that the Fresnel surface was of the fourth order and class with sixteen double
points and sixteen double planes. These turned out to be characteristic properties
of the special quartics that Kummer first described in "Ueber die Flächen vierten
Grades mit sechzehn singulären Punkten."

Constructing Infinitely Thin Pencils of Rays

Already in "Theory of systems of rays,"[15] his first study of ray systems, Hamilton
dealt with infinitely thin pencils of lines, a topic that would recur throughout much
of the nineteenth century.[16] He returned to this topic in his "Third Supplement,"[17]
where he described three types of structures that can occur when considering rays
of lines infinitely close to a given line. In his earlier work, Hamilton focused his
attention entirely on so-called normal systems of rays, those that satisfy the theorem of Malus-Dupin. For such systems, i.e. rays that undergo simple reflection
and refraction, only the first type of structure can occur. On a given line ℓ, the thin
pencil consists of an infinitesimal subset within a given congruence of lines that
fill all of space. This pencil determines three special points M, F_1, F_2 on ℓ, the
midpoint and two foci, such that $\overline{MF_1} = \overline{MF_2}$. In addition, there are two focal
planes, Π_1 and Π_2, containing ℓ, which determine two infinitesimal lines ε_1, ε_2

---

[12] See: Thomas L. Hankins, *Sir William Rowan Hamilton* (Baltimore: Johns Hopkins University Press, 1980), 88–95.

[13] Cited from Hankins, *Sir William Rowan Hamilton*, 95.

[14] George Salmon, "On the degree of a surface reciprocal to a given one," *The Cambridge and Dublin Mathematical Journal* 2 (1847): 65–73.

[15] William Rowan Hamilton, "Theory of systems of rays," *Transactions of the Royal Irish Academy of Sciences* 15 (1828): 69–174.

[16] Atzema, "The Structure."

[17] Hamilton, "Third supplement."

lying, respectively, in Π_1, Π_2, perpendicular to ℓ, and passing through F_1 and F_2. Hamilton then deduced that the infinitely thin pencil centered around ℓ can be constructed as the set of lines that intersect both ε_1 and ε_2. He noted, further, that in the case of a normal congruence the two focal planes will always be perpendicular to one another. This corresponds to the first type of pencil, whereas for the second type the planes Π_1 and Π_2 will no longer be perpendicular. Hamilton's third type of pencil no longer has real foci F_1, F_2; instead these are conjugate imaginary points.

Few of those who afterward studied infinitely thin pencils of lines took notice of Hamilton's general theory, as pointed out in Kummer's "Allgemeine Theorie der gradlinigen Strahlensysteme,"[18] which seems to have been the first study to deal with it. One of Kummer's main interests concerned the angle between the two focal planes Π_1 and Π_2, which would vary within the congruence. He was also interested to consider how the second and third types of pencils arose in uniaxial and biaxial crystals, where the wave fronts form, respectively, ellipsoids of revolution and Fresnel surfaces. Only in the latter case does Hamilton's third type of pencil come into play. In conjunction with this study, Kummer produced three string models for visualizing the structure of each type, describing these in "Über drei aus Fäden verfertigte Modelle der allgemeinen, unendlich dünnen, gradlinigen Strahlenbündel."[19] The models are meant to show the surface of lines that bounds the infinitely thin pencil. Kummer constructed this surface by taking a small circle Ω in the plane passing through M perpendicular to ℓ. The lines that meet ε_1, ε_2, and Ω then form a fourth-order surface that bounds the pencil. These models served as the prototype for others that were produced in Göttingen by Wilhelm Apel, whose father had earlier founded a workshop for building scientific instruments in Göttingen (see Fig. 6).

Apel's three models were among the many on display in Munich in 1893. In the catalog, the Munich geometer Sebastian Finsterwalder briefly explained their construction.[20] Kummer's motivation[21] was at least partly physical. He formulated a general theorem, valid for any homogeneous transparent medium, that generalized what he had found for the Fresnel wave surface. In the latter case, the two focal planes are correlated and cut out reciprocal curves which are tiny conics, familiar in differential geometry as the Dupin indicatrix. Kummer asserted that the same type of correlation will occur in general. He also claimed that Hamilton's three types of pencils are the only possible cases, excepting the special types with a conical or cylindrical structure, arising when the lines pass through a point $P \in \ell$, which will be a cylinder if $\overline{PM} = \infty$.

---

[18] Ernst Eduard Kummer, "Allgemeine Theorie der gradlinigen Strahlensysteme," *Journal für die reine und angewandte Mathematik* 57 (1860): 189–230.

[19] Ernst Eduard Kummer, "Über drei aus Fäden verfertigte Modelle der allgemeinen, unendlich dünnen, gradlinigen Strahlenbündel," *Monatsberichte der Königlich Preußischen Akademie der Wissenschaften zu Berlin*, 30. Juli 1860, 469–74.

[20] Dyck, ed., *Katalog*, 280.

[21] As described in Kummer, "Über drei aus Fäden verfertigte Modelle."

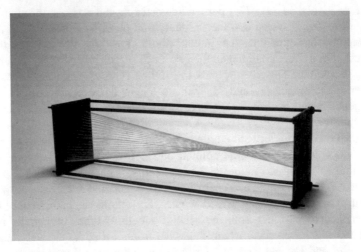

Fig. 6 Model by Wilhelm Apel of Hamilton's second type of infinitely thin pencil of lines (Göttingen Collection of Mathematical Models and Instruments, model 281). © Mathematical Institute, Georg-August-University Göttingen, all rights reserved

Kummer Surfaces

Since the Fresnel wave surface arose from a physical problem—to account for double refraction in biaxial crystals—its metrical properties guided the initial investigations. Geometers had, however, long recognized that algebraic surfaces can be studied from a purely projective standpoint. This gradually led to the recognition that the Fresnel surface belonged to a large class of quartics, culminating with Kummer's fundamental paper "Ueber die Flächen vierten Grades mit sechzehn singulären Punkten." In the 1840s, Arthur Cayley took up the study of special quartic surfaces he called tetrahedroids.[22] These have the property that the four planes of a tetrahedron meet the surface in pairs of conics that pass through four singularities of the surface. The Fresnel surface corresponds to the special case where, in each of the four planes, one of the two conics is a circle. These pairs of curves intersect in four points, which are real in only one of the four planes. The other twelve singular points are imaginary, so altogether there are sixteen. In the 1860s, Kummer used elaborate, yet conventional algebraic machinery to explore the properties of quartic surfaces with the maximum possible number

---

[22] Hudson, *Kummer's Quartic Surface*, 89–92.

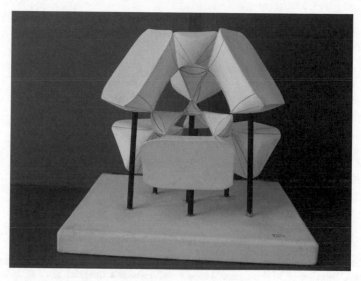

Fig. 7 Karl Rohn's model of a Kummer surface with sixteen real nodes (Göttingen Collection of Mathematical Models and Instruments, model 124). © Mathematical Institute, Georg-August-University Göttingen, all rights reserved. See also: Karl Rohn, *Drei Modelle der Kummer'schen Fläche* (Darmstadt: Brill, 1877)

of singularities, namely sixteen.[23] At the same time, he set about producing models of various types of quartics, though, unlike cubics, quartic surfaces proved too plentiful and complicated to succumb to a general classification.[24]

The sixteen singular points of a Kummer quartic form an interesting and much studied spatial configuration. These points lie in groups of six in sixteen singular planes, or tropes (see Fig. 7). Each of these sixteen tropes is a tangent plane to the surface that touches it along a conic section which contains six of the sixteen singularities. These sets of six points can thus be seen to lie in a special position, since only five coplanar points in general position determine a conic. In fact, the singular points and planes of a Kummer surface form a symmetric (16,6) configuration, so six singular planes also pass through each of the singular points of the surface. Since all these singularities can be real, this case posed an obvious challenge to model makers, beginning with Kummer himself. The model shown in Fig. 7 was produced in Munich by Klein's student, Karl Rohn; he described it along with two others in *Drei Modelle der Kummer'schen Fläche*. Soon afterward, these models proliferated widely, since they were marketed by the Darmstadt firm of Ludwig

---

[23] Ernst Eduard Kummer, "Ueber die Flächen vierten Grades mit sechzehn singulären Punkten," *Monatsberichte der Königlich Preußischen Akademie der Wissenschaften zu Berlin*, 18. April 1864, 246–60.

[24] David E. Rowe, "Mathematical Models as Artefacts for Research: Felix Klein and the Case of Kummer Surfaces," *Mathematische Semesterberichte* 60, no. 1 (2013): 1–24.

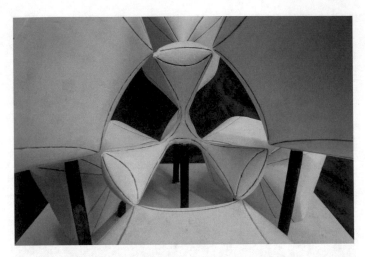

Fig. 8 Close-up of Rohn's model showing a trope containing a conic that passes through six nodes (Göttingen Collection of Mathematical Models and Instruments, model 124). © Mathematical Institute, Georg-August-University Göttingen, all rights reserved

Brill (Series II, Nr. 1).[25] In 1911, these models cost between 21 and 32.50 Mark, which today would be roughly 105 to 160 Euro. Brill's brother was Klein's colleague at the TH München, Alexander Brill (1842–1935), one of the key figures who promoted model-making during the era. Below, we briefly indicate how the model facilitates the visualization of the (16,6) configuration of singularities on a Kummer surface.[26]

This model represents a finite portion of a highly symmetric Kummer quartic situated in projective 3-space. All of its singularities are visible in the model, which can be seen to consist of eight tetrahedral-like pieces: an inner tetrahedron with four others attached to each vertex, plus three outer tetrahedra that join at infinity. These last three tetrahedra have been split and thus truncated into two pieces by planar sections, which is why the model has six outer pieces. By imagining the opposite pieces to extend through infinity, they would then join to form the three outer tetrahedra.

If we now consider the three nearest outer pieces (Fig. 8), we see that each contains two vertices that belong to two of the three tetrahedra associated with the front face of the inner tetrahedron. This gives us six double points, and from the model we can see that these six points are coplanar. Evidently, this is one of the sixteen tropes that touches the surface along a conic, which has been drawn on the model. Similarly, one can visualize three other tropes by rotating the model to

[25] For photos, see: Gerd Fischer, ed., *Mathematische Modelle*, 2 vols. (Berlin: Akademie Verlag, 1986), 34–37.

[26] David E. Rowe, "On Models and Visualizations of Some Special Quartic Surfaces," *Mathematical Intelligencer* 40, no. 1 (2018): 59–67.

bring into view another set of three half-tetrahedra lying in a plane parallel to one of the three formerly invisible faces of the inner tetrahedron. This procedure produces four of the sixteen tropes. Notice, furthermore, that the vertices of each of the three outer tetrahedra are shared by one vertex on each of the four middle tetrahedra. This accounts for the full symmetry of the (16,6) incidence configuration of singularities on the surface. The remaining twelve tropes can now be brought into view by using projective mappings that interchange the inner tetrahedron with one of the three outer tetrahedra, whereby four new tropes become visible, thus yielding sixteen altogether. This model of a highly symmetric Kummer surface also served as a starting point for Klein's models of Plücker's complex surfaces, which he later related to Rohn's important paper "Transformation der hyperelliptischen Functionen $p = 2$ und ihre Bedeutung für die Kummer'sche Fläche."[27]

An earlier version of a similar model was built in 1869, when Klein was in Berlin. There he met with his friend Albert Wenker, who designed it for him. Although this Wenker model has disappeared, it should be remembered for the role it played shortly afterward when Klein and Sophus Lie were collaborating for the first time. After meeting during the fall of 1869 in Berlin, where they both attended Kummer's seminar, they continued to work together the following spring in Paris. During this time, Lie found his famous line-to-sphere mapping, a contact transformation with many interesting properties.[28] Lie used it to map families of lines and the surfaces enveloped by them to families of spheres and their corresponding envelopes.

The simplest example arises when the lines are the generators of a hyperboloid of one sheet (Fig. 9), in which case the spheres envelope a Dupin cyclide (Fig. 10). One can picture this most easily by taking three skew lines in space, which then map to three spheres. The set of all lines that meet these mutually skew lines form the generators of a quadric surface, and since Lie's mapping is a contact transformation, these generators go over to a one-parameter family of spheres tangent to three fixed spheres. Since the second system of generators has the same property, the image is a Dupin cyclide, since these arise as surfaces enveloped by two families of spheres. Moreover, Lie's mapping has a special property of great importance for differential geometry: it sends the asymptotic curves of one surface to the curvature lines of another. In this simple case, the generators themselves are the asymptotic curves on a hyperboloid, and these are mapped to circles of tangency on the Dupin cyclide.[29]

---

[27] Karl Rohn, "Transformation der hyperelliptischen Functionen $p = 2$ und ihre Bedeutung für die Kummer'sche Fläche," *Mathematische Annalen* 15 (1879): 315–54.

[28] Sophus Lie, "Ueber Complexe, insbesondere Linien- und Kugelcomplexe, mit Anwendung auf die Theorie partieller Differential-Gleichungen," *Mathematische Annalen* 5 (1872): 145–256.

[29] See: Sophus Lie and Georg Scheffers, *Geometrie der Berührungstransformationen* (Leipzig: Teubner, 1896), 470–75.

Fig. 9 Historical model of a hyperboloid of one sheet by Théodore Olivier. © Musée des arts et métiers-Cnam, Paris/photo S. Pelly, all rights reserved

Fig. 10 Historical model of a cyclide from the Brill collection (Mathematical Institute, University of Oxford). © Photo: Adam Barker, all rights reserved

In 1864, Gaston Darboux and Theodore Moutard had begun to work on generalized cyclides, which they studied in the context of inversive geometry.[30] Klein and Lie learned about this new French theory when they met Darboux just before the outbreak of the Franco-Prussian War. These cyclides are special quartic surfaces with the property that they meet the plane at infinity in a double curve, namely the imaginary circle that lies on all spheres. Darboux also found that their lines of curvature are algebraic curves of degree eight. This finding set up one of Lie's earliest discoveries, communicated to Darboux at that time. This came from Lie's line-to-sphere mapping, when he considered the caustic surface enveloped by lines in a congruence of the second order and class.[31] This was precisely the context that had led Kummer to his discovery of Kummer surfaces. Lie found that these Kummer quartics map to generalized cyclides of Darboux, and since the curvature lines of the latter were known, he immediately deduced that the asymptotic curves on a Kummer surface must be algebraic of degree sixteen.

Lie communicated these findings to the Norwegian Scientific Society in Christiania in early July 1870, but this note was only published by Ludwig Sylow in 1899, the year of Lie's death.[32] In Paris, Lie and Klein discussed this breakthrough in detail. At first, Klein took a skeptical view of Lie's claim, but then he quickly realized that he had already come across these same curves in his own work without recognizing that they were asymptotic curves. Little more than a week later, however, Klein had to flee from Paris. Staying at his parents' home in Dusseldorf, he continued to think about the paths of these asymptotic curves as well as their singularities. He did so by tracing them on the physical model of a Kummer surface (like Fig. 11) made by his friend Albert Wenker. After a short time, Klein realized that what he had told Lie earlier in Paris about the singularities of these asymptotic curves was, in fact, wrong. After giving the necessary corrections, he wrote:

> I came across these things by means of Wenker's model, on which I wanted to sketch asymptotic curves. To give you a sort of intuitive idea how such curves look, I enclose a sketch (Fig. 12). The Kummer surface contains hyperboloid parts, like those sketched; these are bounded by two of the six conics (K_1 and K_2) and extend from one double point (d_1) to another (d_2). Two of the curves are drawn more boldly; these are the two that not only belong to linear complexes but also are curves with four-point contact. They pass through

[30] On Darboux's early geometrical work, see: Barnabé Croizat, "Gaston Darboux: naissance d'un mathématicien, genèse d'un professeur, chronique d'un rédacteur," (PhD diss., Université Lille 1, 2016).

[31] David E. Rowe, "Klein, Lie, and the Geometric Background of the Erlangen Program," in *The History of Modern Mathematics: Ideas and their Reception*, ed. David E. Rowe and John McCleary, vol. 1 (Boston: Academic Press, 1989), 209–73.

[32] Sophus Lie, *Gesammelte Abhandlungen*, ed. Friedrich Engel and Poul Heegaard, vol. 1 (Leipzig: Teubner, 1934), 86–87.

Fig. 11 Felix Klein's model of a Kummer surface designed in 1871 (Göttingen Collection of Mathematical Models and Instruments, model 95). © Mathematical Institute, Georg-August-University Göttingen, all rights reserved

d_1 and d_2 readily, whereas the remaining curves have cusps there. This is also evident from the model. At the same time, one sees how K_1 and K_2 are true enveloping curves.[33]

By the 'hyperboloid parts' on a Kummer surface, Klein meant those places where the curvature was negative. Only in these regions are the asymptotic curves real and hence visible. Klein later reproduced the same figure in the note that he and Lie sent to Kummer for publication in the *Monatsberichte* of the Prussian Academy.[34] Afterward, this picture became a standard part of the growing literature on Kummer surfaces.[35] Klein had a special fondness for it, too. When five years later he married Anna Hegel, granddaughter of the famous philosopher, he ordered a wedding dress decorated with these arabesque curves. No doubt he was happy to explain their significance to any would-be listener.

---

[33] Felix Klein to Sophus Lie, 29 July 1870, in *Letters from Felix Klein to Sophus Lie, 1870–1877*, ed., Eldar Straume, (Heidelberg: Springer-Verlag), to appear.

[34] Felix Klein and Sophus Lie, "Über die Haupttangenten-Curven der Kummerschen Fläche vierten Grades mit 16 Knotenpunkten," *Monatsberichte der Königlich Preußischen Akademie der Wissenschaften zu Berlin*, 15 December, 1870, 891–99.

[35] As for example in: Rohn, "Transformation," 340, and Hudson, *Kummer's Quartic Surface*, 119.

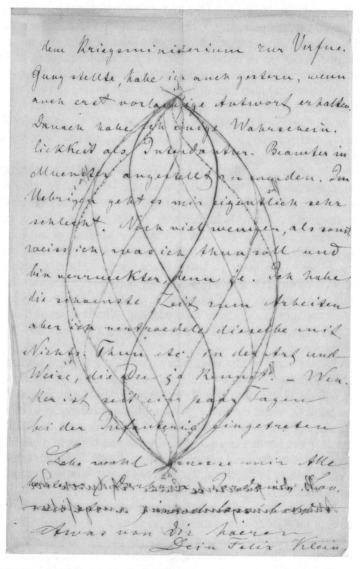

Fig. 12 Felix Klein's sketch of the asymptotic curves between two double points on a Kummer surface from his letter to Sophus Lie, 29 July 1870 (National Library of Norway, Brevs. 289). CC BY-NC-ND 4.0 (https://creativecommons.org/licenses/by-nc-nd/4.0/)

Plücker's Complex Surfaces

During the last years of his life, Plücker devoted a great deal of attention to studying and building the special quartics he called *complex surfaces*. Their name comes from line geometry because these quartics are enveloped by subsets of lines in a quadratic line complex. These are the central objects of interest in a field of inquiry he largely invented: line geometry.[36] In classical studies in geometry, the fundamental elements are points, whereas in projective geometry the principle of duality puts points and planes on an equal footing. Plücker's line geometry, on the other hand, took the manifold of all lines in space as fundamental. These, however, require four parameters rather than only three, as with points and planes. One can introduce different systems of line coordinates for this purpose, but in any case, the space of all lines will be four-dimensional. In order to appreciate Plücker's motivation for studying complex surfaces, a few basic concepts from classical line geometry are needed.

Using line coordinates, an algebraic equation corresponds to a 3-parameter family of lines, a *line complex*. For degree 2, a simple, yet instructive case comes by taking the lines tangent to a nonsingular quadric surface F_2, for example an ellipsoid. Here the local structure is immediately obvious: for a given point $P \notin F_2$, the lines through P tangent to F_2 form a quadratic cone (real or imaginary). This cone, however, collapses to a tangent plane whenever $P \in F_2$, which means that F_2 is the associated *singularity surface* S of the complex. Dually, a typical plane π will cut out a conic, $\pi \cap F_2 = C_2$, whose tangents are also tangents to F_2. The only exceptions are planes $\pi = T_P$ tangent at $P \in F_2$, where the lines enveloping the conic then collapse to a pencil of lines centered at P. This shows S is self-dual, since the same surface arises for points as for planes, a property that holds in all quadratic complexes.

The special character of this elementary quadratic line complex can be seen from its local structure. In the general case, the cone of lines associated with a point $P \in S$ will degenerate into two planar pencils of lines, one centered at P and another at a point Q, where the line PQ is a double line belonging to both pencils. Similarly, a singular plane T_P is one in which the conic degenerates into two point pencils, one at P and another at a point Q. The tangent cones to a quadric surface F_2 thus correspond to the case where the points P and Q fall together forming a single pencil of lines in the plane T_P, counted twice. Similarly, the singularity surface S is the degenerate quartic obtained by doubling F_2, so F_2^2. In the general case, S is a Kummer quartic, which depends on eighteen parameters, not just nine, as in the case of an F_2. Moreover, the tangents to an F_2 completely determine the complex, so it, too, has only nine parameters. A generic quadratic complex has nineteen because, as Klein showed in "Zur Theorie der Liniencomplexe des ersten und zweiten Grades,"[37] any given Kummer surface is the singularity surface for a one-parameter family of quadratic complexes. The same holds for another special case, the tetrahedral complexes $T(\lambda)$, which depend on 13-parameters. Given any four planes in general position, $T(\lambda)$ is the complex determined by those lines whose four intersection points with this tetrahedron have a fixed cross ratio λ. The

[36] Julius Plücker, *Neue Geometrie des Raumes gegründet auf die Betrachtung der geraden Linie als Raumelement*, vol. 1 (Erste Abtheilung) (Leipzig: Teubner, 1868).

[37] Felix Klein, "Zur Theorie der Liniencomplexe des ersten und zweiten Grades," *Mathematische Annalen* 2 (1870): 198–226.

Fig. 13 A Plücker complex
surface showing four pinch
points on its double line.
Vizualization by Oliver Labs.
© Oliver Labs, all rights
reserved

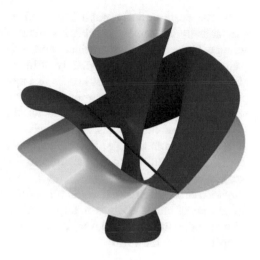

singularity surface S of $T(\lambda)$ is the tetrahedron itself, and by varying λ one gets a one-parameter family of quadratic complexes.

The totality of lines in a quadratic complex is nearly impossible to visualize, except in special cases like the tangents to a quadric surface or those that belong to a tetrahedral complex. Since Plücker's work was strongly guided by *Anschauung*,[38] he studied the local structure of quadratic complexes from the standpoint of lines, the fundamental objects in line geometry.[39] His idea was to view the lines in a complex K_2 with respect to a fixed line g, where in most cases $g \notin K_2$. More precisely, he considered the lines in $K_2 \cap K_1(g)$, where $K_1(g)$ is the first-degree complex consisting of those lines in space that meet g. The intersection of these two complexes produces a 2-parameter family of lines, a line congruence of the second order and class. This type of ray system was familiar from geometrical optics, the background for Kummer's work in the 1860s. Such congruences of lines will envelope a caustic surface (*Brennfläche*), which in this case will be a surface of the fourth order and class. Unlike Kummer surfaces, however, Plücker surfaces have a double line g and other singularities, but never more than eight nodes (thus, Fig. 2 shows the maximal case). Plücker's complex surfaces thus represent a degenerate type of Kummer surface where the nodal line g forms a double line. This double line contains four higher point singularities (so-called pinch points) where the leaves of the surface meet. These pinch points also demarcate the boundaries between real and imaginary portions of the surface as illustrated in Fig. 13.

---

[38] Alfred Clebsch, "Zum Gedächtniss an Julius Plücker," *Abhandlungen der Königlichen Gesellschaft der Wissenschaften zu Göttingen* 16 (1872), 1–40, here 2.
[39] Julius Plücker, *Neue Geometrie des Raumes gegründet auf die Betrachtung der geraden Linie als Raumelement*, vol. 2 (= Zweite Abtheilung), ed. Felix Klein (Leipzig: Teubner, 1869).

In the posthumously published *Neue Geometrie des Raumes* (1869), which was actually written by Plücker's student, Felix Klein, one finds a classification scheme for seventy-eight distinct types of so-called equatorial complex surfaces (see below). This scheme is mainly of interest because it shows how Plücker went about constructing the various cases in a systematic fashion. Plücker also designed prototypes for models of certain types of complex surfaces, both equatorial and meridianal, which were then manufactured by a company in Cologne.

For many of his models, Plücker chose the line $g \subset E_\infty$, the plane at infinity. The family of planes $E(g)$ that contain such a nodal line g are then parallel, and the quartic S_4 enveloped by the lines in $K_2 \cap K_1(g)$ is then an *equatorial surface* (examples are shown in Figs. 1 and 2). All other cases are known as *meridianal surfaces*. Since each $E(g) \cap S_4$ is a quartic curve that contains g as a double line, these curves break up into a family of conics C_2 together with g. In the case of equatorial surfaces, Plücker showed that the equation for the surface in point coordinates has the form

$$\frac{y^2}{Ex^2 + 2Ux + C} + \frac{z^2}{Fx^2 - 2Rx + B} + 1 = 0.$$

This coordinate system is chosen so that the x-axis coincides with the diameter of the complex K_2. This ensures that the conics that form the latitudinal curves (*Breitenkurven*) $E(g) \cap S_4$ have centers that lie on the x-axis. Plücker next shows how to relate the lengths of their semi-diameters to two other fundamental conics in the xy- and xz-planes. The equations for their respective intersections with S_4 are then

$$y^2 + Ex^2 + 2Ux + C = 0; \quad z^2 + Fx^2 - 2Rx + B = 0.$$

One notes that the zeros of the quadratics in x in the denominators correspond to two singular lines that lie in each of the two coordinate planes. Ignoring these, Plücker calls the two remaining conics the *characteristic curves* of the surface. Depending on their reality and relative position, he shows that there are seventeen possible cases.

To illustrate how these characteristic curves can be used to construct the latitudinal curves of S_4, he discusses the case shown in Fig. 14. A graphic of the resulting surface is given in Fig. 15. (Original models exist in the collections in Göttingen (Nr. 110) and in Munich (Nr. 9).) To obtain the plane figure, Plücker rotates one of the two planes around the x-axis so that it coincides with the other. He thus obtains two coplanar conics with four points (real or imaginary) of intersection with this axis. These correspond to the places where the latitudinal curves transition from one type of conic to a different type. The dotted hyperbolas indicate these transitions, so the conics $E(g) \cap S_4$ that lie between A and C will be ellipses, those between A and A' and C and C' are hyperbolas, and those above A' and below C' are imaginary hyperbolas. The points of intersection of the characteristic curves marked K correspond to two circular cross sections, whereas the unmarked

Fig. 14 Julius Plücker's figure of the characteristic curves for an equatorial surface. From: Julius Plücker, *Neue Geometrie des Raumes gegründet auf die Betrachtung der geraden Linie als Raumelement*, Erste Abtheilung (Leipzig: Teubner, 1868), 351

igen *x* entsprechend, in der Zeichnung eine
ttellen die beiden Abschnitte auf derselben,
ttiken und ihre Ergänzungs-Curven bestimmt

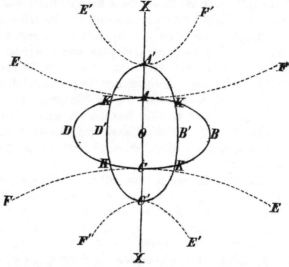

zwei Kreise. Das elliptische Flächenstück
lbe schliessen sich, von *A* bis *A'*, bezüglich
trbolische Flächenstücke der zweiten Art.
ie gleichseitige Hyperbeln enthalten, sind
chschnitt der Ellipse *A' B' C' D'* mit der

intersection points of the inner hyperbola with the elongated ellipse yield cross sections which are rectangular hyperbolas.

Fig. 15 An equatorial surface with 4 real singular points. Vizualization by Oliver Labs. © Oliver Labs, all rights reserved

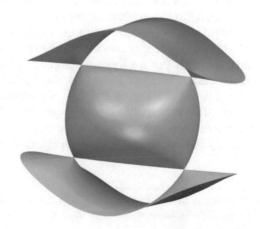

Plücker's entire classification involved seventy-eight cases in all, but the methodology he employed is of far more interest and can be sketched as follows. For each of his seventy-eight cases, Plücker introduced a special symbol based on his construction method. The case just considered (Nr. 9) has the Plücker symbol $I_1H_2E_1H_2I_1$, where E, H, and I indicate the alternation between elliptic, hyperbolic, and imaginary latitudinal curves. The suffix 1 signifies that the bounding singular lines are parallel, 2 that they lie in perpendicular directions. If the two characteristic curves are real ellipses, then two other neighboring cases arise. Nrs. 8 and 10 occur when the intersection points of the ellipses with the x-axis lie outside each other or, respectively, the points of intersection alternate between the two curves. In the latter case 10, the Plücker symbol is $I_2H_2E_2H_2I_2$, which differs only slightly from case 9 in that the suffixes are identically 2. So, imagining that we view the cross sections kinematically by passing from one type of curve to the next, then each singular line encountered lies in the opposite direction from the one preceding.

From these seventeen canonical cases, Plücker goes on to derive systematically all the other degenerate types. He starts by considering cases where two of the singular lines coincide. For example, case Nr. 9 with symbol $I_1H_2E_1H_2I_1$ can pass over into surface Nr. 23 with symbol $I_1H_2H_2|I_1$, where | denotes a double singular line. Here the latitudinal conics within the visible segment of the x-axis are hyperbolas, where at one end of the segment the double singular line appears. As before, the curves above and below the segment are imaginary. Another type of degeneracy occurs when the two characteristic curves have coincident intersection points on the x-axis. An instance of this is Nr. 34 with Plücker symbol $I \times EH_2I$, where the symbol $\times$ indicates that the latitudinal curve consists of two perpendicular lines. This case can arise by continuous deformation starting with either of the two cases 9 or 10.

On Deforming Quartics

Soon after his teacher's death, Klein supplemented Plücker's collection with four additional models to illustrate the connection between Kummer surfaces and three principal types of Plücker surfaces. Beginning with a Kummer surface, Klein produced these models by considering the relation of the double line g to the quadratic line complex. For $g \notin K_2$, the complex surface has eight double points and four pinch points, whereas for $g \in K_2$ there will be only four double points. Finally, if g is a singular line in K_2, then the Plücker surface will have just two double points. This eventually led to a program for classifying Plücker surfaces by means of deformation techniques, an important part of research on real quartics later pursued by Klein's student Karl Rohn (1855–1920).[40] It seemed evident to Klein that

---

[40] Rohn, "Transformation;" Karl Rohn, "Die verschiedenen Gestalten der Kummer'schen Fläche," *Mathematische Annalen* 18 (1881): 99–159.

one should start with a Kummer surface, carrying out deformations that systematically lowered the number of double points. Thus, the double line g on a Plücker surface in effect absorbed eight of the sixteen double points on a Kummer surface. Only eight then remain as the others pass over into the four pinch points on g. Klein's first model (Fig. 11) proved very useful for picturing the configuration of singularities on a Kummer surface. Its connection with the model shown in Fig. 2, on the other hand, was much less clear. One of the main difficulties in visualizing how a Kummer surface can, via deformation, acquire a double line is that the Plücker quartics also contain four other lines as tangential singularities. So, in the process of absorbing eight of the sixteen double points to form four pinch points on the double line g, four lines determined by the four tropes that pass through g need to come into view.

In the late 1870s, Rohn found a way to link the Plücker and Kummer surfaces by deformations. Instead of starting with a general Kummer surface, he began with a highly degenerate case of a quartic surface, namely a quadric counted twice (see Fig. 16). A familiar quadric surface F_2 is a hyperboloid of one sheet, which contains two systems of real generators, or rulings by lines. These one-parameter sets of lines arise naturally in line geometry simply as the intersections of three first-degree line complexes.

Rohn next took four lines each from the two rulings, which then produces the checkerboard configuration of lines on the surface of the hyperboloid shown in Fig. 7. This pattern of lines yields sixteen double points, and the shaded portion

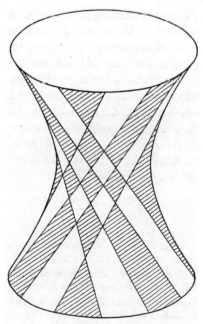

Fig. 3.

can then be treated as the region corresponding to the real points on a degenerate Kummer surface, each piece of which corresponds to a flattened image of the tetrahedra in Rohn's model, discussed above. Passing through infinity, we then see that there are exactly eight pieces, in agreement with what we observed with the model in Fig. 7. So, Rohn could now obtain a general Kummer surface by blowing up the flattened tetrahedra while leaving their vertices fixed.

Rohn's approach made it far easier to visualize the passage from a Kummer surface to a Plücker quartic with a double line. One can carry out this deformation by starting with the degenerate Kummer surface, then what happens is that two of the four lines in, say, the first ruling gradually fall together. As they do so, the eight singular points on these two lines combine in pairs to form four pinch points on the double line. After this, one only has to blow up the flattened pieces, which are now five instead of eight in number. So, by this simple means Rohn was able to show how these degenerate quartics made the relationship between the Kummer and Plücker surfaces more transparent. This insight made it possible to visualize the relationship between the singularities on these two types of special quartic surfaces. Klein later added a schematic drawing to illustrate how the double line with four pinch points is formed when a Kummer surface passes over into a Plücker surface (Fig. 17).

The Kummer surfaces—as well as those quartics related to them but with various other singularities—turned out to be of central importance for the classification of quadratic line complexes, the topic of Klein's doctoral dissertation. Klein only dealt with the generic case, however, whereas a detailed classification requires using Weierstrass' theory of elementary divisors to analyze all possible degeneracies among the eigenvalues of a 6×6 matrix. Five years later, Klein's student Adolf Weiler gave the first detailed analysis of forty-eight different types of quadratic line complexes by making use of their singularity surfaces.[41] In the most general case these are Kummer surfaces, which then pass over to Plücker quartics if the complex contains double lines. In several cases, Weiler found other types of quartic surfaces that had been studied earlier by Luigi Cremona,[42] Arthur Cayley, Jakob Steiner, Sophus Lie, and Ludwig Schläfli. Thus, his classification scheme drew on much recent knowledge from algebraic surface theory. Ten years later, a third doctoral dissertation, written by Corrado Segre, presented still another, even more refined classification of quadratic line complexes; Segre's work has remained the last word on this topic.[43]

[41] Adolf Weiler, "Ueber die verschiedenen Gattungen der Complexe zweiten Grades," *Mathematische Annalen* 7 (1874): 145–207.

[42] Luigi Cremona, "Sulle superficie gobbe di quarto grado," *Memorie dell'Accademia delle Scienze dell'Istituto di Bologna*, series 2, vol. 8 (1868): 235–50.

[43] David E. Rowe, "Segre, Klein, and the Theory of Quadratic Line Complexes," in *From Classical to Modern Algebraic Geometry: Corrado Segre's Mastership and Legacy*, ed. Gianfranco Casnati Alberto Conte, Letterio Gatto, Livia Giacardi, Marina Marchisio and Alessandro Verra (Basel: Birkhäuser, 2016), 243–63.

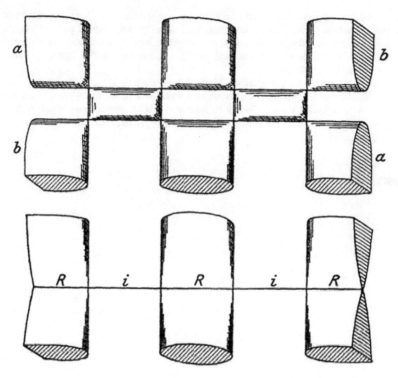

Fig. 17 A Kummer surface with eight nodes passing into a Plücker surface. From Felix Klein, *Gesammelte Mathematische Abhandlungen*, vol. 2: *Anschauliche Geometrie. Substitutionsgruppen und Gleichungstheorie. Zur mathematischen Physik* (Berlin: Springer, 1922), 8, Fig.1; 9, Fig. 2 © All rights reserved

Modeling Parallel Transport

Tilman Sauer

Introduction

This paper is about one particular set of models, a set of three material models of simple curved surfaces dating from 1918. What makes these three surface models special in the context of a history of mathematical modeling is that they carry an additional layer of information, i.e. these plain surface models serve to illustrate a new concept that was painted onto the model surfaces. The basic models were most likely purchased from a standard collection of models that had been commercially available for a long time. The interesting feature is what was painted onto them, and this embellishment happened in 1918. These models were turned into illustrations of a new abstract concept by the Dutch geometer Jan Arnoldus Schouten (1883–1971). He wanted to illustrate or visualize the new geometric concept of parallel transport. Photographs of his illustrated models were included in a paper that introduced this new concept. Identical copies of these photographs appeared in four different publications up until 1924, but it appears that they have never been republished or referenced again after that.

I will argue that these three illustrative models represent a certain transitional stage in a process of conceptual development of differential geometry and it is this transitional nature that makes this case interesting. We also see here that there were two levels to these models at work because Schouten took existing geometric models and added a visual layer, so as to illustrate and visualize an emerging new abstract concept. Such a two-layered use of models was not unusual, but here we see the significance of using the basic models as background for more abstract concepts very clearly.

T. Sauer (✉)

Institute of Mathematics, Johannes Gutenberg-Universität Mainz, Staudinger Weg 9, 5. OG, 55128 Mainz, Germany

e-mail: tsauer@uni-mainz.de

© The Author(s) 2022

M. Friedman and K. Krauthausen (eds.), *Model and Mathematics: From the 19th to the 21st Century*, Trends in the History of Science, https://doi.org/10.1007/978-3-030-97833-4_5

Fig. 1 Jan A. Schouten's
model for geodesic motion of
a frame over a spherical cap
as an example of a surface of
positive curvature. From Dirk
J. Struik, *Grundzüge der
mehrdimensionalen
Differentialgeometrie in
direkter Darstellung* (Berlin:
Springer, 1922), 47. ©
Springer, all rights reserved

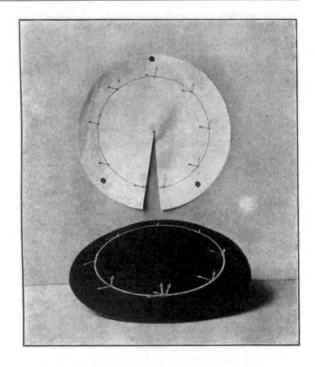

Historical Context: Localization of the Models in Space and Time

The models were first referenced and depicted in Schouten's monograph-length
contribution to the *Verhandelingen der Koninklijke Akademie van Wetenschapen*,
entitled *Die direkte Analysis zur neueren Relativitätstheorie*. Photographs of the
models were reproduced in the paper.[1] Max von Laue (1879–1960) then reproduced the same pictures in the second volume of his textbook on relativity that
was one of the first textbook expositions in Germany of the new general theory
of relativity.[2] Dirk Struik (1894–2000) then reproduced them again in his 1922
monograph on *Grundzüge der mehrdimensionalen Differentialgeometrie in direkter
Darstellung*[3] (see Figs. 1 and 2). The very same photographs were reproduced

---

[1] Jan A. Schouten, *Die direkte Analysis zur neueren Relativitätstheorie* (Amsterdam: Johannes
Müller, 1918), 48, 70.
[2] Max von Laue, *Die Relativitätstheorie*, vol. 2: *Die allgemeine Relativitätstheorie und Einsteins
Lehre von der Schwerkraft* (Braunschweig: Vieweg & Sohn, 1921), 110–11.
[3] Dirk J. Struik, *Grundzüge der mehrdimensionalen Differentialgeometrie in direkter Darstellung*
(Berlin: Springer, 1922), 47–48.

Fig. 2 Jan A. Schouten's model for geodesic motion of a frame over two examples of surfaces of negative curvature. From Dirk J. Struik, *Grundzüge der mehrdimensionalen Differentialgeometrie in direkter Darstellung* (Berlin: Springer, 1922), 48. © Springer, all rights reserved

once more in an introduction to modern differential geometry jointly authored by Schouten and Struik.[4]

When Laue republished Schouten's photographs in his textbook, he did not give any information about the origin and creator of these models. In fact, Struik found it necessary to pass on proper credit for his teacher Schouten and through this comment we learn about the origin and original location of these models. He added in a footnote: "These models were made first by Prof. J. A. Schouten and are part of the model collection of the Technical University Delft."[5] We do have, in fact, accounts of that larger collection which appears to have been used intensely

---

[4] Jan A. Schouten and Dirk J. Struik, "Einführung in die neueren Methoden der Differentialgeometrie," *Christiaan Huygens. International mathematisch Tijdschrift* 1 (1922), 333–53, and 2 (1923), 1–24, 155–71, 291–306; For the images see the reprint in: Jan A. Schouten and Dirk J. Struik, *Einführung in die neueren Methoden der Differentialgeometrie* (Groningen: Noordhoff, 1924), 31.

[5] "Diese Modelle sind zuerst von Prof. J.A. Schouten hergestellt worden und befinden sich in der Modellsammlung der Technischen Hochschule Delft." Struik continued by mentioning von Laue: "Die Abbildungen sind der Arbeit von Schouten, 1918, 10, S. 48 u. 70 entnommen und sind auch (Fig. 1 teilweise) von v. Laue in sein Buch über die Relativitätstheorie aufgenommen worden." ("The images were taken from the work by Schouten, 1918, 10, 48 and 70, and they are also included (Fig. 1 in parts) by v. Laue in his book on the theory of relativity.") (Struik, *Grundzüge*, 48). Incidentally, there is another example of Laue's reproducing a figure that was not of his own making without giving proper credit. On page 226, he reproduced a diagram showing possible particle trajectories in a Schwarzschild space–time, which he had taken from Hilbert's lecture course

Fig. 3 David van Dantzig during one of his lectures at Delft Polytechnic with geometric models in 1938 (Photo: Archiv Gerard Alberts; reproduced in Irene Polo Blanco, "Physical models for the learning of geometry," *Nieuwe Wiskrant* 31, no.1 (September 2011), 36; also reproduced in Irene Polo-Banco, "Theory and history of geometric models," (Ph.D. diss., Groningen, 2007, 8).

also in later times: Fig. 3 shows David van Dantzig (1900–1959) during one of his lectures at Delft Polytechnic in 1937 engaged with a geometric thread model.

The specific interest of these three models derives from the combination of the material geometric surface together with the inscription on them, painted onto the surface with white on black.

The geometric model surfaces were, in fact, rather standard and not in any way exotic shapes, a segment of a sphere, a hyperbolic paraboloid and a pseudosphere, i.e. the rotational surface of a tractrix. Interest in the shape of a sphere and a pseudosphere derived from the properties that they display constant positive or constant negative curvature. Plaster models of these shapes could have been ordered readily from Schilling's catalogue (see also Fig. 4).[6] Spheres were available in different

---

on General Relativity of the winter semester 1916/17 (Tilman Sauer and Ulrich Majer, eds., *David Hilbert's Lectures on the Foundations of Physics: 1915–1927* (Dordrecht: Springer, 2009), 277). But contrary to Schouten, Laue did give credit to Hilbert in the preface, where he wrote: "Vor allem hat D. Hilbert die Ausarbeitung seiner Vorlesung über die Grundlagen der Physik zur Verfügung gestellt, aus der sehr viel in unser Buch übergegangen ist." ("Above all, Hilbert has made available to us the worked out lecture notes of his course on the Foundations of Physics, from which a lot was transferred into our book.") (Laue, *Die Relativitätstheorie*, VII). Schouten, on the other hand, is mentioned only once in the book, together with Levi–Civita and Hessenberg as originators of the concept of parallel transport (Laue, *Die Relativitätstheorie*, 257), citing Schouten, *Die direkte Analysis*, as a general reference.

[6] See: Martin Schilling, ed., *Catalog mathematischer Modelle für den höheren mathematischen Unterricht*, 7th ed. (Leipzig: Martin Schilling, 1911), 115, 114, 144, for a sphere, a hyperbolic paraboloid, and a pseudosphere, respectively.

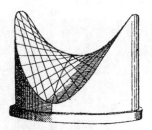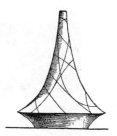

Fig. 4 The basic shapes of Jan A. Schouten's models could be purchased from Schilling's catalogue. From Martin Schilling, ed., *Catalog mathematischer Modelle für den höheren mathematischen Unterricht*, 7th ed. (Leipzig: Martin Schilling, 1911), 115, 114, 144. Similar pictures can be
found in Walther Dyck, ed., *Katalog mathematischer und mathematisch-physikalischer Modelle,
Apparate und Instrumente* (Munich: C. Wolf und Sohn, 1892), 259, 292

sizes "with black board paint and a wooden stand" ("mit schwarzem Tafelanstrich
und Holzuntersatz") so that they could be drawn on with chalk as on a regular
blackboard.[7]

Spheres were available with diameters of 35, 14, and 10 cm, with price tags
of Mk. 43.20, 12.20, or 9.50, respectively. The size of the paraboloid was 15
by 13 cm and its cost was Mk. 4.50 or 8.00 depending on whether it had the
lines of horizontal cuts, i.e. equilateral hyperbolas imprinted on them or not. The
production of the pseudosphere was credited to "stud. math. Bacharach" from
Munich, its size was given as 25 by 18 cm and its price tag was Mk. 11.00.[8]
Examples of the hyperbolic paraboloid and of the pseudosphere are depicted also
in Gerd Fischer's *Mathematische Modelle*[9] (see Fig. 5a, b) and in *Digitales Archiv
mathematischer Modelle*.[10]

The materiality of the models reflects a culture of preparing modeling clay of
a kind especially suitable for the making of geometric models. Fischer cites a
contemporary recipe for making modeling clay, an elaborate procedure involving
various ingredients, specific temperatures, kneading, and patience.[11]

---

[7] "Die drei Kugeln in den verschiedenen Größen gestatten wie auf einer Wandtafel leicht die
Anwendung von Kreide und Schwamm zum Zeichnen." ["The three spheres of different sizes
allow the easy application of chalk and sponge for drawing as on a blackboard."] (Schilling, ed.,
Catalog, 115).

[8] Schilling, ed., *Catalog*, 144.

[9] Gerd Fischer, ed., *Mathematische Modelle: Aus den Sammlungen von Universitäten und Museen*,
vol. 1 (Braunschweig: Vieweg & Teubner, 1986), 8, 77.

[10] www.mathematical-models.org (accessed November 23, 2021).

[11] Fischer, ed., *Mathematische Modelle*, vol. 1, VIII.

a   b

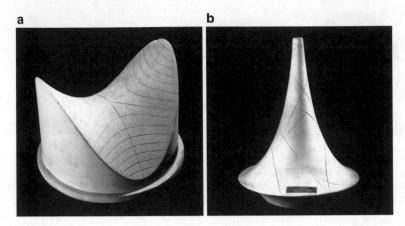

Fig. 5 Photographs of typical specimens of models of the hyperbolic paraboloid and the pseudosphere. Reprinted by permission from Springer Nature: Gerd Fischer, ed., *Mathematical Models*,
vol. 1, 2nd ed. (Wiesbaden: Springer Spektrum, 2017), 8, photo 7; 77, photo 82. © Springer Nature,
all rights reserved

The Notion of Parallel Transport

The completion of the general theory of relativity in late 1915 triggered a renewed
interest in higher-dimensional differential geometry. Thus, Struik wrote in a 1922
monograph: "Einstein's theory of relativity, which had made use of the Ricci Calculus since 1913, has kindled an interest in this method of calculation and in
the differential geometry of higher manifolds in broader circles of society and
has stimulated new investigations."[12] As an example he mentions the "geometric
meaning of covariant differentiation."

Similarly, Ludwig Berwald (1883–1942) wrote in his chapter on differential
invariants for the *Encyklopädie der mathematischen Wissenschaften (Encyclopedia
of Mathematical Sciences):*

> In recent years the interest in Riemannian manifolds has been given new incentive from Ein-
> stein's gravitation theory, a boost that has also resulted in an important principal theoretical
> advance: the introduction of the concept of parallelism in a [Riemannian manifold] V_n by
> Levi-Civita.[13]

---

[12] "Die Einsteinsche Relativitätstheorie, die sich seit 1913 des Ricci-Kalküls bedient hat, hat
das Interesse für diese Rechnungsmethode und für die Differentialgeometrie höherer Mannigfaltigkeiten in weiteren Kreisen wachgerufen und zu neuen Arbeiten angeregt." ["Einstein's relativity theory, which had made use of Ricci's calculus since 1913, kindled interest in this method
of computation and in the differential geometry of higher-dimensional manifolds in a wider public
and instigated further work in these fields."] (Struik, *Grundzüge*, 4).

[13] "In den letzten Jahren hat das Interesse für die Riemannschen Mannigfaltigkeiten durch die
Gravitationstheorie A. Einsteins einen neuen Aufschwung genommen, der auch einen wichti-
gen prinzipiellen Fortschritt in der Theorie zur Folge hatte: die Einführung des Begriffes des

Einstein had worked out the general theory of relativity in the years prior to its completion in late 1915 in a purely analytic way using only concepts from invariant theory. The transition from a scalar theory of gravitation to a metric one was taken in 1912 and involved the introduction of a metric tensor and a differential line element defined by it. The theory of differential invariants was the main resource for Einstein and his collaborator Marcel Grossmann (1878–1936) when they began developing a relativistic theory of gravitation. In the analytic tradition, the field had been established by Elwin Bruno Christoffel (1829–1900) and others. To be sure, there was an implicit geometric meaning, which was more explicit in Riemann's work and the tradition based on his work[14] and which came to the fore when the general theory was restricted to binary forms in two variables that could then be interpreted in terms of Gaussian surface coordinates. But Christoffel did not emphasize this geometric implication nor did Einstein and Grossmann make any use of it. Instead, Grossmann explicitly denied the advantage of geometric conceptualization in their endeavor, writing: "I have purposely not employed geometrical aids because, in my opinion, they contribute very little to an intuitive understanding of the conceptions of vector analysis."[15]

When Einstein and Grossmann were searching for field equations of generalized covariance, the geometric intuition of the two-dimensional case may have been distracting from the general case of a four-dimensional space–time with the added complication of a Lorentz signature metric or an imaginary time coordinate. But with the establishment of the general theory, and especially after the publication of Einstein's gravitational field equations in November 1915, mathematicians and physicists began to explore the geometric implications introduced by the new theory. General Relativity had introduced a curvature in space–time, a consequence that was only by and by extracted from the differential equations.

Parallelismus in einer V_n durch T. Levi–Civita." (Ludwig Berwald, "Differentialinvarianten in der Geometrie: Riemannsche Mannigfaltigkeiten und ihre Verallgemeinerungen," in *Encyklopädie der mathematischen Wissenschaften mit Einschluss ihrer Anwendungen,* ed. Wilhelm Franz Meyer and Hans Mohrmann, vol. 3, bk. 3: *Geometrie* (Leibniz: Teubner, 1902–1927), 73–181, here 124–25).

[14] Alberto Cogliati, "Riemann's *Commentatio Mathematica*: A Reassessment," *Revue d'histoire des mathematiques* 20, no. 1 (2014): 73–94.

[15] "Dabei habe ich mit Absicht geometrische Hilfsmittel beiseite gelassen, da sie meines Erachtens wenig zur Veranschaulichung der Begriffsbildungen der Vektoranalysis beitragen." (Albert Einstein and Marcel Grossmann, "Entwurf einer verallgemeinerten Relativitätstheorie und einer Theorie der Gravitation," *Zeitschrift für Mathematik und Physik* 62, no. 3 (1913): 225–61, here 244. English translation: *The Collected Papers of Albert Einstein,* ed. Martin. J. Klein, Anne J. Kox, Jürgen Renn, and Robert Schulmann, vol. 4: *The Swiss Years, 1912–1914* (Princeton: Princeton University Press, 1995), Docs. 13, 26, 325), see also Tilman Sauer, "Marcel Grossmann and his contribution to the general theory of relativity," in *Proceedings of the 13th Marcel Grossmann Meeting on Recent Developments in Theoretical and Experimental General Relativity, Gravitation, and Relativistic Field Theory,* ed. Robert T. Jantzen, Kjell Rosquist, and Remo Ruffini (Singapore: World Scientific, 2015), 456–503.

In the course of this early elaboration of Einstein's new theory, a conceptual distinction between the metric and the affine connection was introduced.[16] The introduction of the related concept of parallel transport allowed the geometric interpretation of the new theory of space–time in terms of parallel transport of vectors, etc. After the introduction of general relativity, it was Tullio Levi-Civita (1873–1941), Jan Arnoldus Schouten, Gerhard Hessenberg (1874–1925), Hermann Weyl (1885–1955), and others who realized the implications.[17]

The Context of the History of Mathematics

The notion of a covariant derivative generalizes the notion of ordinary differentiation for the case when invariance of a differential form was required for arbitrary (smoothly differentiable) transformations of the basic variables. This concept was readily available to Einstein and Grossmann in the so-called absolute differential calculus of Gregorio Ricci-Curbastro and Tullio Levi-Civita.[18] That calculus was designed and presented from the outset for manifolds of an arbitrary number of dimensions.

From the outset, the problem of parallel transport, therefore, was one of interpreting an analytical concept for an arbitrary number of dimensions. It was only when the geometric meaning of the generalized n-dimensional algebraic concept became an object of study that the geometric interpretation for the case of a two-dimensional manifold (a surface) embedded in Euclidean three-dimensional space was reconsidered.

Gauss, indeed, had introduced the notion of the intrinsic curvature of a surface, a notion of curvature that was independent of the specific way the surface was embedded in space. Riemann had begun to generalize this notion of intrinsic curvature from surfaces to n-dimensional spaces.

For spaces with intrinsic curvature, the notion of parallelism needs to be generalized: how can one define a notion of parallelism or equivalently of the transport of a vector such that the transported vectors remain parallel to each other in a curved space, especially if the notion of the parallelism of the embedding space can no longer be drawn upon? The answer is provided by the concept of an affine connection and its associated notion of parallel transport.

---

[16] John Stachel, "The Story of Newstein or: Is Gravity Just Another Pretty Force," in *The Genesis of General Relativity*, ed. Jürgen Renn, vol. 4 (Dordrecht: Springer, 2007), 1041–78.

[17] Karin Reich, "Levi-Civitasche Parallelverschiebung, affiner Zusammenhang, Übertragungsprinzip: 1916/17–1922/23," *Archive for History of Exact Sciences* 44, no. 1 (1992): 77–105.

[18] Gregorio Ricci-Curbastro Ricci and Tullio Levi–Civita, "Méthodes de calcul différentiel absolu et leurs applications," *Mathematische Annalen* 54, no. 1–2 (1900): 125–201. Reprinted in Tullio Levi–Civita, *Opere Matematiche: Memorie e note*, vol. 1 (Bologna: Nicola Zanichelli, 1954), 479–559.

The Levi-Civita Connection

Nowadays, we call a Levi-Civita connection the uniquely defined affine connection that is symmetric and compatible with the metric. For most practical purposes, it is *the* affine connection used routinely in standard applications of general relativity. At the time, however, it was a pioneering concept that established the link between the algebraic formulation of field equations in Einstein's original formulation with a geometric interpretation.

Let me briefly recapitulate the early history of the notion of an affine connection.[19] Levi-Civita's notion of affine connection was presented in a paper entitled "Nozione di parallelismo in una varietà qualunque e conseguente specificazione geometrica della curvature Riemanniana" that was published in *Rendiconti del Circolo Matematico di Palermo*, presented on 24 December 1916.[20] In it, Levi-Civita explicitly stated his motivation to further develop what he called the "embryo" of Riemann's ideas.[21] In the sequel, Levi-Civita argued very explicitly in geometric language and with the notion of parallelism. The latter concept was established by way of embedding the manifold in a Euclidean reference space, which can always be obtained by increasing the number of dimensions (at most you need $n(n+1)/2$ according to a theorem going back to Ludwig Schläfli (1814–1895) if n is the dimension of the original manifold). Nevertheless, in Levi-Civita's original paper the reference to the embedding space is only for convenience, it is not a conceptual necessity.[22]

Another relevant paper that even predates Levi-Civita's "Nozione" paper, although it was published later, was written by Gerhard Hessenberg and is entitled "Vektorielle Begründung der Differentialgeometrie." It is dated "June 1916" but apparently was issued only later. This paper of 32 pages begins with a general remark on the relevance of differential forms for relativity theory. It states that the aim of the paper was to establish a "connection between the theory of differential forms and differential geometry." Hessenberg explicitly refers to Christoffel's 1869 paper and to the 1901 paper by Ricci and Levi-Civita. He further mentioned work by Knoblauch, Maschke, Wright, Gauss, Riemann, Grassmann, Pfaff, and Beltrami.

---

[19] For the following, I am drawing on Reich, "Levi-Civitasche Parallelverschiebung" and Alberto Cogliati, "Schouten, Levi–Civita and the notion of parallelism in Riemannian geometry," *Historia Mathematica* 43, no. 4 (2016): 427–43.

[20] On Levi–Civita, see: Judith Goodstein, *Einstein's Italian Mathematicians: Ricci, Levi–Civita, and the Birth of General Relativity* (Providence: American Mathematical Society, 2018).

[21] Tullio Levi–Civita, "Nozione di parallelismo in una varietà qualunque e consequente specificazione geometrica della curvature Riemanniana," *Rendiconti del Circolo Matematico di Palermo* 42 (1917): 173–204, here 173.

[22] Iurato and Ruta recently looked at Levi–Civita's paper and claim that the mechanical principle of virtual work plays a significant role in its formulation. See: Giuseppe Iurato and Giuseppe Ruta, "On the role of virtual work in Levi–Civita's parallel transport," *Archive for History of Exact Sciences* 70, no. 5 (2016): 553–65.

Although Hessenberg wants to give a geometric formulation, the word 'parallel' does not appear in the paper. Instead, his program is captured by the following quote:

> The path we shall follow will lead us to Christoffel's three index expressions via the equations (12) and (13) [i.e. the vanishing of the covariant derivative of the metric and the symmetry of the Christoffel symbols]. It will become clear that for the invariant relations that we will get, also for the curvature tensor, only the equations (12) are relevant, while the Christoffel symmetry (13) can be discarded. Their meaning can be summarized by saying that in the geometry of an n-dimensional manifold the 'straightest' lines are also the 'shortest.'[23]

Let us now look at Schouten's work. His so-called "direct analysis" aimed at giving a formulation of geometric quantities that was explicitly independent of coordinates and already in its form displayed the relevant geometric characteristics. Schouten's 1918 paper, in this sense, extended earlier work of his from 1914 which had been motivated by the wish to look at geometric quantities from an engineer's point of view.[24]

Schouten's notion of a geodesically transported frame was constituted in a technically elaborate way, a detailed discussion that would go beyond the framework of this paper. Schouten relied on a so-called 'symbolic method' which was widely used in invariant theory at the time and which allowed an algebraic handling of invariance and transformation properties. The method introduced so-called ideal vectors, which do not have a simple interpretation from a modern understanding, for the representation of the metric tensor and other invariant objects. In more modern terms, the basic presuppositions here were a manifold with a differentiability structure and a metric field defined on it that allows the intrinsic determination of distances within the manifold. It therefore also allowed the definition of the notion of shortest lines, i.e. lines between two points A and B along which the integrated distance is minimal. Schouten then looked at infinitesimally small rigid frames, i.e. linearly independent, small vectors at each point of the manifold. Since a notion of parallel transport was not available, a clear distinction between the manifold and its tangent space in a point was only emerging and was not made explicit. Geodesic transport of frames was then defined in terms of ideal vectors which were introduced in such a way that the resulting relations did not depend on them explicitly.

---

[23] Gerhard Hessenberg, "Vektorielle Begründung der Differentialgeometrie," *Mathematische Annalen* 78 (1917): 187–217, here 192: "Der Weg, den wir im Folgenden einschlagen werden, führt zu den Christoffelschen Dreizeigergrößen über die Formeln (12) und (13). Dabei wird sich zeigen, daß für die abzuleitenden Invarianzen, auch diejenige des Krümmungstensors, nur die Formeln (12) wesentlich sind, während die christoffelsche Symmetrie (13) völlig ausgeschieden werden kann. Ihre Bedeutung läßt sich dagegen in der Aussage zusammenfassen, daß in der Geometrie der betrachteten n-dimensionalen Mannigfaltigkeit die '*geradesten*' Linien zugleich die '*kürzesten*' sind."

[24] Jan A. Schouten, *Grundlagen der Vektor-und Affinoranalysis* (Leipzig: Teubner, 1914).

Schouten then developed the notion of parallel transport by arguing that one needs a geodesically co-moving coordinate system that follows a geodesic line and at each point of the geodesic line is uniquely determined in its orientation. At this point, he illustrated this idea with the above-mentioned models of surfaces of positive and negative curvature. The illustration consists, on the one hand, in drawing frames of reference onto the curved surfaces, i.e. by depicting the moving frames in this curved two-dimensional manifold directly as a pair of orthogonal vectors in the local tangent spaces. For the case of the sphere, he also showed a piece of paper that can be laid out, without tearing or wrinkling, onto the sphere along the connecting line (not the geodesic) where we follow the rigid frame. The picture thus illustrated the ability to develop the notion of geodesic transport.[25] Schouten also created two examples of parallel transport on surfaces with negative curvature (Fig. 2): he painted geodesically moving frames onto models of a hyperbolic paraboloid and on a model of a pseudosphere, i.e. a rotated tractrix. Together with the illustration of the geodesic transport on the spherical surface (Fig. 1), these models seem to be the first visual illustration of the concept of parallel transport, i.e. a two-dimensional illustration of a vector parallel transported along a path in a two-dimensional curved space. If the transport is done along a closed loop, e.g. along a closing parallelogram, the difference between the initial and the final orientation of the frame is a measure of the integrated curvature of the enclosed area. Such a loop is nowadays often referred to as a Levi-Civita parallelogram.

Before continuing with our story, we should mention also Hermann Weyl's concept of infinitesimal geometry. His work was published together with the first edition of his highly influential book entitled *Space-Time-Matter*, the preface of which was dated "Easter, 1918." In this preface Weyl wrote that he could make use of Levi-Civita's paper, but that Hessenberg's paper had appeared only just before his book was going to press. Weyl clearly formulated the problem in terms of his philosophy of a "pure near geometry" ("reine Nahegeometrie"). Perhaps more importantly, Weyl realized very clearly in this context that the affine connection is an independent concept that can be defined without recourse or reference to the notion of a metric. Schouten listed and discussed Hessenberg's paper, and he also refers to Levi-Civita's work. However, it appears that Schouten's creation of the notion of geodesic transport was an independent discovery. In fact, Dirk Struik, who was a collaborator of Schouten at the time, later reminisced:

> Schouten and Levi-Civita had thus obtained the same result, but there were differences in the way each of them introduced parallelism. Schouten's method was entirely intrinsic, whereas that of Levi-Civita utilized a surface embedded in space. He also had derived his result only for the case $n = 2$, although it was clear that it was intrinsic and valid for all values of n. The main difference, however, insofar as influence was concerned, was that Levi-Civita's text was elegant and employed his absolute differential calculus (the tensor

---

[25] As Struik emphasized, the part of the illustration that shows the piece of paper for the spherical case was left out in Laue's reproduction of the illustration.

calculus with which mathematicians all over the world were becoming familiar) whereas Schouten's work was difficult to read due to its unfamiliar notation. And, of course, Levi-Civita also had priority of publication, so that the discovery has since become known as the 'parallelism of Levi-Civita.'[26]

As to the question of priority, Struik remarked:

> Schouten published this work in 1918. Although he translated some of his formulas into the language of the Ricci-Einstein tensor calculus, his theory was so overloaded with symbols that it proved next to impossible to follow. Direct analysis is fine for vectors, when only two multiplications · and x are involved, but for higher systems one gets lost in the maze of the dots, hooks, and crosses necessary to perform the various multiplications.
> Despite all this, Schouten had succeeded in giving a geometrical interpretation of the covariant derivative, an important accomplishment. Circumstances conspired against him, however, preventing him from being credited with a major mathematical achievement. It was nearing the climax of the war, so that communications with Italy were most difficult.
> Thus Schouten was totally unaware of Levi-Civita's work. I still remember how Schouten came running into my office one day waving a reprint he had just received of Levi-Civita's paper. 'He has it, too!' he cried out.[27]

Struik's recollection many years later may be subject to the skepticism that all such reminiscences should evoke. But we do have contemporary evidence to back up some of Struik's claims. Schouten refers to Levi-Civita's "Nozione" paper in his 1918 article. In a footnote, he wrote:

> T. Levi-Civita conceived of the concept of parallelism in a general space already in 1917 in a work of which I only received an offprint by friendly mediation after finalizing this manuscript due to the political circumstances. The geodesically comoving frame successively takes on positions, which are parallel in the sense of Levi-Civita, and the notion of geodesic motion therefore is contained in the notion of parallelism. The relationship between the covariant differential and the geodesically co-moving frame, and the fundamental importance, which the motion of such a system carries for the geometric properties of space [...] Levi-Civita, however, has not yet been made aware of.[28]

A Mechanical Model of Parallel Transport

It is interesting to note that Schouten, perhaps due to his engineering background, also thought about other ways of illustrating his concept of geodesic transport. In a footnote to his 1918 paper, he wrote:

---

[26] Dirk J. Struik, "Schouten, Levi–Civita, and the Emergence of Tensor Calculus," in *The History of Modern Mathematics*, ed. David E. Rowe and John McCleary, vol. 2, *Institutions and Applications* (Boston: Academic Press, 1989), 98–105, here 104.

[27] Struik, "Schouten," 103.

[28] Schouten, *Die direkte Analysis*, 46.

For surfaces the geodesically transported frame can sometimes be realized by a three-dimensional mechanism. A Foucault pendulum which travels around any line of constant latitude along the earth thought to be at rest and perfectly spherical, will always remain oriented with respect to the geodesically co-moving coordinate system, and the same will hold when the pendulum runs along any arbitrary curve on the surface. We can also construct a two-wheeled mechanism by means of a differential wheel, which rolled over the surface will show at each point the geodesically co-moving system.[29]

Schouten's reference to the Foucault pendulum references a discussion about realizing what he called a *Kompaszkörper*, i.e. a rigid solid whose motion realizes the geodesic transport of a frame. The question whether an extended rigid solid has enough degrees of freedom to realize geodesic motion in an arbitrary curved manifold was clarified in discussion with Adriaan D. Fokker (1887–1972). The problem is whether different points of a rigid frame can each follow the geodesic trajectories of curved space while at the same time preserving their mutual distances, a more detailed discussion of this problem by Fokker[30] was communicated to the Amsterdam Academy a year later. The problem was still relevant for the well-known Gravity Probe B experiment of the early twenty-first century.[31]

In our context, the other reference is equally interesting. Schouten did not explicate his idea any further than this, but it seems clear that what he refers to is the same as what was and is known as a *south-pointing chariot*.[32] Such a mechanical device is said to have been known already in ancient China and to have been built in that period, although the documentary evidence is thin. Later reconstructions and rebuilds nevertheless abound. Figure 6 shows a picture of a reconstruction of such a device that was erected in front of the National Museum in Taipeh. Figure 7 shows a model built using Fischer-Technik and was designed by Thomas Püttmann.

The idea is to create a mechanism that is devised in such a way that on a plane surface any difference in path length between the left and the right wheels is compensated by a differential gear mechanism such that a flag or pointer connected to the mechanism will always point in the same direction, even if the carriage is

---

[29] Schouten, *Die direkte Analysis*, 50: "Für gewöhnliche Flächen kann das geodätisch mitbewegte Bezugssystem manchmal durch einen dreidimensionalen Mechanismus realisiert werden. Ein Foucault'sches Pendel welches irgend einen Breitenkreis entlang die ruhend und kugelförmig gedachte Erde umkreist, bleibt stets zu einem geodätisch mitbewegten Koordinatensystem orientiert und das Selbe gilt, wenn das Pendel irgend eine beliebige Kurve auf der Oberfläche durch läuft. Auch mit Hilfe eines Differentialrades ließe sich ein zweirädriger Mechanismus konstruieren, welcher, rollend über die Oberfläche geführt, in jedem Punkte das geodätisch mitbewegte System anzeigt."

[30] Adriaan D. Fokker, "On the equivalent of parallel translation in non-Euclidean space and on Riemann's measure of curvature," *Proceedings of the Section of Sciences. Koninklijke Akademie van Wetenschappen te Amsterdam* 21, no. 1 (1919): 505–17.

[31] Egbertus P. J. de Haas, "The geodetic precession as a 3D Schouten precession and a gravitational Thomas precession," *Canadian Journal of Physics* 92, no. 10 (2014): 1082–93.

[32] See, for example: Dierck-E. Liebscher, "Mit dem Kompasswagen über den Globus," *Der mathematisch-naturwissenschaftliche Unterricht* 52 (1999): 140–44.

Fig. 6 Model of a south-pointing chariot in front of the National Museum in Taipeh. From Dierck-E. Liebscher, "Mit dem Kompasswagen über den Globus," *Der mathematisch-naturwissenschaftliche Unterricht* 52 (1999): 140–144, Fig. 4. All rights reserved

pushed along curved paths. Naturally, the device works on curved surfaces as well and will then indicate the path difference between the two wheels on the curved ground. Pushing the carriage along a closed loop, i.e. along a Levi-Civita loop, the difference between the initial and final position of the pointer will not in general coincide and will instead be a measure of the local curvature integrated over the loop.

It is unclear whether Schouten had anything like this in mind when he added his footnote about the mechanical device, but it appears quite possible, if not likely given his intellectual background as an engineer.

Fig. 7 Model of a south-pointing chariot realized with Fischer-Technik according to a design by Thomas Püttmann. © Photo: Tilman Sauer, all rights reserved

Later History

It seems a notable fact that the very same photographic images of Schouten's models for geodesic frame transport were reproduced three more times between 1921 and 1924 after the initial publication in 1918. I have not seen later instances of reproduction. In fact, I have not come across later instances of *material* models of curved surfaces for the purpose of illustrating the notion of parallel transport of vectors. But I cannot claim that such material models do not exist.

Graphical representations of parallel transport of vectors or connections of tangent spaces to two-dimensional surfaces embedded in three-dimensional space, however, have become rather common, both in textbooks on general relativity as on differential geometry, although they by no means accompany each and every analytical discussion.

As an example of such common illustrations of parallel transport, Fig. 8 illustrates the definition of the covariant derivative of a vector field v along a curve $P(\lambda)$ in an abstract sense. Figure 9 then illustrates the notion of connection coefficients by geodesic transport of two-dimensional frames over a spherical surface.

These illustrations appear in the widely used, highly influential textbook on relativity and gravitation published by Charles W. Misner, Kip S. Thorne, and John A. Wheeler in 1973, a textbook that became a standard source of reference for generations of physicists for many reasons, one of them being its emphasis on physical intuition and visual illustration. In fact, the authors quite explicitly reflected on the necessity of what they called a "pictorial treatment" of geometry: "Gain the power […] to discuss tangent vectors, 1-forms, tensors in curved spacetime; gain the power […] to parallel-transport vectors, to differentiate them, to discuss geodesics; use this power […] to discuss geodesic deviation, to define

Fig. 8 Graphical Representation of the covariant derivative of a vector field along a curve. From Charles W. Misner, Kip S. Thorne, and John A. Wheeler, *Gravitation* (New York: Freeman, 1973), 209, Fig. 8.2. © Freeman, all rights reserved

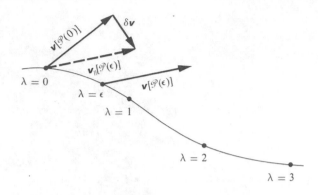

Fig. 9 Graphical Representation of geodesic transport of a frame over a curved spherical surface. From Charles W. Misner, Kip S. Thorne, and John A. Wheeler, *Gravitation* (New York: Freeman, 1973), Fig. 8.3. © Freeman, all rights reserved

curvature, [...] But full power this will be only if it can be exercised in three ways: in pictures, in abstract notation, and in component notation."[33]

Another example of how Schouten's illustrative model of parallel transport has found its way onto modern textbooks is shown in Fig. 10, taken from V.I. Arnold's, *Mathematical Methods of Classical Mechanics*.[34] Although the vector that is being parallel-transported here points off the surface, the illustration almost looks like a graphical representation of Schouten's original model.

---

[33] Charles W. Misner, Kip S. Thorne, and John A. Wheeler, *Gravitation* (New York: Freeman, 1973), 198.

[34] Vladimir I. Arnold, *Mathematical Methods of Classical Mechanics* (New York: Springer, 1989).

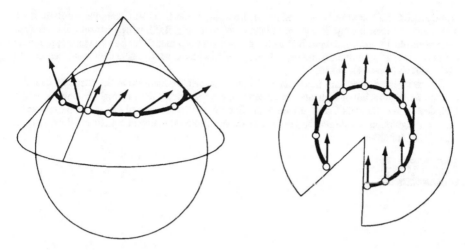

Fig. 10 Graphical representation of parallel transport on the sphere as shown in Vladimir I. Arnold, *Mathematical Methods of Classical Mechanics* (New York: Springer, 1989), 302, Fig. 231. © Springer, all rights reserved

Concluding Remarks

The transitional nature of Schouten's model of parallel transport is indicative of the primacy of the ideal in mathematics. Parallel transport, even though it can be regarded as a genuine geometric concept, is defined and represented exclusively in analytic terms for an arbitrary number of dimensions. Geometric intuition is derived from finite objects and their properties in three-dimensional Euclidean space. Our intuition for higher dimensions, curvature, or extensions to infinity, even though genuinely geometrical, must be assisted by and eventually based on other means. Schouten's model for geodesic transport was mostly and from the beginning merely illustrative. It built on a well-established tradition of models that were used for teaching. Yet it was illustrative for a conceptual problem that was at the time still in the process of being explored and discovered. It investigated a notion that is inherently n-dimensional and associated with curvature, and it helps our understanding of the abstract concept by providing an interpretation of the special and intuitive case of a curved two-dimensional surface embedded in three-dimensional Euclidean space. His mechanical devices, like the Foucault pendulum or the south-pointing chariot are also illustrative but in a different sense. Their physical properties help us understand the necessary implications of the abstract concept and may indeed have played an important role in Schouten's heuristics of concept development.

The Great Yogurt Project: Models and Symmetry Principles in Early Particle Physics

Arianna Borrelli

Introduction: The Coral Gables Conferences on "Symmetry Principles at High Energy" and the Yogurt Project

Between the years 1964 and 1968 five conferences on "Symmetry Principles at High Energy" were held at the University of Miami in Coral Gables.[1] Behram Kursunoglu, particle theorist and professor in Miami, who, in 1965, became founding director of the local Center for Theoretical Studies, organized the meetings.[2] As explained by Kursunoglu and his co-editor Arnold Perlmutter in the short preface to the proceedings of the first conference, the motivation for the initiative was to reflect and discuss theoretical methods that had begun to be employed in particle

---

[1] Behram Kursunoglu and Arnold Perlmutter, eds., *Coral Gables Conference on Symmetry Principles at High Energy* (San Francisco: Freeman, 1964); Behram Kursunoglu, Arnold Perlmutter, and Ismail Sakmar, eds., *Coral Gables Conference on Symmetry Principles at High Energy. Second Conference* (San Francisco: Freeman, 1965); Arnold Perlmutter, Joseph Wojtaszek, George Sudarshan, and Behram Kursunoglu, eds., *Coral Gables Conference on Symmetry Principles at High Energy. Third Conference* (San Francisco: Freeman, 1966); Arnold Perlmutter and Behram Kursunoglu, eds., *Coral Gables Conferences on Symmetry Principles at High Energy. Fourth Conference* (San Francisco: Freeman, 1967); Arnold Perlmutter, Behram Kursunoglu, and C. Angas Hurst, eds., *Coral Gables Conferences on Symmetry Principles at High Energy. Fifth Conference* (New York: Benjamin, 1968).

[2] Behram Kursunoglu, "The Launching of the Coral Gables Conferences on High Energy Physics and Cosmology and the Establishment of the Center for Theoretical Studies at the University of Miami," in *High-Energy Physics and Cosmology: Celebrating the Impact of 25 Years of Coral Gables Conferences*, ed. Behram Kursunoglu, Stephan L. Mintz, and Arnold Perlmutter (Boston: Springer, 1997), 5–18.

A. Borrelli (✉)
Technische Universität Berlin, Sekretariat H 23, Straße des 17. Juni 135, 10623 Berlin, Germany
e-mail: aborrelli@weatherglass.de

M. Friedman and K. Krauthausen (eds.), *Model and Mathematics: From the 19th to the 21st Century*, Trends in the History of Science,
https://doi.org/10.1007/978-3-030-97833-4_6

physics "during the past few years" and which were "sufficiently novel to warrant frequent gatherings of experts in particle physics."[3] These methods, referred to as "symmetry principles," were largely based on the use of group theoretical structures.

Born and raised in Turkey, Kursunoglu jokingly inserted in the proceedings of the second conference a series of anecdotes about Nasreddin Hoja, a traditional jester figure from Turkish folklore. Among them was this story:

> One morning a woodcutter saw Hoja by the edge of a lake, throwing quantities of yeast into the water. 'What the devil are you doing, Hoja?' he asked. Hoja looked up sheepishly and replied, 'I am trying to make all the lake into yogurt.' The woodcutter laughed and said, 'Fool, such a plan will never succeed.' Hoja remained silent for a while, and stroked his beard. Then he replied, 'But just imagine if it should work!'[4]

In 1968, at the fifth and last conference of the series, Israeli theorist Yuval Ne'eman was asked to hold a final talk summing up the results of all five meetings and, at the end of his speech, he expressed his appreciation for Hoja:

> I also express general sentiment when I say that we have all especially appreciated the proceedings of the second conference, with Nasreddin Hoja's contributions. I wonder whether Kursunoglu can sometime tell us where Hoja learned his physics. I particularly appreciated Hoja's Yogurt project.[5]

Ne'eman did not explain why he especially liked the yogurt story among the various Hoja anecdotes scattered in the 1965 proceedings, but it certainly suggests a disillusioned attitude towards the research program brought forward in the first five Coral Gables conferences: a dream many theorists had believed in, but that so far had not paid out. This impression is strengthened by the words immediately following the mention of the yogurt project:

> I only have to add now a piece of information which may not be known to you, and this is that I am not only the summarizer of these five conferences but I am also the undertaker responsible for their lying in peace forever. The series of conferences on 'Symmetry Principles at High Energy' is hereby closed. I think Prof. Kursunoglu has felt that they represented such a nice achievement that one should not risk stretching it too much.[6]

---

[3] Behram Kursunoglu and Arnold Perlmutter, "Preface," in *Coral Gables Conference on Symmetry Principles at High Energy*, ed. Behram Kursunoglu and Arnold Perlmutter (San Francisco: Freeman, 1964), [i].

[4] Kursunoglu, Perlmutter, and Sakmar, eds., *Coral Gables,* 1965, 5.

[5] Yuval Ne'eman, "Progress in Five Coral Gables Conferences on Symmetry Principles at High Energy," in *Coral Gables Conferences on Symmetry Principles at High Energy. Fifth Conference*, ed. Arnold Perlmutter, Behram Kursunoglu, and C. Angas Hurst (New York: Benjamin, 1968), 355–76, here 375–76.

[6] Ne'eman, "Progress," 376.

Ne'eman explained that the next Coral Gable conference would be devoted to a new topic: "Fundamental Interactions."[7] From today's point of view it seems difficult to account for Ne'eman's downbeat mood, or for the decision to put an end to the conferences on symmetries, since the 1960s appear in retrospect as the time in which symmetry principles, especially in the form of group theory, became one of the main driving forces behind theoretical research in particle physics. Indeed, recollections of theorists, reflections of philosophers, and even some works by historians of science have expressed the conviction that, in the 1950s and 1960s, symmetry principles and group theoretical methods, which at the time were not usually regarded as equivalent, proved to be an effective heuristic tool to mathematically represent the many newly observed particle phenomena and allow for the prediction of as yet unobserved ones.[8] From this background, theoretical developments in early particle theory have often been framed in terms of a progressing, successful application of abstract principles of mathematical invariance to the representation and explanation of particle phenomena. As I have shown elsewhere for the early 1950s, however, a closer analysis of primary historical sources does not support this claim.[9] In the present paper I will focus on the late 1950s and early 1960s, showing how in that period, too, the relationship between group theory and the mathematical conceptualization of phenomena was not as straightforward as usually presented when we look at it today.

---

[7] Ibid.

[8] Ronald Anderson and Girisch Joshi, "Interpreting Mathematics in Physics: Charting the Applications of SU(2) in 20th Century Physics," *Chaos, Solitons & Fractals* 36, no. 2 (April 2008): 397–404; Katherine Brading, Elena Castellani, and Nicholas Teh, "Symmetry and Symmetry Breaking," *The Stanford Encyclopedia of Philosophy* (Winter 2017), https://plato.stanford.edu/archives/win 2017/entries/symmetry-breaking/ (accessed September 23, 2020); Murray Gell-Mann, "Particle Theory from S-Matrix to Quarks," in *Symmetries in Physics (1600–1980): Proceedings of the 1st International Meeting on the History of Scientific Ideas held at Sant Feliu de Guíxols, Catalonia, Spain, September 20–26, 1983*, ed. Manuel G. Doncel, Armin Hermann, Louis Michel, and Abraham Pais (Barcelona: Universitat Autònoma de Barcelona, 1987), 473–97; Yuval Ne'eman, "Hadron Symmetry, Classification and Compositeness," in *Symmetries in Physics (1600–1980): Proceedings of the 1st International Meeting on the History of Scientific Ideas held at Sant Feliu de Guíxols, Catalonia, Spain, September 20–26, 1983*, ed. Manuel Doncel, Armin Hermann, Louis Michel, and Abraham Pais (Barcelona: Universitat Autònoma de Barcelona, 1987), 499–555; Andrew Pickering, *Constructing Quarks: A Sociological History of Particle Physics* (Chicago: University of Chicago Press, 1984), 56–57; Silvan S. Schweber, "Quantum Field Theory: From QED to the Standard Model," in *The Cambridge History of Science*, ed. Mary Jo Nye, vol. 5: *The Modern Physical and Mathematical Sciences* (Cambridge: Cambridge University Press, 2003), 375–93, here 386–87; Steven Weinberg, "Changing Attitudes and the Standard Model," in *The Rise of the Standard Model: Particle Physics in the 1960s and 1970s*, ed. Lillian Hoddeson, Laurie Brown, Michael Riordan, and Max Dresden (Cambridge: Cambridge University Press, 1997), 36–44; Steven Weinberg, "Symmetry: A key to Nature's Secrets," *New York Review of Books* 58, no. 16 (October 2011).

[9] Arianna Borrelli, "The Making of an Intrinsic Property: 'Symmetry Heuristics' in Early Particle Physics," *Studies in History and Philosophy of Science* A 50 (April 2015): 59–70.

The main thesis of this paper is that theorists engaging in the construction of models or theories of particle phenomena in the 1950s and early 1960s did not in general regard abstract group theoretical constructs as particularly effective tools for the job. Some authors made no use of the concepts of group theory at all, while others only employed them in a very selective way, recurring only to those group structures that could be related to the invariances of space or space-time, like rotations, mirroring or translations. Interestingly, authors following the latter approach tended to employ the term 'model' to indicate purely descriptive constructs, and 'theory' to characterize those constructs they hoped might eventually be seen as explaining the laws behind particle phenomena. It was in building such 'theories' that they believed space-time invariances would provide an effective guideline. Thus, in the 1950s and early 1960s notions of group theory were primarily employed in particle physics by a specific group of theorists who sought not just to fit observation, but also to mathematically capture the features of fundamental microphysical interactions. Incidentally, their efforts were largely unsuccessful. I will support my claim by discussing two historical constellations: proposals made in the years 1953–1956 aimed at expanding the Gell-Mann-Nishijima model of particle classification into a 'theory' whose invariances were similar to those of space-time, and the systematic attempts to connect internal particle symmetries (e.g. isospin, strangeness) to space-time invariances which were discussed at the Coral Gable Conferences in the early 1960s.

In the next two sections of this paper, I will address some methodological issues regarding, first, the terms 'model' and 'theory' and, second, the distinction between different mathematical practices. In context of the second topic I will sketch the practices linked to rotation phenomena in quantum physics, and the way in which the formalism of non-relativistic spin provided the template first in the 1930s for describing nuclear interaction and then in the 1950s for formulating the first empirically successful scheme of particle classification: the so-called Gell-Mann-Nishijima model. After that, I will discuss the (ultimately unsuccessful) attempts made during the 1950s of either replacing or embedding that model in a broader mathematical construct based on invariances similar to those of relativistic space-time. In section "The Path from SU(2) to SU(3), or: Did Particle Physicist Know Group Theory?," I will summarize the way in which, in the early 1960s, the Gell-Mann-Nishijima model was embedded in an empirically successful classification based on the group SU(3) (the eightfold way and the quark model), and that this happened through a complex path essentially unrelated to the projects described in section "The Search for a Theory of Isospin and Strangeness in the 1950s." Finally, sections "Beyond SU(3)—The Mathematical Marriage of Space-Time and Internal Symmetries" and "The Rise and Fall of SU(6)" discuss the research program unfolding in the five Coral Gables conferences on "Symmetry Principles at High Energy," following the (ultimately unsuccessful) attempts to expand the eightfold way and the quark model by combining them with the symmetries of relativistic space-time.

'Models' and 'Theories' as Actors' Categories in Early Theoretical Particle Physics

There is in philosophy of science a long tradition of trying to characterize both the distinction and the mutual relationship between models and theories, and in particular their specific functions within scientific practice.[10] No consensus exists so far on whether and how a distinction between the two may be drawn, and in the present study I will assume that all theoretical practices analyzed are equally well characterized as modeling or theorizing. However, I will endeavor to trace an (admittedly vague) distinction made by some, though by far not all, historical actors between 'models' (or 'schemes') on the one hand and 'theories' on the other. I will argue that, while both 'models' and 'theories' were seen as mathematical constructions whose prediction were expected to fit observation, the latter could be associated with an additional explanatory function, or at least with the hope that the mathematical structures involved in some way captured features of natural order beyond observable phenomena. Theorists endorsing this position in general did not attempt to bring forward arguments to support it, but usually simply expressed their belief that certain mathematical principles had a closer relationship to fundamental physical laws.

Theorists who drew the distinction sketched above between 'models' and 'theories' were also the ones who made explicit use of group theoretical structures, and more in general of abstract mathematical notions. However, what is particularly interesting is that they did not regard all group-theoretical concepts as equally relevant from a physical point of view, but rather pinned their hopes on symmetries linked to invariances of three-dimensional space or relativistic four-dimensional space-time. Thus, the focus was for example on the group O(3) (rotations and mirror transformations in three dimensions), on the Lorenz group (rotations in four-dimensional space-time), or on the Poincaré group (rotations and translations in four-dimensional space-time). The authors did not offer any particular reason why these specific mathematical structures should be epistemically preferred and, in the end, their efforts in constructing theories along these lines were largely unsuccessful. Yet the research program which Ne'eman, and later on also Kursunoglu, compared to Hoja's yogurt project was in so far heuristically fruitful, that it generated a number of new, more limited theoretical approaches to studying high energy phenomena by means of group-theoretical methods.

In conclusion, if one asks what role abstract mathematical notions, and more specifically group theory played in theoretical practices of early particle physics, the answer is that they were predominantly used by authors who made a distinction between 'models' and 'theories,' and expected their constructs to be 'theories'

---

[10] For recent perspectives on these topics, see: Roman Frigg and Stephan Hartmann, "Models in Science," *The Stanford Encyclopedia of Philosophy* (Summer 2018), https://plato.stanford.edu/archives/sum2018/entries/models-science/ (accessed September 23, 2021); Mary Morgan and Margaret Morrison, eds., *Models as Mediators: Perspectives on Natural and Social Sciences* (Cambridge: Cambridge University Press, 1999).

which expressed fundamental features of nature. However, these theorists did not regard all invariances as fit to serve that aim, and rather focussed on group theoretical structures that could be associated to the symmetries of space-time. As we shall see, from this point of view empirically successful constructs were perceived as in some sense arbitrary if they did not conform to the expectation that all of the symmetry in nature should be similar to that occurring in space-time. Other theorists instead, although in practice employing the same mathematical methods, did not refer to them in terms of abstract concepts like group representations. Because of this, it is important in the analysis to make a methodological distinction between the performance of certain mathematical operations and the explicit use of the relevant abstract notions. This is the topic of the next section.

Mathematical Practices of Rotations and the Emergence of the Gell-Mann-Nishijima Model of Particle Classification

In the following pages I will often speak not of mathematics, but rather of mathematical practices. This choice allows for much more flexibility when trying to reconstruct what historical actors were doing (or not doing) when making use of mathematical formalisms and notions, and what aim they were (or were not) pursuing. An example of this distinction that is of relevance for our topic is the rotations of objects in three-dimensional space. Rotations of solid objects around themselves were known and had been studied since Antiquity, and at the latest in the early modern period geometrical and analytical formalisms for modeling them were developed.[11] During the nineteenth century this formalism allowed both to study in greater detail rotating bodies, and to connect those experiences to more abstract notions of rotations, such as the interplay of electric and magnetic forces.[12] With the emergence of vector analysis around the end of the nineteenth century, the rules of rotation for solid bodies were interpreted as pertaining to the mathematical objects labeled as vectors, while other rules were valid for axial vectors or tensors.[13] Another mathematical perspective was added by group theory, which also emerged in that period, and in whose context the rules for rotations

---

[11] Arianna Borrelli, "Angular Momentum Between Physics and Mathematics," in *Mathematics Meets Physics*, eds. Karl-Heinz Schlote and Martina Schneider (Frankfurt am Main: Harri Deutsch, 2011), 395–440.

[12] Olivier Darrigol, *Electrodynamics from Ampère to Einstein* (Oxford: Oxford University Press, 2003), 137–76.

[13] Michael J. Crowe, *A History of Vector Analysis: The Evolution of the Idea of a Vectorial System* (Notre Dame: University of Notre Dame Press, 1967).

of vectors, tensors and other objects could be conceptualized as different representations of a more abstract notion: the group of rotations in three-dimensional space.[14]

It is neither necessary nor possible to discuss here the details of these complex historical developments, but it is important to note that, despite the emergence of all these new mathematical reflections, the rules for rotating a vector in space remain quite simple and can be employed without any knowledge of vector analysis or group theory. Moreover, even if a person has (some) knowledge of those fields of mathematics, this does not imply that she or he is conceiving the practice of analytically representing the rotation of a force or a velocity as the employment of a representation of the group of rotations in three-dimensional space. This rather trivial observation becomes relevant for our topic when we consider the subject of particle spin, which is of paramount importance for the history of particle theory.

During the early 1920s, in the context of emerging quantum physics, the spectroscopic and chemical properties of atoms were described in terms of a new degree of freedom attributed to electrons, which came to be referred to as 'spin,' as it was initially tentatively thought of as linked to a rotation of the electron around its axis.[15] On the basis of spectroscopic results, Wolfgang Pauli formally represented the new degree of freedom by means of a two-component object (Pauli spinor). Electrons were spin doublets, since they only had two possible spin states: up or down. The rules according to which Pauli spinors transformed under space rotations were expressed in the form of the so-called Pauli matrices, and Hermann Weyl later recognized these as a representation of the group SU(2).[16] With the emergence of a Paul Dirac's quantum relativistic theory of electrons, Pauli spinors came to be interpreted as non-relativistic versions of the transformation properties of the quantum fields associated to electrons (and other particles) under relativistic space-time rotations. These in turn could be thought of in terms of an 'intrinsic angular momentum' which formally took the value 1/2 for some particles (e.g. electrons, protons, neutrons) and 0 or 1 for others (e.g. pions, photons). In the non-relativistic limit, the intrinsic angular momentum determined the way in which particles transform with respect to rotations in three-dimensional space. Thus, what had started as a simple, ad-hoc formalism to represent spectroscopic data had eventually been interpreted as a manifestation of mathematically more

---

[14] Hans Wussing, *Die Genesis des abstrakten Gruppenbegriffes: Ein Beitrag zur Entstehungsgeschichte der abstrakten Gruppentheorie* (Berlin: VEB Deutscher Verlag der Wissenschaften 1969), 144–84.

[15] For a short but exhaustive overview on the development of non-relativistic and relativistic spin, with special attention to the interplay between physics and mathematics and the seminal contribution of Bartel L. van der Waerden, see: Martina Schneider, *Zwischen zwei Disziplinen: B. L. van der Waerden und die Entwicklung der Quantenmechanik* (Berlin: Springer, 2011), 39–42 and 126–36.

[16] Schneider, *van der Waerden*, 40. The special unitary group SU(n) is the group of $n \times n$ unitary matrices with determinant 1. The group SU(2) is therefore the group of 2×2 matrices with determinant 1, which can all be represented as linear combination of three basis matrices used by Pauli to represent the three components of electron spin, and today usually known as the Pauli matrices.

complex invariance properties of relativistic space-time. However, at first only mathematicians took interest in the more refined group-theoretical implications of these developments, while physicists only employed those mathematical tools strictly necessary for their work. More specifically, from the 1940s onward, scientists working in atomic, molecular and nuclear physics became fully familiar with the Pauli spinor formalism, which played an important role in their fields, but they did not necessarily know or care much about the relevant group theoretical notions, such as SU(2). In turn, the Pauli formalism provided the template for isospin, a new particle property originally introduced in nuclear physics the 1930s and extended to the many new particles discovered in cosmic ray and accelerator experiments from the late 1940s onward.

The formalism of isospin had been developed during the 1930s and 1940s in formal analogy to intrinsic angular momentum to mathematically represent interactions in atomic nuclei.[17] To this end, the proton and the neutron were represented as two components of an 'isospinor' formally, but not physically, analogous to a Pauli spinor. This formalism proved useful in nuclear physics, although until the 1950s it was not regarded as expressing a physical quantity. Later on, the three pions, which had positive, negative and zero charge, were assumed to form an isospin triplet transforming like a vector under rotations in 'isospace.' Of course, rotations in 'isospace' had nothing to do with rotations in real space, yet they were formally fully analogous to them, so that, to manipulate isospin, it was not necessary to know anything about group theory, but it sufficed to be familiar with the formalism of vectors and (two-component) spinors commonly used in quantum physics to represent angular momentum and spin. In the early 1950s it was confirmed that further particles existed and were capable of strong interactions. They were variously known as V-particles, tau-mesons, and later on even as 'curious' or 'strange' particles, and experimental results on their possible charges, masses and decay modes were in constant flux.[18] Nonetheless, some theorists attempted to come up with schemes describing their properties and interactions, and in this context some authors tentatively attributed isospin to them.

The first proposals of this kind were made in 1951 by Japanese theorists, among them Kazuhiko Nishijima, and their approach was soon taken up by U.S. American physicists, among them Murray Gell-Mann.[19] For the most part, these authors tried to fit the (rapidly shifting) experimental results on decay modes of the new particles by grouping them into isospin doublets, triplets or singlets without making

---

[17] For a detailed discussion of the emergence of isospin and its changing epistemic status, see: Arianna Borrelli, "The Uses of Isospin in Early Nuclear and Particle Physics," *Studies in History and Philosophy of Science Part B: Studies in History and Philosophy of Modern Physics* 60 (November 2017): 81–94.

[18] Francesca Bresolí Català, *Descobriment experimental de les partícules estranyes i construcció teòrica del concepte d'estranyesa (1947–1957)* (Universitat Autònoma de Barcelona, 2006), http://www.tdx.cat/handle/10803/3383 (accessed September 23, 2021).

[19] Arianna Borrelli, "Constructing Strangeness: Exploratory Modeling and Concept Formation," *Perspectives on Science* 29, no. 4 (August 2021): 388–408.

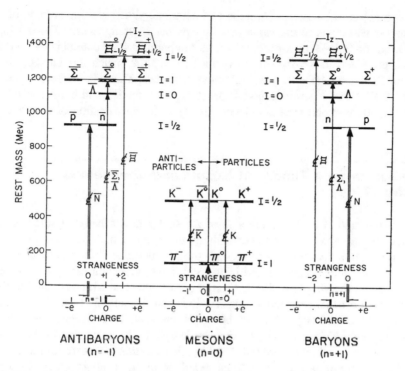

Fig. 1 Diagram representing the properties of the "strange" particles in: Murray Gell-Mann and Arthur H. Rosenfeld, "Hyperons and Heavy Mesons (Systematics and Decay)," *Annual Review of Nuclear Science* 7 (December 1957): 407–78, here 415, Fig. 1; permission conveyed through Copyright Clearance Center Inc. © Annual Reviews, Inc., all rights reserved

use of abstract group-theoretical notions. In the years 1955–56, an approach developed independently by Nishijima and by Gell-Mann turned out to be empirically quite successful.[20] Gell-Mann was the first one to make use of graphic representations of the model, placing the symbols he had introduced for labeling the new particles in a diagram with mass and electric charge displayed on the vertical and horizontal scales (see Fig. 1).[21] Gell-Mann described his construct by stating that the new particles formed isospin multiplets which were 'displaced' with respect

---

[20] Murray Gell-Mann, "The Interpretation of the New Particles as Displaced Charge Multiplets," *Il Nuovo Cimento (1955–1965)* 4, no. S2 (April 1956): 848–66; Kazuhiko Nishijima, "Charge Independence Theory of V Particles," *Progress of Theoretical Physics* 13, no. 3 (March 1955): 285–304.

[21] Murray Gell-Mann and Arthur Rosenfeld, "Hyperons and Heavy Mesons (Systematics and Decay)," *Annual Review of Nuclear Science* 7 (December 1957): 407–78.

to those of the old particles (protons and neutrons, pions) and the amount of displacement was equal to the value of a new property: strangeness.[22] For our topic, the details of Gell-Mann's proposal are not important: what counts is the fact that it was presented in a very simple mathematical form which left unchanged the isospin formalism, only 'displacing' some multiplets. This scheme became known as the 'Gell-Mann-Nishijima model' or 'strangeness model' and in the second half of the 1950s was accepted as an empirically valid classification of new and old particles.

The Search for a Theory of Isospin and Strangeness in the 1950s

Both before and after the emergence of the strangeness scheme, a small number of theorists attempted to employ considerations based on mathematical invariances and group-theoretical notions to classify the new particles.[23] In this section I will discuss a few such proposals and in particular the motivation brought forward by their authors, showing how they preferred groups identical with or related to those of (relativistic or non-relativistic) space-time transformations. One of the earliest and most persistent supporters of this approach was Abraham Pais who, already in 1953, sought a way not only to attribute isospin to the new particles, but also to take the isospin variable "seriously."[24] Significantly, this meant for him to connect isospin to space-time transformations described by the Lorentz group. However, he noted, "the exploration of the irreducible representations of the Lorentz group shows that there is no such freedom left,"[25] so he opted for regarding the isospin formalism as part of a three-dimensional "ω-space" in which, using the relevant group-theoretical structures, the formal equivalents of both angular momentum and parity could be defined. Pais described his proposal as a "theory"[26] and explained his key assumption so:

> The element of space-time is not a point but is a manifold ("ω-space") which is carried into itself by all transformations of a three-dimensional real orthogonal group. Now this is a fancy way to talk about a simple manifold namely a two-dimensional sphere. Yet the phrase

[22] The term 'strangeness' was introduced by Gell-Mann. Nishijima, who had been the first one to introduce the relevant variable, called it 'v-spin.' Nishijima, "Charge Independence Theory of V Particles," 285–304; see also: Borrelli, "Strangeness."

[23] Borrelli, "Symmetry Heuristics;" Arianna Borrelli, "Symmetry, Beauty and Belief in High-Energy Physics," *Approaching Religion* 7, no. 2 (November 2017): 22–36; Borrelli, "Uses of Isospin;" Arianna Borrelli, "The Weinberg-Salam Model of Electroweak Interactions: Ingenious Discovery or Lucky Hunch?" *Annalen der Physik* 530, no. 2 (February 2018), online: https://doi.org/10.1002/andp.201700454 (accessed September 23, 2021).

[24] Abraham Pais, "Isotopic Spin and Mass Quantization," *Physica* 19 (1953): 869–87, here 873.

[25] Ibid.

[26] Ibid., 870.

is used advisedly to direct attention to the group rather than to the sphere seen as a metric object or as being embedded in a three-dimensional Euclidean space.[27]

The details of Pais' proposal cannot be summarized here, yet we note how, on the one hand, he underscored the abstract notion of group as opposed to the intuitive image of a two-dimensional sphere, while on the other hand presenting his "ω-space" as something physical, and not as a purely formal construct, as it had been usually characterized so far. However, stating that each point in space-time is associated to a manifold did not add anything from the mathematical point of view: It was only a way to take isospin "seriously," attributing to it an epistemic status analogous to that of space-time and thus, at least in Pais' view, constructing not just an empirically successful 'model,' but an explanatory "theory" of elementary interactions. As we shall see later on, the explicit employment of certain group-theoretical structures in theoretical practices was often related to the wish of connecting to space-time. In other words, Pais' approach was as empirically motivated as those of Gell-Mann and Nishijima, but made use of more refined mathematical tools, which Pais tried to interpret as expressing underlying physical structures.

Pais's paper was among the earliest proposals on how to attribute isospin to the new particles, and some of its features served as a starting point for the schemes of Nishijima and Gell-Mann, yet those authors only took up specific points of Pais' theory, and not its general group-theoretical approach. In the same year, Pais developed his ideas further, both to relate to new experimental results and to explore the physical–mathematical implications of his earlier theory.[28] He presented his reflections as an initial move to approach a form of theorization guided by mathematical invariances: "The present work must, therefore, be viewed as a first step in employing new invariance principles. If this direction of approach proves fruitful it should be followed, by further refinement."[29] As in the first paper, Pais spoke of his "theory,"[30] and described the proposals by other authors, like Nishijima or Gell-Mann as "models."[31]

In 1954 Pais brought forward yet another, expanded version of his theory of ω-space. He explained that the previous scheme was "too narrow,"[32] and that ω-space should have not three, but four dimensions, corresponding to the orthogonal group O(4) of rotations in four-dimensional Euclidean space, which Pais noted was closely related to the Lorentz group of rotations in relativistic space-time.[33] In fact, at the end of the paper Pais mentioned attempts to relate space-time with isospin

[27] Ibid., 873.

[28] Abraham Pais, "On the Baryon-Meson-Photon System," *Progress of Theoretical Physics* 10, no. 4 (October 1953): 457–69, here 461.

[29] Ibid.

[30] Ibid., 462.

[31] Ibid., 460.

[32] Abraham Pais, "On the Program of a Systematization of Particles and Interactions," *Proceedings of the National Academy of Sciences* 40, no. 6 (June 1954): 484–92, here 484.

[33] Pais, "Systematization," 491.

more fundamentally then what he had proposed, suggesting that his efforts might provide indications on the properties of a "more complete theory."[34] Interestingly, in this paper he referred both to his own scheme and to the simpler construct of Gell-Mann as "models,"[35] using the word theory only to describe the "more complete theory" he was still looking for. In conclusion, Pais saw his proposals not just as attempts to describe observations, but also as first steps in constructing a theoretical structure expressing fundamental features of nature using as a guideline invariance principles and group-theoretical structures similar to those of space-time. In conclusion, Pais, Gell-Mann and Nishijima followed the same research program, whose primary goal was to come up with a mathematical classification of observed particle phenomena. Pais, however, tried to interpret some of the mathematical tools used in this process as related to underlying physical features, while Gell-Mann and Nishijima did not raise any such claims. The interesting point is that, for some reason which is today difficult to grasp, both Pais and later authors saw as potentially physically significant only those group-theoretical structures linked to space-time transformations.

Another author who, like Pais, attempted early on to employ space-time symmetries to classify the new particles was the Polish theorist Jerzy Rayski, who in 1954 proposed that "for every irreducible representation of the group of rotations and reflections corresponds a particle type."[36] If this hypothesis should be correct, he explained, "the situation will be more satisfactory from a group-theoretical viewpoint, moreover, we shall possess a clue for understanding the existence of various types of particles and their properties."[37] Like Pais' ideas, Rayski's proposal was quite short-lived and soon, as described in the previous section, the Gell-Mann-Nishijima strangeness model was recognized as empirically successful. Yet the strangeness scheme was not regarded by physicists as expressing fundamental properties of nature, and in 1956 Robert Oppenheimer, in his introduction to a theoretical session at the sixth Rochester conference on "High Energy Nuclear Physics," made a parallel between that model and the development of non-relativistic and relativistic spin from spectroscopy:[38]

---

[34] Ibid.

[35] Ibid., 487–88.

[36] Jerzy Rayski, "On a Systematization of Heavy Mesons and Hyperons," *Il Nuovo Cimento (1943–1954)* 12, no. 6 (December 1954): 945–47, here 946.

[37] Ibid., 945.

[38] The Rochester Conferences on High Energy Nuclear Physics (from 1958 onward on High Energy Physics) were the main yearly meeting point for the growing community of theoretical and experimental physicists studying the interactions of particles at high energies. Robert E. Marshak, "Scientific Impact of the first decade the Rochester Conferences (1950–1960)," in *Pions to Quarks: Particle Physics in the 1950s*, ed. Laurie Brown, Max Dresden, and Lillian Hoddeson (Cambridge: Cambridge University Press, 1989), 645–67.

Perhaps, using an analogy, one may say that [with the strangeness model] we are at a stage corresponding to the finding of the duplexity of atomic spectra, but not yet at the point of the discovery of electron spin, and certainly not at the stage of Dirac's theory of the electron.[39]

In the late 1950s, various authors attempted to develop the Gell-Mann-Nishijima model into a construct expressing the structure of subatomic physics, and, once again, group-theoretical considerations were employed to that end. However, just like in the case of Pais and Rayski, the groups chosen as guideline in this enterprise were those connected to space-time invariances. One might assume that, as suggested by Oppenheimer, the reflections of theorists were shaped by the example of spin, which, as we saw above, had emerged as an ad-hoc formalism to be revealed as expressing transformation properties of relativistic space-time.

The proposal by Pais served as the starting point for a new approach by Rayski,[40] as well as for the work of two theorists, the French Bernard d'Espagnat and French-Polish Jacques Prentki, both based at CERN at that time, who published a paper entitled "Mathematical Formulation of the Model of Gell-Mann."[41] In their paper, they compared the "theory" by Pais with the "model" by Gell-Mann, explaining how the first one was mathematically satisfactory, but experimentally problematic, while the latter one described experimental data well, but lacked a theoretical basis.[42] To solve this problem their idea was to employ the same isospin space as Gell-Mann, but consider also mirror transformations in it, and then use the requirement of invariance of strong interactions with respect to this larger group to

---

[39] Robert Oppenheimer as quoted in: Joseph Ballam, Val L. Fitch, Thomas Fulton, Kerson Huang, R. R. Rau, and Sam Bard Treiman, eds., *High Energy Nuclear Physics: Proceedings of the Sixth Annual Rochester Conference on High Energy Nuclear Physics 1956* (New York: Interscience, 1956), here VIII-1. The reference here is to Dirac's relativistic equation for electrons, which is written in terms of four-component (i.e. relativistic) spinors and was the final results of the development started with Pauli's notion of two-component electron spin.

[40] Jerzy Rayski, "Bilocal Field Theories and Their Experimental Tests.—I," *Il Nuovo Cimento (1955–1965)* 4, no. 6 (December 1956): 1231–41; Jerzy Rayski, "Bilocal Field Theories and Their Experimental Tests—II," *Il Nuovo Cimento (1955–1965)* 5, no. 4 (April 1957): 872–85.

[41] Bernard d'Espagnat and Jacques Prentki, "Formulation Mathématique du Modèle de Gell-Mann," *Nuclear Physics* 1, no. 1 (1956): 33–53.

[42] "En résumè, nous sommes en présence de théories dont l'une au moins, celle de Pais, est mathématiquement satisfaisante mais est difficile á concilier avec les données expérimentales tandis que d'autres, celle de Gell-Mann, en particulier, sont plutôt des modèles qui décrivent trés correctement les faits observé jusqu'á présent mais souffrent d'un manque de bases théoriques. Dans ces conditions il a paru interessant de rechercher une formulation plus mathèmatique du modèle de Gell-Mann et de tenter par là de le libérer du reproche d'arbitraire qui lui a été fait." ["In short, we are in the presence of theories of which at least one, that by Pais, is mathematically satisfactory but difficult to reconcile with experimental data, while others, in particular the one by Gell-Mann, are rather models which describe very correctly the facts observed so far, but suffer of a lack of theoretical foundations. In this situation it has seemed interesting to search for a more mathematical formulation of Gell-Mann's model and trying thus to free it from the accusation of arbitrairness which has been made against it."] d'Espagnat and Prentki, "Formulation Mathématique," 37 (translation by A.B.).

determine all possible particle fields, deducing the Gell-Mann model from invariance principles. In this way, they concluded, Gell-Mann's "model" would receive the support of a "theory." The two authors motivated their research by characterizing Gell-Mann's model as "arbitrary" and in need of a "more mathematical formulation."[43] However, they gave no reason why the model should be regarded as arbitrary or not mathematical enough, since it was quite efficient in fitting phenomena. It was purely their conviction that an explanatory construct should have specific mathematical features that led them to dismiss Gell-Mann's model as arbitrary.

The year 1956 also saw the publication of a paper by Julian Schwinger on a "Dynamical Theory of K Mesons" in which yet another extension of isospin space to four dimensions was proposed.[44] Starting from the Gell-Mann-Nishijima model, Schwinger listed experimental results which in his opinion supported the view that the (approximate) invariance under rotations in the usual three-dimensional isospin space was a leftover of a rotational symmetry in a higher, four-dimensional Euclidean space, a symmetry broken by the presence of pion interactions.[45] Another attempt at expanding the Gell-Mann-Nishijima classification was made by the Brazilian theorist Jayme Tiomno, who spoke of the "scheme" by Gell-Mann and Nishijima, referring to his construct as a "theory":

> It is shown that the usually accepted Gell-Mann-Nishijima scheme is not unique and that another scheme [...] is possible. A theory is developed based on general symmetry principles, with this new scheme, and it is shown that [...] it is equivalent to Schwinger's 4-dimensional (in isotopic spin space) theory.[46]

These examples should have shown how, during the 1950s, there was a small but active group of theorists who employed abstract group-theoretical concepts in their work, and who were convinced that the structures related to space-time invariances provided a privileged means not just to fit observation, but also to better understand fundamental interactions, It was not the first, but rather the latter goal which motivated the use of group theory. At the same time, though, empirically successful approaches to fit phenomena arose from the work of theorists like Gell-Mann or Nishijima, who neither made use of group-theoretical notions, nor expressed a belief that any specific invariance would possess a special epistemic status in the exploration of microphysical interactions.

---

[43] d'Espagnat and Prentki, "Formulation Mathématique," 37.

[44] Julian Schwinger, "Dynamical Theory of K Mesons," *Physical Review* 104, no. 4 (November 1956): 1164–72.

[45] "[T]he internal symmetry space is a four-dimensional Euclidean manifold which is reduced to the aspect of the three-dimensional isotopic spin space through the operation of the [pion] interactions." Schwinger, "Dynamical Theory," 1166.

[46] Jayme Tiomno, "On the Theory of Hyperons and K-Mesons," *Il Nuovo Cimento (1955–1965)* 6, no. 1 (July 1957): 69–83, here 69.

Meanwhile, in 1956, a very important event had taken place: the discovery of the violation of left–right invariance in weak interactions.[47] This development plays a role in our story because it prompted theorists to devote more attention to both symmetry and symmetry-breaking, and because it lowered the (so far extremely high) status of space-time invariances, letting it appear more plausible that internal symmetries like isospin and strangeness might be somehow connected to space-time transformations.[48] Thus, approaches like those of Pais or d'Espagnat and Prentki became more popular in the theoretical community and, as we shall see in the following pages, in the early 1960s the main agenda of group-theoretically-minded theorists became the establishment of a close connection between internal symmetries, like isospin and strangeness, and space-time invariances.

The Path from SU(2) to SU(3), or: Did Particle Physicist Know Group Theory?

Among the proposals made in the 1950s for expanding the Gell-Mann-Nishijima model only a few made use of group-theoretical notions. A different, more success-ful approach was to tentatively regard strongly interacting particles as composites of a small number of more fundamental objects. In 1956, the Japanese theorist Shoichi Sakata made the most influential proposal in this direction regarding only protons, neutrons and the newly discovered Lambda particle as elementary par-ticles.[49] The "Sakata model", as it was (and still is) usually referred to, had an implicit group-theoretical structure of type SU(3), which Sakata's colleagues and students from the Nagoya school soon explicitly discussed in attempts to expand the model and compute its implications.[50] These results were presented at the tenth Rochester Conference on High Energy Physics held at CERN in 1960, so that at the latest at that point they became well known to the international particle community.[51]

---

[47] Richard Dalitz, "K-Meson Decays and Parity Violation," in *Pions to Quarks: Particle Physics in the 1950s*, ed. Laurie Brown, Max Dresden, and Lillian Hoddeson (Cambridge: Cambridge University Press, 1989), 434–57; Allan Franklin, "The Nondiscovery of Parity Nonconservation," in *Pions to Quarks: Particle Physics in the 1950s*, ed. Laurie Brown, Max Dresden, and Lillian Hoddeson (Cambridge: Cambridge University Press, 1989), 409–33.

[48] Borrelli, "Beauty and Belief;" Borrelli, "Weinberg-Salam Model."

[49] Shoichi Sakata, "On a Composite Model for the New Particles," *Progress of Theoretical Physics* 16, no. 6 (December 1956): 686–88.

[50] Mineo Ikeda, Shuzo Ogawa, and Yoshio Ohnuki, "A Possibile Symmetry in Sakata's Model for Bosons-Baryons System," *Progress of Theoretical Physics* 22, no. 5 (November 1959): 715–24; Yoshio Ohnuki, "Models for Elementary Particles and the Nagoya School 1955–1973," *Progress of Theoretical Physics* 122, no. 1 (July 2009): 23–30. As noted above, the special unitary group SU(3) is the group of 3×3 unitary matrices with determinant 1. Unlike SU(2), SU(3) cannot be linked to transformations of space of space-time.

[51] Marshak, "Rochester Conferences," 660–61.

Because of the interest in the Sakata model, attention was drawn to the SU(3) structure, and in 1961 Gell-Mann and Ne'eman independently proposed a classification scheme for strongly interacting particles based on that group.[52] The scheme combined the isospin singlets, doublets and triplets of the Gell-Mann-Nishijima model into two 'octets' and Gell-Mann, who was always good in coming up with catchy labels, called it "the eightfold way" in reference to Zen Buddhism. It is interesting to note that neither Gell-Mann nor Ne'eman represented their model in graphic diagrams. In fact, the earliest diagrams of the eightfold way I could locate appeared in the proceedings of the first Coral Gables Conference on Symmetry Principles at High Energy Physics (see Fig. 2).[53] The eightfold way is still valid today, albeit as part of a more complex overarching theory of strong interactions, and is represented diagrammatically in most manuals of particle physics as in Fig. 2. Both Gell-Mann and Ne'eman later claimed to have arrived at SU(3) fully independently from the Sakata model, although, as we saw, in 1961 that structure was well known to theorists.[54] Apart from issues of priority, though, Gell-Mann and Ne'eman in their papers indeed introduced SU(3) through a path which had little to do with Sakata's idea of a small number of elementary building blocks, and was instead inspired by a new approach to invariance which had been developed in the later 1950s and was referred to as 'local gauge invariance.' It is beyond the scope of this paper to discuss Gell-Mann's and Ne'eman's route to SU(3), but a short description of local gauge invariance is necessary, as it was yet another kind of symmetry principle striving to combine internal degrees of freedom with space-time variables.

Although Pais had spoken of each point in space-time as being an isospace manifold, he and other authors had conceived the symmetries of particles as made out of two mutually independent rotations, one in space-time and the other one in isospace. In 1954 Chen Ning Yang and Robert Mills instead proposed an invariance with respect to rotations of the isospin variable dependent on the space-time

---

[52] Murray Gell-Mann, "The Eightfold Way. A Theory of Strong Interaction Symmetry," California Institute of Technology, Pasadena. Synchrotron Lab, Report CTSL-20 (March 1961), online: https://doi.org/10.2172/4008239 (accessed September 23, 2021); Yuval Ne'eman, "Derivation of Strong Interactions from a Gauge Invariance," *Nuclear Physics* 26, no. 2 (August 1961): 222–29.

[53] Robert Adair, David Barge, W. T. Chu, and Lawrence Leipuner, "The Hunting of the Quark," in *Coral Gables Conference on Symmetry Principles at High Energy*, ed. Behram Kursunoglu and Arnold Perlmutter (San Francisco: Freeman, 1964), 36–44, here 43 and Sidney Meshkov, "Comparison of Experimental Reaction Cross Sections with Various Relations Obtained from SU(3)," in *Coral Gables Conference on Symmetry Principles at High Energy*, ed. Behram Kursunoglu and Arnold Perlmutter (San Francisco: Freeman, 1964), 104–22, here 114.

[54] Gell-Mann, "Particle Theory;" Yuval Ne'eman, "Patterns, Structures and Their Dynamics: Discovering Unitary Symmetry and Conceiving Quarks," *Proceedings of the Israel Academy of Sciences and Humanities – Section of Sciences* 21 (1983): 1–26; Ne'eman, "Hadron Symmetry."

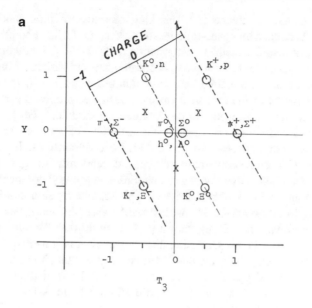

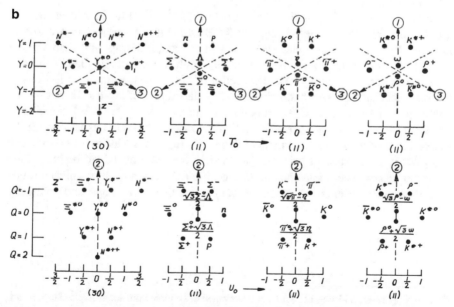

Fig. 2 Diagrams of strongly interacting particles as multiplets of SU(3) from two contributions to the proceedings of the first Coral Gable Conference on Symmetry Principles in High Energy Physics. **a** is from Robert Adair, David Barge, W. T. Chu, and Lawrence Leipuner, "The Hunting of the Quark," in *Coral Gables Conference on Symmetry Principles at High Energy*, ed. Behram Kursunoglu and Arnold Perlmutter (San Francisco: Freeman, 1964), 36–44, here 43. © Macmillan Publishers, all rights reserved. **b** is from Sidney Meshkov, "Comparison of Experimental Reaction Cross Sections with Various Relations Obtained from SU₃," in *Coral Gables Conference on Symmetry Principles at High Energy*, eds., Behram Kursunoglu and Arnold Perlmutter (San Francisco: Freeman, 1964), 104–22, here 114. © Macmillan Publishers, all rights reserved

coordinate ("local gauge invariance").[55] This kind of transformation was called "local" because it changed with space-time location. Yang and Mills had proposed a local version of the usual isospin transformation with its SU(2) structure, but this approach turned out to be too simple to accommodate the variety of particles and interactions observed at the time, and later authors expanded it, for example through local versions of the extended isospace variables proposed earlier on, or adding local rotations with respect to the strangeness variable.[56] However, the groups employed in local gauge invariance were still those of rotations in space, space-time or in Euclidean spaces with more than three dimensions. Instead, in 1961, Gell-Mann and Ne'eman contemporarily but independently brought forward proposals in which the local gauge invariance had the same group-theoretical structure as the Sakata model: SU(3). They arrived at this scheme by setting aside the idea of turning isospin into a non-local space-time-like variable, and instead looking for a local gauge invariance fitting known particle multiplets. Because of this closeness to observation, it is not surprising that they arrived at a similar solution as Sakata, though not exactly the same one.[57] However, the main aim of their plan did not work out, and local gauge invariance soon dropped out of the model, to make a comeback only a few years later, in a quite different form. What remained, were the SU(3) octets: the eightfold way.

Today, when looking at the octet structure of SU(3) (Fig. 2) and at the diagram drawn in 1957 by Gell-Mann and Rosenfeld (Fig. 1), one is tempted to read into the 1957 image the octet structure of the eightfold way, and then wonder how come no-one had made the connection before Ne'eman and Gell-Mann did. This retrospective, anachronistic perspective is often found in the recollections of the historical actors, who blamed themselves and their colleagues for not having come up earlier with the SU(3) structure. Ne'eman explicitly marveled that his colleagues had not deployed group-theoretical knowledge already available since the nineteenth century, noting that most of them had attended the lectures Giulio Racah had given on that topic at Princeton in 1951.[58] Ne'eman explained he knew little of group theory, and spent the summer of 1960 in systematically going through lists of group representations searching for the one fitting his scope, and eventually

[55] Chen Ning Yang and Robert L. Mills, "Conservation of Isotopic Spin and Isotopic Gauge Invariance," *Physical Review* 96, no. 1 (October 1954): 191–95. For an overview of the development of local gauge theories in particle physics up to 1961, see: Pickering, *Constructing Quarks*, 160–65. The term 'gauge invariance' refers to invariance with respect to multiplication by a phase, and the adjective 'local' indicates that the phase depends on local space-time coordinates.

[56] Jun John Sakurai, "Theory of Strong Interactions," *Annals of Physics* 11, no. 1 (September 1960): 1–48; Abdus Salam and John Ward, "On a Gauge Theory of Elementary Interactions," *Il Nuovo Cimento (1955–1965)* 19, no. 1 (January 1961): 165–70.

[57] Marshak, "Rochester Conferences," 661.

[58] Ne'eman, "Patterns," 10–11.

discovering the octet representation of SU(3).[59] Gell-Mann was among those who had attended Racah's lectures, but explained in his recollections that he could not pay attention to their content because of Racah's strong accent:

> In 1951 I attended the beautiful lectures by Giulio Racah [...] but I did not really understand the material. The reason was not that the lectures were not elegant, or that they were not explicit. The problem was his accent. Of course I understood the words; I have no trouble following English spoken with a foreign accent. But his accent was so remarkable that I could not hear the substance. Every English word was pronounced with a perfect Florentine accent. For example he would say: 'Tay vah-loo-ay eess toh eeg' (The value is too high). So I never learned about Lie algebras, and I had to rediscover them.[60]

Gell-Mann discussed at length how he allegedly rediscovered SU(3) on his own. Moreover, he complained that mathematicians teaching to physicists "give only trivial examples" instead of discussing potentially useful methods.[61] Perhaps because of the colorful descriptions of Gell-Mann's and Ne'eman's alleged path to group-theoretical knowledge, historians have assumed that SU(3) remained hidden so long because physicists knew too little group theory, and its use by Gell-Mann and Ne'eman came to be described as a success of the application of group-theoretical methods to quantum physics.[62]

Yet this explanation is much too simplistic: in the 1950s the idea of employing SU(3) to fit particle phenomena was in no way straightforward, and the initial failure to do so was not due to theorists' ignorance of group theory. As we saw in the previous section, theorists interested in group-theoretical extensions of the isospin schemes were quite knowledgeable and, had they had any interest in SU(3), they could certainly have informed themselves like Ne'eman had done. In 1956 even Racah himself wrote a short paper on d'Espagnat and Prentki's theory, and it is extremely improbable that he did not know about the group SU(3).[63] Therefore, despite all of the claims of what theorists would have used SU(3) after the fact, if only they had known about it, this was not the case: SU(3) was known at least to some of them, who however did not choose to employ it. The reason why none of these authors took interest in SU(3) has rather to be sought in the specific aims and premises of their mathematical practices. As we saw in section "The Search for a Theory of Isospin and Strangeness in the 1950s," they were not trying to fit particles to some symmetric scheme regardless of its type, but attempted to find an invariance which was not only empirically plausible, but also—in their perception—physically satisfying, in that they could regard it as expressing features of

---

[59] Ibid., 9–11. In group theory, groups are abstract sets of transformations that can be implemented in different ways through 'representations.' For example, the group of space rotations can be represented by a vector, a scalar (i.e. a rotationally invariant object), an axial vector or other, more complex constructs.

[60] Gell-Mann, "Particle Theory," 487.

[61] Ibid., 487–88.

[62] Pickering, Constructing Quarks, 56–57; Schweber, "Quantun Field Theory," 386–87.

[63] Giulio Racah, "Remarques sur une formulation mathématique du modèle de Gell-Mann," Nuclear Physics 1, no. 4 (March 1956): 302–3.

reality. For them, this meant a symmetry displaying formal analogies or direct connections to space-time transformations. Since SU(3) had no relationship to those transformations, authors like Pais, d'Espagnat and Prentki, who most probably knew about it, never thought of employing it. On the other hand, Sakata and his collaborators saw mathematical invariances rather as tools to fit phenomena than as means to explain them, and eventually employed SU(3) to classify particles.

So we see how speaking in general of an application of 'symmetry principles' or 'group theoretical methods' to physics is not a heuristically fruitful way to frame the historical-epistemological constellations discussed in the previous sections, as the situation is much more complex. For example, Gell-Mann and Ne'eman, in developing the eightfold way, were indeed following a symmetry principle, namely local gauge invariance, yet it was a thoroughly different one than those that had guided Pais or d'Espagnat and Prentki. All the same, both research programs were shaped by an interest in connecting internal and space-time symmetries, and not of deploying group theory in general to fit phenomena. For theorists in the 1950s and 1960s, not all groups were equal.

A few years after the emergence of the eightfold way, in 1964, the quark model was proposed on the basis of purely theoretical considerations.[64] As shown above, in the eightfold way particles had been grouped into octects (and later also decuplets), which were representations of SU(3). However, so far no particles had been assigned to the smallest SU(3) representation, which is the triplet. Thus, the question arose whether still unobserved particles might exist that corresponded to the triplet and might perhaps take up the role of fundamental building blocks, as Sakata had envisaged.[65] Proposals in this sense were advanced, once more, by Gell-Mann and by George Zweig, a young U.S. American theorist working at CERN at the time.[66] Gell-Mann suggested three 'quarks,' while Zweig spoke of three 'aces,' but the two schemes were otherwise largely equivalent, and eventually became known under Gell-Mann's label as 'quark model.' However, as we shall see in the next sections, the eightfold way and the quark model were seen by many theorists only as a starting point for a long sought after merger of internal and space-time symmetries.

---

[64] "Quark model" (as opposed to "quark theory") was the expression most often employed to indicate the use of the SU(3) triplet to classify strongly interacting particles, e.g. in: Susumo Okubo and Robert E. Marshak, "The Quark Model as a Probe of Higher Symmetries," in *Coral Gables Conference on Symmetry Principles at High Energy. Second Conference*, ed. Behram Kursunoglu, Arnold Perlmutter, and Ismail Sakmar (San Francisco: Freeman, 1965), 128–44; Marco Ademollo and Roberto Gatto, "Weak Non-leptonic Interactions in the Quark Model," *Physics Letters*, 10, no. 3 (June 1964): 339–40; Walter Thirring, "The $\omega \to \pi o + \gamma$ Decay in the Quark Model," *Physics Letters* 16, no. 3 (June 1965): 335.

[65] Pickering, *Constructing Quarks*, 85–124.

[66] Murray Gell-Mann, "A Schematic Model of Baryons and Mesons," *Physics Letters* 8, no. 3 (February 1964): 214–15; George Zweig, "An SU(3) Model of Strong Interaction Symmetries and Its Breaking. Part 2," *CERN* preprint, February 21, 1964.

Beyond SU(3)—The Mathematical Marriage of Space-Time and Internal Symmetries

With the success of the eightfold way, a growing number of theorists took interest in abstract group-theoretical structures and their possible use in high-energy physics. The five conferences on "Symmetry Principles at High Energy" (1964–1968) held at Coral Gables were both evidence of and a meeting point for a growing community of theoretical physicists interested in symmetries. Participants of the meetings included names which already were or would soon become well known in the theoretical high energy physics community, like Asim Barut, Nicola Cabibbo, Gerson and Sulamita Goldhaber, Bernard D'Espagnat, Robert Marshak, Louis Michel, Yoichiro Nambu, Yuval Ne'eman, Robert Oppenheimer, Lochlainn O'Raifeartaigh, Abraham Pais, Abdus Salam, Julian Schwinger, George Sudarsahn and Bruno Zumino. Therefore, the five volumes of proceedings of those conferences provide a good guideline to follow the rise of symmetry principles in high-energy physics and the mathematical practices that were referred to under that expression.

Were abstract group-theoretical notions now finally being regarded as a generally effective tool to fit observations in particle physics? As I will argue in the next pages, the answer to this question is *no*. The participants to the Coral Gable conferences certainly regarded empirical adequacy as a necessary feature of any mathematical conceptualization of particle phenomena, and therefore took as a starting point for all of their reflections the empirically successful eightfold way and quark model and tried to expand them into an equally successful construct. However, their explicit use of group theory was not aimed at improving the fit to phenomena, but rather at ensuring that the mathematical construct produced would also in some way reflect a deeper structure of fundamental interactions. Once again, these authors made a distinction between 'models' and 'theories' which was not necessarily shared by the whole particle physics community. The scientists meeting at Coral Gables expected the symmetries of nature to take a form similar to space-time invariances. The shared tenet of the emerging symmetry community was the idea that internal symmetries like isospin and strangeness should be combined to space-time degrees of freedom. Kursunoglu expressed this goal very forcefully in his contribution to the first conference:

> The study of elementary particle events from the point of view of symmetry properties of their interactions has had both qualitative and quantitative success [...] However, it is also known that the present picture of the description of elementary particles based on the experimentally tested symmetry concepts is not entirely satisfactory. [...] The introduction of fictitious spaces like isotopic spin or unitary spin space, distinct from the space-time structure of elementary events has long been recognized to be quite unsatisfactory for further progress towards a real understanding of the dynamical principles underlying elementary particles interactions.[67]

---

[67] Behram Kursunoglu, "A New Symmetry Group for Elementary Particles," in *Coral Gables Conference on Symmetry Principles at High Energy*, ed. Behram Kursunoglu and Arnold Perlmutter (San Francisco: Freeman, 1964), 20–32, here 20.

Kursunoglu immediately stated that he had no solution for this problem, but went on to present his reflections:

> There are a number of ways of introducing new groups, that is, new quantum numbers to describe some properties of elementary particles. Almost any finite or infinite group can provide some discrete quantum number. The physics of the things almost always emerge from a skillful bookkeeping of the correspondence between these discrete numbers and observed facts. In the absence of basic physical principles it is quite possible that the correlation of facts and some real numbers can be achieved in more than one way. For example if SU(3) is an invariance group for elementary particles, why not the so-called G_2 or SU(4) or for that matter O_7 or any other well suited finite dimensional group?[68]

We see here how Kursunoglu did not consider success in fitting phenomena as a sufficient criterion for regarding a particular group-theoretical structure as physically significant. To that aim, more guidance was needed:

> We must, therefore, seek some guidance from the most basic invariance principles of physics, meaning that any extra quantum degree of freedom for elementary particles must be based on the inhomogeneous Lorentz group. We must establish a bridge between space-time and unitary structure of microphysics.[69]

Although Kursunoglu did not use here the terms model and theory, it is clear that he was making a sharp distinction between "some real numbers" fitting "observed facts" (models) and groups mirroring the principles of natural laws (theories). In a similar vein, Asim Barut started his paper by saying:

> We shall take the point of view that there is an exact symmetry governing the quantum numbers and the mass states of elementary particles. It is clear that this symmetry has to encompass the space-time symmetry and go beyond to include all other internal degrees of freedom or quantum numbers. We shall call this larger symmetry the dynamical symmetry.[70]

Barut explained that physics had proceeded in the past by describing systems in terms of increasingly larger groups, quoting the history of spin as an example that one should search for a symmetry connecting space-time and internal quantum numbers. In a joint paper "On the Origin of Symmetries," Ne'eman, Nathan Rosen and Joe Rosen explored the possibility that internal symmetries might "emerge from space-time itself."[71] They even brought gravity into play and

---

[68] Ibid., 20–21.

[69] Ibid., 21.

[70] Asim Barut, "On Dynamical Symmetry Groups and Mass Spectrum of Elementary Particles," in *Coral Gables Conference on Symmetry Principles at High Energy*, ed. Behram Kursunoglu and Arnold Perlmutter (San Francisco: Freeman, 1964), 81–89, here 81.

[71] Yuval Ne'eman, Nathan Rosen, and Joe Rosen, "On the Origin of Symmetries," in *Coral Gables Conference on Symmetry Principles at High Energy*, ed. Behram Kursunoglu and Arnold Perlmutter (San Francisco: Freeman, 1964), 93–99, here 93.

concluded: "In fact, we got a larger symmetry than the observed one, and contracted it by assuming that first order gravitational curvature should not be left out and conjecturing that it should have just that observed effect."[72] It is of course not possible to discuss here the details of this and other proposals, but it is important to see how, once again, group-theoretical structure linked to space-time—and only those—were seen as guidelines for fundamental reflection more than for fitting phenomena. As Schwinger put it in his contribution: "what I want to talk about is not some modification of existing classification schemes, but rather a fundamental field theory of matter, i.e. of everything."[73]

Despite their focus on explanatory constructs, the participants of the Coral Gable conferences regarded empirical adequacy as a prerequisite for any theory, and so were quite interested in experimental results, to which some talks were devoted.[74] Interestingly, these experimental talks contained the earliest diagrams geometrically representing strongly interacting particles as multiplets of SU(3), as shown in Fig. 2, which is today the standard representations of the eightfold way. In Gell-Mann and Ne'eman's papers no graphic representation of that (or other) kind appeared and, as noted above, the diagrams of the strangeness model developed by Gell-Mann did not express any group theoretical properties, but rather the measurable features of particles, like masses and charges (see Fig. 1). This observation confirms that it was only when the eightfold way became relevant for experimenters that the observed properties of particles started being seen and diagrammatically represented through the mathematical lens of SU(3). Soon, such representations would become a standard means of introducing young physics students to elementary particles and, at that point, it started being assumed that, had one known more group theory, one might have read off the multiplets from experimental reports. Yet it was not so, and the construction of the Gell-Mann-Nishijima model and the eightfold was a highly non-trivial development at the intersection of theory and experimentation.

Now that the SU(3) structure was there, though, all particles appeared as potential bearers of quantum numbers representing higher symmetries, and it is in this sense that they were featured on the cover of the proceedings of the first Coral Gables conference (see Fig. 3). It was there that Nasreddin Hoja made his first appearance, sitting backward on his donkey surrounded by the symbols of the particles making out the eightfold way.[75] Hoja reprised this role on the cover of the

---

[72] Ne'eman, Rosen, and Rosen,"Origin of Symmetries," 97.

[73] Julian Schwinger, "A Ninth Baryon?," in *Coral Gables Conference on Symmetry Principles at High Energy*, ed. Behram Kursunoglu and Arnold Perlmutter (San Francisco: Freeman, 1964), 127–37, here 127.

[74] Adair et al., "Quark;" Gerson Goldhaber, "Experimental Study of Multiparticle Resonance Decays," in *Coral Gables Conference on Symmetry Principles at High Energy. Second Conference*, ed. Behram Kursunoglu, Arnold Perlmutter, and Ismail Sakmar (San Francisco: Freeman, 1965), 34–126.

[75] In the proceedings of the second Coral Gable conference we can read Hoja's explanation to his students of why he rode his donkey backwards: "If I rode with my face straight ahead, you would

Fig. 3 Cover image of:
Behram Kursunoglu and
Arnold Perlmutter, eds.,
*Coral Gables Conference on
Symmetry Principles at High
Energy* (San Francisco:
Freeman, 1964). ©
Macmillan Publishers, all
rights reserved

proceedings of the second conference (see Fig. 4), but this time he was surrounded
by the symbols of various groups that had been or might be used to classify particles. This image represented the work done by theorists during the last year and
expressed their hopes of further progress. The second Coral Gable conference was
held in January 1965, and its proceedings appeared during that same year.[76] As
noted by the editors in their short preface, very important developments had taken
place since the first conference and, from the point of view of the Coral Gables
participants, the most important event had been the proposal to expand the SU(3)
quark model into a classification based on SU(6), which at least partially combined
internal and space-time symmetries:

> The subject of combining internal and space-time symmetries was one of the topics dis-
> cussed at the First Coral Gables Conference, held in January 1964. The issues raised at that
> time have since been considered - both extensively and intensively - by many experts, and

---

be behind me. If, on the other hand, you were to walk in front, you would turn your backs upon
me. I think, therefore, to ride this way is to solve all the problems, and it is more polite!" Behram
Kursunoglu, Arnold Perlmutter, and Ismail Sakmar, "Preface," in *Coral Gables Conference on Symmetry Principles at High Energy. Second Conference*, ed. Behram Kursunoglu, Arnold Perlmutter,
and Ismail Sakmar (San Francisco: Freeman, 1965), [i].

[76] Kursunoglu, Perlmutter, and Sakmar, eds., *Coral Gables*, 1965.

Fig. 4 Cover image of:
Behram Kursunoglu, Arnold
Perlmutter, and Ismail
Sakmar, eds., *Coral Gables
Conference on Symmetry
Principles at High Energy.
Second Conference* (San
Francisco: Freeman, 1965). ©
Macmillan Publishers, all
rights reserved

their efforts have culminated in the proposal that SU(6) be accepted as a possible symmetry group of hadrons.

At this year's conference a major part of the discussion was concerned with the expansion and development of this proposal, and it was the opinion of the participants that considerable progress was achieved.[77]

At the time, approaches based on SU(6) appeared as a great breakthrough. Today this is practically forgotten, and in the following section we will follow its rise and fall in context of the rapid diversification of mathematical practices linked to the search for symmetry principles at high energy.

The Rise and Fall of SU(6)

When the quark model appeared in 1964, the fact that quarks would have electric charge equal to 1/3 of that of the electron raised many doubts about their existence, since so far no objects with fractional charges had ever been observed.[78] However,

[77] Behram Kursunoglu, Arnold Perlmutter, and Ismail Sakmar, "Preface," in *Coral Gables Conference on Symmetry Principles at High Energy. Second Conference,* ed. Behram Kursunoglu, Arnold Perlmutter, and Ismail Sakmar (San Francisco: Freeman, 1965), [i].

[78] Pickering, *Constructing Quarks*, 25–124.

it was noted by a number of theorists that, if one considered quarks with spin up and quarks with spin down as two distinct particles, and not as two states of the same particle, then one would have six independent objects out of which strongly interacting particles might be constructed according to a scheme based on representations of the group SU(6). As it turned out, this classification appeared empirically plausible, with the particles forming 35-dimensional (mesons) and 56-dimensional (hadrons) SU(6) multiplets.[79]

For theorists with an interest in symmetry principles this scheme was extremely attractive, because SU(6) was not simply a product of SU(3) and SU(2), but intimately combined the degrees of freedom of internal symmetries with two-component spin. The scheme was of course non-relativistic, but, as we shall see, Kursunoglu and other theorists were convinced that its relativistic extension was possible and would soon follow. The various symbols surrounding Hoja on the cover of the 1965 proceedings were proposals in that direction.

In his opening remarks at the second Coral Gables conference, Robert Oppenheimer spoke of the "successes of SU(3) and the successes of SU(6)," and stated that a main topic of the conference would be "in what sense, for what purpose and at what cost can one combine relativity with the good things about SU(6)."[80] This topic was indeed central to many talks at the conference. Salam proposed a "Covariant theory of strong interaction symmetries" based on a group with 144 parameters which he labeled as $\tilde{U}(12)$.[81] Bunji Sakita presented a relativistic SU(6) theory, noting at the beginning of his paper that its formulation and results were "almost identical to those of Salam's,"[82] Korkut Bardacki, John Cornwall, Peter Freund, and Benjamin Lee instead argued for an extension to U(6) × U(6),[83] while Yoichiro Nambu assumed the existence of three distinct quark triplets.[84] Some authors discussed more in general the possibilities and problems of the research program: Robert Marshak and Susumu Okubo made an inventory

[79] Jeffrey Mandula, "Coleman-Mandula Theorem," *Scholarpedia* 10, no. 2 (February 2015): 7476.

[80] Robert Oppenheimer, "Remarks on Symmetry Principles," in *Coral Gables Conference on Symmetry Principles at High Energy. Second Conference*, ed. Behram Kursunoglu, Arnold Perlmutter, and Ismail Sakmar (San Francisco: Freeman, 1965), 2–4, here 2–3.

[81] Abdus Salam, "The Covariant Theory of Strong Interaction Symmetries," in *Coral Gables Conference on Symmetry Principles at High Energy. Second Conference*, ed. Behram Kursunoglu, Arnold Perlmutter, and Ismail Sakmar (San Francisco: Freeman, 1965), 6–25. A better known notation for that group is Mathieu group M(12).

[82] Bunji Sakita, "A Phenomenological Approach to Relativistic SU(6) Theory," in *Coral Gables Conference on Symmetry Principles at High Energy. Second Conference*, ed. Behram Kursunoglu, Arnold Perlmutter, and Ismail Sakmar (San Francisco: Freeman, 1965), 354–56, here 354.

[83] Korkut Bardakci, John Michael Cornwall, Peter Freund, and Benjamin Lee, "Intrinsically Broken U(6) X U(6) Symmetry for Strong Interactions," in *Coral Gables Conference on Symmetry Principles at High Energy. Second Conference*, ed. Behram Kursunoglu, Arnold Perlmutter, and Ismail Sakmar (San Francisco: Freeman, 1965), 260–68.

[84] Yoichiro Nambu, "Dynamical Symmetries and Fundamental Fields," in *Coral Gables Conference on Symmetry Principles at High Energy. Second Conference*, ed. Behram Kursunoglu, Arnold Perlmutter, and Ismail Sakmar (San Francisco: Freeman, 1965), 274–83.

of possible higher symmetries embedding the quark model,[85] George Sudarshan discussed "Questions of Combining Internal Symmetries and Lorentz Group,"[86] Louis Michel addressed "The Problem of Group Extensions of the Poincaré Group and SU(6) Symmetry."[87]

A critical note came from Julian Schwinger, who presented a development of his earlier idea of a distinction between "fundamental," non-observable quantum fields and "phenomenological" fields representing observed particles.[88] He could not spell out the details of how to connect the two kinds of fields, as "this problem defies direct dynamical attack, at the moment,"[89] but developed some general considerations on symmetry, and in the end explicitly underscored that they should be regarded as different form the SU(6) approach pursued by most other authors:

> It would be quite erroneous, however, to identify the concepts described in this note with the ideas of the SU(6) workers. The latter view SU(6), or some larger group, as an idealized dynamical invariance group, despite the difficulties raised by relativistic considerations, or, more conservatively, regard it as valid in a non-relativistic limit. [In our theory] [t]he requirement of invariance under the inhomogeneous Lorentz group is met through the machinery of field theory, which was invented for that purpose, and not by forcing an unhappy union between physically incompatible partners.[90]

Yet Kursunoglu pursued further his dream of a union of internal and space time symmetries, wondering whether one might have to consider infinite Lie groups to that aim, and speculating that "[i]n such a theory, one may end up with a single infinite supermultiplet of hadrons."[91]

Not all authors at the conference focused on how to combine space-time and internal symmetries, and some theoretical contribution had a closer link to experiment.[92] Moreover, two papers discussed how SU(6) and other symmetry principles might be connected to an approach which, at the time, was emerging as either

---

[85] Okubo and Marshak, "Quark Model."

[86] George Sudarshan, "Question of Combining Internal Symmetries and Lorentz Group," in *Coral Gables Conference on Symmetry Principles at High Energy. Second Conference*, ed. Behram Kursunoglu, Arnold Perlmutter, and Ismail Sakmar (San Francisco: Freeman, 1965), 147–58.

[87] Louis Michel, "The Problem of Group Extensions of the Poincaré Group and SU(6) Symmetry," in *Coral Gables Conference on Symmetry Principles at High Energy. Second Conference*, ed. Behram Kursunoglu, Arnold Perlmutter, and Ismail Sakmar (San Francisco: Freeman, 1965), 331–50.

[88] Julian Schwinger, "Phenomenological Field Theory," in *Coral Gables Conference on Symmetry Principles at High Energy. Second Conference*, ed. Behram Kursunoglu, Arnold Perlmutter, and Ismail Sakmar (San Francisco: Freeman, 1965), 372–78.

[89] Ibid., 372.

[90] Ibid., 377.

[91] Behram Kursunoglu, "Symmetry and Strong Interaction," in *Coral Gables Conference on Symmetry Principles at High Energy. Second Conference*, ed. Behram Kursunoglu, Arnold Perlmutter, and Ismail Sakmar (San Francisco: Freeman, 1965), 160–75, here 161.

[92] Laurie Brown and Harry Faier, "Meson Decaysand the Sigma Hypothesis," in *Coral Gables Conference on Symmetry Principles at High Energy. Second Conference*, ed. Behram Kursunoglu, Arnold Perlmutter, and Ismail Sakmar (San Francisco: Freeman, 1965), 219–39; Bernard

a complement or an alternative to quantum field theory: the so-called bootstrap theory and the relevant techniques of dispersion relations, Regge poles and current algebra. The bootstrap approach was promoted by Geoffrey Chew as an alternative to quantum field theory, because it did not assume the existence of elementary particles, but only aimed at mathematically representing processes of mutual transformation between particles states, none of which was regarded as more elementary than the others.[93] At the second Coral Gables conference, both Roger Dashen and Ne'eman discussed how far SU(6) and other symmetries might be applied to the bootstrap, and Ne'eman concluded his summarizing talk of the meeting by discussing symmetries in a bootstrap theory "clean of elementary quarks."[94]

Despite the great expectations and the efforts of theorists present (and absent) at the second Coral Gables conference, the research program of a relativistic extension of SU(6) saw little progress in the following year, and at the third Coral Gable conference in January 1966 a large part of the papers were devoted either to the bootstrap approach or to attempts to theoretically fit new experimental results on weak and strong interactions. Once again, Ne'eman held a final, summarizing talk, and this time he devoted some remarks to "a discussion of the 'Philosophy' of what we are doing here, in the context of the general development of 'elementary particle' physics."[95] It is significant that Ne'eman not only put the word philosophy in quotations, but also the expression elementary particles: an indication of his interest in the bootstrap approach. Ne'eman went on to offer his interpretation of the history of chemistry as a development from the grasping of "patterns" in the form of the periodic table of elements to the understanding of "structures" with quantum physics.[96] He then compared this development with the present situation in high-energy physics, which he claimed was still in the pattern identification stage:

---

d'Espagnat and Mary Gaillard, "A Model for PC Violation in the Framework of Unitary Symmetry," in *Coral Gables Conference on Symmetry Principles at High Energy. Second Conference*, ed. Behram Kursunoglu, Arnold Perlmutter, and Ismail Sakmar (San Francisco: Freeman, 1965), 380–400; Harry Lipkin, "Tests of Unitary Symmetry and Possible Higher Symmetries," in *Coral Gables Conference on Symmetry Principles at High Energy. Second Conference*, ed. Behram Kursunoglu, Arnold Perlmutter, and Ismail Sakmar (San Francisco: Freeman, 1965), 202–13; Fredrik Zachariasen and George Zweig, "U(2) and Time Reversal Violation in the Weak Interactions," in *Coral Gables Conference on Symmetry Principles at High Energy. Second Conference*, ed. Behram Kursunoglu, Arnold Perlmutter, and Ismail Sakmar (San Francisco: Freeman, 1965), 362–70.

[93] Pickering, *Constructing Quarks*, 73–78.

[94] Roger Dashen, "Weak and Electromagnetic Currents and Strong Interactions Symmetries in the Bootstrap Program," in *Coral Gables Conference on Symmetry Principles at High Energy. Second Conference*, ed. Behram Kursunoglu, Arnold Perlmutter, and Ismail Sakmar (San Francisco: Freeman, 1965), 179–93, here 422.

[95] Yuval Ne'eman, "Hadron Matrix Mechanics," in *Coral Gables Conference on Symmetry Principles at High Energy. Third Conference*, ed. Arnold Perlmutter, Joseph Wojtaszek, George Sudarshan, and Behram Kursunoglu (San Francisco: Freeman, 1966), 263–74, here 263.

[96] Ne'eman, "Hadron Matrix Mechanics," 263–65.

Some day we may be able to go beyond this pattern identification stage. Quarks and their like may yet land us in a world of subparticle physics with a very different set of basic laws; alternatively we shall perhaps really learn to use the bootstrap as a working piece of physics, not just as an idea".[97]

Although Ne'eman did not speak of models and theories, we may interpret his distinction between "patterns" and "structures" as reflecting the idea often expressed at the time with those terms: an opposition between mathematical constructs which simply fit phenomena (models, patterns), and those expressing the (hidden) principles of their coming-to-be (theories, structures). Once again it must be noted that Ne'eman was expressing the beliefs of the community gathered in Coral Gables—beliefs which were not necessarily shared by all particle theorists at the time (not to mention experimentalists). In general, among high-energy physicists, the term 'model' was used in a rather neutral sense, without implying expectations that at some point a 'theory' would emerge. Therefore, the distinctions made by Ne'eman cannot be taken to reflect a general feature of theoretical practices of early particle physics. Moreover, all theorists agreed that any theoretical construct should be expected to fit phenomena and allow for successful predictions, so that any 'theory' would also be an empirically successful 'model.' In his conclusions Neeman stated:

> The search for 'relativistic' SU(6) has been replaced by the elaboration of a Lorentz-frame dependent structure, relativistic, but not explicitly covariant. Let us hope that the coming year will bring about a more complete understanding of the operator structure involved [...] with better chances of some structural understanding resulting from these developments.[98]

Contrary to Ne'eman's hopes, however, the following year brought little clarity, and the difficulties of linking internal symmetries with space-time invariances were shown to be not just of a technical, but also rather of a fundamental character. At the fourth Coral Gables conference, the final, summarizing talk was held by Nicola Cabibbo, who started with a look back at the development of symmetry principles in particles physics during the last years:

> The discovery of the octet scheme led to a substantial step forward in our understanding of hadrons [...] This was followed by attempts to introduce even higher symmetries. [...] After some ambitious attempts to have an intimate merger of internal and Lorentz symmetries, discussed in 1964 at this conference, we had, later in 1964, the explosive success of non-relativistic SU(6). The initial attempts to make SU(6) into a completely relativistic scheme were frustrated by a series of beautiful impossibility theories, the latest version of which was presented here by Professor Coleman.[99]

---

[97] Ibid., 265.
[98] Ibid., 273.
[99] Nicola Cabibbo, "Symmetry and Elementary Particle Physics," in *Coral Gables Conferences on Symmetry Principles at High Energy. Fourth Conference*, ed. Arnold Perlmutter and Behram Kursunoglu (San Francisco: Freeman, 1967), 181–89, here 170.

The proceedings of the conference do not contain any paper by Sidney Coleman, but in that year a joint paper by Coleman and Geoffrey Mandula appeared, in whose abstract the authors stated: "We prove a new theorem on the impossibility of combining space-time and internal symmetries in any but a trivial way."[100] At the beginning of the paper, they summarized the hope and efforts leading up to the present no-go theorem:

> Until a few years ago, most physicists believed that the exact or approximate symmetry groups of the world were (locally) isomorphic to direct products of the Poincaré group and compact Lie groups. This world-view changed drastically with the publication of the first papers on SU(6); these raised the dazzling possibility of a relativistic symmetry group that was not simply such a direct product. Unfortunately, all attempts to find such a group came to disastrous ends, and the situation was finally settled by the discovery of a set of theorems which showed that, for a wide class of Lie groups, any group which contained the Poincare group and admitted supermultiplets containing finite numbers of particles was necessarily a direct product.[101]

The authors explained that those theorems only dealt with specific cases, and that their own formulation provided a comprehensive argument covering all of them. The Coleman-Mandula theorem is today famous as the negative starting point for supersymmetry, since in the early 1970s a number of authors independently showed how it was, after all, possible to non-trivially extend the Poincare group by introducing transformations of fermions into bosons and vice-versa.[102] Yet in that case, too, no combination of internal SU(3) with space-time invariance was possible.

In 1966 a NATO International Advanced Study Institute on "Symmetry Principles and Fundamental Particles" took place in Istanbul, co-organized by Kursunoglu, and in its proceedings, Coleman started his lectures by explaining the clues that had led to the attempts of expanding SU(6), and then stated:

> We would have been extremely happy to find a relativistic generalization of SU(6), even if all particles in a supermultiplet had the same mass, and even if there were no good perturbative mass formulas. The remaining lectures will be devoted to explaining why even this modicum of happiness is denied us.[103]

It is interesting to note how Coleman half-jokingly remarked that theorists would have been happy even with a 'theory' that provided a quite inadequate fit to observation, or did not allow computing any predictions. Because of the obstacles encountered by the main research line of the previous years, the mathematical

[100] Sidney Coleman and Jeffrey Mandula, "All Possible Symmetries of the S-Matrix," *Physical Review* 159, no. 5 (July 1967): 1251–56, here 1251.
[101] Ibid.
[102] Mandula, "Coleman-Mandula Theorem."
[103] Sidney Coleman, "The Past of a Delusion: Attempts to Wed Internal and Space-time Symmetries," in *Symmetry Principles and Fundamental Particles*, ed. Behram Kursunoglu and Arnold Perlmutter (San Francisco: Freeman, 1967), 3–45, here 7.

practices of symmetry presented at the fourth Coral Gables conference were more varied than in earlier meetings and, to use Ne'eman's terminology, they explored rather the patterns than the structure of particle phenomena. To this end, they employed approaches linked to the bootstrap, like current algebra and dispersion relations.[104] In his summary talk, Cabibbo also noted the role of experiment, stating that "the work on the application of SU(3) has, in the meantime, gone on, in particular through the painstaking experimental search for new resonant states and their possible classification in multiplets of SU(3)."[105] A new, more diverse phase of symmetry practices had begun.

Conclusion: The End of the Yogurt Project?

By 1968, Kursunoglu had accepted that the dream pursued so far in the Coral Gable conferences would not become reality, but that the approaches developed in trying to realize it still constituted powerful heuristic tools for theoretical research. Accordingly, he decided that the fifth Coral Gable conference in January 1968 would be the last one devoted to "Symmetry Principles at High Energy," leaving the stage to meetings studying "Fundamental Interactions at High Energies."[106] In the preface to the proceedings, the editors listed as themes of this last meeting, beside the use of symmetry principles "for the elucidation of the nature of the fundamental laws governing the behavior of elementary particles," also "the collation of experimental data and the prediction of new effects" and "the need to explain the deeper mystery of broken symmetries."[107] The search for a 'theory'

---

[104] James Bjorken, "Current Algebra and Sum Rules at High Momentum Transfer," in *Coral Gables Conferences on Symmetry Principles at High Energy. Fourth Conference*, ed. Arnold Perlmutter and Behram Kursunoglu (San Francisco: Freeman, 1967), 168–76; Sergio Fubini, "Superconvergence, Current Algebra and the Lorentz Group," in *Coral Gables Conferences on Symmetry Principles at High Energy. Fourth Conference*, ed. Arnold Perlmutter and Behram Kursunoglu (San Francisco: Freeman, 1967), 106–19; Seymour Lindenbaum, "Forward Scattering Amplitudes at High Energies, Dispersion Relations, and Asymptotic Predictions," in *Coral Gables Conferences on Symmetry Principles at High Energy. Fourth Conference*, ed. Arnold Perlmutter and Behram Kursunoglu (San Francisco: Freeman, 1967), 122–63; André Martin, "Analytic Properties of Scattering Amplitudes," in *Coral Gables Conferences on Symmetry Principles at High Energy. Fourth Conference*, ed. Arnold Perlmutter and Behram Kursunoglu (San Francisco: Freeman, 1967), 32–43; Yuvan Ne'eman, "Structure of the High Energy Currents," in *Coral Gables Conferences on Symmetry Principles at High Energy. Fourth Conference*, ed. Arnold Perlmutter and Behram Kursunoglu (San Francisco: Freeman, 1967), 2–10; Jun John Sakurai, "Vector Meson Dominance and Current Algebra," in *Coral Gables Conferences on Symmetry Principles at High Energy. Fourth Conference*, ed. Arnold Perlmutter and Behram Kursunoglu (San Francisco: Freeman, 1967), 48–58.

[105] Cabibbo, "Symmetry," 170.

[106] Arnold Perlmutter, Behram Kursunoglu, and C. Angas Hurst, "Preface," in *Coral Gables Conferences on Symmetry Principles at High Energy. Fifth Conference*, ed. Arnold Perlmutter, Behram Kursunoglu, and C. Angas Hurst (New York: Benjamin, 1968), v.

[107] Ibid.

of fundamental symmetries had given way to the use of a broad range of mathematical practices involving invariances and group theory that attempted rather to establish patterns and 'model' phenomena. As I have discussed in more detail elsewhere, techniques and notions of symmetry breaking had in the meantime come to play a main role in this context.[108] However, the contributions to the fifth Coral Gables conference acknowledged the importance of their original program, and Lochlainn O'Raifeartaigh gave a summary of attempts to link internal and space-time symmetries:

> The most dramatic occurrence in the field of internal symmetry-space-time speculations, was, of course, SU(6), and it raised immediately the question of if and how it could be made relativistic. This question gave rise to 'no-go' theorems, this time of the Coleman, Jordan, Weinberg type and to a number of relativistic and semi-relativistic generalizations.[109]

As discussed at the beginning of this paper, Ne'eman gave a final talk summarizing the progress in all five conferences.[110] In a section titled "Matchmakers," he praised O'Raifeartaigh's presentation and reminded the audience of the productivity of the second conference, where so many proposals for combining internal and space-time symmetries had been presented, adding:

> All this material was thrown into the mêlée, it all got caught in some kind of mixmaster and I think that his was perhaps the most successful of the conferences. [...] It was out of this mess that [...] the algebra of the currents in its better understood form grew, so did the infinite component theories.[111]

The image of the mixmaster is helpful to express the way in which the initial, rather simple and fundamental approach for deploying some group-theoretical structures in high energy physics became the starting point of a diversity of mathematical practices linked to symmetry principles which could be physically interpreted either as fundamental theories or as phenomenological models. Symmetry principles went on to a new life, or lives, yet the original 'matchmaking' dream seemed to be at an end. It is in this sense that Ne'eman mentioned his appreciation for Hoja's yogurt project, as both a warning and a tribute to those who had believed—and still believed—that unity of internal and space-time symmetries might be possible.

Almost twenty years later, in 1997, Kursunoglu recalled the launching of the Coral Gables conferences, and, like Ne'eman, he praised espccially the second one, recalling the theories presented there by Salam and adding: "Unfortunately it

---

[108] Arianna Borrelli, "The Story of the Higgs Boson: The Origin of Mass in Early Particle Physics," *The European Physical Journal H* 40, no. 1 (March 2015): 1–52.

[109] Lochlainn O'Raifeartaigh, "Brief Survey of Recent Aigebraic Approaches to Particle Physics," in *Coral Gables Conferences on Symmetry Principles at High Energy. Fifth Conference*, ed. Arnold Perlmutter, Behram Kursunoglu, and C. Angas Hurst (New York: Benjamin, 1968), 49–76, here 50.

[110] Ne'eman, "Progress."

[111] Ibid., 373.

turned out to be another good example of Hodja's [sic] 'yogurt project'."[112] We may thus conclude that the systematic employment of group-theoretical structures in early particle physics has to be seen not as a fruitful practice for fitting phenomena, but rather as an ambitious, ultimately unsuccessful attempt at mathematically grasping the inner structures of nature. While rather simple mathematical methods proved successful in the first task, the use of abstract, more refined mathematical notions failed to achieve the second goal. Those theorists who wished to make a distinction between purely descriptive 'models' and both descriptive and explanatory 'theories' were faced with the choice of either giving up their wish for a deeper understanding of particle interactions, or accepting that the fundamental structures of microphysics could be captured by symmetries only vaguely (or not at all) related to the invariances of space-time.

Looking at the developments of the 1970s, one might argue that some authors chose the second possibility, and that the great yogurt project of symmetry matchmaking lived on in Grand Unified Theories, which were based on SU(5) and other special unitary group, but were eventually ruled out by the failure to observe proton decay. From the 1980s onward, though, a third answer to the theoretical dilemma sketched above emerged, which had already been implicit in Coleman's remark from 1967. It was the possibility that a 'theory' of fundamental interactions would not necessarily need to deliver new, successful empirical predictions, but only needed not to explicitly contradict established experimental results. In this spirit, increasingly speculative proposals of physics 'Beyond the Standard Model' were made, which were assumed to coincide with the Standard Model at low energy, and whose predictions, when falsified, were moved up to energies not yet reached, as in the case of supersymmetry, or even only obtained in an experimentally inaccessible range, as in the case of string theory. Interestingly, theorists bringing forward these research projects usually describe their activities as 'model-building,' and, as I have argued in detail elsewhere, they tend to use the term 'theory' to indicate hypothetical mathematical constructs which are expected to reflect the inner structure of fundamental interactions, yet do not at present exist.[113] However, despite—or perhaps thanks to—their highly speculative character, these theoretical research programs are without doubt still capable of inspiring in mathematically-minded authors the same enthusiasms which led Hoja to exclaim: "But just imagine if it should work!"

Acknowledgements The research presented in this paper was funded by the Institute for Advanced Study on Media Cultures of Computer Simulation (MECS), Leuphana University Lüneburg (DFG grant KFOR 1927) and by the project "Exploring the 'dark ages' of particle physics" (DFG grant BO 4062/2-1).

---

[112] Kursunoglu, "Coral Gables Conferences," 7.

[113] Arianna Borrelli, "The Case of the Composite Higgs: The Model as a 'Rosetta Stone' in Contemporary High-Energy Physics," *Studies in History and Philosophy of Modern Physics* 43, no. 3 (August 2012): 195–214; Arianna Borrelli, "Between Symmetry and Asymmetry: Spontaneous Symmetry Breaking as Narrative Knowing," *Synthese* 198, no. 4 (April 2021): 3919–48.

Interview with Myfanwy E. Evans: Entanglements On and Models of Periodic Minimal Surfaces

Myfanwy E. Evans, Michael Friedman, and Karin Krauthausen

KK: Maybe we can start with a question concerning your training and profession. You're a trained mathematician and you often work in fields which are by no means fully mathematically developed. Hence, you're doing fundamental mathematical research on the one hand. But, on the other hand, you're often classified as an applied mathematician, since a lot of your work is relevant to the materials sciences and biology. How do you see the relationship between pure and applied mathematics? Where do you see the advantages and disadvantages of applied mathematics concerning research, in the sense that, for example, applied mathematics advances mathematics as a whole?

ME: My initial training was in mathematics, but my PhD is actually in physics and mathematics together, that is, in the Department of Applied Mathematics inside the Research School of Physics (at the Australian National University). As such, I was exposed extensively to physics (and a little chemistry and biology) throughout my PhD, even though the content of my research was based in mathematical techniques. I was always very much connected to the idea that these things should be

M. E. Evans
Institut für Mathematik, Universität Potsdam, Campus Golm, Haus 9, Karl-Liebknecht-Str. 24-25, 14476 Potsdam, Germany
e-mail: evans@uni-potsdam.de

M. Friedman (✉)
The Cohn Institute for the History and Philosophy of Science and Ideas, Tel Aviv University, The Lester and Sally Entin Faculty of Humanities, Ramat Aviv, 6997801 Tel Aviv, Israel
e-mail: friedmanm@tauex.tau.ac.il

K. Krauthausen
Cluster of Excellence "Matters of Activity. Image Space Material," Humboldt-Universität zu Berlin, Unter den Linden 6, 10099 Berlin, Germany
e-mail: Karin.Krauthausen@hu-berlin.de

M. Friedman and K. Krauthausen (eds.), *Model and Mathematics: From the 19th to the 21st Century*, Trends in the History of Science, https://doi.org/10.1007/978-3-030-97833-4_7

a   b

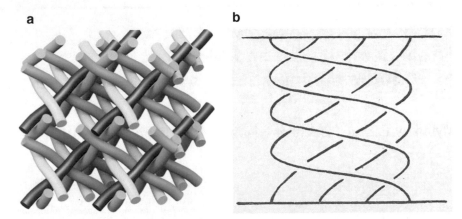

Fig. 1 a Three-dimensional periodic entanglements Reprinted with permission from Myfanwy E. Evans and Roland Roth, "Shaping the Skin: The Interplay of Mesoscale Geometry and Corneocyte Swelling," *Physical Review Letters* 112, no. 038102 (2014): 3, Fig. 3. © 2021 American Physical Society, all rights reserved. **b** A braid with four strings. From Emil Artin, "Theorie der Zöpfe," *Abhandlungen aus dem Mathematischen Seminar der University Hamburg* 4 (1925): 47–72, here 54, Fig. 7. © All rights reserved

useful in physics, but also the physics should be useful for generating new directions in mathematics. The interdisciplinary research that I do, along with many other people in my research community, is an interesting connection, because it's an application directly from pure mathematics to the natural sciences. The typical field of mathematical physics involves a very different type of mathematics and very different type of physics to what I'm looking at. But we've made a connection from pure geometry and topology, subjects which always sit inside the pure mathematics department and never in the applied mathematics one. And this connection was made between these domains and the materials sciences and physics. One might say we're doing applied pure mathematics. We're using these pure mathematical constructions applied in a theoretical context in the natural sciences, such as in theoretical biophysics, theoretical biology, and theoretical chemistry. We're staying on a theoretical level behind all of these different disciplines. In that sense, it's applied from the perspective of a mathematician, but it's not really applied from the perspective of a natural scientist. At no point are we really going into real-world applications of the ideas. We're just stepping between the disciplines on a theoretical level.

MF: I would like to focus on this connection between mathematics and research on materials that characterizes the applied pure mathematics which you just mentioned, and to concentrate on a particular area of your research which looks into mathematical three-dimensional entanglements. The starting point for this research is, surprisingly, the structure of the skin, to which we'll return later. But to concentrate for now on these entanglements, you've worked with visual models created by computer visualization programs, focusing on three-dimensional periodic entanglements (see Fig. 1a). Can you explain what these visual models mean for you?

Are they illustrations? Do you understand them as geometrical models? Or do they serve as a source of inspiration?

ME: When it comes to a three-dimensional entanglement, and particularly a periodic one, we can't say anything (yet) about it mathematically. If I have two different complicated weavings of filaments in space that are sitting side by side, I can't say whether they're the same or different. You could compare their exact geometry, you could try to lay one on top of the other, but when you look at entanglements, you want them to move around. The question you want to ask is: can I take one and deform it a little bit in space, not allowing the strings to cross each other? Can we deform one entanglement into the other? That question is possibly unanswerable. When I look at graphs (or collections of them), there's information that I can write down—how the edges and vertices are connected, for example. When I have a filamentous structure, I'm lacking concrete characteristics that I can write down. So when I employ enumerative techniques to build three-dimensional structures, my end product is simply just the model. The models really are the research in this sense, that is, until we can find a better way to describe the entanglements mathematically. But until then, the models are *it*. So I think they do fulfil a very important role, as they are the research.

MF: Taking these difficulties into account, can one assume that braid theory[1] is not a sufficient tool to research these entanglements?

ME: While being careful about absolute statements, I think that it's unlikely that braid theory (as it stands) will be sufficient for understanding these structures. Braid theory works nicely, because you can project a braid to a plane, examine the pattern of crossings on the plane (see Fig. 1b), and associate to this pattern of crossings an algebraic structure. If I have something that's three-periodic (so truly three-dimensional), I can't project it to a plane in a reasonable way. Two-periodic (like a fabric weaving on a flat page): that works. As soon as it's three-dimensional, I can't do it anymore without distorting or losing essential information. It becomes a purely three-dimensional geometry problem. Without this planar projection, the algebraic techniques of braid theory are simply no longer defined.

KK: We would like to examine a more concrete example: filaments.[2] One of the inspirations for your research is skin. Skin has a highly complicated structure that's in constant exchange with the environment, and therefore undergoes

---

[1] Officially, the research on braid theory began with Emil Artin's paper "Theorie der Zöpfe," published in 1926 (Emil Artin, "Theorie der Zöpfe," *Abhandlungen aus dem Mathematischen Seminar der Universität Hamburg*, 4 (1926): 47–72), though the mathematical interest in braids had already begun in the nineteenth century. See: Michael Friedman, "Mathematical formalization and diagrammatic reasoning: The case study of the braid group between 1925 and 1950," *British Journal for the History of Mathematics* 34, no. 1 (2019): 43–59.

[2] See: Myfanwy Evans, Vanessa Robins, and Stephen Hyde, "Periodic entanglement I: nets from hyperbolic reticulations," *Acta Cryst* A69 (2013): 241–61; Myfanwy Evans, Vanessa Robins, and Stephen Hyde, "Periodic entanglement II: weavings from hyperbolic line patterns," *Acta Cryst* A69 (2013): 262–75; Myfanwy Evans, and Stephen Hyde, "Periodic entanglement III: tangled degree-3 finite and layer net intergrowths from rare forests," *Acta Cryst* A71 (2015): 599–611.

transformations—for example, the swelling of skin in water. Modeling such a changing structure mathematically is a challenge. You've suggested modeling the filaments of the outer layer of the skin by describing the geometry of the swelling of dead skin cells with the help of triply periodic minimal surfaces (TPMSs).[3] How can these surfaces model this geometry? And can you explain how this idea of modeling with these TPMSs came to your mind?

ME: One approach to analyzing the filaments in a three-periodic entangled structure is to consider them as physical structures, like a tube of a given radius (rather than simply curves in space), and try and minimize their energy as a way of representing that structure,[4] giving it a canonical form, and as such giving us a way of comparing it with other structures. That's something that's done in knot theory already, and is termed 'ideal knots.' By finding a canonical form for the structure, you can compare it to other canonical forms as a way of comparing entangled structures. This is typically done computationally, as analytic solutions have proven difficult in most cases. I was extending these ideas of ideal knots to periodic structures, where I take a particular weaving of tubular filaments of a given tube radius and I computationally minimize the length of the filaments within the box, not allowing the tubes to intersect each other. There's a set of filamentous structures that are constructed as decorations of TPMSs, where I found that the filaments weren't straight in the minimized state, which was a big surprise and somewhat counterintuitive. These structures could be realized as symmetric cylinder packings in space, so I assumed that it would minimize their length when they take on the straight cylinder geometry. The actual minimizer is found when I allowed the filaments to curve, which in turn decreases the size of the periodic translations, kind of like compacting the structure down. This compaction actually decreased the length of the structure, and hence was identified as the minimizing structure. This physical property was counterintuitive, and lead to the question: when the structure is in its minimal form, is it somewhat 'spring-loaded'? The simple process of pulling the filaments straight is going to isotropically expand the material. While working on this, I came across a paper by a researcher, Lars Norlén, in Sweden,[5] who had been looking at the structure of skin and had proposed this particular weaving as the internal structure of corneocyte cells. I started to look at the swelling property and was wondering if it could be explained by

---

[3] A minimal surface is a surface having zero mean curvature (which means that the sum of principal curvatures is zero).

[4] In this case, we are defining the energy of a particular embedding of the structure to be the length of the tube divided by the radius of the tube. The constraint on the object is that the tube cannot intersect other parts of the tube. The minimizer of this energy is then pulling the knot of filaments as 'tight' as possible. Other energies can be defined on the configurations, but this notion of tightness is a good starting point due to its simplicity.

[5] Lars Norlén and Ashraf Al-Amoudi, "Stratum corneum keratin structure, function, and formation: the cubic rod-packing and membrane templating model," *Journal of Investigative Dermatology* 123, no. 4 (October 2004): 715–32. See also: Lars Norlén, "Stratum corneum keratin structure, function and formation—a comprehensive review," *International Journal of Cosmetic Science*, 28, no. 6 (December 2006): 397–425.

Fig. 2 The compact (left) and swollen (right) states of filaments. © Myfanwy E. Evans, all rights reserved

our spring-loaded model. Our filament model has this nice deformation mechanism where it can expand and contract, keeping filaments in contact through the process. We started exploring our model in the context of the skin swelling, and it turned out to be a very nice way to model the behavior that we can see in the skin (see Fig. 2). We were coming at it from a purely mathematical perspective, exploring configurations, and every time we started to do calculations, it always matched up with the experimental data from skin. The relationship of the geometric structures to the TPMS in this case is also not coincidental, where the TPMS decoration is likely a biological formation mechanism for the intricate structures in the skin cells.

KK: The history of these periodic minimal surfaces and their discovery is intertwined with the production of material models. The production of the material models of TPMSs started in the last third of the nineteenth century with Hermann Amandus Schwarz and his student Edvard Rudolf Neovius,[6] who not only drew sketches but also built material models of these periodic minimal surfaces (see Fig. 3). This was followed in the twentieth century by Alan H. Schoen, during the 1960s, who built plastic models, but at the same time also did computer drawings in order to explore these kinds of surfaces, discovering in 1970 another type of TPMS, the gyroid.[7] Schoen's work is a reference in your own work. How do

---

[6] Hermann Amandus Schwarz, *Bestimmung einer speciellen Minimalfläche* (Berlin: Königliche Akademie der Wissenschaften, 1871); Edvard Rudolf Neovius, "Bestimmung zweier spezieller periodischer Minimalflächen," *Akad. Abhandlungen* (Helsingfors: J. C. Freckell & Sohn, 1883).

[7] See: Alan H. Schoen, *Infinite Periodic Minimal Surfaces Without Self-Intersections*, (Washington D.C.: National Aeronautics and Space Administration, 1970).

a   b

c

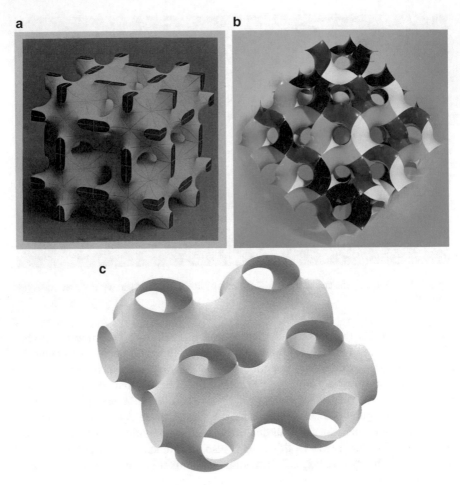

Fig. 3 a Neovius's photo of his model of a triply periodic minimal surface from 1883. From Edvard Rudolf Neovius, "Bestimmung zweier spezieller periodischer Minimalflächen," in *Akad. Abhandlungen* (Helsingfors: J. C. Freckell & Sohn, 1883), Plate 4. **b** The first physical model (1968) approximating the gyroid, made by Alan Schoen (https://schoengeometry.com/e-tpms.html, all rights reserved). It is composed of several lattice fundamental domains (source: https://schoen geometry.com/e-tpms.html, all rights reserved). **c** Approximation of the Schwarz *P* minimal surface. Graphic by Andreas Sandberg, 2012. CC BY-SA 3.0 (https://creativecommons.org/licenses/by-sa/3.0/, all rights reserved).

you consider the models of Schwarz, Neovius, and Schoen? And how are these models different—also compared to your own kind of modeling—from computer visualizations?

ME: I find the spirit in which Alan H. Schoen did research very inspirational: a three-dimensional exploration. To come across the gyroid minimal surfaces, the mathematics is extremely difficult in order to get there. The mathematicians during the nineteenth century never got there, in the sense that they never found those minimal surfaces. Whereas Schoen's more conceptual exploration of networks and

a

b

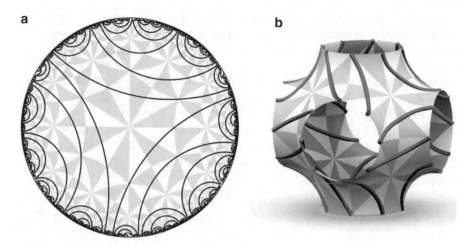

Fig. 4 On the right, a 'weaving' of filaments on a basic unit of one of the TPMSs, induced from a free tiling (or packing of lines) in the hyperbolic plane $\mathbb{H}^2$, modeled via the Poincaré disk model on the left. Reproduced with permission of the International Union of Crystallography from Myfanwy E. Evans, Vanessa Robins, and Stephen Hyde, "Periodic entanglement II," *Acta Crystallographica Section A* 69 (2013): 270, Fig. 14, all rights reserved. Here, the (cut) hyperbolic plane $\mathbb{H}^2$ is thought of as a covering of the basic unit of one of the TPMSs

partitioning of space was a different approach. He was looking at the network channels of the P and the D surfaces—two known types of TPMS already discovered by Schwarz (see e.g. Fig. 3c), which have a simple cubic network structure through the channels, or respectively a diamond network structure. Schoen's observation was that there was a third network in a similar class. The cubic network has degree six vertices, the diamond has degree four vertices; Schoen was aware that there's a third network that has degree three vertices. He started exploring what the surfaces would look like between these networks. This idea of exploring what you can do with three-dimensional structure materially and then diving back into what that means mathematically—that's a really nice way of doing things that are too complicated for mathematics as it currently stands.

These exploratory techniques rely heavily on models, be they real models to hold and explore, or computer models that you can rotate and explore computationally. Observing different features and perspectives of the models brings new intuition for the structures. Virtual reality models are starting to play a bigger role in these explorations too, and to bridge the gap between the physical and computational models.

MF: If we return to your research on entanglements considered as a sort of 'tiling' of TPMSs (see Fig. 4), you also work with hyperbolic geometry, which is associated with this tiling and with the TPMSs. As you note with respect to your research on skin, "keratin filaments of the skin lie almost exclusively inside one of the channels of an appropriately sized Gyroid, leaving the second channel filled with water."[8] As is known, it's impossible to isometrically immerse a complete hyperbolic plane—a surface which has everywhere constant negative Gaussian curvature—in a three-dimensional Euclidean space[9]; that is, a surface with constant negative Gaussian curvature embedded in the Euclidean space $\mathbb{R}^3$ doesn't exist. But the gyroid is a minimal surface with the most uniform negative curvature and with finite average Gaussian curvature; hence, the gyroid can be considered as the 'best' embedding of the hyperbolic plane $\mathbb{H}^2$. In your work you choose a *model* of the hyperbolic plane called the Poincaré disk model. Here though, we're not talking about a material model but we're rather dealing with a model in another sense, in the sense of a realization of a set of axioms. Why did you choose this model? What advantages does it have?

ME: Yes, you're correct that this is not a physical model but rather a mathematical model of the hyperbolic plane. The Poincaré disk allows me to visualize tilings of the hyperbolic plane in a nice way. In particular, this is a good model to use because it's conformal, so it preserves the angles between lines in the plane. I also think that symmetry, reflections and rotations in this model has quite an intuitive feel. The research does require a lot of intuition. Because it's not really clear what techniques should be used, and it may be that this research is in its infancy, you need models that display the properties that you would like to be able to explore.

MF: Can you explain in more detail the connection between the tiling of the Poincaré disk model and the tiling of the TPMSs?

ME: The hyperbolic plane is the universal cover of the minimal surfaces, which means that I can essentially wrap the hyperbolic plane over the minimal surfaces. If I have a cylinder, its universal cover is a flat Euclidean plane. When I have a flat plane (or, more precisely, a rectangular piece of it), I have two independent translation directions, X and Y. If I make one of those translations into a loop, and one of those I leave open, then I get a cylinder. In a similar manner, if I take a particular tiling of the hyperbolic plane, I can cut out a particular portion of this plane (a hyperbolic dodecagon), and then I have six independent translation directions in the hyperbolic plane—so, one can say, there's a lot more space to move in the hyperbolic plane. Out of these six, I make three of them into loops and I leave three of them open. So three of them become closed loops on the resulting surface, and the other three become X, Y, and Z—the translations in three-dimensional space. This is the wrapping process from the hyperbolic plane to the minimal surfaces (see Fig. 4). Everything that I draw in the hyperbolic plane, I can just print onto the surface. That's really analogous to drawing a line in the Euclidean plane, which would wrap to a helix of a given pitch on the cylinder. So a packing of lines in the Euclidean plane gives me helices on a cylinder. In the same way, a packing of lines in the hyperbolic plane—and there are infinitely

[8] Myfanwy E. Evans and Gerd E. Schröder-Turk, "In a material world: Hyperbolic geometry in biological materials," *Asia Pacific Mathematics Newsletter* 5, no. 2, (2015): 21–30, here 22.
[9] David Hilbert, "Über Flächen von konstanter Krümmung," *Transactions of the American Mathematical Society* 2 (1901): 87–99.

many ways to construct these—gives me a set of helices on the minimal surfaces once I wrap up the hyperbolic plane.

MF: If I return to what was noted earlier about the impossibility of isometrically immersing a complete hyperbolic plane in the Euclidean space $\mathbb{R}^3$, this means that one has to have some distortion, seen e.g. in the fact that the curvature of the TPMS is not everywhere negative. How does this distortion affect the research?

ME: There's some kind of distortion of the surface, but I haven't been concerned with it because the resulting structures that we're looking at are the entanglements in three-dimensional space; we're not looking at the precise geometry of the curves (so far). If you wanted to do this more precisely, you could certainly use some approximations to follow geodesic curves on the surfaces rather than geodesic curves on the hyperbolic plane. This could be an interesting mathematical direction, which would involve some advanced differential geometry; however, in the connection to the physical systems, this doesn't seem to be significant.

MF: My next question is a more general one. How does the mathematical theory of entanglement express the connection between geometry, topology, and mechanical and material behavior? Were new mathematical tools and theories developed for this?

ME: The way the project exists at the moment is enumerative. That's to say, here's a technique that we have which has been developed related to various objects that we find in the natural sciences, in particular the minimal surfaces. And so the question arises: how can we add an extra layer of complexity to start looking at more complicated three-dimensional structures using these minimal surfaces as a stepping stone? As with most enumerative techniques, the difficulty is in making sense of how we order and characterize the structures. That's where various forms of mathematics has been—or needs to be—developed, in order to answer questions like: do all possible structures exist in the enumeration, can I uniquely describe each of these structures, and how can we have a sensible complexity ordering? In terms of a comprehensive mathematical theory of entanglement, we're still a long way away from something like that.

MF: I want to make the question a little bit more precise concerning the mechanical behavior, focusing on whether the materiality of entanglement was taken into consideration, and also connecting this to the materiality of the models. You said the tactile models have the advantage that we can touch them. But you also work with filaments and entangled nets which have a certain thickness, that is, with the material itself. But can one say that the mathematical situation is close in some way to the real world? The filaments are not one-dimensional strings, like mathematical knots. Does this materiality have any influence on how to formulate a corresponding mathematical theory?

ME: That's a good question. This type of question is already being tackled from the mathematical perspective in the topic of 'ideal knots' that I mentioned earlier, where the knots are made from a tube of a given thickness. Exact mathematical descriptions of these objects are really difficult to obtain in most cases, but approximate results are already proving useful in applications, like in properties of knotted DNA. Developing a parallel idea for the filamentous structures is what we've done computationally but, again, any kind of exact mathematics on these structures is elusive. So my answer would be that yes, mathematics can deal with the materiality, but we jump to the field of geometric knot theory and geometric analysis rather than the algebraic structure of braids that we saw earlier, and this makes the mathematics significantly more difficult.

KK: The focus on the real entangled object leads me back to the topic of visualization and modeling. Is there a difference for you when you visualize something in your computer or your mind and when you model something?

ME: No, I see them as the same thing. I can't really come up with an example where I differentiate between the two. To visualize something is to model it.

KK: And what kind of programs do you use for these kinds of visualizations?

ME: I use Houdini, which is a proprietary animation software. There are a few research groups using Houdini for mathematical purposes. But with the increasing use of 3D printers, modeling software is now easy to obtain and relatively user friendly. Since we're using specific mathematical formulations as inputs, it doesn't matter too much what kind of visualization software you use in the end.

KK: Is there a link between your visualization programs and the programs that the materials scientists use?

ME: Not really. A materials scientist or a physicist will typically want to simulate a process, and will have specific programs to fulfil this. Basic visualization will often be incorporated into these programs already, and further visualization is rarely done.

KK: If we take a look at the beginning of the twentieth century, together with a certain decline in the production of material models, there was a strong tendency in certain areas of mathematics to distance oneself from intuition, and especially from visualization. I wonder, when I listen to you as you describe a somewhat opposite tendency, how this changed at the end of the twentieth century, as intuition and three-dimensional tangibility seems so important and such a driving force.

ME: I don't know if it has really changed across most of mathematics. I think, if we walked around the corridor here[10] and started to discuss these questions, one would get very different responses. I wonder if this is how the sciences are starting to filter back into mathematics. One may suggest that rigorous mathematics and physics can only describe things up to a certain complexity. Biology, for example, is feeding examples back into the sciences that are far too complex to even consider in that level of detail and rigor. But when you're taking those concrete examples and are trying to say something about them, then you're really inventing new mathematical ideas. In order to know what's worth describing, you need to use intuition gained from the real structure and its function. In general, the typical mathematics problem is very easy to pose but very difficult to solve. But for these interdisciplinary questions it's the other way around: they're relatively easy to solve but extremely difficult to pose. What question do you even want to ask and

---

10 At the mathematical department of the Technische Universität Berlin, where the interview took

what question do you want to answer? That's the most difficult part of the whole process, and it often takes a long time for these research questions to evolve to something reasonable.

MF: So how do you see the connection between the material models and the task of asking the right questions? Since, if I paraphrase what you said earlier, the material models really are the research object.

ME: For me, the models are the catalyst for starting to formulate what kind of research question I might ask. I make a model (using some kind of mathematically based technique), then I look at what's interesting about it: symmetry, mechanics, something curious. These observations then guide the exploration of finding good research questions to ask.

MF: And an example of that would be the filaments?

ME: Definitely. I build filamentous structures using the TPMS; then, using the construction and observation of the structures, I can begin to formulate possible research questions, possible characteristics worth exploring, similarities and differences. There are often really spectacular arrangements of the filaments that arise, that only become spectacular when you look at the model, that become important in comparing and contrasting structures. The skin structure is one such structure, with a very interesting symmetry and entanglement. So it's not surprising at all that nature should choose such a structure.

Myfanwy E. Evans is Professor of Applied Geometry and Topology at the University of Potsdam. She works on geometric and topological problems inspired by physical and biological structures and processes, in particular relating to tangling, surfaces, and tangling on surfaces. She works around the central question of how much of the function of a structure in nature can be described by basic geometric and topological principles. This involves various computational and theoretical studies alongside the study of some really beautiful geometric objects. She studied Mathematics and Physics at the Australian National University in Canberra, also completing her PhD there in 2011 in the Department of Applied Mathematics with the dissertation "Three-dimensional entanglement: Knots, knits and nets." Before moving to Potsdam in 2020, she led an Emmy Noether research group at the Technische Universität Berlin. She is an active member of the Cluster of Excellence "Matters of Activity. Image Space Material" at the Humboldt Universität zu Berlin and the cooperative research center "Discretization in Geometry and Dynamics" at the Technische Universität Berlin.

Acknowledgements The authors acknowledge the support of the Cluster of Excellence "Matters of Activity. Image Space Material" funded by the Deutsche Forschungsgemeinschaft (DFG, German Research Foundation) under Germany's Excellence Strategy–EXC 2025–390648296.

The Dialectics Archetypes/Types (Universal Categorical Constructions/Concrete Models) in the Work of Alexander Grothendieck

Fernando Zalamea

Along with David Hilbert, Alexander Grothendieck (1928–2014) is certainly one of the two greatest mathematicians of the twentieth century. The range (functional analysis, complex variables, algebraic geometry, category theory, number theory, topological algebra) and the depth (nuclear spaces, sheaves, schemes, topos, motives, stacks) of his contributions are simply outstanding. In his work, Grothendieck systematically explored the pendulum of the abstract and the concrete, the universal and the particular, a *back-and-forth* that we can subsume in a dialectics between *archetypes* and *types*. Archetypes are mathematical constructions which act as universals in certain categories, and which are *projected* into the many types living in those categories. On the other hand, types are often embedded into global archetypes, which govern the local behavior of the types. We will thus understand 'archetypes' as *universal categorical constructions*, and 'types' as *concrete models*, and study the correlations between them.

The article is divided in three parts: (1) an exploration of the notion of 'archetype,' and its *projectivity* to 'types,' at the beginning of Grothendieck's algebraic geometry period (the *Tôhoku* paper, published in 1957 and the *Rapport* on classes on sheaves, also published in 1957),[1] (2) a study of the *embedding* of types

---

[1] Alexander Grothendieck, "Sur quelques points d'algèbre homologique," *Tôhoku Mathematical Journal* 9, no. 2 (1957): 119–221; and Alexander Grothendieck, "Classes de faisceaux et théorème de Riemann-Roch," *Séminaire de Géométrie Algébrique* 6 (1957): 20–77 (*Rapport* to Serre, published 1966–67).

F. Zalamea (✉)
Departamento de Matemáticas, Universidad Nacional de Colombia, Oficina 205, Edificio 405, Carrera 45 No. 26-85, 500001 Bogotá, Colombia
e-mail: fernandozalamea@gmail.com

© The Author(s) 2022
M. Friedman and K. Krauthausen (eds.), *Model and Mathematics: From the 19th to the 21st Century*, Trends in the History of Science, https://doi.org/10.1007/978-3-030-97833-4_8

into archetypes, at the end of Grothendieck's topological algebra period (*Pursuing Stacks* (1983)[2] and *Dérivateurs* (1991)[3]), (3) an account of the appearances of the notion of "modèle" and its variants in *Récoltes et semailles* (1986).[4]

Archetypes and Types in the *Tôhoku* and the *Rapport*

The *Tôhoku* introduced sheaves and category theory into the landscape of 'real' mathematics. Along with David Corfield in his 2003 *Towards a Philosophy of Real Mathematics*,[5] we understand by 'real' mathematics, the corpus of the hard-working mathematicians: combinatorics, number theory, abstract algebra, algebraic geometry, topology, complex variables, functional analysis, differential geometry, etc., well beyond sets and classical logic. Jean Leray invented sheaves in 1942. Saunders MacLane and Samuel Eilenberg invented categories in 1945.[6] Nevertheless, their actual use in real mathematics, as signaled by the same MacLane is due to Grothendieck.[7] Grothendieck's contributions to category theory are truly impressive on their own (introduction of sub-objects, adjoint functors, equivalences, representable functors, additive and abelian categories, generators, infinitary axioms), but the central points of the *Tôhoku* paper deal with two kinds

[2] Alexander Grothendieck, *Pursuing Stacks* (1983), c. 600 pp., manuscript distributed, unpublished.

[3] Alexander Grothendieck, *Les Dérivateurs* (1991), c. 2000 pp., manuscript distributed, unpublished.

[4] Alexander Grothendieck, *Récoltes et semailles* (1986), c. 1600 pp., manuscript distributed, unpublished.

[5] David Corfield, *Towards a Philosophy of Real Mathematics* (Cambridge: Cambridge University Press, 2003).

[6] Given a topological space X, a *sheaf F* is a contravariant functor which associates to every open set U in X another set $F(U)$ of objects (e.g. abelian groups, vector spaces, modules over a ring etc.), which satisfy certain axioms. A category C is a collection of objects $Ob(C)$ that between them are morphisms $hom(C)$, when these satisfy certain axioms. For example, the class of all sets (as objects) together with all functions between them (as morphisms) form the category *Set*. For the seminar papers of Leray, see, for example: Jean Leray, "Sur la forme des espaces topologiques et sur les points fixes des représentations," *Journal de Mathématiques Pures et Appliquées* 9, no. 24 (1945): 95–167. Jean Leray, "L'anneau d'homologie d'une représentation," *Comptes rendus de l'Académie des Sciences* 222 (1946): 1366–68. Jean Leray, "Structure de l'anneau d'homologie d'une représentation," *Comptes rendus de l'Académie des Sciences* 222 (1946): 1419–22. For the first works of Saunders MacLane and Samuel Eilenberg, see: Samuel Eilenberg and Saunders MacLane, "Group extensions and homology," *Annals of Mathematics* 43, no. 4 (1942): 757–831; Samuel Eilenberg and Saunders MacLane, "Natural Isomorphisms in Group Theory," *Proceedings of the National Academy of Sciences of the United States of America* 28, no. 12 (December 1942): 537–43; Samuel Eilenberg and Saunders MacLane, "General theory of natural equivalences," *Transactions of the American Mathematical Society* 58 (1945): 231–94.

[7] Ralf Krömer, *Tool and Object. A History and Philosophy of Category Theory* (Basel: Birkhäuser 2007), 158.

of *archetypes* related to mathematical practice: (1) to prove that in an abelian category with a generator and a suitable infinitary axiom, any object ('type') can be embedded into an injective object ('archetype'),[8] (2) to prove that the cohomology of a space, with coefficients in a sheaf ('types'), can be reconstructed as a projection of a formalism of derived functors ('archetypes').[9] In that way, (co)homology, one of the main themes of mathematical development in the period 1895–1950, can be seen as a projection of an even more fundamental theme: a theory of abstract derivators in suitable categories.[10]

Grothendieck's *Tôhoku* strategy is very interesting in terms of 'models.'[11] First, categories, abstract universal properties, are, through a series of *axioms*, held in the diversity of mathematical regions: associativity and identities (all categories), existence of an abelian group structure in the *Hom*-sets (additive categories, *e.g.* category of holomorphic fibered spaces over a Riemann surface),[12] existence of kernels, cokernels, and adequate factorizations through them (abelian categories, *e.g.* category of sheaves with fibers abelian groups),[13] etc. In this approach, the *axiomatic abstract definitions* have their expected *concrete models*. Second, in the abstract setting, that is in general categories which satisfy the axioms, not yet 'reified' in concrete models, archetypes occur in a *natural way* (definitions and properties related to the quantifier 'exists only' ∃!), and only afterwards are they projected into existential objects of concrete categories. Thus, *abstract categories* are 'doubly archetyped' (*inside* the categories free objects appear, *outside* the categories free objects incarnate in apparently very different disguises), while *concrete categories* become just partial modelizations of general categorical transits. For example, an initial object 0 in an *abstract category* is defined through the property that there exists only a single morphism from 0 to any other object of the category. If the *abstract category is modelized in diverse concrete categories*, the initial object 'incarnates' in apparently very different constructions: 0 is the empty set in the category of sets, 0 is a group with one element in the category of groups, 0 is the ring of integers in the category of rings, 0 is the discrete topology in the category of topological spaces, etc. Third, in the abstract settings, the general theorems (*e.g.* (1) and (2) above) occur thanks to *free machinery* that has escaped from the particularity of the concrete models.

---

[8] See: Grothendieck, "Sur quelques points d'algèbre homologique," Chap. 1.

[9] See: Ibid., Chap. 3.

[10] We will look below to an even more general setting, which will try to encompass not only homology, but also homotopy, in Grothendiecks's *Les Dérivateurs* (1991). See the section below, "Types and archetypes in *Pursuing Stacks* and *Dérivateurs*."

[11] ... though, it must be noted that Grothendieck himself did not use the term 'model' in the *Tôhoku* paper.

[12] See: Grothendieck, "Sur quelques points d'algèbre homologique," 127.

[13] See: Ibid., 129.

For Grothendieck, abstraction is never artificial,[14] but, on the contrary, has the very precise purpose of *smoothing* many of the obstructions that live along the diverse particular regions of mathematics. *Modelizing* through axioms and categories allows for the *integration* of many possible differences. Archetypes (universal categorical constructions) allow to *unify smoothly* a diversity of types (concrete models). It is a simple matter of elevation and perspective: while, in the maze of the concrete, walls surround us, when we elevate ourselves to the general, we are afforded the possibility of viewing from a wider panorama. Orientation is obtained at the top of the mountain, if we can escape the underlying jungle.

Grothendieck's extension of the Riemann-Roch theorem is a good example of the *abstract orientation* advantages obtained through categorical modelization. To recall, the Riemann-Roch theorem unifies two very different approaches to obtain a *natural invariant* for a complex surface. On one hand, the genus g of the surface is obtained by counting the number of holes of the surface, or, equivalently, the number of cuts (minus one) with which the surface becomes disconnected (*e.g.,* the genus of the sphere is 0, the genus of the torus is 1, etc.) On the other hand, one can think about the 'good' functions (holomorphic) and the 'bad' functions (meromorphic) definable on the surface. If we fix n points on the surface, consider the (vector) space *Hol* of holomorphic functions with zeroes on those points, and the (vector) space *Mer* of holomorphic functions with poles on those points. The Riemann-Roch theorem says that

$$n - g + 1 = \dim(Mer) - \dim(Hol).$$

In this way, an *intrinsic* geometric invariant (the genus) is related to an *extrinsic* complex variable invariant (harmonic difference between *Mer* and *Hol*). The Riemann-Roch theorem marks the beginning of modern algebraic geometry at the very heart of profound connections between geometry, topology, complex variables, differential geometry and algebra.

To examine Grothendieck's approach in more detail, recall that *coherent sheaves* may be understood as the simplest, well-behaved sheaves, emerging from ring representations.[15] For a variety X with enough smooth conditions, Grothendieck imagined the *free group K(X)* generated by *all* coherent sheaves on X. Coherence, freeness and totality—diverse forms of simplicity—help to understand why

---

[14] The repeated prejudice that Grothendieck's work, or category theory in general, can be reduced to some 'abstract nonsense,' is a good example of a poor utterance by mathematicians who have never read Grothendieck. Instead, all his work is filled with *examples*, and the *back-and-forth* between the concrete and the abstract always comes back to the real mathematics (functional analysis, complex variables, algebraic geometry, topological algebra) in which he was a Master.

[15] Given a sheaf whose fibers are rings, one can iterate the construction and consider a sheaf over the space of those rings, whose fibers are now ring modules. If the vertical structural transitions behave well (modules of *finite type*) and the horizontal homomorphic transitions also do so (kernels of *finite type*), we have a *coherent sheaf* (Jean-Pierre Serre, "Faisceaux algébriques cohérents," *Annals of Mathematics* 61, no. 2 (1955): 197–278). The sheaf of germs of holomorphic functions is coherent.

K(X) may behave like an *archetype*. The main theorem in Grothendieck's K-theory assures that it is the case. *K* becomes a functor related to the homology functor *H*, through a natural transformation *C* given by Chern's classes.[16] Then, an *obstruction* to commutativity (*CK* does not equal *KC*) occurs, but it can be factored through Todd's classes,[17] producing an extended Riemann-Roch-Serre-Hirzebruch formula. The particular case of the Riemann-Roch formula is obtained from the general case, specifying a variety in one point. Thus, the general archetype (universal categorical construction *K(X)*), while projected on a trivial homology, captures the specific type (concrete equation on the surface

$$X, n - g + 1 = \dim (Mer) - \dim (Hol).$$

Here, the modelization does not occur though axioms, but it uses instead a very *soft* categorical environment, where little is specified. Well-behaved coherence, freeness, and universality, are the only ingredients used in the construction of the K-theory group *K(X)*. *Categorical abstraction provides an archetypal guide, and concrete modeling provides its associated types.* The fact that Riemann-Roch's wonderful connection between magnitude (genus) and number (harmonic difference) can be seen as a projection from an *abstract sheaf's behavior* shows the depth of Grothendieck's approximation. By way of sheaves, geometry and algebra become unified, something, which further developments, will also underline forcefully.[18]

Types and Archetypes in *Pursuing Stacks* and *Dérivateurs*

Both in the *Tôhoku* paper and the *Rapport*, both from 1957, precedence is given to archetypes (derived functors, group of the K-theory), which then become projected into types (homology constructions). In Grothendieck's work in the 1980s, the method is somewhat *inverted*: a study of many types (*e.g.*, moduli spaces, that is, isomorphism classes of Riemann surfaces, and the absolute Galois group $Gal(\overline{\mathbb{Q}}, \mathbb{Q})$) gives rise to an emergence of archetypes (*e.g.*, anabelian conjectures, stacks, derivators). In this section I will explore some of these situations in *Pursuing Stacks* (1983) and *Dérivateurs* (1991).[19]

---

[16] See: Grothendieck, "Classes de faisceaux," 40–41, 63–64.

[17] See: Ibid., 72.

[18] It is the case with the emergence of schemes (1958) and topos theory (1962) in Grothendieck's famous *Séminaire de Géométrie Algébrique*. Some types of sheaves (schemes) generalize algebraic varieties (discretion), while other kind of collections of sheaves (topos) generalize topological spaces (continuity). Space and number become unified again through sheaf theory.

[19] $\overline{\mathbb{Q}}$ is the algebraic closure of the field of rational numbers, $\mathbb{Q}$ (i.e. it is the set of all roots of a non-zero polynomial in one variable with rational coefficients). To recall, given an algebraic, normal extension E of a field F, $Gal(E, F)$ is the group of all isomorphisms f of E s.t. $f(x) = x$ for every $x \in F$.

Grothendieck's *La longue marche à travers la théorie de Galois* (1981)[20] establishes a new research program, after his works of the first (1949–1957) and
second (1958–1970) decades. His attention is now turned to 'low-level' complexity objects, that is, surfaces in general and Riemann surfaces in particular, modular
curves, complex towers, and combinatorial approximations to the absolute Galois
group Gal($\overline{\mathbb{Q}}$, $\mathbb{Q}$). Given an algebraic extension X of $\mathbb{Q}$,, one can ask if his algebraic group $\pi_1^{alg}(X)$ (profinite completion of the fundamental group $\pi_1^{top}(X)$)
fully characterizes X: if the answer is positive, Grothendieck says that we have
an *anabelian* variety ('an' stands for 'non,' since the homotopy groups involved
are strongly non-commutative). One of his main conjectures (still open) states that
the anabelian varieties over $\mathbb{Q}$ are essentially the isomorphism classes of Riemann
surfaces. In this way, an *archetypal* property (anabelianity) is expected to capture
some of the main types of modern mathematical thought, the Riemann surfaces.

Pursuing stacks (1983) opens the stage for a profound connection between
homotopy and homology. A *hierarchy* of concepts, objects and techniques is
fundamental to be able to delve into the many levels of the homological and
homotopical groups. In that sense, an *n*-stack can be understood as an *n*-truncated
homotopy type, which in turn can be imagined as an *(n + 1)*-sheaf. The *itera-
tion of types* gives rise to ∞-categories, where all information on *n*-morphisms
(morphisms, functors, natural transformations, etc.) is piled-up. ∞-categories act
as very large archetypes, which cover huge varieties of models. *Pursuing stacks*,
whose first episode is called "histoire de modèles" (modelizing story),[21] studies
the many ways in which adequate functors in *Cat* (the category of all categories)
can model fragments of homotopical and homological constructions. In particular, the *localizers* of a category through weak equivalences cover a lot of concrete
realizations. We can observe here the action of *both* grothendieckian strategies
(i) bottom-up, and (ii) top-down. In fact, with the first strategy (i) motivated by
the visualization of concrete objects (Riemann surfaces, homotopy groups), the
abstract anabelian conjectures emerge, while, on the other hand, with the second (ii) motivated by the desire to unify homotopy and homology, the concrete
modelizers in *Cat* appear.

Dérivateurs (1991) proposes several *axiomatic* treatments of the constructions
first imagined in *Pursuing stacks*. Axioms are initially imposed on classes W of
morphisms in *Cat*, to try to model some properties of weak equivalences in the
category *Hot* of CW-complexes with homotopies between them. The program consists of attempting to elevate types in *Hot* to archetypes in *Cat*. Axioms, as we have
mentioned, *free* the lower level constructions (say, in *Hot*), in order to get *smooth-
ness* properties at higher levels (say, in *Cat*). Other axioms are afterwards imposed
on functors $Cat \to Cat$: $C \to C\ W^{-1}$, in order to capture the good structural properties of localizations. Such well-behaved functors are called *derivators*, since they

---

[20] Grothendieck, *La longue marche à travers la théorie de Galois* (1981), c. 1600 pp., manuscript
distributed, unpublished.
[21] See: Grothendieck, *Pursuing Stacks*, 4 (table of contents).

generalize, *in a canonical way*, the nice properties of derived functors (*Tôhoku*). In this way, in an abstract, free, archetypal environment, derivators capture some of the main constructions of homotopy (coming from topoi and homotopy groups) and homology (coming from abelian categories and homology groups).

Models in *Récoltes et Semailles*

Récoltes et semailles (1986) is considered without doubt as the most profound text ever written[22] on the emergence of mathematical creativity. Beyond unnecessary (but comprehensible) quarrels with the mathematical community, *Récoltes et semailles* provides, through the lenses of a mathematician of the highest caliber, an extremely detailed analysis of his mathematical career. The result illuminates the many polarities, forces and layers that govern mathematical thought. If we follow in *Récoltes et semailles* the term "modèle" and some related concepts ("unité," "universel"), we obtain another complementary perspective on the archetypes/types dialectics that we have quickly perused in Grothendieck's mathematical work.

The term "modèle" appears in approximately sixty pages of the manuscript. Four sorts of uses of "modèle" are on display: (i) "modèle" as a concrete mathematical model, that is, a structure with some well determined properties (e.g., modelizers, homotopical constructions, motivic Galois groups, *étale* topos, set-theoretic models, Chern classes),[23] (ii) "modèle" as an abstract mathematical framework of ideas (e.g., algebraic geometry, Euclidean geometry, Newtonian mechanics, Einsteinian relativity, coherent duality),[24] (iii) "modèle" as a philosophical trend of thought (e.g., mixturing continuity/discreteness, foundations, *yin/yang*),[25] (iv) "modèle" as a human model of conduct (e.g., simplicity, perfection, Guru and Krishnamurti behavior, maternity).[26] One can see how Grothendieck explores, as usual, *many layers and many different perspectives, which, afterwards, become unified*. In fact, unity and universality (circa 50 appearances each in *Récoltes et semailles*) become closely connected with modelization, since any of the four reference "modèle" categories (i)-(iv) provides a *common ground*, in order to understand, in correlative, reticular ways, either mathematical structures, mathematical thought, philosophy, or ethics.

---

[22] Henri Poincaré, "L'invention mathématique," *Bulletin de l'Institut Général de Psychologique* 8, no. 3 (1908) is certainly another of the profound texts written around on the subject of mathematical creativity. Nevertheless, the range and depth of Grothendieck's *Récoltes et semailles* far exceeds Poincaré's brief, albeit wonderful, reflections.

[23] See: Grothendieck, *Récoltes et semailles*, 1.V, 1.VIII, 1.117, 2.177, 2.200, 2.211, 2.304, 2.326, 2.411, 4.941, 4.947, 4.1205, for example (*Récoltes et semailles* is divided in a preface (P), four parts (1–4), and an Appendix (A)).

[24] See: Grothendieck, *Récoltes et semailles*, 29, 56–58, 60, 4.976, 4.992, for example.

[25] See: Ibid., 58, 2.392, 3.694–96, for example.

[26] See: Ibid., 1.III, 2.383, 3.538–41, 3.654, for example.

Conclusion

The archetype/type dialectics between categorical constructions and structured models *enriches* considerably the panorama of mathematical practice. *Both levels*—the abstract and the concrete, the universal and the particular—are necessary to *nurture mathematical imagination.* Mathematics has to explore all possible worlds, beyond mere actuality, but, on the other hand, many of its greatest achievements occur when a general architecture illuminates an actual conjecture.[27] The *back-and-forth* between types and archetypes, elevated to a method in Grothendieck's work, helps to understand the importance of a *hierarchy of models* in mathematical thought.

Acknowledgements I thank Michael Friedman for his suggestion that I track the term 'modèle' in Grothendieck's *Pursuing Stacks* and *Récoltes et semailles.*

---

[27] It is the case of Fermat's theorem illuminated by Taniyama-Shimura's identification of modularity and ellipticity, the Weil conjectures illuminated by Grothendieck's *étale* topos of a scheme, or, perhaps (time will tell), the Riemann Hypothesis illuminated by Connes' arithmetic topos.

Part II
Epistemological and Conceptual Perspectives

'Analogies,' 'Interpretations,' 'Images,' 'Systems,' and 'Models': Some Remarks on the History of Abstract Representation in the Sciences Since the Nineteenth Century

Moritz Epple

Despite the rapidly growing interest in the history of 'thinking in models', a detailed historical understanding of the different modes and forms of such thought-forms is still some way off. This is mainly due to the fact that, when considered retrospectively, not enough attention is paid to the at times surprisingly diverse epistemological language used by the actors in this history. Accordingly it is not uncommon that not only the language use but also the thought-forms themselves are characterised with some vagueness in historical accounts.

This can be illustrated by a frequently quoted passage from the introduction to Heinrich Hertz's *Prinzipien der Mechanik*, published in 1894. Here Hertz remarks:

> We form for ourselves images [*innere Scheinbilder*] or symbols of external objects; and the form which we give them is such that the necessary consequents of the images [*Bilder*] in thought are always the images of the necessary consequents in nature of the things pictured [*abgebildeten Gegenstände*]. In order that this requirement may be satisfied, there must be a certain conformity between nature and our thought. Experience teaches us that the requirement can be satisfied, and hence that such a conformity does in fact exist. When

A German version of this paper has been published under the title "'Analogien,' 'Interpretationen,' 'Bilder,' 'Systeme' und 'Modelle': Bemerkungen zur Geschichte abstrakter Repräsentationen in den Naturwissenschaften seit dem 19. Jahrhundert," *Forum interdisziplinäre Begriffsgeschichte* 5, no. 1 (2016): 11–30. This English version has been slightly revised and corrected. Many thanks are due to the translator Benjamin Carter and the editors of this volume for their great help and patience.

M. Epple (✉)
Goethe-Universitat Frankfurt, Historisches Seminar – Wissenschaftsgeschichte, IG-Farben-Haus, Norbert-Wollheim-Platz 1, 60323 Frankfurt am Main, Germany
e-mail: epple@em.uni-frankfurt.de

M. Friedman and K. Krauthausen (eds.), *Model and Mathematics: From the 19th to the 21st Century*, Trends in the History of Science, https://doi.org/10.1007/978-3-030-97833-4_9

from our accumulated previous experience we have once succeeded in deducing images of the desired nature, we can then in a short time develop by means of them, as by means of *models* [*wie an Modellen*], the consequences which in the external world only arise in a comparatively long time, or as a result of our own intervention. We are thus enabled to be in advance of the facts, and to decide as to present affairs in accordance with the insight so obtained. The images [...].[1]

Around two decades later Ludwig Wittgenstein, on whom Hertz's reflections on images of external reality had made a strong impression, would—with a barely perceptible alteration—form from this the famous proposition 2.12 of his "Tractatus logico-philosophicus": "The image is a model of reality."[2]

The combination of these two texts and motifs has occasionally been made the centrepiece of a genealogy of thinking in abstract and, particularly, mathematical models in scientific modernity. In Wittgenstein's formulation, however, a difference is lost that, if read carefully, is still found in Hertz's text: namely that between images that are 'like models' and the models themselves. *After* Wittgenstein it was possible—although initially still uncommon—to speak also in the sciences of abstract, immaterial models of real relations. Until Hertz—and still in the language of his introduction—the models that the images *resembled* without being *identical* with them were not abstract but concrete, such as the models that had been used for centuries in a range of artisanal, artistic and academic contexts, from architecture to painting, and had recently acquired a certain prominence in the sciences in the form of the material structures representing scientific objects that can still be found in great quantities in our collections and museums. Some examples selected at random from an abundance of similar ones are the once commercially distributed plaster or string models of mathematical objects (Figs. 1 and 2) and the material models of biological and chemical phenomena (Fig. 3).[3]

---

[1] Heinrich Hertz, *Die Prinzipien der Mechanik* (Leipzig: Geest u. Portig, 1894), 1–2: "Wir machen uns innere Scheinbilder oder Symbole der äußeren Gegenstände, und zwar machen wir sie von solcher Art, dass die denknotwendigen Folgen der Bilder stets wieder die Bilder seien der naturnotwendigen Folgen der abgebildeten Gegenstände. Damit diese Forderung überhaupt erfüllbar sei, müssen gewisse Übereinstimmungen vorhanden sein zwischen der Natur und unserem Geiste. Die Erfahrung lehrt uns, dass die Forderung erfüllbar ist und dass also solche Übereinstimmungen in der That bestehen. Ist es uns einmal geglückt, aus der angesammelten bisherigen Erfahrung Bilder von der verlangten Beschaffenheit abzuleiten, so können wir an ihnen, wie an Modellen, in kurzer Zeit die Folgen entwickeln, welche in der äußeren Welt erst in längerer Zeit oder als Folgen unseres eigenen Eingreifens auftreten werden; wir vermögen so den Thatsachen vorauszueilen und können nach der gewonnenen Einsicht unsere gegenwärtigen Entschlüsse richten. Die Bilder [...]." English translation: Heinrich Hertz, *The Principles of Mechanics* (London: Macmillan and Co., 1899), 2 (emphasis M.E.).

[2] In the original: "Das Bild ist ein Modell der Wirklichkeit." Ludwig Wittgenstein, "Tractatus logico-philosophicus," *Annalen der Naturphilosophie* 14 (1921): 185–262, Proposition 2.12.

[3] On mathematical models as they were understood at the time, see *inter alia*: Gerd Fischer, ed., *Mathematische Modelle*: *From the Collections of Universities and Museums*, 2 vols. (Braunschweig and Wiesbaden: Vieweg, 1986); Herbert Mehrtens, "Mathematical Models," in *Models: The Third Dimension of Science*, ed. Soraya de Chadarevian and Nick Hopwood (Stanford: Stanford University Press, 2004), 276–306; Anja Sattelmacher,

Fig. 1 Model of a space curve with singular points (Sammlung mathematischer Modelle, Institute of Discrete Mathematics and Geometry, TU Wien). © Research group Differential Geometry & Geometric Structures, TU Wien, all rights reserved

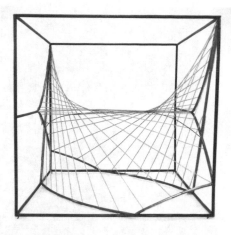

Fig. 2 Plaster model of a third-order surface (Sammlung historischer mathematischer Modelle, Martin-Luther-University Halle-Wittenberg. All rights reserved

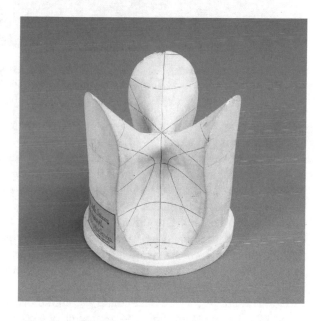

"Geordnete Verhältnisse: Mathematische Anschauungsmodelle im frühen 20. Jahrhundert," *Berichte zur Wissenschaftsgeschichte* 36, no. 4 (December 2013): 294–312; Anja Sattelmacher, *Anschauen, Anfassen, Auffassen: Eine Wissensgeschichte Mathematischer Modelle* (Wiesbaden: Springer Spektrum, 2021). See also: the overview at http://www.universitaetssamml ungen.de/modelle (accessed November 13, 2021). The architectural model in particular has a long history in the context of academic teaching. It was used—as was said for example in the eighteenth century—in the 'practical' part of academic training.

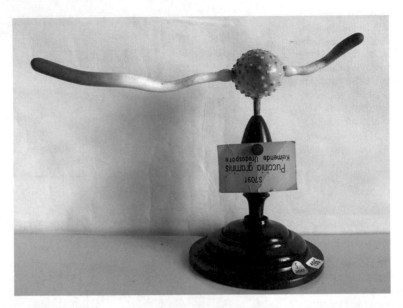

Fig. 3 Model of a spore of the fungus Puccinia graminis (grain rust) (Botanisches Museum Greifswald). Photo: David Ludwig. CC BY-SA 3.0 (https://creativecommons.org/licenses/by-sa/3.0/)

The historiography of thought-forms and epistemological motifs that are related to what scientific modernity—following Wittgenstein and many others—first designated as (theoretical as well as mathematical) 'models,' stands before a fork in its potential path at this point. A history of models in the strict sense would have to carry out a detailed study of the aforementioned *material* objects and their uses. While some contributions to this first way of approaching a detailed history of models are beginning to emerge, it, too, remains largely unwritten. However, the sciences of the nineteenth century and of earlier periods were of course also familiar with different forms of more or less abstract, more of less *non-material* representations. For this second strand in the history of models in science, Heinrich Hertz's 'images' are only *one* late example.

Besides 'images,' nineteenth-century scientific discourses also spoke of a broad range of other ways in which to represent objects of knowledge, even without being (material) models as these were understood at the time. In particular the notions of 'analogies,' 'interpretations' or 'systems' of real or mental objects were widespread in this function. The examples associated with these notions have frequently been of substantial importance for the development of science. However, as I hope to suggest in the following, they often stand for quite different forms and functions of abstract representation. Here I use the expression 'abstract representation' somewhat vaguely and naïvely as a generic term for diverse ways of describing a set of scientifically interesting objects or facts by means of something else, and to thematise this for scientific practice *without* thereby referring to material, tangible objects—as was the case with the notion of 'model' in the language of the nineteenth century. At the same time (pace Wittgenstein and notwithstanding the inflationary use of the term 'model' beginning in the mid-twentieth century) it is important to avoid premature talk of 'thinking in models'. As we shall see, the abstract representations in question occasionally had very *concrete* epistemic functions. Therefore the word 'abstract' should not be overstated. In the following I set out to characterise in particular the respective *epistemic situation*—i.e. the specific circumstances of knowledge and knowledge practices—in which recourse to a form of abstract representation took place and seemed promising to those involved.

I understand these remarks as prolegomena in two respects: first, historically, as preliminary remarks on a history of thinking in abstract representations that is oriented to *actor's categories* and that sees in the variation of these contemporary categories a field of important historical *and* epistemological differences that should not be levelled out too quickly; second, epistemologically, as preliminary remarks contributing to a differentiated analysis of the forms and functions of abstract representations in the sciences of modernity. On both levels I aim in particular to show that a reduction of the discussion of the historical function of abstract representations to the problem of perspectivity or relativity (a set of facts or phenomena can have many—and even inequivalent—abstract representations/models, which for a thinking in models ultimately calls into question the category of the real) would fall short. The recognition of this problem has rightly been regarded as an essential signum of scientific modernity, and it received its first thorough treatment in Gaston Bachelard's *Le nouvel esprit scientifique*, in which Bachelard set out to develop his 'non-Cartesian' epistemology.[4] In the historical material, however—as I hope to make plausible—the issue of perspectivity or of the potential multitude of abstract representations of the same segment of 'reality' is by no means the only noteworthy aspect, and sometimes one finds almost the opposite epistemological tendencies to this development. This does not speak against Bachelard's thesis; rather it speaks against an underestimation of the complexity of the epistemological history of the modern sciences.

Dynamical Analogies, Physical/Mechanical Analogies, Mathematical Analogies

I shall begin with a form of abstract representation that was frequently referred to in the mid-nineteenth century and that was of great importance particularly in mathematically-based areas of physics and still played a terminologically fixed role in Hertz's mechanics: namely 'analogy.' More accurately, this form of abstract representation should be divided into a number of sub-varieties, since it was referred to sometimes as 'physical analogy,' sometimes as 'mathematical analogy' and

---

[4] Gaston Bachelard, *Le nouvel esprit scientifique* (Paris: Les Presses universitaires de France, 1934).

sometimes (in the case of Hertz) as 'dynamical analogy.'[5] An early explicit epistemological treatment is found in James Clerk Maxwell's famous 1856 text "On Faraday's Lines of Force."[6]

The text begins with a description of the problematic epistemic situation in the area of physics that Maxwell was engaged in: "The present state of electrical science seems peculiarly unfavourable to speculation."[7] For Maxwell there were several reasons for this: While the laws of static electricity and some parts of the mathematical theory of magnetism were known, in other parts there was still insufficient experimental data. There were mathematical formulae for the flow of currents in conducting materials as well as for their mutual attraction, but the relationship of these formulae to other areas of the theory of electricity and magnetism remained unclear, and so on. In this situation the development of a theory of electricity required that the various known parts be correlated with mathematical precision and that proposals be made for unknown areas *without*, however, making "physical hypotheses,"[8] that is to say without making uncertain assumptions about the precise nature of the facts underlying electrical and magnetic phenomena. Given the partial knowledge of experimental data at the time, such hypotheses were always subject to error. Hence a 'physical theory'—i.e. a realistic, causal explanation of electricity and magnetism—was still unattainable, not in principle but owing to the particular conditions of knowledge at the time.

Maxwell concluded that in this situation of highly incomplete and partial knowledge it was requisite "to obtain physical ideas without adopting a physical theory."[9] And in order to do that, he continued,

> we must make ourselves familiar with the existence of physical analogies. By a physical analogy I mean that partial similarity between the laws of one science and those of another which makes each of them illustrate the other.[10]

---

[5] The history of thinking in analogies (in a precise sense of the term) can of course be traced back a long way. The scientific tradition in antiquity knew analogy principally as a *terminus technicus* in mathematics. It was introduced as a relation of (an epistemically illuminating) proportionality $A : B = C : D$ between four terms, of which three were generally known and one unknown, albeit determined by the analogy (literally: the sameness of ratios). In Latin the term was rendered as *proportio*. Through metaphorical extension analogy was introduced into many domains of knowledge—as has been shown not least by Michel Foucault in his description of pre-classical thought (see: Michel Foucault, *Les mots et les choses: Une archeologie des sciences humaines* [Paris: Gallimard, 1966], Chap. 2). The fate of thinking in analogies in this sense in the seventeenth and eighteenth centuries still needs to be studied in more detail. In the context of the physical sciences, analogies are found in a similar function in texts predating the texts discussed here; see, for example: note 12 below.

[6] James Clerk Maxwell, "On Faraday's Lines of Force," in *The Scientific Papers of James Clerk Maxwell*, vol. 1, ed. William Davidson Niven (Cambridge: Cambridge University Press, 1890), 155–229.

[7] Ibid., 155.

[8] Ibid.

[9] Ibid., 156.

[10] Ibid.

The examples that Maxwell went on to provide—particularly one taken from the earlier work of his colleague William Thomson (later Lord Kelvin)—clearly show that with the "partial similarity between the laws" of two sciences he meant the agreement of their mathematical form. Of note in the example taken from Thomson was that certain mathematical formulae that regulated the flow of heat in bodies were consistent with those describing the attraction of bodies under the influence of a reciprocal force that was inversely proportional to the square of their distance. Both cases could be described by means of so-called potentials—i.e. by solutions of Laplace's differential equation given certain boundary conditions:

> We have only to substitute source of heat for centre of attraction, flow of heat for acceler-ating effect of attraction at any point, and temperature for potential, and the solution of a problem in attractions is transformed into that of a problem in heat.[11]

Maxwell pointed out that the physical and causal circumstances of both cases were very different and *not* analogous. In the case of heat, a phenomenon in a continuous material medium was concerned; in the case of attraction, discrete material bodies interacted with each other through forces across an empty space. It was merely the partially correlating mathematical form that conveyed the analogy. Hence the analogy between the two domains was a matter of translation, of symmetry and of a common formal structure. Unlike the modern model, analogy here was not an asymmetrical relationship between representation and represented but a relation of reciprocal representation. Moreover, it was not the mathematical form *itself* that was declared to be the representation of a domain of physical phenomena; rather Maxwell's 'physical analogy' was a symmetrical relation between different domains of physical phenomena placed on the same level, each representing and illustrating the other. When, as in the example taken from Thomson, one of the domains belonged to mechanics, Maxwell spoke of a 'mechanical analogy.' This was a particularly attractive class of analogies for the explanation of nature, since mechanics had to be regarded provisionally as the best-known area of physics. Once, through Thomson and others, 'dynamical theory' became the catchphrase of a more abstract theoretical and mathematized version of mechanical theory, the expression 'dynamical analogy' was also frequently found in the texts of English-language natural philosophers.[12]

The physics of the second half of the nineteenth century discussed a broad range of additional examples of such analogies. In the following years Maxwell

---

[11] Ibid., 157. Thomson probably owed his example to Michel Chasles, "Mémoire sur l'attraction d'une couche ellipsoïdale infiniment mince," *Journal de l'Ecole Polytechnique* 15 (1837), 266–316 (I owe this reference to Jesper Lützen, *Joseph Liouville, 1809–1882, Master of Pure and Applied Mathematics* (Heidelberg and New York, 1990), 141).

[12] On 'dynamical analogies,' see *inter alia*: Ole Knudsen, "Mathematics and Physical Reality in William Thomson's Electromagnetic Theory," in *Wranglers and Physicists: Studies on Cambridge Physics in the Nineteenth Century*, ed. Peter M. Harman (Manchester: Manchester University Press, 1985), 149–79 as well as (with a focus on Maxwell) the contributions by Daniel M. Siegel, Jed Z. Buchwald and Norton M. Wise.

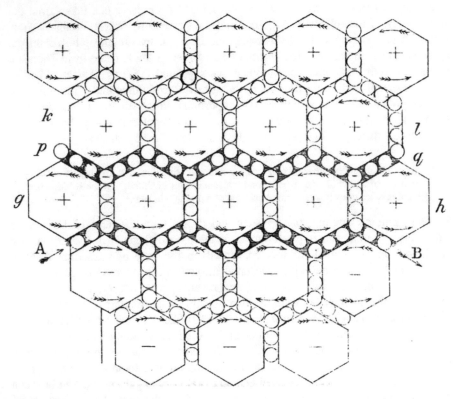

Fig. 4 James Clerk Maxwell's mechanical analogy. From James Clerk Maxwell, "On Physical Lines of Force," in *The Scientific Papers of James Clerk Maxwell*, vol. 1, ed. William Davidson Niven (Cambridge: Cambridge University Press, 1890), 451–513, here 489, Fig. 2

himself endeavoured to find analogies for electromagnetism. Among other things, he developed a 'mechanical analogy' for the dynamical interaction between electrical and magnetic fields, as illustrated in Fig. 4. (If the secondary literature has sometimes described this analogy as a 'mechanical model,' this is strictly speaking yet another anachronism).

In the epistemic situation in which Maxwell made use of this analogy it was difficult to interpret it as anything other than an attempt, with the help of mechanical analogy and controlled by mathematical equations, to find out something about the microstructure of reality itself, even if this should turn out *not* to be mechanically structured. Moreover Maxwell presented his attempt not least as an offer of dialogue to those who considered magnetism a micromechanical phenomenon.[13]

---

[13] See, for example: Maxwell's reference in "On Physical Lines of Force," 453, to a paper published in 1847 by his colleague William Thomson entitled "On a Mechanical Representation of

A particularly interesting example of physical analogy was introduced into the literature by Hermann Helmholtz in 1858. This was the analogy between hydrodynamics—i.e. the continuum mechanics of fluids—and the static theory of magnetic fields generated by electric currents:

> This [hydrodynamic] problem leads to a peculiar analogy between the vortex motions of water and the electro-magnetic effects of electric currents. Thus, if in a simply connected space filled with a moving fluid there is a velocity potential, then the velocities of the water particles are equal and in the same direction as the forces exerted on a magnetic particle in the interior of the space by a certain distribution of magnetic masses on its surface. If, on the other hand, vortex filaments exist in such a space, then the velocities of the water particles are to be set equal to the forces on a magnetic particle by closed electric currents that in part flow through the vortex filaments in the interior of the mass, in part in its surface, and whose intensity is proportional to the product of the cross section of the vortex filaments and the velocity of rotation.
> In the following I shall therefore frequently avail myself of the fiction of the presence of magnetic masses or of electric currents, simply in order to obtain a briefer and more vivid representation of the nature of functions that are the same kind of functions of the coordinates as the potential functions or attractive forces that attach to those masses or currents with respect to a magnetic particle.[14]

Here too the aim was to demonstrate analogical formal-mathematical relationships between different subject areas. It should be borne in mind, however, that Helmholtz brought epistemic asymmetry into play in two respects: on the one hand electromagnetism was used here to explain hydrodynamics, not mechanics to explain electromagnetism; on the other—and this asymmetry is even more striking—the electromagnetic side of the analogy also explains the abstract mathematical objects involved: namely the (in certain situations multivalued) potential

---

Electric, Magnetic, and Galvanic Forces." *Cambridge and Dublin Mathematical Journal*, vol. 2 (1847): 61–64.

[14] Hermann Helmholtz, "Ueber Integrale der hydrodynamischen Gleichungen, welche den Wirbelbewegungen entsprechen," *Journal für die reine und angewandte Mathematik* 55 (1858), 25–55, here 27: "Diese [hydrodynamische] Aufgabe führt zu einer merkwürdigen Analogie der Wirbelbewegungen des Wassers mit den elektromagnetischen Wirkungen elektrischer Ströme. Wenn nämlich in einem einfach zusammenhängenden, mit bewegter Flüssigkeit gefüllten Raume ein Geschwindigkeitspotential existirt, sind die Geschwindigkeiten der Wassertheilchen gleich und gleichgerichtet den Kräften, welche eine gewisse Vertheilung magnetischer Massen an der Oberfläche des Raumes auf ein magnetisches Theilchen im Innern ausüben würde. Wenn dagegen in einem solchen Raume Wirbelfäden existiren, so sind die Geschwindigkeiten der Wassertheilchen gleichzusetzen den auf ein magnetisches Theilchen ausgeübten Kräften geschlossener elektrischer Ströme, welche theils durch die Wirbelfäden im Innern der Masse, theils in ihrer Oberfläche fliessen, und deren Intensität dem Product aus dem Querschnitt der Wirbelfäden und ihrer Rotationsgeschwindigkeit proportional ist. Ich werde mir deshalb im Folgenden öfter erlauben, die Anwesenheit von magnetischen Massen oder electrischen Strömen zu fingiren, blos um dadurch für die Natur von Functionen einen kürzeren und anschaulicheren Ausdruck zu gewinnen, die eben solche Functionen der Coordinaten sind, wie die Potentialfunctionen oder Anziehungskräfte, welche jenen Massen oder Strömen für ein magnetisches Theilchen zukommen." English translation: Herman von Helmholtz, "On Integrals of the Hydrodynamic Equations That Correspond to Vortex Motions," *International Journal of Fusion Energy* 1 (1978), 41–68, here 43.

functions that Helmholtz used to describe vortex motions.[15] Felix Klein would later make this (or a closely related) epistemic asymmetry the starting point for a more comprehensive 'physical' representation of certain epistemic objects in mathematics—a representation which he attributed not to Helmholtz, however, but to Bernhard Riemann.[16]

To sum up, the epistemic functions of physical analogies provided, firstly, a very fruitful heuristic and, secondly, a way out of the research-practice dilemma of uncertain "physical hypotheses," as Maxwell wrote. In this respect one can also see in them an initial distancing from a strong (and at least temporarily unattainable) version of realism in the physical sciences. Of course the recourse to physical analogies also involved a clear tendency to emphasize formal similarities in the physical world and to identify uniform mathematical forms in the diversity of the physical world. This tendency was clearly directed *against* a relativistic epistemology. Indeed it may even have worked directly towards a realism of physics at a more abstract level. Where, then, are 'physical analogies' to be classified on a scale between concrete and abstract thought-forms? Here the evidence is clear. Analogies were principally a matter of constructing bridges between different concrete imaginations, possibly even of the provision of concrete imaginations that could represent abstract mathematical objects (such as multivalued potential functions). Such concrete imaginations as those developed in the analogies could and should be used for the development of new 'physical ideas,' as can be traced in particular in the physics of ether in the second half of the nineteenth century.[17] The explanatory side of analogies thus gave rise—even against Maxwell's intention and assertion—to tentative imaginations about the unknown physical nature of the phenomena at hand. Used in this way, analogy also helped, in spite of everything, an early form of perspectivist explanation of physical reality.

Interpretations of Non-Euclidean Geometry

I shall now discuss another central chapter in the development of the exact sciences in the nineteenth century, an episode that retrospectively and anachronistically has often been described as a shift to a thinking in mathematical models: namely the emergence and propagation of so-called non-Euclidean geometry. For Bachelard it constituted the paradigmatic development in relation to which the formation of the new scientific spirit of modernity could be most clearly understood, at least in

---

[15] For a detailed discussion of this analogy, see: Moritz Epple, "Topology, Matter, and Space, I: Topological Notions in 19th-Century Natural Philosophy," *Archive for the History of Exact Sciences* 52 (1998), 297–392; Moritz Epple, *Die Entstehung der Knotentheorie: Kontexte und Konstruktionen einer modernen mathematischen Theorie* (Wiesbaden: Vieweg, 1999), Chap. 4.

[16] Felix Klein, *Über Riemann's Theorie der Algebraischen Functionen und ihrer Integrale* (Leipzig: Teubner, 1882).

[17] See: Helge Kragh, "The Vortex Atom: A Victorian Theory of Everything," *Centaurus* 44 (July 2002), 32–114.

some of its important respects, and which also represented the earliest episode of this transformation.[18]

On closer inspection, however, it turns out once again that the actors for good reason did *not* speak of models (and if they did, then again in the traditional sense; see below)—and nor did they refer to the physical analogies discussed above. Instead the Italian mathematician Eugenio Beltrami, in an important step in this development, drew on the notion of 'interpretation'—and more precisely on the interpretation of a system of basic propositions or principles of geometry. Here a certain form of pluralism came into play: such a system of basic propositions—this was the thrust of Beltrami's intervention—could admit *multiple* different interpretations. Already before Beltrami the complementary question had been raised: How many possible systems of basic propositions (principles) of geometry are there? This indicates that another epistemological concept was in play: namely that of 'systems' (of basic propositions of geometry). Moreover it is in this episode that we reencounter the notion of 'image', invoked at the beginning of this article, in the historical material. Again I shall briefly sketch the epistemic situation in which Beltrami made his attempt to provide an interpretation of non-Euclidean geometry (as the Italian title of his essay suggests).[19]

Even if well known in its general outline, the gradual, dispersed beginnings of the development that would eventually be summarised under the name non-Euclidean geometry still continue to occupy historians. Here I shall limit myself to a few cursory remarks that are essential for understanding the principal point, and ask the more initiated reader to be patient.[20]

The first thing to note is that during the 1820s and early 1830s a number of mathematicians had come to the conclusion that, in addition to the traditional system of geometry that could look back on a continuous development since antiquity and was considered the perhaps clearest example of a science that was both demonstrative and descriptive, there was still (at least) one other system of geometry that could (and should) be developed mathematically. Both mathematical outsiders, such as János Bolyai, and established scientists, such as Nikolai Lobachevsky in Kazan and Carl Friedrich Gauss in Göttingen, agreed that a geometry could be developed in which Euclid's much-discussed postulate on three intersecting lines (usually, but somewhat misleadingly called 'parallel postulate') was *not* valid (Fig. 5).

In the eighteenth century several geometers (in particular Girolamo Saccheri in a book published in 1733, shortly before his death, with the telling title *Euclides ab*

---

[18] See: the first chapter ("Les dilemmes de la philosophie géométrique") in Bachelard, *Le nouvel esprit*.

[19] See: Eugenio Beltrami, "Saggio di interpretazione della geometria non-euclidea," *Giornale di Mathematiche* VI (1868), 285–315.

[20] For the following, see, for example: Jeremy J. Gray, *Ideas of Space: Euclidean, Non-Euclidean, and Relativistic* (Oxford: Oxford University Press, 1989).

Fig. 5 Euclid's 'parallel postulate.' If two straight lines intersect a third in such a way that the sum of the inner angles between them on one side is less than two right angles, then the two lines on this side will meet if extended indefinitely

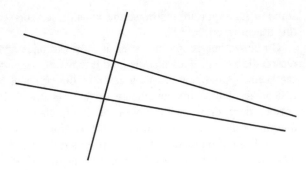

Fig. 6 Girolamo Saccheri's quadrangle

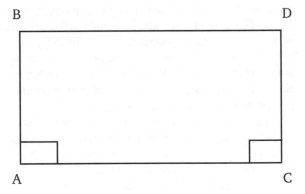

omni naevo vindicatus) discussed logical alternatives to this postulate or axiom[21] with the intention and hope of showing that these alternatives were absurd, if the other axioms of Euclid's geometry were retained. This discussion yielded a tripartition of possible cases, which can be illustrated for example as in Fig. 6: Suppose two line segments of equal length AB and CD are perpendicular to a third segment AC, while the free end points are connected by a fourth segment BD.

Three cases are then possible:

1. The angles ∢ABD and ∢CDB are equal and smaller than a right angle
   (acute angle hypothesis).
2. The angles ∢ABD and ∢CDB are both equal to a right angle
   (right angle hypothesis; this hypothesis is equivalent to the parallel axiom).
3. The angles ∢ABD and ∢CDB are equal and greater than a right angle
   (obtuse angle hypothesis).

---

[21] For reasons of simplicity I skip a discussion of the tradition of Euclidean geometry to distinguish between axioms (or common notions) in a more traditional sense and postulates; it is not relevant for my purposes here.

After a few intermediate steps the third case led to a contradiction with the assumption that lines could be extended to infinity, which while not being an explicit axiom of traditional geometry was nonetheless taken for granted by most participants in the discussion. The second case was that of traditional geometry. For the first case—as the eighteenth century geometers who were occupied with this question believed—a contradiction could also be found. Remarkably enough, however, this came to light only after a lengthy and rather subtle argumentation, the weaknesses of which were obvious to experts.

It was precisely here that the aforementioned authors announced their disagreement. Bolyai, Lobachevsky, Gauss and a few others believed that the acute angle hypothesis could *not* be refuted but, on the contrary, could be developed into a valid system of geometry. This system—and this is the decisive epistemic point— was different in content from the traditional system of geometry. Thus in the new geometry, given a line and a point not on this line, there were *several* lines in the plane defined by the line and the point, passing through this point that do not intersect the given line (and not just a single line, as in traditional [Euclidean] geometry); also in this new geometry the sum of the angles of a triangle was always less than 180 degrees, and so on. Both systems of geometry could not be regarded as simultaneously correct descriptions of the structure of physical space and be true in this sense—an unprecedented epistemic situation and a completely new problematic for the venerable science of geometry.

Lobachevsky and Gauss regarded this problematic as an empirical question to be decided for instance on the basis of astronomical observations, a position that a few decades later would strongly promote empiricism in the natural sciences. However, for the time being it remained impossible (for a number of reasons) to confidently make such an empirical decision. The 'young radical' Bolyai, in turn, believed he possessed speculative proof that the new deviant form of geometry was in fact the true one—unfortunately this 'proof' has not come down to us.

Important for the historical situation in which Beltrami wrote was certainly also the fact that the overwhelming majority of the mathematically educated were still firmly convinced of the truth of the traditional geometry of Euclid and his successors.[22] The epistemological irritation caused by the alternative geometry, later so celebrated (not least by Bachelard), initially had very little effect. Indeed, who could be sure that the acute angle hypothesis did not simply conceal a more remote error that had not yet been detected? Gauss remained cautious, expressing himself on the subject only in letters to fellow scientists, never in print. Bolyai was virtually unknown. While Lobachevsky was a professor and rector at the prestigious University of Kazan, this was far from the centres of European learning.

Beltrami's intervention in 1868 occurred precisely in this epistemic situation in geometry. In the years immediately before, a number of Gauss's statements in his

---

[22] Compare for example the bibliography of the writings on Euclid's parallel axiom in Paul Stäckel and Friedrich Engel, *The Theory of Parallel Lines from Euclid to Gauss* (Leipzig: Teubner, 1895). The vast majority of publications listed until about 1860 sought to prove the truth of traditional geometry.

Fig. 7 Surface of revolution
of a tractrix with constant
negative curvature, designed
by student of mathematics
Isaak Bacharach, Munich
1877 (Göttingen Collection
of Mathematical Models and
Instruments, model 188). ©
Mathematical Institute,
Georg-August-University
Göttingenn, all rights
reserved

correspondence had been posthumously edited, and interest among mathematicans
in the obscure texts of Lobachevsky and Bolyai had grown somewhat. The main
question, however, was what to do with them. Beltrami decided to try to *interpret*
them. Precisely what he had in mind is the point that now needs to be addressed.

Beltrami's principal interest was in mathematical objects in ordinary, traditional
geometry that had internal relations corresponding to, or at least approximating, those of Bolyai-Lobachevskian geometry (henceforth BL geometry). Beltrami
found several such objects. In the first step of his argumentation he referred to
curved surfaces in the three-dimensional space of ordinary Euclidean geometry, in
particular to surfaces that could be obtained through the revolution of a suitable
curve around a spatial axis that had surfaces with constant negative curvature.[23]
Figure 7 shows a slightly later model (in the nineteenth century sense of the
term) of such a surface. If the points on a surface of constant negative curvature were interpreted as points on a plane of BL geometry (henceforth BL plane)
and the 'shortest lines' (geodesic lines, lines of least curvature) on such a surface as lines on the BL plane, the BL geometry used for these was the usual
length and angle measurement in certain limited areas. However, there was still
a difficulty: all known concrete surfaces of constant negative curvature in three-
dimensional Euclidean space were unable to represent the entire BL plane (this
led, among other things, to multiple self-intersections of geodesic lines, and generally the regions of a BL plane outside these limited areas could no longer be
represented).[24]

---

[23] At about the same time, as a result of Gauss among others, surface curvature had become an
interesting object of research of geometry in which alternative geometries were also first articulated. Of course there were other solid reasons for this: namely the geodesic work of this period.
See: Erhard Scholz, *Geschichte des Mannigfaltigkeitsbegriffs von Riemann bis Poincaré* (Basel:
Birkhäuser, 1980).

[24] In 1901, David Hilbert would prove that the complete BL plane could not be "represented in
the Beltramian way" ("in der Beltramischen Weise zur Darstellung gebracht werden") by an analytical surface of constant negative curvature in three-dimensional Euclidean space; see: David

Hence Beltrami went a step further in his argumentation and considered the interior of a circle with a length and angle measurement that *deviates* from the (Euclidean) norm. Such an 'auxiliary circle,' with the correct specification of its metric determinations, could also *represent* a surface of constant negative curvature that was no longer thought of as being in three-dimensional Euclidean space, and that *in this sense* could be imagined as an abstract surface of constant negative curvature. And this new object (or imagination) in turn could also be interpreted as a BL plane. Thus Beltrami's interpretation combined two stages of representation: the auxiliary circle (to which appropriate determinations of length and angle measurement were assigned) represented an (unlimited) surface of constant negative curvature, which in turn represented a (complete) plane of BL geometry. As he went on to summarise:

> It follows from the above that the geodesic lines [of the BL plane regarded as a surface of constant negative curvature, ME] are represented in their total (real) development by the chords of the limit circle, whereas the extensions of these chords outside this circle have no (real) representation. On the other hand two real points of the [BL] plane are represented by two likewise real points inside the limit circle which determine *one* chord of this circle. It can thus be seen that two *arbitrarily* chosen real points of the [BL] plane *always determine one geodesic line*, which is represented on the auxiliary plane by the chord passing through their corresponding points.
> [...] What is more, the theorems [of non-Euclidean planimetry] are only accessible to concrete interpretation when one relates them not to the [Euclidean] plane but precisely to these surfaces [of constant negative curvature], as we shall demonstrate in detail below.[25]

Hilbert, "Über Flächen von konstanter Gaussscher Krümmung," *Transactions of the American Mathematical Society* 2, no. 1 (1901): 87–99. Today's mathematical textbooks render this statement in yet another form. One should keep in mind, though, that none of these later clarifications were available to Beltrami or his contemporaries.

[25] This translation is based on the contemporary French translation of Beltrami's "Saggio" (1868) by Charles Hoüel. See: Eugenio Beltrami, "Essai d'interprétation de la géométrie non euclidienne," *Annales scientifiques de l'école normal supérieure* 6 (1869), 251–88, here 259 : "De ce qui précède il résulte que les lignes géodésiques de la surface sont représentées, dans leur développement total (réel), par les cordes du cercle-limite, tandis que les prolongements de ces cordes en dehors de ce même cercle n'ont aucune représentation (réelle). D'autre part, deux points réels de la surface sont représentés par deux points, également réels, intérieurs au cercle-limite, lesquels déterminent *une* corde de ce cercle. On voit donc que deux points réels de la surface, *choisis d'une manière quelconque,* déterminent *toujours une ligne géodésique unique,* qui est représentée sur le plan auxiliaire par la corde passant par leurs points correspondants. [...] Il y a plus, ces théorèmes, en grande partie, ne sont susceptibles d'une interprétation concrète que si on les rapporte précisément à ces surfaces, au lieu du plan, comme nous allons le démontrer tout à l'heure avec détail." For the original formulation, see: Beltrami, "Saggio," 290: "Dalle cose precedenti risulta che le geodetiche della superficie sono rappresentate nel loro totale sviluppo (reale), dalle corde del cerchio limite, mentre i prolungamenti di queste corde fuori del cerchio stesso sono destituiti d'ogni rappresentanza (reale). D'altronde due punti reali della superficie sono rappresentati da due punti, parimenti reali, interni al cerchio limite, i quali individuano *una* corda del cerchio stesso. Si vede dunque che due punti reali della superficie, *scelti in modi qualunque,* individuano sempre *una sola e determinata linea geodetica,* che è rappresentata nel piano ausiliare dalla corda passante pei loro punti corrispondenti. [...] Anzi questi teoremi non sono in gran parte suscettibili di concreta interpretazione,

Fig. 8 A (paper) model of
non-Euclidean geometry by
Eugenio Beltrami, ca.
1869–1872. © Dipartimento
di Matematica, Università di
Pavia, all rights reserved. The
photograph was kindly
provided by Professor
Maurizio Cornalba at the
Dipartimento di Matematica,
Università di Pavia

By means of this two-stage representation Beltrami now also established an epistemic symmetry for his interpretation. All elements of BL geometry (both figures
and relations between them) now found a counterpart in *both* stages of the interpretation: in the (abstract) surface of constant negative curvature as well as in
the interior of the auxiliary circle with its metric determinations. Just as in the
physical analogies, now also in the interpretation of non-Euclidean geometry two
imaginations *mutually* explained each other. In order to achieve this epistemic
symmetry Beltrami was prepared to forgo a degree of imaginative concretion: His
unlimited surface of constant negative curvature was an imagination that (unlike
the surfaces of revolution with constant negative curvature with which he had
begun his considerations) no longer found a place in the world of traditional geometric imaginations. It is an example of one of the early epistemic objects of
mathematics *beyond* the sphere of ordinary spatial intuition that would become
so characteristic of mathematical modernity.[26] Nevertheless Beltrami insisted that
even this imagination was still concrete, as is shown by the final sentence of the
quote above.

Indeed Beltrami's interest in the concretisation of the problematic epistemic
object 'non-Euclidean plane' is documented in yet another way. He made several models (in the literal sense of the time) of the non-Euclidean plane from
paper, which he frequently described in his correspondence—at least one version
of which has survived and can today be found at the Istituto Matematico of the
University of Pavia (Fig. 8).[27]

---

se non vengono riferiti precisamente a queste superficie anziché al piano, come ora procediamo a
diffusamente dimostrare."

[26] See: Epple, *Entstehung der Knotentheorie*, Chap. 7.

[27] See: Luciano Boi, Livia Giacardi, and Rossana Tazzioli, *La découverte de la géométrie non
euclidienne sur la pseudosphère: Les lettres d'Eugenio Beltrami à Jules Houle (1868–1881)*
(Paris: Albert Blanchard, 1998). On Beltrami's paper models, see: Antonio C. Capelo and
Mario Ferrari, "La 'cuffia' di Beltrami: storia e descrizione," *Bollettino di storia delle scienze*

About twenty years later, in 1889, it was incidentally also Beltrami who brought Saccheri's aforementioned book, *Euclides ab omni naevo vindicatus*, back to the attention of a wider audience of mathematicians and historians of science.

Beltrami's successful attempt at an interpretation of non-Euclidean geometry—along with two other texts published in 1868: Riemann's previously unpublished inaugural lecture "Über die Hypothesen, welche der Geometrie zu Grunde liegen" and Helmholtz's "Ueber die Thatsachen, welche der Geometrie zum Grunde liegen"—substantially contributed to the reception of this obscure object of growing mathematical desire in wider mathematical circles. In contemporary mathematical literature in other languages, his notion of *interpretazione* was often rendered with the help of the term 'image'.

Thus Felix Klein in his 1871 paper "Ueber die sogenannte Nicht-Euklidische Geometrie," in which he turned his attention to the new subject, initially introduced Beltrami's motif of interpretation to then consistently use the terminology of the 'image' for the representation of non-Euclidean geometries, as can be seen in the following quotation that links Beltrami's construction with a technique of projective geometry developed by the English geometer Arthur Cayley:

> For the ideas of hyperbolic geometry, in the light of the above, we immediately obtain an image if we draw an arbitrary real conic section basing it on a projective determination of measure.[28]

The expressions *Abbild* (roughly: image of a mathematical mapping) and *anschauliches Bild* (intuitive image) are also frequently found in the autographed 1893 lectures on non-Euclidean geometry, which introduced Klein's ideas on non-Euclidean geometry to a new generation of mathematicians.[29] In these lectures, and in line with his general mathematical orientation, Klein was explicitly concerned with the *Versinnlichung* (the making available to the senses) of the new geometries—another frequently used catchword for the epistemic function of the images of non-Euclidean geometries. As with Beltrami here too the choice of language reflects the main epistemological interest in the *material* models of this period.

---

matematiche 2, no. 2 (1982): 233–47; Livia Giacardi, "Problematiche emergenti dalla corrispondenza inedita Beltrami-Hoüel. Parte prima: Eugenio Beltrami artigiano della pseudosfera," *Quaderno di Matematica, Università di Torino* 77 (1984): 1–31. Thanks to Rossana Tazzioli.

[28] Felix Klein, "Ueber die sogenannte Nicht-Euklidische Geometrie," *Mathematische Annalen* 4 (1871): 573–625, here 611: "Für die Vorstellungen der hyperbolischen Geometrie erhalten wir nach dem Vorstehenden sofort ein Bild, wenn wir einen beliebigen reellen Kegelschnitt hinzeichnen und auf ihn eine projectivische Massbestimmung gründen."

[29] Felix Klein, *Nicht-euklidische Geometrie: Vorlesung, gehalten während des Wintersemesters 1889–90. Ausgearbeitet von Fr. Schilling*, 2 vols. (Göttingen: [without publisher imprint], 1893). The lectures were held in the winter semester 1889–1890 and in the summer semester 1890. These notes are not to be confounded with Klein's later *Vorlesungen über nicht-euklidische Geometrie*.

When Klein's 1893 lectures appeared in print for the first time posthumously in 1928 in a completely revised version by Walter Rosemann, one found in addition to the still widely used terminology of the 'image,' 'representation' (*Abbild*) etc. the word 'model' used in the modern sense. Thus in an anachronistic semantic shift one reads for example about the "model of planar hyperbolic geometry already known to Beltrami,"[30] where "model" no longer refers to Beltrami's paper model, which was presumably unknown to Rosemann, but to his 'interpretation' BL geometry. The precise source from which Rosemann adopted the modern usage of the term 'model' must remain an open question here. Felix Hausdorff's inaugural lecture 'Das Raumproblem,' discussed below, and Hermann Weyl's lectures on space, time and matter (first published in 1918), both of which adopt the new usage at least to some extent, are two possibilities, however.[31]

Thus the epistemic functions of Beltrami's 'interpretation' and the 'images' of non-Euclidean geometry before the semantic shift of the notion of 'model'— just like the 'physical analogies' of the time—mainly comprised a heuristic of new research objects with unclear status. And as with analogies, it is also very clear that the representations called an 'interpretation' added concrete imaginations of more familiar mathematical objects to a still problematic system of geometric principles. This step could well be interpreted as a means of safeguarding the plausibility or 'existence' of non-Euclidean geometry (or geometries); this function would, however, shift historically. The images and/or interpretations of non-Euclidean geometry led to a *relativisation* of the real to the extent that in the course of a long debate involving many authors it gradually became clear that the correspondence of the epistemically symmetrical, distinct images of geometries with the *faits accomplis* of experience was not *fully determining* such images, and thus that even mathematically inequivalent 'images' of geometric relations—or different systems of basic geometric propositions, in Beltrami's terms—could be made to correspond to the same empirical facts (see below).[32]

As we have seen, in the construction of the interpretations and images of non-Euclidean geometry, too, the main aim of the researchers involved was the production of epistemic symmetries. What was valid and existed in an interpretation or an image also existed and was valid in the domain of thought that was interpreted or pictured (*abgebildet*). The geometers of the last third of the nineteenth century (besides Klein one should at least mention Wilhelm Killing and Henri Poincaré) outdid each other in creating more and more 'images' of non-Euclidean geometries and in studying the 'mappings' (*Abbildungen*) between them. This gave rise to an

---

[30] Felix Klein, *Vorlesungen über nicht-euklidische Geometrie: Für den Druck neu bearbeitet von W. Rosemann* (Berlin: Julius Springer, 1928), 309: "Modell der ebenen hyperbolischen Geometrie, das bereits Beltrami bekannt war."

[31] I thank Erhard Scholz for the reference to Hermann Weyl's book.

[32] For a concise summary of these debates, see my historical introduction: Moritz Epple, "Felix Hausdorffs Erkenntniskritik von Zeit und Raum," in Felix Hausdorff, *Gesammelte Werke,* vol. 6: *Geometrie, Raum und Zeit,* ed. Moritz Epple (Berlin: Springer, 2020), 1–207, here Sect. 2.3.

increasingly broad range of different possible representations of the 'same' geometric relations by means of different images. Under the title "Interprétation des géométries non euclidiennes" Poincaré for instance explicitly invoked the obvious idea of the translation between such images with a matching lexicon, again underlining epistemic symmetry.[33]

As a consequence of these developments, the plurality of geometric systems possessed (at least) two dimensions at the end of the nineteenth century. Not only were there now different, mathematically inequivalent geometries, but also each of these geometries in turn admitted several interpretations or images between which there was epistemic symmetry. In this way, in particular the (symmetrical and asymmetrical) relations between various possible representations of different possible geometric systems themselves came to light. Bachelard would later say that it was not *one representation* of spatial relations that constituted the subject of geometry but *the entire collection of interrelated, not completely equivalent representations* of such relations; geometry was not the science of one unequivocally determined mathematical form of spatiality but the science of the multiplicity of possible mathematical forms of spatiality and their complex interrelationships.[34]

Systems, *Spielräume, Euklidische Modelle*: Some Remarks by Felix Hausdorff, Ca. 1900

In order to take a closer look at how far the discussion of the representations of epistemic objects in geometry at the turn of the century could go beyond the limits of traditional intuition and imagination, and in order to discuss one of the earliest uses of the term 'model' in a non-material (albeit still limited) sense in mathematics, I shall now look briefly at how the mathematician and epistemologist Felix Hausdorff dealt with the subject of non-Euclidean geometry. At this point in time Hausdorff, who was born in 1868, was a *außerplanmäßiger Extraordinarius*, i.e. a professor by title but without a regular salary, at the University of Leipzig. It was only later when he was made a *planmäßiger Extraordinarius* (a salaried professor) at the University of Bonn in 1910 and through his groundbreaking monograph *Grundzüge der Mengenlehre*, published in 1914, that he eventually received professional recognition as a mathematician. During his Leipzig years, alongside his mathematical career, he also tried his hand as a Nietzschean writer—under the pseudonym Paul Mongré—and radical advocate of scientific and, in particular, mathematical modernism.[35] In his epistemological monograph *Das Chaos in kosmischer Auslese*, published in 1898, Mongré introduced the still novel language of

---

[33] See: Henri Poincaré, *La Science et l'Hypothese* (Paris: Flammarion, 1902), 57–58.

[34] See: Bachelard, *Le nouvel esprit*, Chap. 1.

[35] On the notion of mathematical modernity, see: Herbert Mehrtens' important monograph *Moderne—Sprache—Mathematik* (Frankfurt am Main: Suhrkamp, 1990); and more recently Jeremy J. Gray, *Plato's Ghost: The Modernist Transformation of Mathematics* (Princeton: Princeton University Press, 2008).

Georg Cantor's theory of infinite sets into the world of literary metaphors, among other things, by using the idea of an arbitrary map between sets as a tool to describe variations of temporal and spatial relations. In the monograph, his main emphasis was a radical and comprehensive critique of any and all kinds of metaphysical ideas about time, and an illustration of the consequences of this critique in other fields from the natural sciences to morals.[36]

In 1903 Hausdorff gave his inaugural lecture as associate professor in Leipzig on the problem of space, published in the same year and followed in the winter semester of 1903–1904 by a lecture course on time and space, carefully drafted in a manuscript which remained unpublished until very recently.[37] In these papers Hausdorff first identified himself publicly as Mongré, and the mathematician now engaged in an open exchange with the epistemologist. Around the same time he also wrote an undated paper, "Nichteuklidische Geometrie," aimed at a non-specialist audience; it was intended for publication in a natural science journal but never made it to press.[38] In this paper Hausdorff—not least in reference to the publication a few years earlier of David Hilbert's *Grundlagen der Geometrie*—drew on the notion of 'system' already used by Beltrami in order to describe the still new, deviant geometries:

> By a single non-Euclidean geometry we understand any system of geometric propositions that deviates in any more or less important respect from a particular system of Euclidean geometry. Non-Euclidean geometry as a mathematical discipline sets itself the task of testing and comparing all these single systems.[39]

It was precisely this shift that prompted Bachelard to speak in relation to this scientific field of a new scientific spirit. Hausdorff now also pushed ahead with the epistemological as well as ontological problem associated with this shift:

---

[36] See: Moritz Epple, "Felix Hausdorff's Considered Empiricism," in *The Architecture of Modern Mathematics: Essays in History and Philosophy*, ed. José Ferreiros and Jeremy J. Gray (Oxford: Oxford University Press, 2006), 263–89; Moritz Epple, "Spielräume des Denkens: Felix Hausdorff und Paul Mongré," in *Das bunte Gewand der Theorie: Vierzehn Begegnungen mit philosophierenden Forschern*, ed. Astrid Schwarz and Alfred Nordmann (Freiburg: Karl Alber, 2009), 235–62. A full biography of Hausdorff can be found in Felix Hausdorff, *Gesammelte Werke*, vol. 1B: *Biographie*, ed. Egbert Brieskorn and Walter Purkert (Berlin: Springer, 2018).

[37] Felix Hausdorff, "Das Raumproblem" (originally published in *Ostwald's Annalen der Naturphilosophie* 3 (1903), 1–23) and "Zeit und Raum." Both texts have been published in Felix Hausdorff, *Gesammelte Werke*, vol. 6, ed. Moritz Epple (Berlin: Springer, 2020), 279–303 and 391–450.

[38] His manuscript has also been published for the first time in Felix Hausdorff, "Nichteuklidische Geometrie," in Felix Hausdorff, *Gesammelte Werke*, vol. 6, ed. Moritz Epple (Berlin: Springer, 2020), 347–89.

[39] Hausdorff, "Nichteuklidische Geometrie," 350: "Unter einer einzelnen nichteuklidischen Geometrie verstehen wir jedes System geometrischer Sätze, das in irgend einer mehr oder minder belangreichen Beziehung von einem bestimmten System, der euklidischen Geometrie abweicht; nichteuklidische Geometrie, als mathematische Disciplin, stellt sich die Prüfung und vergleichende Betrachtung aller dieser einzelnen Systeme zur Aufgabe."

But that particular single system, the underlying normal system as it were, compared with which all the rest are already in the naming mere negations, abnormalities and varieties, is logically speaking no more just, natural or necessary than the others. In this respect mathematics cannot object strongly enough to the prejudice popular among many philosophers that wishes to deduce Euclidean space with all its peculiarities as the a priori 'product of logical positing.'[40]

If traditional [Euclidean] geometry ceased to appear natural or necessary, then the problem of the *empirical validity* of geometric systems became just as pressing as that of understanding the unavoidable *Spielräume*—the spaces of play or ranges of freedom—in the construction of such systems. In his inaugural lecture of 1903, Hausdorff distinguished the kinds of such *Spielräume*— there was a *Spielraum* of experience, of intuition, and of thought.[41] Given these ranges of play in the construction of geometrical systems, almost every specific, concrete determination of geometric space lost its binding character, as Hausdorff continued in his popular article on non-Euclidean geometry:

As strangely reactionary as it may sound, even propositions such as those that speak of the infinity and limitlessness of the world in space and time are strictly speaking very premature conclusions from the centre of the universe to its fringes and have about the same value as the opinions of a jellyfish in the Atlantic Ocean about the shape of the American coast.[42]

Like other representatives of mathematical modernism Hausdorff experienced the rejection of such strictures as a liberation. This was primarily due to the liberation of *language*, which had been practised not least by the extensive use (and translation) of the various analogies, interpretations and images of new mathematical objects that had accumulated over the preceding decades:

Modern mathematics owes its essential and enlightening insights to the radicalism with which it has proceeded [in the uses mathematical language], but which for the uninitiated observer has something arbitrary and harmful to his or her dearest habits. Here one calculates with 'numbers', where 2a can be equal to a [...], the shortest lines on a [curved] surface are designated 'straight' according to need, a straight line of ordinary space is interpreted as a 'point' of a four-dimensional space, and so on. Why does mathematics do that?

---

[40] Ibid.: "Jenes bestimmte einzige System aber, das sozusagen als Normalsystem zu Grunde gelegt wird und dem sich bereits in der Namensgebung alle übrigen als bloße Negationen, Abnormitäten, Spielarten gegenüberstellen, ist logisch betrachtet um kein Jota berechtigter, natürlicher, denknothwendiger als die andern; in dieser Hinsicht kann die Mathematik nicht scharf genug dem bei vielen Philosophen beliebten Vorurtheil widersprechen, das den euklidischen Raum mit all seinen speciellen Eigenthümlichkeiten als 'Product logischer Setzung' a priori deduciren will."

[41] The distinction between these three types of *Spielräume* in the construction of mathematical spatial concepts can be found in Hausdorff's "Das Raumproblem;" see on this: Epple, "Spielräume des Denkens."

[42] Hausdorff, "Nichteuklidische Geometrie," 351: "So seltsam reactionär dies klingen mag: selbst Sätze wie diejenigen, welche von der Unendlichkeit und Unbegrenztheit der Welt in Raum und Zeit reden, sind doch streng genommen sehr voreilige Schlüsse aus der Mitte des Universums auf seine Ränder und haben etwa den Werth, den die Meinungen einer Qualle im Atlantischen Ozean über die Gestalt der amerikanischen Küste beanspruchen könnten."

[...] [B]ecause an appropriate name can reveal the most important connections between seemingly distant domains, while an inappropriate one can obscure them. The transfer of a language use to unusual, deviant cases serves in particular to make visible one by one the numerous presuppositions that float unresolved in the ordinary sphere of concepts. It is almost the 'logical experiment' [...], the chemical analysis of concepts.[43]

At the same time the free use of language in mathematics was able to call forth *intuitive ideas* insofar as mathematics—as in the interpretations and images of non-Euclidean geometries—"seeks to intuitionally imagine its pure thought-figures and herein proceeds with the free use of the elements of reality."[44] Of course Hausdorff did not intend to suggest a solid fundament for mathematical language here through the back door of intuition. Unlike Klein, for example, Hausdorff emphasised that "the word 'intuitive' means too many things, and indeed to each person something different."[45] Nevertheless the use of this intuition, imagination or fantasy, while always an individual use, was a legitimate means of gaining access to the thought-objects of mathematics. On Beltrami-Cayley's image (*das Beltrami-Cayley'sche Bild*) of pseudospherical geometry Hausdorff commented: "For these relations [of the formal system of pseudospherical geometry, ME] we can obtain an intuitive analogy which at present will only prove a helpful symbol, but later will prove to be the true equivalent of the matter."[46] The image 'versinnlichte' (made available to the senses)[47] while the object itself was the formal system of pseudospherical (or BL) geometry. Moreover, Beltrami's image provided an additional epistemic function for non-Euclidean geometry: Assuming that traditional, Euclidean geometry was free of internal contradictions, Beltrami's image allowed to conclude that pseudospherical (or BL) geometry was equally free of internal contradictions, since any such contradiction would have to manifest itself as a contradiction in Beltrami's *image*:

---

[43] Ibid., 352: "Gerade die moderne Mathematik verdankt wesentliche und aufklärende Einsichten dem Radicalismus, mit dem sie [im mathematischen Sprachgebrauch] verfahren ist, und der für den uneingeweihten Betrachter allerdings etwas Willkürliches, die liebsten Gewohnheiten verletzendes hat. Da wird mit 'Zahlen' gerechnet, bei denen $2a = a$ sein kann [...]; die kürzesten Linien auf einer Fläche werden nach Bedürfniss als 'Gerade' bezeichnet, eine Gerade des gewöhnlichen Raums als ein 'Punkt' eines vierdimensionalen Raums gedeutet u.s.w. Ja warum thut das die Mathematik? [...] [W]eil ein sachgemässer Name die wichtigsten Zusammenhänge zwischen scheinbar entfernten Gebieten aufdecken, ein unsachgemässer sie verschleiern kann. Die Übertragung eines Sprachgebrauchs auf ungewöhnliche, abweichende Fälle dient insbesondere dazu, die zahlreichen Voraussetzungen einzeln sichtbar zu machen, die in der gewöhnlichen Begriffssphäre ungelöst schwimmen; sie ist geradezu das 'logische Experiment' [...], die chemische Analyse der Begriffe."

[44] Ibid., 353: "nämlich ihre reinen Gedankengebilde auch anschaulich vorzustellen sucht und hierin mit freier Benutzung der Wirklichkeitselemente verfährt."

[45] Ibid.: "weil das Wort 'anschaulich' zu vielerlei und eigentlich bei Jedem etwas Anderes bedeutet."

[46] Ibid., 360: "Wir können uns für diese Verhältnisse eine anschauliche Analogie verschaffen, die, gegenwärtig nur als hülfreiches Symbol, sich später als das wahre Äquivalent der Sache erweisen wird."

[47] Ibid., 361.

From the consistency of Euclidean geometry one will be able to conclude the consistency of a non-Euclidean [geometry] if, to speak somewhat vaguely and generally, one is able to produce a *Euclidean image* of this geometry, i.e. if one can set up a correspondence between non-Euclidean elements (points, lines, surfaces, etc.) and Euclidean elements in such a way that the non-Euclidean relations between the former elements are represented by the Euclidean relations between the latter.[48]

Despite this emphatic plea for the freedom of mathematical expression, imagination and thought, and despite the epistemological twist just mentioned, the term 'model' was not used by Hausdorff in this manuscript, written probably between 1900 and 1903. The term appeared just once in it, and not as part of Hausdorff's own text, but in a longer passage quoted from an earlier article by Helmholtz in which the latter had pointed out that even in traditional, Euclidean geometry there were topics that challenged the abilities of spatial intuition, such as complicated knotted threads in space, many-surfaced 'crystal models,' or complex architectural drawings.[49] Such 'models' were still the traditional, tangible models of the nineteenth century, understood as material aids to overcome the weaknesses of spatial intuition (which, as Hausdorff insisted, could never serve as a foundation for mathematical truth anyway).

In his inaugural lecture "Das Raumproblem" of 1903, and in the lecture course on "Zeit und Raum" (given in the winter term 1903/1904), Hausdorff revised his terminology in a slight but telling fashion. Now, in a very specific sense, he spoke of "Euclidean models" of certain geometrical systems where he had used the term "images" before. Again, the immediate context was the proof of consistency of geometrical systems:

The absence of an internal contradiction has been proved directly by suitable mappings of non-Euclidean geometries to Euclidean models, and of Euclidean geometry to pure arithmetics.[50]

---

[48] Ibid., 370: "Von der Widerspruchlosigkeit der euklidischen Geometrie wird man nun wieder auf die einer nichteuklidischen schliessen können, wenn man, allgemein und etwas ungenau ausgedrückt, ein *euklidisches Bild* derselben herstellen kann, d.h. wenn man den nichteuklidischen Elementen (Punkten, Linien, Flächen u.s.w.) Elemente der euklidischen Geometrie so zuordnen kann, dass die nichteuklidischen Beziehungen zwischen jenen durch die euklidischen zwischen diesen dargestellt werden."

[49] Ibid., 353: "Wir stossen übrigens auf ganz ebenso grosse und ähnliche Schwierigkeiten des Vorstellens, wenn wir uns die Führung eines in einem verwickelten Knoten geschlungenen Fadens, oder ein vielflächiges Krystallmodell oder ein Gebäude von verwickeltem Bauplan vorstellen wollen, ohne es gesehen zu haben, obgleich die Anschaubarkeit dieser letztern Gebilde durch thatsächliche Anschauung erwiesen werden kann." Hausdorff quotes this passage from Hermann von Hemholtz: "Ueber den Ursprung und Sinn der geometrischen Sätze: Antwort gegen Herrn Professor Land," in Hermann Helmholtz, *Wissenschaftliche Abhandlungen*, vol. 2 (Leipzig: Barth, 1883), 640–61, here 645.

[50] Hausdorff, "Das Raumproblem," 283: "Die Abwesenheit eines Widerspruchs ist durch geeignete Abbildungen der nichteuklidischen Geometrien auf euklidische Modelle und der euklidischen Geometrie auf die reine Arithmetik direkt bewiesen worden."

Similarly, and explicitly, Hausdorff introduced the term 'Euclidean model' (*euklidisches Modell*) in some passages of his lecture course on time and space, given in 1903/1904. Once again, the immediate context was a proof of the consistency of non-Euclidean geometries. We can see, therefore, that the epistemic role of what now was an immaterial 'model' was still to provide a *familiar*, concrete imagination to an *unfamiliar* system of geometrical axioms, and in this sense epistemologically similar to the material models of earlier decades, and to Beltrami's 'interpretation.' As far as I can see Hausdorff never used the term 'model' in these years except in this limited sense of 'Euclidean models' or 'images' for non-Euclidean geometries.[51]

Underlying Hausdorff's remarks on non-Euclidean geometries was not merely a plea for a liberated construction of mathematical thought-objects. The purpose of the "chemical analysis of concepts" based on a free use of language and on the exploration of the (conceptual, intuitional and experiential) *Spielräume* for providing mathematical descriptions of reality was ultimately also a "self-criticism of science."[52] In order to do science—understood as an attempt of the human intellect to find rational order in experienced reality—, Hausdorff insisted, one had to be aware of the irreducible indeterminacy of any given attempt to describe reality in mathematical terms, be it in geometry or in any other field of mathematized science. Mongré's *Das Chaos in kosmischer Auslese* thus ended with the following passage—which is obviously in need of further explanation[53]:

> In the passage to the transcendent the whole wonderful and richly articulated structure of our cosmos collapsed into chaotic indeterminacy. Thus on returning to the empirical even the attempt to establish the simplest forms of consciousness as the necessary incarnations of appearance fails. As a result the bridges that in the fantasy of every metaphysician pass between chaos and cosmos are destroyed and *the end of metaphysics* is declared—of the acknowledged one no less than the hidden one. And the natural science of the next century will not be spared the task of extracting the latter from its current structure.[54]

---

[51] For a full discussion of Hausdorff's writings on non-Euclidean geometry and the epistemological concerns adressed in the last paragraphs see Epple "Hausdorff's Erkenntniskritik von Zeit und Raum," Sects. 4.2 and 4.3.

[52] Felix Hausdorff, *Sant Ilario—Gedanken aus der Landschaft Zarathustras* (Leipzig: C. G. Naumann, 1897), 342: "Selbstkritik der Wissenschaft."

[53] For this explanation, see: Epple, "Considered Empiricism," Epple, "Spielräume des Denkens," and Epple, "Felix Hausdorff's Erkenntniskritik von Zeit und Raum," Sect. 3.

[54] Felix Hausdorff, *Das Chaos in kosmischer Auslese* (Leipzig: C. G. Naumann, 1898), 209: "Die ganze wunderbare und reichgegliederte Structur unseres Kosmos zerflatterte beim Übergang zum Transcendenten in lauter chaotische Unbestimmtheit; beim Rückweg zum Empirischen versagt dementsprechend bereits der Versuch, die allereinfachsten Bewusstseinsformen als nothwendige Incarnationen der Erscheinung aufzustellen. Damit sind die Brücken abgebrochen, die in der Phantasie aller Metaphysiker vom Chaos zum Kosmos herüber und hinüber führen, und ist das *Ende der Metaphysik* erklärt,—der eingeständlichen nicht minder als jener verlarvten, die aus ihrem Gefüge auszuscheiden der Naturwissenschaft des nächsten Jahrhunderts nicht erspart bleibt."

Images and Dynamical Models: Heinrich Hertz Once Again

In the last step of my considerations I shall briefly look again at Hertz's use of the terms 'image' and 'model'.[55] Remember that for Hertz always several 'images' of the same segment of the real were possible, with the significant restriction that each of these images had to stand in an 'inference equivalence' (*Folgenäquivalenz*) to the causality of the real. This epistemological structure was closely related to that of the images of non-Euclidean geometry. In the domain of geometry, the equivalence between on the one hand the inference relationships (*Folgerungs-beziehungen*) within a geometric system and on the other those in its images (or interpretations) was a logico-mathematical one. Hertz, however, was concerned with the causal relationships between physical causes and effects. In both cases the images assisted in the comprehension of something that was unfamiliar and difficult for the human mind to access.

In this context Hertz also developed a very idiosyncratic, technical sense of the notion of 'model' that broke away from the still common understanding of a model as a concrete, material representation of a scientific object (as we saw at the beginning Hertz was not unfamiliar with this sense). Once more the notion of 'system' or of a variety of systems was essential. In this new, technical sense Hertz stated:

Definition. A material system is said to be a dynamical model of a second system when the connections of the first can be expressed by such coordinates as to satisfy the following conditions:

1. That the number of coordinates of the first system is equal to the number of the second.
2. That with a suitable correspondence of the coordinates in both systems the same defining equations for the systems hold.
3. That by this correspondence of the coordinates the expression for the magnitude of a displacement agrees in both systems.[56]

And he concluded:

Corollary 1. If one system is a model of a second, then, conversely, the second is also a model of the first. If two systems are models of a third system, then each of these systems

---

[55] See: Jesper Lützen, *Mechanistic Images in Geometric Forms: Heinrich Hertz's Principles of Mechanics* (Oxford: Oxford University Press, 2005).

[56] Hertz, *Die Prinzipien der Mechanik*, 197: "Definition. Ein materielles System heißt dynamisches Modell eines zweiten Systems, wenn sich die Zusammenhänge des ersteren durch solche Koordinaten darstellen lassen, dass den Bedingungen genügt ist: / 1. dass die Zahl der Koordinaten des ersten Systems gleich der Zahl der Koordinaten des andern Systems ist, / 2. dass nach passender Zuordnung der Koordinaten für beide Systeme die gleichen Bedingungsgleichungen bestehen, / 3. dass der Ausdruck für die Größe einer Verrückung in beiden Systemen bei jener Zuordnung der Koordinaten übereinstimme." English translation: Hertz, *The Principles of Mechanics*, 175.

is also a model of the other. The model of a model of a system is also a model of the original system.[57]

Just as in the earlier contexts of physical analogies the point here was also a—definitionally enforced, complete—reflexivity, symmetry and transitivity of the model relationship. Being models of each other was a relation of equivalence between "material systems." Even Hertz's technically defined "models" were, therefore not models in the modern, reductive and relativizing sense. And, as in the case of the proliferation of images of non-Euclidean geometries, Hertz was also aware that there could always be, or indeed were, *many* models of the same system:

> Corollary 3. A system is not completely determined by the fact that it is a model of a given system. An infinite number of systems, quite different physically, can be models of one and the same system. Any given system is a model of an infinite number of totally different systems.[58]

Limited by this requirement of equivalence there was, however, a certain measure of perspectivity in Hertz' models. This was desirable and useful principally from an epistemological point of view, since the central 'proposition' was that the dynamical conditions of systems that were models of each other were in mutual correspondence[59] so that insights into the dynamics of a system could be obtained by examining an (equivalent) model system. In the middle of these definitional considerations Hertz once again brought the concept of mental images of material processes used in the introduction to his book into play in an observation that can hardly be described as anything other than cryptic:

> Observation 2. The relation of a dynamical model to the system, of which it is regarded as the model, is precisely the same as the relation of the images which our mind forms of the things to these things themselves. For if we regard the state of the model as the representation of the state of the system, then the consequents of this representation, which according to the laws of this representation must appear, are also the representation of the consequents which must proceed from the original object according to the laws of this original object. The agreement between mind and nature may therefore be likened to the agreement between two systems which are models of one another, and we can even account for this agreement by assuming that the mind is capable of making actual dynamical models of things, and of working with them.[60]

---

[57] Ibid.: "Folgerung 1. Ist ein System Modell eines zweiten Systems, so ist auch umgekehrt das zweite System Modell des ersten. Sind zwei Systeme Modelle eines dritten, so sind sie auch Modelle voneinander. Das Modell des Modells eines Systems ist auch Modell des ursprünglichen Systems." English translation: ibid.

[58] Ibid.: "Folgerung 3. Ein System ist noch nicht vollständig bestimmt dadurch, dass es Modell eines gegebenen Systems ist. Unendlich viele, physikalisch gänzlich verschiedene Systeme können Modelle eines und desselben Systems sein. Ein System ist Modell unendlich vieler, gänzlich verschiedener Systeme." English translation: ibid., 176.

[59] See: ibid., 198. English translation: ibid.

[60] Ibid., 199: "Anmerkung 2. Das Verhältnis eines dynamischen Modells zu dem System, als dessen Modell es betrachtet wird, ist dasselbe, wie das Verhältnis der Bilder, welche sich

If we take Hertz's consideration seriously, then it is an ambitious, even if empirically entirely unsupported hypothesis of cognitive science—a sort of mechanics of the mind.

Despite its avowed perspectivity, Hertz's discussion of images and models therefore remains bound to a rigorous idea that is ultimately deeply rooted in the presented epistemological traditions of the nineteenth century—the idea of a correspondence between causally structured reality, the activity of the mind and scientific theory formation. Hertz believed in this correspondence on the basis of his consistent emphasis of the epistemic symmetry of the relationships in 'images' or 'models' (just as Maxwell had believed in the epistemic symmetry of his 'analogies'). The epistemological rupture that geometry had already undergone or was in the process of undergoing, as we have seen particularly in the case of Hausdorff, was not embraced by Hertz.

Epilogue: The Rise of (Modern) Mathematical Models

The rise (and inflation) of the modern concept of the mathematical model as an actor's category is a mid-twentieth century affair that I cannot go into here. It was only conceivable *after* the break noted by Bachelard that brought about the *nouvel esprit scientifique* of the twentieth century. To tell the history of this rise in detail it would be necessary not only to examine further how the concept of models in geometry began to acquire its modern, abstract meaning, the beginnings of which we have traced in Hausdorff and, among other places, in Rosemann's pseudo-Klein from 1928, but also in particular to trace the epistemological shifts in mathematical economics of the same period. There too one finds in numerous early-twentieth-century texts mechanical analogies in the style of the nineteenth century[61] *before* the modern sense of the term 'model' came into general use. Several historians of mathematical economics refer to Jan Tinbergen's 1935 paper "Quantitative Fragen der Konjunkturpolitik" (Quantitative Issues of Economic Policy)[62] as the place

---

unser Geist von den Dingen bildet, zu diesen Dingen. Betrachten wir nämlich den Zustand des Modells als eine Abbildung des Zustandes des Systems, so sind die Folgen des Zustandes, welche nach den Gesetzen dieser Abbildung eintreten müssen, zugleich die Abbildung der Folgen, welche sich an dem ursprünglichen Gegenstand nach den Gesetzen dieses ursprünglichen Gegenstandes entwickeln müssen. Die Übereinstimmung zwischen Geist und Natur lässt sich also vergleichen mit der Übereinstimmung zwischen zwei Systemen, welche Modelle voneinander sind, und wir können uns sogar Rechenschaft ablegen von jener Übereinstimmung, wenn wir annehmen wollen, dass der Geist die Fähigkeit habe, wirkliche dynamische Modelle der Dinge zu bilden und mit ihnen zu arbeiten." English translation: ibid., 177.

[61] See *inter alia*: Mary S. Morgan, *The World in the Model: How Economists Work and Think* (Cambridge: Cambridge University Press, 2012).

[62] Jan Tinbergen, "Quantitative Fragen der Konjunkturpolitik," *Weltwirtschaftliches Archiv* 42 (1935): 366–99.

where the concept of the mathematical model was first introduced into economic theory, a concept that then spread rapidly in this field.[63]

For as long as the epistemological history of the shifts in the understanding and use of abstract representations in the natural sciences remains unwritten—one that ultimately led to the repression of the earlier and the rise of the modern concept of models—the historian can only cautiously recollect the epistemic functions of analogies, interpretations, images and, not least, tangible models of the natural sciences and mathematics of the nineteenth century.

What for us today has become an irreversible insight into the perspectivity and relativity of scientific representations as well as into the unavoidable and epistemically *not* indifferent reduction of the complexity of the real in any theoretical model—and even in any biological model system—did *not* obtrude on the thought of the physical sciences of the century before last. At that time the focus was on other epistemic functions of analogies, images and material models: namely their collective or individual heuristic value, the relief or replacement that they could offer in epistemic situations where causal hypotheses were not (yet) available, the epistemic symmetry or equivalence between different concrete representations of the same supposedly more abstract structures, the development of uniform mathematical structure of different segments of nature.

However, something else came into play besides such functions in the interpretations, systems and images of non-Euclidean geometry: namely the task of dealing with *inequivalent* mathematical representations of something real (physical space) that, for that very reason, had to be interpreted, *versinnlicht* and evaluated anew as free imaginations. Hence wherever the question of the relationship between such imaginations and a scientifically explored segment of reality was posed, a persistent epistemological problematic was raised that ultimately forced a "self-criticism of science"[64]—one that only the "natural science of the next century" would carry through.[65]

Translated by Benjamin Carter and Nathaniel Boyd.

---

[63] See: Gerard Alberts, *Jaren van berekening: Toepassingsgerichte initiatieven in de Nederlandse wiskundebeoefening 1945–1960* (Amsterdam: Amsterdam University Press, 1998); Marcel Boumans, *How Economists Model the World Into Numbers* (London and New York: Routledge, 2005).

[64] Hausdorff, *Das Chaos in kosmischer Auslese*, 209.

[65] Hausdorff, *Sant Ilario*, 342.

Mappings, Models, Abstraction, and Imaging: Mathematical Contributions to Modern Thinking Circa 1900

José Ferreirós

1921, Vienna and Leipzig: the journal *Annalen der Naturphilosophie*, edited by chemist Wilhelm Ostwald,[1] publishes a hard-to-classify paper, an essay written in aphorisms about knowledge and its limits, about the world and "what cannot be spoken of," polished during the Great War by Ludwig Wittgenstein.[2] In this terse and difficult writing, barely 78 pages, one finds an attempt to definitively clarify main philosophical questions. At the core of this reflection is logic and its link with the world, a topic articulated around the notions of *Bild*, image or figure, and *Abbildung*, figuration or representation, mapping. Wittgenstein's entire attempt is unequivocally modernist in spirit, an exemplary specimen of a cultural shift intensified by the horrendous war. In those years one started to talk about *structures* in different fields, not least within mathematics, and the new architectonic efforts look for purity of lines, a purist functionalism linked (surprisingly or not) to an intense search for transcendence. This can be applied, e.g., to Wassily Kandinsky's

---

[1] Wilhelm Ostwald was a Nobel prize winner and tried to promote a scientific, 'monist' worldview; also famous for promoting the history of science.

[2] Ludwig Wittgenstein, "Tractatus logico-philosophicus," *Annalen der Naturphilosophie* 14 (1921): 185–262. Proposition 7 reads: "Wovon man nicht sprechen kann, [...]."

[3] In what follows, to clarify my discourse, I will often employ the relevant German words. The reader should keep in mind that *Bilder* is the plural of *Bild*, from which the verb *abbilden* is derived (to represent pictorially) and the substantive *Abbildung* (representation, mapping).

English version of: José Ferreirós, "Matemáticas y pensamiento: en torno a imágenes, modelos, abstracción y figuración," in *Rondas en Sais: ensayos sobre matemáticas y cultura contemporánea*, ed. Fernando Zalamea (Bogotá: Editorial de la Universidad Nacional de Colombia, 2013), 15–40.

J. Ferreirós (✉)
Facultad de Filosofía, Universidad de Sevilla, C. Camilo José Cela, s/n, 41018 Sevilla, Spain
e-mail: josef@us.es

© The Author(s) 2022
M. Friedman and K. Krauthausen (eds.), *Model and Mathematics: From the 19th to the 21st Century*, Trends in the History of Science,
https://doi.org/10.1007/978-3-030-97833-4_10

canvases or the scores of Anton Webern but applies in no lesser degree to the beautiful "Tractatus logico-philosophicus."

Let us consider more closely Wittgenstein's pictorial or figural theory (*Bildtheorie*) of meaning. The sentences of a language are an expression of thoughts, which in turn are images, *Bilder*, of states of affairs in the world. At the basis of everything is the making of an image or *Bild* of a fact, a model of reality–figuration or representation, *abbilden* ("Tractatus," 2.1: "Wir machen uns Bilder der Tatsachen"). Of course, Wittgenstein, and his predecessors, employ these words in a wide and somewhat abstract sense.[3] "The logical picture of the facts is the thought," he wrote in the "Tractatus" (3.: "Das logische Bild der Tatsachen ist der Gedanke"). Even more, logic itself is, according to Wittgenstein, a picture of the world;[4] as he said in a memorable phrase, "Die Logik ist [...] ein Spiegelbild der Welt," logic is a mirror image of the world.[5]

For all of its novelty and its modernist style, the logico-philosophical work of the Viennese is also a knot that links receptively a number of proposals and developments that have already gone a long way. Few people are cited in the text, with the exceptions of Gottlob Frege, Bertrand Russell, and Heinrich Hertz. Moreover, it is well-known that the work of Hertz, a physicist, includes passages that emphasize the notions of image and representation, *Bild* and *Abbildung*, as key to understanding the function of scientific thinking and its links with reality (see section "Helmholtz and Hertz").

The purpose of this essay will be to establish a partial genealogy of these notions, following them through an undulating path that will traverse the territories of science, mathematics, and philosophy. It should be well-known: frontiers are a work of humans, an artifice; they exist only in our social life and our imagination—regardless of how hard and consistent they may seem to be.

Generalities

Academic writers often do not pay attention to such trespassing of frontiers, like the cases I just indicated above. A source of such lack of sight is what might be called 'disciplinary blindness': the mathematician looks back in the search for mathematical contributions that can be recognized as relevant to the present context of mathematics. The same is true for the philosopher or the physicist. It is easy to perceive that such disciplinary blindness goes together with a certain kind of Whiggism, which is more usual among academics than in the world of art (compensated perhaps by an increase of rigor in their statements). Ideally, to understand changes in the world of thought like the ones I will discuss, a highly interdisciplinary approach is required.

---

[4] This thesis is polemical and very open to discussion, but we cannot enter into the question here. See: José Ferreirós, "The Road to Modern Logic – An Interpretation," *The Bulletin of Symbolic Logic* 7 (2001): 441–84.

[5] Wittgenstein, "Tractatus," 6.13.

Chance or not, the authors I will talk about in the following pages belong to German speaking contexts. Between 1800 and 1940 such contexts turned out to be a great breeding ground for exchanges between philosophical and scientific thought. Hybrid figures of great interest, the scientist-philosopher or the philosopher-scientist, are easy to find in that period and context—from Immanuel Kant to Albert Einstein or Hermann Weyl, to mention just some paradigmatic cases at the beginning and end of that era. In the following pages some other examples will emerge—Johann F. Herbart, Bernhard Riemann, Hermann von Helmholtz, Richard Dedekind, Hertz, and Wittgenstein.

That same richness complicates the task of tracing faithfully the maze of influence and affluence. It may be worthwhile to mention a case that I encountered early on, when I was starting to work on the history of mathematical thought. Studying Dedekind's works, highly interesting for the change of (architectonic) style in mathematics, I found many indications of the possible influence of the philosophical ideas of Gottfried W. Leibniz and Kant. A first natural impulse was to follow them in detail and trace their genealogy in particular works of those philosophers. Even today I think it is highly likely that Dedekind may have read the *Critique of Pure Reason* (*Kritik der reinen Vernunft*), the *New Essays on Human Understanding* (*Nouveaux Essais sur l'entendement humain*), etc., and for the case of Leibniz I found an elaborate argument concerning the exact how and where of those supposed influences. Yet the critical exercise of historical research led me to realize that none of my arguments was certain enough. I came to understand that Kantian and Leibnizian ideas were so widely disseminated in the German-speaking nineteenth century that they could flow in many ways, through the writings not only of philosophers (Jakob Fries, Hermann Lotze—who taught Dedekind at Göttingen) but also of mathematicians (Carl Friedrich Gauss, Riemann, William R. Hamilton, Hermann Grassmann) or scientists (Wilhelm Weber and Helmholtz). Thus, I became circumspect and I learnt a valuable lesson about direct and indirect sources of cultural transmission.

Let me return to representations or 'mappings,' what in German today are called *Abbildungen*. Dedekind coined the term in an 1888 work,[6] in the wake of the more fragmentary suggestions of Gauss and Riemann, his masters. To put it in simple terms, close to Dedekind's own, a representation or *Abbildung* φ is "a law" that correlates objects of some domain D with those of another C, the counterdomain, in such a way that to each object in D there corresponds one and only one object in C. Being d an element of D, its correlate $\varphi(d)$ is called the *image*, *Bild*, of d; analogously, one says that d is the *original* of $\varphi(d)$. Moreover, the whole set of elements in C that are images of elements of D can be called "the image of D" and denoted by $\varphi(D)$; which corresponds to the whole canvas depicting an original scene.

---

[6] Richard Dedekind, *Was sind und was sollen die Zahlen?* [*What are numbers, and what for?*], of which the reader should employ the revised English version in *From Kant to Hilbert*, ed. William B. Ewald, vol. 2 (Oxford: Claredon Press, 1996), 787–833; see Sect. 2 on p. 799, "Representation [*Abbildung*] of a set."

The terminology is obviously inspired by painting, and Dedekind himself wrote in a letter that the mapping or representation φ is "the painter who paints" ("der abbildende Maler").[7] But it is clear that the idea can be applied to the relation between symbols in a pentagram and musical sounds, or to any other language including scientific representations.

The moment when that elegant idea was established marks an epoch insofar as it denotes the transition from 'classical' to 'modern,' twentieth century mathematics. Dedekind immediately applied the idea to structure-preserving mappings, what would later be called *morphisms*. This transition involved a transfiguration of mathematics, it has often been claimed that the nineteenth century was a true Renaissance for mathematics. On the one hand, Dedekind's mappings or representations link back with the concrete functions that were studied in 'classical' algebra or analysis. But he precisely abandoned the concrete element (definition by means of an analytic, explicit formula) in favor of a simplicity, generality and abstraction that is characteristically modern. Those mappings, and especially the morphisms, inaugurate the path toward new levels of conceptual abstraction that will end up in the *arrows* and *functors* of category theory. Dedekind can thus be acknowledged as a landmark in the path toward contemporary mathematics, an example being his reconceptualization of Galois theory in 1894,[8] and as such he was acknowledged by mathematicians like Emmy Noether and Emil Artin.

The coinage of the modern notion of *Abbildung* must be seen as a key moment in a transformation of *longue durée*, which has profoundly affected scientific and philosophical thought. I mean the transition from a substantial-causal conception of phenomena, characteristic of ancient and medieval thought, to a relational-functional conception that is typical of modern and contemporary thought. Substantial and causal models have been found incomplete and insufficient for the understanding of physical, human and social phenomena.[9] Simultaneously, led by physics and mathematical analysis, the idea of function came to the foreground in the context of admitted scientific models. All of science today aims to formulate relational-functional models that allow exact prediction

---

[7] Richard Dedekind: "Letter to Keferstein" [1890], in *From Frege to Gödel. A Source Book in Mathematical Logic 1879–1931*, ed. Jean van Heijenoort (Cambridge: Harvard University Press, 1967), 98–103, here 102.

[8] The theory of Évariste Galois, a true cornerstone of modern algebra, was born in the writings of the famous Frenchman around 1830, but Dedekind was the first to sketch the modern formulation in terms of groups of field automorphisms (in Appendix XI to the 1894 edition of *Vorlesungen über Zahlentheorie*, see: José Ferreirós, "Dedekind's Map-theoretic Period," *Philosophia Mathematica* 25, no. 3 (2017): 318–40). This very abstract version was developed by Emil Artin thirty-five years later.

[9] Note for philosophers: the idea of *causality* admits several definitions, and there is a widespread tendency to call current scientific models 'causal'—yet this linguistic usage is anachronistic when applied to the acme of causal thinking. Thus, it seems confusing to me. I employ the word 'causality' for the simple cause-effect relation that occupied Hume in his famous criticism, which Kant tried to defend by postulating it a priori as one of the categories.

of diverse kinds of phenomena, just like gravity theory (be it Newtonian or relativistic) describes interactions mediated by gravitation, like quantum theories (elementary or field-theoretic) describe the phenomena due to electromagnetism and nuclear forces.

The triumph of functional thinking, in terms of mapping or representation, surfaced during the eighteenth century and the beginning of the nineteenth, the time that today would be called—from the viewpoint of mathematics—the era of Leonhard Euler and Gauss. Since then it has only consolidated, in spite of and throughout the many transformations that have happened in science and in mathematics. I will come back to this.

The Riemann Inflexion

I will now consider some facets of the work and thought of Bernhard Riemann. This impressive intellectual figure represents a turning point in the history of mathematical and scientific thought. Of course, mathematics is collective work, and almost nothing of importance can be reduced to the contribution of a single 'genius' (a notion that is today reviled by historians, perhaps exaggeratedly). Yet some mathematicians concentrate ideas and trends in such a way that they amaze and prompt the question as to what might have been without their proposals. The work of Riemann, as one important twentieth century mathematician put it, is "full of almost cryptic messages to the future" (Lars V. Ahlfors in 1953).[10]

Relationalism and the idea of *Abbildung* play key roles in several places in Riemann's thought, but let me start here in the domain of philosophy. It should be briefly noted that Riemann studied under Gauss, as did Dedekind, who regarded his somewhat older colleague as a true master. An important characteristic of the mathematicians at Göttingen was emphasized by Gauss's collaborator, the physicist Wilhelm Weber: "With Dirichlet and Riemann, Göttingen remained the plantation of a deeply philosophical orientation in mathematical research, which it had become under Gauss."[11]

The *topos* of relations had become more and more relevant in philosophical reflections towards the period that I marked out above. It seems no coincidence that Leibniz, inventor of calculus, was a pioneer of this turn and one who insisted on relations as key for the understanding of reality. From his metaphysics, where each monad contains a (more or less partial) representation of the universe, to his scientific contributions, we see that notion appear in diverse forms. He even applied

[10] Quoted in: Samuel J. Patterson, "Reading Riemann," in *Exploring the Riemann Zeta Function: 190 Years from Riemann's Birth*, ed. Hugh Montgomery, Ashkan Nikeghbali, and Michael Th. Rassias (Cham: Springer, 2017), 265–85, here 265.

[11] From a University report, 1866; quoted in Pierre Dugac, *Richard Dedekind et les fondements des mathématiques*, (Paris: Vrin, 1976), 166: "Die Pflanzstätte der tiefer philosophischen Richtung im mathematischen Forschen, die Göttingen durch Gauß geworden, ist es unter Dirichlet und Riemann geblieben […]."

it to the matrix ideas of space and time, offering a relationalist approach that, in spite of Einstein, has never been elaborated in a totally satisfactory way. This topos followed its course with authors as central as Kant (cf. his categories of relation) and it seems to have reached a peak during the nineteenth century. In this period, a post-Kantian author who took inspiration from Leibniz, a philosopher-scientist named Herbart, went so far as to say that all our knowledge is knowledge of relations, in perception as much as in conceptual knowledge: "we live amid relations and need nothing else."[12] What there is, both in the ultimate reality postulated by metaphysics and in the contents of our mental acts,[13] is above all relations and records of relations.

Johann Friedrich Herbart is of interest because Riemann regarded him as a master in philosophical questions. But he is not among the usual canon of philosophers, partly due to the disciplinary blindness I mentioned above.[14] Herbart elaborated an interesting theory of knowledge, and here, perhaps for the first time in a philosopher, we find an overtly hypothetico-deductive epistemology, coherent with the experimental approach of the natural sciences. He insisted that our theories emerge from the gradual modification of prior ideas, from the dialectics of *experience* [*Erfahrung*] and *reflection* [*Nachdenken*], that is to say, from the reflective effort to accommodate at the level of theory the experiential data that contradict the old ideas. He rejected outright Kantian apriorism, saying that the "hypothesis" of two forms of intuition was "completely superficial, devoid of content and inadequate," since they were merely "a couple of empty infinite recipients into which the senses must pour their sensations, without any reason for the ordering and configuration."[15] As regards the categories of the understanding, which according to Kant regulate, among other things, our understanding of substances and forces, he wrote:

> the *multitude* of past mistakes [in the history of science] concerning substances and forces prove in fact that the *corresponding concepts are not fixed and determinate in the human*

---

[12] "Wir leben einmal in Relationen, und bedürfen nichts weiter." Johann F. Herbart, *Allgemeine Metaphysik* (Königsberg: Unzer, 1829), 415.

[13] Herbart developed a novel psychology, which figures in histories of the discipline as transitional towards scientific psychology. He is also very relevant in the history of pedagogy. See: Erhard Scholz, "Herbart's influence on Bernhard Riemann," *Historia Mathematica* 9, no. 4 (1982): 413–40.

[14] Twentieth century philosophers have projected their ideals back, tracing their own genealogy in histories that prefer to choose, as representative of the nineteenth century *Zeitgeist*, authors linked to the lines Georg W. F. Hegel–Karl Marx or Arthur Schopenhauer–Friedrich Nietzsche.

[15] "[…] aber ein paar unendliche leere Gefäße hinzustellen, in welche die Sinne ihre Empfindungen hineinschütten sollten, ohne irgend einen Grund der Anordnung und Gestaltung, das war eine völlig gehaltlose, nichtssagende, unpassende Hypothese." (Johann Friedrich Herbart, "Psychologie als Wissenschaft: neu gegründet auf Erfahrung, Metaphysik, und Mathematik. Erster, synthetischer Theil. 1824," in *Johann Friedrich Herbart's Sämtliche Werke in chronologischer Reihenfolge*, ed. Karl Kehrbach, vol. 5 (Langensalza: Hermann Beyer, 1890), 177–434, 428).

spirit, that they are not at all categories or innate concepts, but mutable products of a reflective thought [Nachdenken] *stimulated by experience and altered by all kinds of opinions.*[16]

The mathematician, philosopher and scientist Riemann accepted wholeheartedly this anti-apriorist point of view, the evolutionary gradualism regarding scientific theories, and the hypothetico-deductive scheme of 'experience' and 'reflection.' Such a viewpoint, together with his noteworthy independence of thought, allowed him to advance ideas that would take decades before they seemed reasonable to other scientists. In particular, he went beyond the positivism that was dominant in his lifetime and dared to suggest that the theory of gravitation—then regarded as definitive—would be abandoned for other more adequate theories. Furthermore, he proposed extremely novel geometric ideas that would precisely establish the mathematical framework for Einstein's theory of gravitation (General relativity).

Riemann did not follow Herbart in all areas of philosophy. He rejected in particular his psychological theory and his metaphysics, to develop viewpoints that indeed turn out to emphasize relationalism even more. (E.g., Herbart established a version of the monadology, considering those immaterial, essentially active beings as substances, but Riemann replied that the supposed substantial nature of Herbart's 'monads,' the *Realen,* is contradicted by the central attributes the philosopher himself assigns to them.) Although he rejected important parts of the Herbartian conception of space and the continuum, it is beyond doubt that Herbart transmitted to him a relationalist conception of space, far from that of Kant, and that is related to the great contribution of Riemann to geometry.

Here, however, we are interested primarily in the notion of *Abbildung,* as Riemann elaborated it in his function theory, and its impact upon a pictorial or modelistic conception of scientific representation. Two themes where there seems to be a link between the author's mathematics and his philosophical ideas.

Herbart regarded mathematics as particularly close to philosophy, considering it quite possible to give a philosophical treatment of mathematics, so much so that "treated philosophically, it [mathematics] becomes itself a part of philosophy."[17] To some extent, Riemann's contributions can be regarded as a realization of such a viewpoint, a truly deep realization, quite different from what Herbart might have been able to imagine. He devoted himself to the search for basic concepts around which to restructure and reorganize whole areas of mathematics; concepts which allowed him to dig deeper into its foundations. He was convinced that mathematical research "starting from general concepts" could make a decisive contribution to scientific thinking, preventing it from "being hampered by too narrow views," so

---

[16] "Denn die *Mannigfaltigkeit* der Irrthümer über Substanzen und Kräfte beweist factisch, *daß die Begriffe hievon im menschlichen Geiste nicht vest stehn, daß sie keinesweges Kategorien oder angeborne Begriffe sind, sondern wandelbare Erzeugnisse eines durch die Erfahrung aufgeregten, durch allerley Meinungen umhergeworfenen, Nachdenkens*" (Johann Friedrich Herbart, "Psychologie als Wissenschaft: neu gegründet auf Erfahrung, Metaphysik, und Mathematik. Zweiter, analytischer Theil. 1825," in *Johann Friedrich Herbart's Sämtliche Werke in chronologischer Reihenfolge,* ed. Karl Kehrbach, vol. 6 (Langensalza: Hermann Beyer, 1892), 1–338, 198).

[17] See: Scholz, "Herbart's Influence on B. Riemann," 425, orig. German on 437.

that "progress in knowledge of the interdependence of things [may not be] checked by traditional prejudices."[18] So it was that he came to propose ideas related to the modern concepts of set and mapping, which in turn he tracked down and followed to the roots of scientific thinking.

As a mathematician, Riemann became famous early in life due to his contributions to 'function theory,' that is, complex analysis.[19] A few words have to be said on this topic in order to understand the links with the pictorial or modelistic conception of knowledge. Riemann thought it necessary in function theory to get away from a calculational mathematics based on formulas, opting instead for a highly conceptual approach, of the kind I have previously called abstract. The general concept that would offer the key to a new foundation of function theory was that of *analytic function*, more specifically what we call a holomorphic function.[20] This concept was defined by means of a very general characteristic property, the simple differentiability of the function around any given point (given by the Cauchy-Riemann equations). From that starting point, in order to characterize each particular function from a 'bird's eyes' view, Riemann developed a highly original approach employing geometric elements (actually topological, a discipline that he pioneered) together with elements inspired by mathematical physics. The whole approach was *conceptual*, insofar as the particular formulas employed by analysts to determine the functions (infinite series, integrals) ought to appear only at the end, as a *result* of the general abstract theory. With this, Riemann marked a turning point along the path toward twentieth century abstract mathematics.

All of that had another consequence that caught Riemann's attention: if the domain of the complex variable is conceived as a plane, the Gaussian plane, the function establishes a correspondence between two planes.[21] And holomorphic functions establish, precisely, a *conformal mapping* (*Abbildung*), that is, they "apply the minimal parts of a surface onto the other, so that the image is similar to the original in its minimal parts."[22] In other words, an infinitesimal triangle has

---

[18] Riemann, "On the Hypotheses which Lie at the Foundation of Geometry" (1854/1868); trans. William Kingdon Clifford (revised) in *From Kant to Hilbert*, ed. William Ewald, vol. 2 (Oxford: Claredon Press, 1996), 652–61, here 652.

[19] Based on the numbers $a + bi$, with i the famous complex unit such that $i^2 = -1$; the set of all complex numbers can be seen as a 2-dimensional geometric system, the complex plane.

[20] 'Holomorphic' was a term employed by Charles Briot and Jean-Claude Bouquet, students of Cauchy, in their *Théorie des fonctions doublement périodiques* (Paris, 1859); it stems from Greek ὅλος (*holos*), 'whole,' and μορφή (*morphē*), 'form' or 'appearance.'

[21] If the function $w = f(z)$ is continuous, says Riemann in his dissertation, "it will be possible to conceive that dependence of magnitude w from z as a mapping [*Abbildung*] of the plane A onto plane B;" in German: "Man wird sich also diese Abhängigkeit der Grösse w von z vorstellen können als eine Abbildung der Ebene A auf der Ebene B." (Bernhard Riemann, "Grundlagen für eine allgemeine Theorie der Functionen einer veränderlichen complexen Grösse," in *Bernhard Riemann's Gesammelte mathematische Werke und wissenschaftlicher Nachlass*, 2nd ed., ed. Heinrich Weber (Leipzig: Teubner, 1876), 3–47, here 5).

[22] This is a phrase from the title of a paper by Gauss published in 1825, which inspired his disciple; see: ibid., "Ueber diesen Gegenstand sehe man: 'Allgemeine Auflösung der Aufgabe:

as its image, in the other plane, another triangle with the same angles and proportional sides. We find here, in a geometric and highly pictorial version, the ideas of mapping (*Abbildung*), original and image (*Bild*)—the function is the *pictor*, who paints and conforms. There is an 'isomorphism' between the complex variable and its holomorphic correlate, the forms are preserved. Related to this topic, Riemann proposed and proved, though not with sufficient rigor, his celebrated *mapping theorem* (that there is a conformal mapping between any simply connected domain and the unit disk), one of the most important results in complex analysis.

Seen from a higher standpoint, *mutatis mutandis*, something analogous occurs with any structure-preserving function or mapping, even if the correlation may eliminate all traces of geometric form. Even so, the very existence of a continuous functional relation will establish profound structural analogies between one and the other domain. The idea is very philosophical, but without a doubt it occupied Riemann and inspired some of his epistemological considerations. It also became dear to Dedekind, encouraging him to amplify the pictorial terminology of *Bild* and *Abbildung* and use it for any function or mapping in general.

We can now go on to see how the idea finds application in Riemann's epistemology. The hypothetico-deductive method of Riemann is based on the notion of truth, understood along classical lines (following Aristotle) as the correspondence with facts. Thus, he writes:

> I. When is our conception of the world true?
> 'When the connection between our mental representations [*Zusammenhang unserer Vorstellungen*] corresponds with the connection between things.'
> II. How can the connection between things [*Zusammenhang der Dinge*] be established?
> 'Starting from the connections between phenomena [*Erscheinungen*].'[23]

Even though the ideas expressed above may seem well known, at this level too Riemann manages to introduce some innovative views that have interesting counterparts in the twentieth century. In a Herbartian vein, he indicates that the interesting correspondence is not between simple elements in our conceptual systems and the simple real elements but between their *interrelations*. The relations between elements in our image of the world must faithfully reflect the relations between things. Commenting upon no. I he writes:

'Die Theile einer gegebenen Fläche so abzubilden, dass die Abbildung dem Abgebildeten in den kleinsten Theilen ähnlich wird, von C. F. Gauss. [...].'" A conformal mapping preserves angles and forms around each point, but it may deform figures and affect their size.

[23] Bernhard Riemann, "Fragmente philosophischen Inhalts," in *Bernhard Riemann's Gesammelte mathematische Werke und wissenschaftlicher Nachlass*, 2nd ed., ed. Heinrich Weber (Leipzig: Teubner, 1892), 507–38, here 523. "I. Wann ist unsere Auffassung der Welt wahr? 'Wenn der Zusammenhang unserer Vorstellungen dem Zusammenhang der Dinge entspricht.' [...] II. Woraus soll der Zusammenhang der Dinge gefunden werden? 'Aus dem Zusammenhange der Erscheinungen.'"

The elements in our image of the world [*Bild von der Welt*] are wholly different from the corresponding elements of the represented real. They are something in us, the real elements are something outside us. But the relations [*Verbindungen*] between the elements in the image and in the represented [*Elementen im Bilde und im Abgebildeten*] must coincide if the image is to be true. The truth of the image is independent of its degree of finesse, it does not depend on whether the image-elements represent greater or lesser amounts of the real. But their relations must correspond, in the image one ought not assume an immediate action [*unmittelbare Wirkung*] between two elements, when in reality there is only a mediated one. If so, the image would be false and need rectification.[24]

Riemann's explanations on this point are highly reminiscent of the well-known *Bildtheorie* of thought and language proposed by Hertz, or also that of Wittgenstein at the beginning of his *Tractatus*. Both employ the very Riemannian terminology of 'Bild' and 'Abbildung' (just quoted), which was first published in 1876. It would be interesting to know whether Wittgenstein, besides being influenced by Hertz, could have read with some care Riemann's *Gesammelte mathematische Werke* (1876, 1st ed.). Probably not. What seems highly unlikely is that Hertz did *not* read them: among other indications, in his *The Principles of Mechanics* (*Die Prinzipien der Mechanik*, 1894) Hertz introduces a "geometry of systems of points" ("Geometrie der Punktsysteme") that is a form of *n*-dimensional Riemannian geometry.[25]

Reflections in Science and Mathematics … and New Flashes

The highly pictorial terminology of image and representation, *Bild* and *Abbildung*, produces at the same time an inversion—a semantic displacement which distances

---

[24] Riemann, "Fragmente," 523: "Die Elemente unseres Bildes von der Welt sind von den entsprechenden Elementen des abgebildeten Realen gänzlich verschieden. Sie sind etwas in uns; die Elemente des Realen etwas ausser uns. Aber die Verbindungen zwischen den Elementen im Bilde und im Abgebildeten müssen übereinstimmen, wenn das Bild wahr sein soll. Die Wahrheit des Bildes ist unabhängig von dem Grade der Feinheit des Bildes; sie hängt nicht davon ab, ob die Elemente des Bildes grössere oder kleinere Mengen des Realen repräsentiren. Aber die Verbindungen müssen einander entsprechen; es darf nicht im Bilde eine unmittelbare Wirkung zweier Elemente auf einander angenommen werden, wo in der Wirklichkeit nur eine mittelbare stattfindet."

[25] See: Jesper Lützen, "Geometrising Configurations. Heinrich Hertz and his Mathematical Precursors," in *The Symbolic Universe: Geometry and Physics 1890–1930*, ed. Jeremy Gray (Oxford: Oxford University Press, 1999), 25–46, also for qualifications (p. 39–40) concerning Hertz's conception of geometry, which seems to have been rather Kantian and not Riemannian. Heinrich Hertz, *Die Prinzipien der Mechanik* (Leipzig: J. A. Barth, 1894), 36: "Die Sammlung und Ordnung aller hier auftretenden Beziehungen gehört in die Geometrie der Punktsysteme und die Entwickelung dieser Geometrie hat eigenen mathematischen Reiz; wir verfolgen dieselbe aber nur soweit als es der augenblickliche Zweck der physikalischen Anwendung erfordert. Da ein System von *n* Punkten eine *3n*fache Mannigfaltigkeit der Bewegung darbietet, welche aber durch die Zusammenhänge des Systems auch auf jede beliebige Zahl vermindert werden kann, so entstehen viele Analogien mit der Geometrie eines mehrdimensionalen Raumes, welche zum Teil so weit gehen, daß dieselben Sätze und Bezeichnungen hier und dort Bedeutung haben können."

it from any naïve conception. In fact, Riemann and Dedekind, following a form of Leibnizian and Herbartian relationalism, insist upon the idea that the image representing some real element does not need to bear resemblance to it. The form that is preserved in adequate representations is not the form of the elements but only of the relations between them. Thus, the suggestive terminology, seemingly so close to familiar ideas, contains a new idea that is far from common sense: the symbolic, not iconic nature of scientific images,[26] the conventionality of models and scientific representations, will reappear in several scientific achievements of the time, which in turn will promote it.

Two historical experiences of great epistemological depth came together in the nineteenth century: in mathematics the move beyond the Euclidean geomet ric framework, in physics the abandonment of mechanistic images of the world. Both times we find some German-speaking authors centrally involved, whom I will invoke soon, i.e., Riemann, Helmholtz, Hertz (together, of course, with others of various nationalities, Gauss, Nikolai Lobachevsky, János Bolyai, Michael Faraday, James C. Maxwell, Ernst Mach, Hendrik A. Lorentz, Henri Poincaré, Einstein). A third achievement not entirely unrelated to these two, which I can only mention here, is a deeper reflection on sensations and perceptual experience upon new experimental bases, in the hands of Helmholtz, Mach and others.

If there is a mathematical or scientific sign that we regard as prototypically iconic, this is the geometric figures, the "triangles, circles" and other "characters" which Galileo Galilei deemed elements of the language in which the great book of nature is written.[27] The geometrization of the physical image of the world, due to Johannes Kepler and Galilei, seemed thus to offer a direct route to the inner side of things, an intimate familiarity with real forms. Precisely for this reason, perhaps the most pronounced change in scientific epistemology, during the nineteenth century, was the abandonment of such naïve views. In the terms we have established, one can say briefly and concisely: the key was the very novel idea that *geometric images of physical reality are not iconic but symbolic.* At least when moving away from the middle ranges of experience, characteristic of our bodies and common experience, and going towards the very small (atomic) or the very large (cosmic range).

The problem was implicated already in the idea of a straight line, so hard to make conceptually precise since the essential point in the behavior of straight lines—expressed in the parallel postulate—is not a local property but a global

---

[26] I follow here the terminology of Charles Sanders Peirce, who employs 'sign' as a generic term, under which icons, indices and symbols fall; the icon, unlike the symbol, bears resemblance with the represented. The reader should know that Riemann's work antedates Peirce, and also Dedekind's is independent.

[27] Galileo Galilei, *Il Saggiatore* (1623), in *Opere*, vol. VI, 232. Yet their relation with the designated is far from trivial, since geometric icons are operated upon, not as they are depicted but as conceived; thus they may lack empirically given referents (see: José Ferreirós, "Ancient Greek Mathematics: A Role for Diagrams," Chap. 5 in *Mathematical Knowledge and the Interplay of Practices* (Princeton: Princeton University Press, 2015), 112–52).

one, which involves the infinitely large. Gauss already saw this problem and realized that it was an empirical matter to determine whether it is Euclid's geometry, or perhaps that of Lobachevsky, that most adequately symbolizes physical relations in the largest scales. Riemann was his successor here, and at the same time he left Gauss astonished with the richness of geometric possibilities exposed with the introduction of Riemannian manifolds, i.e., with spaces of variable curvature (tensor), in the famous lecture "On the Hypotheses which Lie at the Foundation of Geometry" ("Über die Hypothesen, welche der Geometrie zu Grunde liegen," 1854, published 1868). Within this framework, *geodesics* (lines of minimal length) play the role of straight lines, and if the space curvature is equal at all points, we come down to more intuitive geometries like those of Euclid and Lobachevsky (and the so-called 'elliptical' geometry of Riemann). Eventually, in the limit case when curvature is always zero, we find Euclidean geometry, and geodesics coincide with the straight lines of our intuition. The dialectics between local and global, which has played such a central role in mathematical thought ever since, was a fundamental issue for Riemann—maybe the first mathematician to take this step. Given that I cannot enter into this topic any further, I refer the reader to Riemann's mathematical works.[28]

The question of the local or extremely small was crucial for nineteenth century science, since the core of mathematics was formed by real and complex analysis, developed around infinitesimal calculus, and at the core of physics one found laws expressed by differential equations.[29] Once more, Riemann saw this very clearly, and went as far as to say that "truly elementary laws can only occur in the infinitely small, only for points in space and time."[30] At the atomic scale "the concepts of a solid body and of a ray of light" cease to be valid, but they had been the basis for our usual metric and geometric determinations, and for that reason the geometry valid in the "infinitely small" may not be Euclidean. And so it should be assumed "if we can thereby obtain a simpler explanation of phenomena."[31] The issue was of utmost relevance for physical theory then, since the central decades of the nineteenth century saw great efforts to find a correct theory of electromagnetism. This theory, linked with the name of Maxwell, became the greatest step forward in physics after Newton and before the quantum theories.

---

[28] English edition: Bernhard Riemann, *Collected papers*, trans. Roger Baker, Charles Christenson, and Henry Orde (Heber City: Kendrick Press 2004); German edition: *Bernhard Riemann's Gesammelte mathematische Werke und wissenschaftlicher Nachlass*, 2nd ed., ed. Heinrich Weber (Leipzig: Teubner, 1892). Many titles deal with such questions, among them e.g. the elegant book *Poetry of the Universe* by Robert Osserman (New York: Anchor Books, 1995). See also: Detlef Laugwitz, *Bernhard Riemann 1826–1866* (Basel: Birkhäuser, 1996).

[29] Partial differential equations like the wave equation, those of Laplace and Hamilton in classical mechanics, the electromagnetic equations of Ampère and later Maxwell's, etc.

[30] Cited in Thomas Archibald, "Riemann and the Theory of Electrical Phenomena: Nobili's Rings," *Centaurus* 34, no. 3 (1991): 247–71, here 269.

[31] Riemann, "On the Hypotheses," 661.

There developed in this field a complex dialectics between the requirement of intelligibility and exploitation of the new theoretical freedom implied in the pictorial-abstract conception of models, a freedom made explicit in the new mathematical tools. Riemann too, like Faraday before and Maxwell after him, was in favor of field theories instead of distance-action theories. They tried to avoid something that was hardly compatible with physical intuition, namely "that one body may act upon another at a distance through a vacuum without the mediation of anything else"—an idea that Newton himself adopted only as a useful expedient, but which he regarded as "so great an absurdity" as to be unacceptable to a person "who has in philosophical matters any competent faculty of thinking."[32] Riemann and Maxwell put in its place the notion of local action between ethereal points, in accordance with some partial differential equations. The celebrated four equations of Maxwell (which do not correspond to what he actually published) enjoyed unprecedented success not only in the explanation of electricity, magnetism and their interrelations but also because they reduced light waves to an electromagnetic phenomenon, and they predicted the existence of radio waves. Interestingly, Riemann assisted Weber and Rudolf Kohlrausch in some experiments where the measurement of an electromagnetic constant led to a result very close to the speed of light. Subsequently, he speculated on its theoretical significance, writing a paper for the Göttingen Scientific Society where he communicated his discovery of a connection between electricity and light.[33]

As well known, it was precisely Helmholtz's disciple Hertz who managed in 1888 to build an apparatus that emitted and detected electromagnetic waves, setting the basis for a technology that dominates our lives today. Like Maxwell and Helmholtz, like Isaac Newton himself, Hertz too was both a great theoretician and a great experimenter. We shall see in a moment that he followed Riemann's path in adopting the abstract-pictorial conception of scientific models. But now it will be interesting to consider another side of his work.

In the search for a theory, Maxwell employed relatively simple mechanical models to back his equations. It was the age of mechanicism, the belief in the unlimited power of attractive and repulsive mechanical actions to explain all phenomena, from planetary movements and light to nerve stimuli. Maxwell himself was aware of the fact that his imaginary mechanisms could not account fully for the actions described in the equations he was handling. In effect the problem was endemic: the ether hypothesis adopted in wave theories of light (from Augustin J. Fresnel onwards) had proven to be intractable—if the medium was mechanical, luminiferous ether seemed to enjoy inconsistent properties.

Hertz inherited this problem through Helmholtz's work on electromagnetism, and within a few years he came to the conviction that Maxwell's theory was right

---

[32] Third letter to Bentley, mentioned by Riemann (see: Riemann, "Fragmente," 534).

[33] Bernhard Riemann, "Ein Beitrag zur Electrodynamik (1858)," in *Bernhard Riemann's Gesammelte mathematische Werke und wissenschaftlicher Nachlass*, ed. Heinrich Weber (Leipzig: Teubner, 1876), 270–75, here 270.

and had put an end to the dominance of Newtonian action at a distance, inaugurating the reign of physical fields. Actually, Hertz became convinced of the need to go one step further, a small step from the technical point of view, but huge as it involved a big shift towards abstraction, with obvious epistemological implications. He abandoned the hope to find mechanistic interpretations of the field models and expressed the idea in eloquent terms: "To the question 'what is Maxwell's theory?' I know of no shorter and more concrete answer than the following: Maxwell's theory is the system of Maxwellian equations." "[... They must be] considered as hypothetical assumptions, so that their probability depends upon the great number of natural laws which they encompass."[34]

With these words he was taking note of the new status attained by physical theory around 1900; mechanicism was abandoned, the age of *Anschauung* or visualizable models based on mechanics had ended, the era of mathematical abstraction was beginning.[35] Hertz himself acknowledged that, as a result, the theory acquires a very "abstract and colourless" appearance.[36]

The situation can be regarded, with certain qualifications, as highly Pythagorean. Weyl has remarked how attempts to provide physical equations with an immediate physical sense, through the means of a suggestive language, do not manage to go farther than more or less fortunate metaphors, analogies which are always a bit unsatisfactory, vague, imprecise, superficial. Hence, the common experience that physical theory can only be adequately understood by those who have command of its mathematical structure (together with the way it is interpreted and applied in paradigmatic laboratory cases, in real and also imaginary experiments).

Helmholtz and Hertz

Let me go back to our story about the pictorial approach (*Bildtheorie*) to scientific theorizing. In the case of Wittgenstein, Hertz seems to have been the crucial influence, although we know that other authors (like Ludwig Boltzmann) may have played a role. Some experts have indicated that Hertz inherited his conception

---

[34] Heinrich Hertz, *Electric Waves*, trans. Daniel Evan Jones (New York: Dover, 1962), 21, 139; German: "Auf die Frage 'Was ist die Maxwell'sche Theorie' wüsste ich also keine kürzere und bestimmtere Antwort als diese: Die Maxwell'sche Theorie ist das System der Maxwell'schen Gleichungen." "[Es erscheint folgerichtig, dieselben] als eine hypothetische Annahme zu betrachten und ihre Wahrscheinlichkeit auf der sehr grossen Zahl an Gesetzmässigkeiten beruhen zu lassen, welche sie zusammenfassen." (Heinrich Hertz, *Untersuchungen über die Ausbreitung der elektrischen Kraft* (Leipzig: Barth, 1892), 23, 148).

[35] A great popular exposition of this topic, which explains it without a single mathematical formula, can be found in Albert Einstein and Leopold Infeld, *The Evolution of Physics* (Cambridge: Cambridge University Press, 1938).

[36] Hertz, *Electric Waves*, 28.

partially from the polymath Helmholtz, his master at the Friedrich-Wilhelms-Universität Berlin.[37] However, the truth may be different, and the influence of Riemann on Hertz seems quite likely.

Curiously, Riemann and Helmholtz form a couple that affords much insight into the deep tendencies of scientific thinking in the nineteenth century. Philosophically both were heirs to reactions toward Kantian epistemology, they enjoyed deep knowledge of physics and mathematics, and their conceptions pointed in different directions, but both would be of great influence in the twentieth century.

Helmholtz represents a kind of scientific neo-Kantianism, oriented towards empiricism, and since decades he has been hailed as a predecessor of logical empiricism. Riemann was a Herbartian and thus radically against Kantian apriorism, but he was no less scientific; his tendency was to emphasize the hypothetical nature of any scientific theorizing, its character as a 'symbolic construction' of the world. To put it in simplistic terms: if Helmholtz was a predecessor of the Vienna Circle, Riemann opened the way towards positions such as Karl Popper's, or also Weyl's (despite matters of emphasis and nuance).

The divergence was already quite clear in the reactions of the one scientist to the core contributions of the other. One of them is found in the last article Riemann was preparing, a reply to Helmholtz's great book *The Theory of the Sensations of Tone as a Physiological Basis for Musical Theory* (*Die Lehre von den Tonempfindungen als physiologische Grundlage für die Theorie der Musik*, 1863).[38] Another, inversely, took place after the publication of Riemann's epoch-making "On the Hypotheses which Lie at the Foundations of Geometry" ("Ueber die Hypothesen, welche der Geometrie zu Grunde liegen," 1868). Helmholtz was quick to reply in the same year, and his paper was named "On the Facts which lie at the Foundations of Geometry" ("Ueber die Thatsachen, die der Geometrie zum Grunde liegen," 1868).[39] *Tatsachen* versus *Hypothesen*—*facts* that would speak for geometries of constant curvature, against the daring *hypotheses* that opened up the world of Riemannian manifolds. And yet Helmholtz's supposed facts would be contradicted by a new empirically supported theory, Einstein's general relativity, in which Riemann's geometries of varying curvature featured centrally.

---

[37] See: Gregor Schiemann "The Loss of World in the Image: Origin and Development of the Concept of Image in the Thought of Hermann von Helmholtz and Heinrich Hertz," in *Heinrich Hertz: Classical Physicist, Modern Philosopher*, ed. Davis Baird, Richard I. G. Hughes, and Alfred Nordmann (Dordrecht: Kluwer, 1998), 25–38, and Michael Heidelberger, "From Helmholtz's Philosophy of Science to Hertz's Picture-Theory," in *Heinrich Hertz: Classical Physicist, Modern Philosopher*, ed. Davis Baird, Richard I. G. Hughes, and Alfred Nordmann (Dordrecht: Kluwer, 1998), 9–24.

[38] See: Bernhard Riemann, "Mechanik des Ohres," in *Bernhard Riemann's Gesammelte mathematische Werke und wissenschaftlicher Nachlass*, ed. Heinrich Weber (Leipzig: Teubner, 1876), 316–28. See also: Hermann von Helmholtz, *Die Lehre von den Tonempfindungen als physiologische Grundlage für die Theorie der Musik* (Braunschweig: Vieweg und Sohn, 1863).

[39] Hermann von Helmholtz, "Ueber die Thatsachen, die der Geometrie zum Grunde liegen," *Nachrichten von der Königlichen Gesellschaft der Wissenschaften und der Georg-Augusts-Universität zu Göttingen* 9 (1868): 193–221.

Those who think that Hertz's *Bildtheorie* stems from his master cite some passages from an important conference that Helmholtz gave in 1878, "The Facts in Perception" ("Die Tatsachen in der Wahrnehmung"). He insists upon an idea that, no doubt, is linked with the topics we have been revisiting:

> Our sensations are indeed effects produced in our organs by external causes: and how such an effect expresses itself naturally depends quite essentially upon the kind of apparatus upon which the effect is produced. Inasmuch as the quality of our sensation gives us a report of what is peculiar to the external influence by which it is excited, it may count as a *sign* [*Zeichen*] of it, but not as an *image* [*Abbild*]. For from an image one requires some kind of alikeness with the object of which it is an image—from a statue alikeness of form, from a drawing alikeness of perspective projection in the visual field, from a painting alikeness of colors as well. But a sign need not have any kind of similarity at all with what it is the sign of. The relation between the two of them is restricted to the fact that like objects exerting an influence under like circumstances evoke like signs, and that therefore unlike signs always correspond to unlike influences. [...] Since like things are indicated in our world of sensations by like signs, an equally regular sequence will also correspond in the domain of our sensations to the sequence of like effects by law of nature upon like causes.[40]

We find here an emphasis on the idea of functional correspondence between stimulus and sensation, and at the same time insistence on the non-resemblance between the elements, i.e., the 'conventional' or dissimilar nature of secondary qualities.[41] But Helmholtz chose his terminology in such a manner that he rejects the pictorial conception—they can be regarded as signs (*Zeichen*), he says, but not as images, not as *Abbilder*. And it is so because he insists (to explain it this way) on the old

---

[40] Hermann von Helmholtz, "The Facts in Perception," in *From Kant to Hilbert*, ed. William Ewald, vol. 2 (Oxford: Claredon Press 1996), 689–727, here 695–96; German: "Unsere Empfindungen sind eben Wirkungen, welche durch äussere Ursachen in unseren Organen hervorgebracht werden, und wie eine solche Wirkung sich äussert, hängt natürlich ganz wesentlich von der Art des Apparats ab, auf den gewirkt wird. Insofern die Qualität unserer Empfindung uns von der Eigenthümlichkeit der äusseren Einwirkung, durch welche sie erregt ist, eine Nachricht giebt, kann sie als ein *Zeichen* derselben gelten, aber nicht als ein *Abbild*. Denn vom Bild verlangt man irgend eine Art der Gleichheit mit dem abgebildeten Gegenstande, von einer Statue Gleichheit der Form, von einer Zeichnung Gleichheit der perspectivischen Projection im Gesichtsfelde, von einem Gemälde auch noch Gleichheit der Farben. Ein Zeichen aber braucht gar keine Art der Aehnlichkeit mit dem zu haben, dessen Zeichen es ist. Die Beziehung zwischen beiden beschränkt sich darauf, dass das gleiche Object, unter gleichen Umständen zur Einwirkung kommend, das gleiche Zeichen hervorruft, und dass also ungleiche Zeichen immer ungleicher Einwirkung entsprechen. [...] Da Gleiches in unserer Empfindungswelt durch gleiche Zeichen angezeigt wird, so wird der naturgesetzlichen Folge gleicher Wirkungen auf gleiche Ursachen, auch eine ebenso regelmässige Folge im Gebiete unserer Empfindungen entsprechen." (Hermann von Helmholtz, *Die Tatsachen in der Wahrnehmung* (Darmstadt: Wissenschaftliche Buchgesellschaft, 1959), 18–19).

[41] This old theme was discussed by Galileo and Descartes, but it can be found already in Greek philosophy. Democritus says: "By convention sweet and by convention bitter, by convention hot, by convention cold, by convention color; but in reality atoms and void. [And the senses reply:] Unhappy mind! You obtain your beliefs from us, and you want to end us? Our fall will be your ruin." (Hermann Diels and Walther Kranz, *Die Fragmente der Vorsokratiker*, vol. 2 (Dublin: Weidmann, 1972–74), fragment 125).

style of figurative painting, not imagining the turn towards abstraction that will come. Specialists thus have to talk about Helmholtz's *Zeichentheorie*, which is not the same as the *Bildtheorie* we find in Riemann and Hertz.

Riemann's option is the opposite one, more forward looking, which is what Hertz will follow. And the point is that this new opposition between Riemann and Helmholtz seems to me quite coherent with the tension between the hypothetico-deductivism of the former and (the relative) empiricism of the latter. And it is also interesting to realize that Hertz did not share his Berlin master's tendency to present everything as derived from 'facts,' insisting instead on the hypothetical character of physical principles. One should make clear that Riemann's epistemological fragments became public in 1876, with the first edition of his *Werke*, and the related ideas of Dedekind were published in a purely logico-mathematical context in 1888. All of it years before the key text by Hertz (1894).

I will now quote the celebrated passage of Hertz's introduction to *The Principles of Mechanics* (1894), where he speaks of theories and scientific hypotheses as images, *Bilder*:

> The most direct, and in a sense the most important, problem which our conscious knowledge of nature should enable us to solve is the anticipation of future events, so that we may arrange our present affairs in accordance with such anticipation. As a basis for the solution of this problem we always make use of our knowledge of events which have already occurred, obtained by chance observation or by prearranged experiment. In endeavoring thus to draw inferences as to the future from the past, we always adopt the following process. We form for ourselves inner images [*innere Scheinbilder*] or symbols of external objects; and the form which we give them is such that the necessary consequents of the images [*Bilder*] in thought [*denknotwendigen Folgen*] are always the images of the necessary consequents in nature [*naturnotwendigen Folgen*] of the things pictured [*abgebildeten Gegenstände*]. In order that this requirement may be satisfied, there must be a certain conformity between nature and our thought. Experience teaches us that the requirement can be satisfied, and hence that such a conformity does in fact exist.[42]

---

[42] Heinrich Hertz, *The Principles of Mechanics, Presented in a New Form* (New York: Dover, 1956), 1; German: "Es ist die nächste und in gewissem Sinne wichtigste Aufgabe unserer bewußten Naturerkenntnis, daß sie uns befähige, zukünftige Erfahrungen vorauszusehen, um nach dieser Voraussicht unser gegenwärtiges Handeln einrichten zu können. Als Grundlage für die Lösung jener Aufgabe der Erkenntnis benutzen wir unter allen Umständen vorangegangene Erfahrungen, gewonnen durch zufällige Beobachtungen oder durch absichtlichen Versuch. Das Verfahren aber, dessen wir uns zur Ableitung des Zukünftigen aus dem Vergangenen und damit zur Erlangung der erstrebten Voraussicht stets bedienen, ist dieses: wir machen uns innere Scheinbilder oder Symbole der äußeren Gegenstände, und zwar machen wir sie von solcher Art, daß die denknotwendigen Folgen der Bilder stets wieder die Bilder seien von den denknotwendigen Folgen der abgebildeten Gegenstände. Damit diese Forderung überhaupt erfüllbar sei, müssen gewisse Übereinstimmungen vorhanden sein zwischen der Natur und unserem Geiste. Die Erfahrung lehrt uns, daß die Forderung erfüllbar ist und daß also solche Übereinstimmungen in der That bestehen." (Heinrich Hertz, *Die Prinzipien der Mechanik in neuem Zusammenhange dargestellt* (Leipzig: Barth, 1894), 1).

The key sentence, as one can see, combines some terminology close to Helmholtz (symbols) with Riemann's pregnant terminology, which has become famous while its origin is forgotten. Maybe it is difficult to grasp on a first read what Hertz is expressing: "the necessary consequents of the images in thought are always images of necessary natural consequents," but it is an elementary mathematical idea, typical of mappings and analyzed by Dedekind in full precision. Call a and b the two natural things that are one consequence of the other, and call $f(a)$ and $f(b)$ their correlated images. As Hertz states, the key condition is: If $f(a)$ follows from $f(b)$ in the image, then a causes b in nature.[43] (R).

He continues insisting that the images he is talking about are our mental representations of things [*Vorstellungen*], and that they are in conformity with things "in a single important respect," namely that they satisfy requirement (R). There need not exist any other kind of conformity and in effect we are not in a position to know whether our mental representations may exhibit any other kind of conformity. It is the same idea Riemann insisted on.

In fact, things are even more complex, as scientists and philosophers of science would discover in the twentieth century.[44] The relations (be they causal or not) that are ascertained by empirical means, "by chance observation or by prearranged experiment," are not precisely the relations between the elements that conform to abstract theory. The elements in this abstract theory, its objects (assumptions) and principles, form a complex whole. Considering this *holos*, the whole, often in combination with one or more auxiliary theories, one establishes particular *models* for certain kinds of concrete systems. These models act as mediators, and it is them that we use to infer, by logical or mathematical or computational means, some predictions that concern the relations one ought to expect to hold among natural phenomena. The validity of requirement (R)—in Hertz's or maybe better in Riemann's version—for these predictions derived from models does not quite imply conformity between the relations of high theory and the relations between natural things.

Many scientists, in particular theoretical physicists, have a strong tendency to think that their theories are even more real than phenomena themselves. This tendency stems in the last analysis from a somewhat primitive epistemology, from a lack of acknowledgment of the subtle relations between the theories (that they handle so well in practice), the models subsequently established, and the data obtained from their experimental colleagues. It is the theoretician's *hybris* and a form of residual Platonism (ideas being more real than appearances), paradoxically

---

[43] Riemann, as you will remember, thought it this way: If $f(a)$ is related with $f(b)$ in the image, then a is related with b in nature; and if a is not related immediately with b, then $f(a)$ will not be related immediately with $f(b)$.

[44] The locus classicus for methodological issues related to holism is Pierre Duhem's work, *The Aim and Structure of Physical Theory* (Princeton: Princeton University Press, 1954; orig. 1906); also relevant is Willard Van Orman Quine, "Two Dogmas of Empiricism," *Philosophical Review* 60, no. 1 (1951): 20–43. But the topic reappears in many other twentieth century authors, like Thomas Kuhn for instance.

the outcome of a too pragmatic, hence, naïve orientation. The subtle equilibrium that scientists like Hertz or Maxwell managed to attain, probably thanks to their mastery of both theoretical and experimental practices, and their acquaintance with conceptual and philosophical difficulties, seems to have been lost in our age of hyper-specialized scientists.

Longue Durée

In 1910, Ernst Cassirer published *Substanzbegriff und Funktionsbegriff*, in simplified translation *Substance and Function*, an important book which aimed to account for a fundamental transformation in the logic of scientific thinking, involving all fields of knowledge and especially the natural sciences. Chapter 1 explains the general idea, before proceeding to trace its course through several sciences, and begins by talking about the Aristotelian way of representing world phenomena, focusing on "generic concepts" (basically for classification) whose metaphysical counterpart are substances;[45] to end up emphasizing the new "logic of the mathematical concept of function." Cassirer wrote:

'Every mathematical function represents a universal law, which, by virtue of the successive values which the variable can assume, contains within itself all the particular cases for which it holds.' If, however, this is once recognized, a completely new field of investigation is opened for logic. In opposition to the logic of the generic concept, which, as we saw, represents the point of view and influence of the concept of substance, there now appears the *logic of the mathematical concept of function*. However, the field of application of this form of logic is not confined to mathematics alone. On the contrary, it extends over into the field of the knowledge of nature; for the concept of function constitutes the general schema and model according to which the modern concept of nature has been molded in its progressive historical development.[46]

---

[45] Ernst Cassirer, *Substance and Function* (New York: Dover, 1980), 8. The comment (ibid., 8) is noteworthy that all determinations of being were subordinated to the substances they inhere in, and this affected in particular the Aristotelian category of relation, "forced into a dependent and subordinate position;" relations can only be modifications of substances, they cannot alter their real "nature."

[46] Cassirer, *Substance and Function*, 20–21; German: "'[...] Denn jede Funktion stellt ein allgemeines Gesetz dar, das vermöge der successiven Werte, welche die Variable annehmen kann, zugleich alle einzelnen Fälle, für die es gilt, unter sich begreift.' Wird dies aber einmal anerkannt, so eröffnet sich damit zugleich für die Logik ein völlig neues Gebiet der Untersuchung. Der Logik des Gattungsbegriffs, die, wie wir sahen, unter dem Gesichtspunkt und der Herrschaft des Substanzbegriffs steht, tritt jetzt die *Logik des mathematischen Funktionsbegriffs* gegenüber. Das Anwendungsgebiet dieser Form der Logik aber kann nicht im Gebiet der Mathematik allein gesucht werden. Vielmehr greift hier das Problem sogleich auf das Gebiet der *Naturerkenntnis* über: denn der Funktionsbegriff enthält in sich zugleich das allgemeine Schema und das Vorbild, nach welchem der moderne Naturbegriff in seiner fortschreitenden geschichtlichen Entwicklung sich gestaltet hat." (Ernst Cassirer, *Substanzbegriff und Funktionsbegriff. Untersuchung über die Grundfragen der Erkenntniskritik* (Berlin: Bruno Cassirer, 1910), 27).

Cassirer's idea seems basically correct and highly perceptive. On the one hand, the evolution of scientific research over several centuries, from Copernicus's *De Revolutionibus* (1543) onward, involved a constant distancing from the conception of phenomena in terms of substance and accident, to the point that the substantialist conception—if not entirely superseded—was increasingly marginalized, less and less visible in the details of scientific theories and models. On the other hand, it is also true that the impact of functional thinking on logic itself has been decisive: modern logic emerged from the effort to understand the logical relations put forward in mathematics, and it is well known that Frege's logic was founded on the abandonment of the subject-predicate scheme (correlate of the notions of substance and attribute, or substance and accident) in favor of the function-argument scheme, which Frege came to regard as fundamentally logical in nature (and which became the basis for his metaphysics of functions and objects).[47]

Another side of the question, which Cassirer did not research into, but which was a topic for contemporaries like Mach and Carnap, was the dissolution of causal thinking in its primeval form. This aspect has been less reflected upon and less incorporated in our cultural consciousness, to the point that it would not be strange if the reader is surprised to read about it. The truth is that, to paraphrase Cassirer, the concepts of relation and function have given form to the general scheme and pattern according to which the contemporary conception of nature has been molded, especially since around 1850. The old and familiar scheme of cause and effect has died out. To be sure, it remains familiar and intuitive because it is the way agents like us represent the possibilities of action in their immediate environment (manipulative causation). Of course, there are many particular circumstances in which it makes sense to ask for causes, and we still do so in scientific practice, but the question of the cause is essentially pragmatic, context-dependent, and does not point toward a complete explanation.

If you ask, why is the sun eclipsed, the answer may be 'because of the moon,' and if you ask why the blood moves, we will answer 'thanks to the heart pumping;' or, why does the orbit of Uranus present some deviations, 'because of the force imparted by Neptune.' But these are only partial indications, contextually relevant given the information previously available, which do not account for the phenomenon in anything like a complete explanation. Causal explanations are very often demanded in practice,[48] but a causal explanation will never be a full explanation of the phenomenon—for which we would resort instead to a

---

[47] Aristotelian propositions express substantial or accidental attributions (some men are small, all humans are mortal), but they are unable to analyze elementary mathematical inferences; here is a related example given by Leibniz, of an inference which escapes Aristotle: Mary was the mother of Jesus of Nazareth, therefore Mary was the mother of God. One could go much deeper into this topic: the logic of quantifiers $\forall x$, $\exists x$ emerged in relation with critical reflections upon the foundations of analysis, a theory which focuses on the study of functions (mappings); see: José Ferreirós, "The Road to Modern Logic," *Bulletin of Symbolic Logic* 7, no. 4 (2001): 441–84.

[48] The pragmatic theory of explanation has been defended by Bas van Fraassen, *The Scientific Image* (Oxford: Clarendon Press, 1980).

relational-functional model. E.g., a three-body model complying with Newton's equations for the system sun-earth-moon, or a complex model of the autonomous nervous system and the movements of the heart and blood, and so on.

The ultimate implications of these deep changes may not yet be clear. By giving primacy to relations and functions, the question arises to what extent the *relata* must antecede relations, a topic in between science and metaphysics. Is the world made of 'bricks' (atoms, particles) or is it fundamentally made of relations? It has been philosophers, above all, who dared to suggest the radical priority of relations over *relata*.[49] Common sense still recommends the opposite idea that *relata*, the objects entering into relations, are prior, no matter how much the sciences belie the notion of permanent objects (in a strong sense) and promote the thesis of the processual and relational character of a reality whose principles are expressed through functional thinking. Maybe this antinomy, which I cannot further develop here, holds the key to a solution of many perplexities in physical theory.

Other Reflections

There are some other related reflections and bright sparks that I could comment upon here. Indeed, the topics I have revisited suggest connections that, more than once, may leave us perplexed. I am thinking above all about how to understand the links between the rise of a pictorial-abstract approach in science and mathematics, and the contemporary (albeit somewhat delayed) emergence of several forms of non-figurative painting. A flat and somewhat reductionistic perspective on the relations between art and science would insist, perhaps, in how photographic techniques emptied the meaning of traditional forms of painting, forcing painters to leave aside realistic representation and to reconsider work on the canvas in a much more autonomous way. This may well be true, but it seems only one factor among several (the function is in several variables). The developments I have reviewed suggest deeper and more philosophical lines of connection, perhaps a bit subterranean but certainly powerful.

Paul Cézanne was a contemporary of Dedekind and Hertz whose exploration of the inner space of the canvas, his free speculation about the reciprocal relations between forms and colors,—which great masters of the twentieth century would consider pioneering and paradigmatic—has many analogies with the architectonic exploration of structural possibilities that guided the most avant-garde

---

[49] Examples can be found in the neo-Kantians Paul Natorp and Hermann Cohen, in Herbart (see above), but apparently also in a physicist like Wilhelm Eduard Weber; see: M. Norton Wise, "German Concepts of Force, Energy, and the Electromagnetic Ether: 1845–1880," in *Conceptions of Ether: Studies in the history of ether theories 1740–1900*, ed. Geoffrey N. Cantor and Michael J. S. Hodge (Cambridge: Cambridge University Press, 1981), 269–307.

mathematicians of his time.[50] In the same period, the sculptures of Auguste Rodin problematize the limits of the image, speculating with the relations between frame and figure, between matter and sculptural form, opening up a network of more complex relations, more autonomous and problematic than in the previous tradition. The final decades of the nineteenth century, and the early ones of the twentieth, saw the transition from the space-frame characteristic of figurative painting (analogous to the ambient space of Cartesian and Newtonian physics and geometry) to the space-network characteristic of the avant-garde (analogous in turn to the relativistic geometry of Riemann and Einstein). There is a surprising parallelism between the evolution of geometries linked with non-Euclidean theories and the abstract explorations in the world of art, immediately afterwards.[51] It can hardly be mere coincidence.

One could go deeper into the contributions of Einstein, the works of Kandinsky, the writings of Wittgenstein, or the new and multiple structures emerging in experimental music around the same years. But limits of space, and of course my own limits, prevent me from continuing down these paths.

Just one last remark. Contemporaneous with the process I have been studying, there was quite a lot of interest among mathematicians in models (actual, 3D models) and model making. They were employed as tools for education, highly recommended by Klein, Alexander Brill and others (plaster models thus made their way to leading universities around the world), but they were also artefacts for research.[52] Indeed, the notions of *Bild* and *Abbild*, and the relation to *Vorstellung*, are often mentioned by those mathematicians, when they literally speak about material models. Was there any close relation? One would tend to say, at first sight, that the two processes are parallel and disconnected: one is the Riemannian line of abstract thinking about *Abbildung*, another the Kleinian line of emphasis on actual hands-on experience and 3D models. Yet this is not quite convincing, if only because Klein regarded himself as a conceptual thinker and direct heir to Riemann. Moreover, one can find intermediary cases, like the 'models' for the non-Euclidean plane in Euclidean space (Minding, Beltrami, etc.). The question certainly deserves further attention, but I leave it open here.

---

[50] Even some analogies in life and mental orientations, so to say, between Paul Cézanne and Dedekind are salient and noteworthy. See my forthcoming paper: José Ferreirós, "Paradise Recovered? Some Thoughts on *Mengenlehre* and Modernism," in *Science as Cultural Practice, Vol. II: Modernism in the Sciences, ca. 1900–1940*, ed. Moritz Epple and Falk Müller (Berlin: Akademie Verlag, forthcoming).

[51] On this topic see: Capi Corrales Rodrigáñez, "Dallo spacio come contenitore allo spazio come rete," in *Matematica e Cultura 2000*, ed. Michele Emmer (Milano: Springer Italia 2000), 123–38. "Local–Global in Mathematics and Painting," in *The Visual Mind*, ed. Michele Emmer, vol. 2 (Cambridge, MA: The MIT Press, 2005), 273–94.

[52] David Rowe, "Mathematical Models as Artefacts for Research: Felix Klein and the Case of Kummer Surfaces," *Mathematische Semesterberichte* 60, no. 1 (2013): 1–24.

Thinking with Notations: Epistemic Actions and Epistemic Activities in Mathematical Practice

Axel Gelfert

The Applicability 'Problem'

In a text that has become an instant favorite among physicists and mathematicians alike, Eugene P. Wigner expressed his sense of wonder at "the enormous usefulness of mathematics in the natural sciences," which he likened to "something bordering on the mysterious," for which "there is no rational explanation."[1] This phenomenon, for which Wigner coined the phrase "the unreasonable effectiveness of mathematics in the natural sciences" (which also serves as the title of his original lecture in 1960), brings into sharp focus one of the core issues at the intersection of philosophy of mathematics and philosophy of science: the applicability of mathematics to the world at large. For as long as the 'Book of Nature' was thought—at least within the Western intellectual tradition—to have been authored in the language of mathematics (as Pythagoreans, in their various guises, would have it), and Nature was seen as bearing the hallmarks of a divine Being's rationality, the natural world's amenability to mathematical description did not constitute a salient problem, but was simply assumed. In turn, no special 'applicability problem' arose for mathematics—which, after all, was thought to be omnipresent in Nature 'by design.' Yet, once "numbers and geometrical forms [were] no longer assumed to be inherent as such in Nature," as Michael Polanyi has argued, "'pure' mathematics, formerly the key to nature's mysteries, became

---

[1] Eugene Wigner, "The Unreasonable Effectiveness of Mathematics in the Natural Sciences," *Communications on Pure and Applied Mathematics* 13, no. 1 (1960): 1–14, here 2.

A. Gelfert (✉)
Technische Universität Berlin, Straße des 17. Juni 135, 10623 Berlin, Germany
e-mail: a.gelfert@tu-berlin.de

© The Author(s) 2022
M. Friedman and K. Krauthausen (eds.), *Model and Mathematics: From the 19th to the 21st Century*, Trends in the History of Science,
https://doi.org/10.1007/978-3-030-97833-4_11

strictly separated from the *application* of mathematics to the formulation of empirical laws."[2] Marveling at the divinely mandated concordance between mathematics and Nature gradually gave way to a sense of puzzlement at the very possibility of applying mathematics to the empirical world.

Mathematics, of course, underwent significant transformations itself, and it may well be argued that it was precisely by outstripping the domain of what was needed in order to explain natural phenomena, that the proliferation of mathematical theories and subfields contributed to this sense of puzzlement. Indeed, this seems to have been Wigner's main point: If certain mathematical concepts are useful for describing natural phenomena, who is to say that other—hitherto undiscovered—mathematical frameworks could not be even more 'unreasonably effective'? As Wigner argues, "because we do not understand the reasons for their usefulness, we cannot know whether a theory formulated in terms of mathematical concepts is uniquely appropriate."[3] Wigner's version of the applicability problem, then, must be seen against the backdrop of the vast proliferation of mathematical frameworks since the nineteenth century and their subsequent uptake in early twentieth-century physics, notably of Riemannian geometry in general, relativity theory, group theory in particle physics, etc. Yet throughout some of its most formative periods, mathematicians and philosophers hardly reflected on the applicability problem and related foundational issues. Around 1800, it was quite common to speak of mathematics, as Leonhard Euler did, as "nothing more than a *science of magnitudes.*"[4] The term 'magnitudes' here refers to the act of measuring physical quantities: perhaps the amount of water used to irrigate a field, or the total weight of the harvest it generated. The thought was not that mathematics should study the qualitative *physical* features of the world, but rather that it deals with the *relations* in which real-world magnitudes stand to one another. At the same time, its applicability to the physical world was, again, simply presupposed.

In the late nineteenth/early-twentieth century, the tables initially seemed to turn. Foundational issues had slowly moved to the center of mathematics, not least due to attempts to formalize mathematics and give it a programmatic dimension. Thus, David Hilbert had this to say about the relation between mathematics and the physical world:

---

[2] Michael Polanyi, *Personal Knowledge* (London: Routledge & Kegan Paul, 1962), 8.

[3] Eugene Wigner, "The Unreasonable Effectiveness," 2. Elsewhere I have described this worry in terms of a novel form of *underdetermination*, viz. "underdetermination by (a multiplicity of conceivable) mathematical frameworks" (Axel Gelfert, "Applicability, Indispensability, and Underdetermination: Puzzling Over Wigner's 'Unreasonable Effectiveness of Mathematics,'" *Science & Education* 23, no. 5 (2014): 997–1009, here 1002).

[4] Leonhard Euler, "Vollständige Anleitung zur Algebra," in Leonhard Euler *Opera Omnia*, vol. 1. (Leipzig: Teubner 1911 [1771]), 1–498, here 15.

We are confronted with the peculiar fact that matter seems to comply well and truly to the formalism of mathematics. There arises an unforeseen unison of being and thinking, which for the present we have to accept like a miracle.[5]

We can see here a nascent—if unresolved—concern with 'bridging the gap' between mathematical knowledge and knowledge of the physical world, formulated from *within* the field of mathematics. At the same time, the rise of logical empiricism in philosophy largely denied the existence of any special problem of the applicability of mathematics. As Torsten Wilholt has convincingly argued, the rise of Carnap-style logical empiricism largely neutralized the philosophical significance of the applicability problem—"not because [the applicability problem] was considered to have received an answer, but because all matters of applicability of frameworks were considered practical questions."[6]

Where the logicist tradition *within* mathematics had at least kept alive the hope of a resolution by grounding applicability in a general theory of how *any* concept can apply to the world, for logical empiricism, mathematics was simply to be equated with whatever conceptual framework was empirically found to be most "serviceable"[7] *for the time being*. Arguably, such a philosophical view left little room for 'problematizing' the fact that mathematics was, indeed, found to be a useful resource in describing the world. Yet once again, developments in mathematics pulled in a different direction. As José Ferreirós observes, in mid-twentieth century mathematics, "it was usual to emphasize that 'all mathematical theories can be considered as extensions of the general theory of sets'"[8] ("as everyone knows"[9]). This recreated, or at least widened, the gulf between mathematics—now understood as a set-theoretic enterprise whose ontology followed from the existential assumptions implicit in the axioms of set theory—and the empirical world, which seemed to conform to it. If the applicability problem has in recent years

---

[5] David Hilbert, *Natur und mathematisches Erkennen: Vorlesungen, gehalten 1919–1920 in Göttingen*, ed. David E. Rowe (Basel: Birkhäuser, 1992), 69; English translation: Torsten Wilholt, "Lost on the Way from Frege to Carnap: How the Philosophy of Science Forgot the Applicability Problem," *Grazer Philosophische Studien* 73, no. 1 (2006): 69–82, here 69: "Wir stehen da der merkwürdigen Tatsache gegenüber, daß anscheinend die Materie sich ganz und gar dem Formalismus der Mathematik fügt. Es zeigt sich hier ein unvorhergesehener Einklang zwischen Sein und Denken, den wir vorläufig wie ein Wunder hinnehmen müssen."

[6] Wilholt, "Lost on the Way," 76.

[7] Ibid., 75.

[8] José Ferreirós, *Mathematical Knowledge and the Interplay of Practices* (Princeton: Princeton University Press, 2016), 36.

[9] Nicolas Bourbaki, "Foundations of Mathematics for the Working Mathematician," *The Journal of Symbolic Logic* 14, no. 1 (1949): 1–8, here 7.

regained some of its prominence,[10] then this is as much a reflection of the 'fading' of (certain commitments of) logical empiricism, as it is the long-term effect of intra-mathematical developments in the second half of the twentieth century.

Whatever the merits and varied fortunes of the applicability problem in the philosophy of mathematics, one aspect of the debate is immediately obvious: it invariably construes applicability in 'global' terms, as the challenge of relating a domain of abstract objects to the empirical world-at-large. The applicability problem, thus understood, refers to the difficulty of specifying the global conditions of possibility for how abstract entities can manifest themselves in a world of concrete events and facts. Such a perspective on the 'applicability' of mathematics, however, is far removed from any actual—concrete—actions of *applying* mathematics. Applying mathematics—proving theorems, deriving corollaries, performing calculations, etc.—takes work and requires significant cognitive effort. Which acts of inferring, deriving, representing, manipulating, or rearranging (e.g. of formulas and equations), are appropriate, and when? Such questions cannot be answered with reference to the global relationship between the realm of mathematical entities and the empirical realm, but instead require attention to mathematical actions and activities.

Mathematical actions and activities, I wish to argue in this paper, are typically mediated by symbol systems, notations, and formalisms, in a way that not only allows for the seamless expression of mathematical concepts, but also *actively shapes mathematical practice*, both at the level of individual derivations and at the collective level of mathematics as a discipline. That is, notations and formalisms do not play a merely auxiliary role, in that they give neutral expression to the underlying foundational concepts, but they are themselves constituents of mathematical practice, without which certain connections and developments within mathematics (and certain ways of applying mathematics to the natural world) would not be possible. The realization that mathematical notations and formalisms are more than just passive 'vessels' for the transmission of mathematical concepts has gradually been catching on in contemporary philosophy of mathematics. Thus, James Robert Brown argues that "the source of the attraction of formalism [as a general position in the philosophy of mathematics] stems from the evident power of notations themselves," and less from its "nominalistic hostility to abstract entities" (which he deems "silly");[11] Mark Colyvan notes "that the notation can help

---

[10] E.g., through the work of Mark Steiner, *The Applicability of Mathematics as a Philosophical Problem* (Cambridge: Harvard University Press, 1998), Mark Colyvan, *An Introduction to the Philosophy of Mathematics* (Cambridge: Cambridge University Press, 2001), and Alan Baker, "Mathematical Explanation in Science," *The British Journal for the Philosophy of Science* 60, no. 3 (2009): 611–33; Alan Baker "Are There Genuine Mathematical Explanations of Physical Phenomena?," *Mind* 114, no. 454 (2005): 223–38.

[11] James Brown, *Philosophy of Mathematics: A Contemporary Introduction to the World of Proofs and Pictures* (Abingdon: Routledge, 2008), 86.

reveal hitherto unknown mathematical facts;"[12] finally, Helen De Cruz and Jan De Smedt (more about whom below) have gone so far as to claim that "symbols are not merely used to express mathematical concepts," but are "constitutive of the concepts themselves."[13]

While the coalescence of such sentiments into a self-proclaimed *philosophy of mathematical practice* is a fairly recent development (see next section), there are multiple historical precursors. I wish to conclude this introductory survey of the debate about the applicability problem with an especially illustrative example from Edmund Husserl's *Philosophie der Arithmetik* (1891, transl. *Philosophy of Arithmetic*, 2003). Among other things, Husserl chides nominalists such as Hermann von Helmholtz and Leopold Kronecker for putting the cart before the horse when they take ordinal numbers to be the most natural starting point for our number concept. On this view, the number 'five' would be "nothing other than a sign for the totality of the signs 'first,' 'second,' 'third,' 'fourth,' 'fifth'"[14]—a view Husserl rejects as fanciful. What is telling is Husserl's analysis of what has gone wrong in this hasty identification of the "*number* of objects" with the "entirety of the designations" that went before (as Kronecker puts it). What "these great mathematicians" have misunderstood is the *autonomy* of "the process of symbolic enumeration, which we carry out as a blind routine,"[15] and which precisely allows us to set apart any considerations of content and treat symbol sequences as mnemonic devices—which need to be given a conceptual interpretation only at the final step of the process. We do not 'grasp,' in the final step of an enumeration, the totality of conceptual designations that went before; instead we generate, with certainty, a symbolic output which, thanks to the rule-governed operation we have been following, is assured to have a definite mathematical content. As Husserl puts it later in his treatise, not every derivation "is an *essentially conceptual operation*;" rather, much mathematical activity takes the form of "*essentially sense perceptible*" operations "which, utilizing the system of number signs, deriv[e] sign from sign according to fixed rules."[16] For Husserl, the latter—sense-perceptible—approach has considerable advantages over the 'method of concepts':

---

[12] Mark Colyvan, *An Introduction to the Philosophy of Mathematics* (Cambridge: Cambridge University Press, 2012), 133.

[13] Helen De Cruz and Johan De Smedt, "Mathematical Symbols as Epistemic Actions," *Synthese* 190, no. 1 (2013): 3–19, here 4.

[14] Edmund Husserl, *Philosophy of Arithmetic*, trans. Dallas Willard (Dordrecht: Kluwer, 2003), 186. "Sollte also Fünf wirklich nichts anderes sein als ein Zeichen für den Inbegriff der Zeichen: Erster, Zweiter, Dritter, Vierter, Fünfter?" (Edmund Husserl, *Philosophie der Arithmetik: Psychologische und Logische Untersuchungen*, vol. 1 (Halle: C. E. M. Pfeffer, 1891), 197).

[15] Husserl, *Philosophy of Arithmetic*, 186. "Die Quelle der merkwürdigen Misverständnisse, in welche die beiden berühmten Forscher [...] verfallen sind, liegt nun, [...], in der Misdeutung des symbolischen Zählungsprocesses, den wir blind-gewohnheitsmässig üben." (Husserl, *Philosophie der Arithmetik*, 197).

[16] Husserl, *Philosophy of Arithmetic*, 272. "Entweder es ist diese Herleitung eine im Wesentlichen begriffliche Operation, bei welcher die Bezeichnungen eine nur untergeordnete Rolle spielen; oder sie ist eine im Wesentlichen sinnliche Operation, welche auf Grund des Zahlzeichensystems nach

The method of concepts is highly abstract, limited, and, even with the most extensive practice, laborious. That of signs is concrete, sense perceptible, all-inclusive, and it is, already with a modest degree of practice, convenient to work with. [...] Thus, it makes the conceptual method entirely superfluous, its use being no longer suited to the scientific state of mind, but only to a childish backward one instead.[17]

One need not endorse Husserl's broadside against conceptual approaches in order to agree with his characterization of the immense usefulness of notations, formalisms, and sign systems that allow us to 'externalize' reasoning processes—especially in contexts where, due to the complexity of the issues at hand, we cannot trust our own intuitions about whether or not we have, indeed, grasped the "conceptual substrata"[18] correctly and have managed to track our conceptual operations successfully over time. Indeed, it seems only a small step to extend Husserl's insight concerning "essentially sense-perceptible operations" also to the use of other representational formats and media, such as the three-dimensional material models that were common in late 19th and 20th-century mathematics—especially in instructional settings—and which, through their concrete realizations, facilitated the haptic and visual examination of mathematical concepts and relationships. As we shall see, it is through paying attention to the interplay of models, notational systems, actions, and operations on the part of its practitioners, that much of mathematical activity can be reconstructed as meaningful, in ways that a purely foundationally oriented approach in the philosophy of mathematics would be likely to miss.

Philosophies of Mathematical Practice

It has become customary to distinguish, among the various conceivable approaches to the philosophy of mathematics, between '*mainstream*' and so-called '*maverick*' approaches.[19] The mainstream has its origin in well-known foundational approaches—such as Platonism, intuitionism, and formalism—and subsequently gave rise to analytic approaches, such as the ones associated with logical positivism, which mainly focused on ontological and epistemological issues. In

---

festen Regeln Zeichen aus Zeichen herleitet, um erst das Resultat als die Bezeichnung eines gewissen, des gesuchten, Begriffes zu reclamiren." (Husserl, *Philosophie der Arithmetik*, 291). For a discussion of Husserl's philosophy of mathematics in its historical context, see: Stefania Centrone, *Logic and Philosophy of Mathematics in the Early Husserl* (Dordrecht: Springer 2010).

[17] Husserl, *Philosophy of Arithmetic*, 272. "Die Methode der Begriffe ist höchst abstract, beschränkt und selbst bei grösster Uebung mühsam; die der Zeichen concret-sinnlich, allumfassend und schon bei mässiger Uebung bequem zu handhaben. [...] So macht sie die begriffliche Methode ganz überflüssig, deren Anwendung nicht mehr dem wissenschaftlichen, sondern nur dem kindlich zurückgebliebenen Geisteszustande angepasst ist." (Husserl, *Philosophie der Arithmetik*, 291–92).

[18] Husserl, *Philosophiy of Arithmetic*, 273 ["begrifflichen Anwendungssubstraten" (Husserl, *Philosophie der Arithmetik*, 292).]

[19] See: William Aspray and Philip Kitcher, *History and Philosophy of Modern Mathematics* (Minneapolis: University of Minnesota Press 1988), 17.

opposition to these discussions, which basically align well with core questions of philosophy in general—pertaining to realism, foundationalism, ontology—'maverick' approaches are marked by anti-foundationalism, anti-logicism, and generally a greater attention to mathematical practice.[20] What these maverick commitments entail is a view that considers mathematical knowledge not a matter of being acquainted with mathematical objects or structures that exist independently, but rather as the fallible product of practices—such as conjecturing, proving, refuting, etc.—that cannot be adequately captured by the inferential rules of logic. Imre Lakatos, in the introduction to his *Proofs and Refutations* (1976), gives a vivid characterization of the motivation behind such early practice-oriented ('maverick') approaches:

> The history of mathematics, lacking the guidance of philosophy, has become *blind*, while the philosophy of mathematics, turning its back on the most intriguing phenomena in the history of mathematics, has become *empty*.[21]

However, these early 'maverick' approaches, it has been argued, have "not managed to substantially redirect the course of philosophy of mathematics,"[22] and the balance of philosophical work continued to be in the ontological and epistemological tradition outlined earlier.

At the same time, the growth of interest in the role of mathematical practice over time led to a proliferation of methodological approaches, just as the so-called 'practice turn' did in the philosophy of science more generally. At one end of the spectrum are those who are keen to highlight the continuity of mathematics with all sorts of practical concerns—such as measuring, estimating, and counting, which have historically developed across various occupations, such as agriculture, crafts, and trades. This, for example, is the preferred approach in the historical parts of the *Oxford Handbook of the History of Mathematics* (2008). By almost exclusively looking for commonalities between mathematics and other practical projects, it is perhaps not surprising that such an approach risks losing track of anything specific that might characterize mathematical practice. At the other end of the spectrum, we find approaches such as that of Philip Kitcher, who conceives of a *mathematical practice* as a quintuple $< L, M, S, R, Q >$, consisting of a language L in which statements, S, are derived by reasoning processes R as answers to questions Q, all informed by a metamathematical framework M.[23] Specific practices—such as the production of proofs, the conjecturing of mathematical relationships, the development of new notations in mathematics, etc.—are all lumped together in a single, overarching, paradigm-like meta-mathematical framework M. As Ferreirós

---

[20] Paolo Mancosu, *The Philosophy of Mathematical Practice* (Oxford: Oxford University Press 2008), 5.

[21] Imre Lakatos, *Proofs and Refutations* (Cambridge: Cambridge University Press, 1976), 2.

[22] Mancosu, *The Philosophy of Mathematical Practice*, 5.

[23] Philip Kitcher, *The Nature of Mathematical Knowledge* (Oxford: Oxford University Press 1985), 164.

rightly criticizes, this renders Kitcher's account "rather abstract"—resulting in a notion of "practice without practitioners."[24]

In recent years, a small, but growing number of scholars have begun to develop what one might call a 'meso-level' approach to the philosophy of mathematical practice: that is, a perspective that is not content with merely positing over-arching paradigms (while holding on to a largely abstract view of the content of what mathematics is about), nor with assimilating mathematical practice to other—non-mathematical—practical projects. Instead, this intermediate perspective acknowledges, amongst others, that mathematics is both a tool and offers de facto foundations for many scientific pursuits, while also maintaining that it differs from a mere "science of magnitudes" by its "concentration on conceptual and theoretical issues *independently* of their potential role as models for physical phenomena."[25] In other words, mathematics—and mathematical knowledge—cannot be reduced to formal and symbolic systems, but must be understood, in Ferreirós's apt phrase, as a "network of practices"[26]: more accurately, of practices that are *specific to* mathematics—such as conjecturing, deriving proofs (as well as explicating them), demonstrating, deriving, formalizing, and so forth. Similar to the rapprochement between practice-oriented approaches and foundational methodologies in more recent philosophy of science, philosophy of mathematics, too, stands to flourish "under the combined influence of both general methodology and classical metaphysical questions [...] interacting with detailed case studies."[27] Other sources of inspiration for this recent emergence of an intermediate-level philosophy of mathematical practice include the growing literature in cognitive science, especially in relation to the range of inferential abilities that underpin all mathematical reasoning, and recent work on cultural techniques such as written representations and their "notational iconicity."[28] It is the professed hope, at least of some of those engaged in this enterprise, that such a practice-based approach "ceases to pose the problem of the 'applicability' of mathematics as external to mathematical knowledge itself," instead recognizing it as "internal to its analysis."[29]

Given the richness of mathematics as a discipline, where can one hope to gain a theoretical foothold in order to make sense of mathematical practice? One place to look—not the only one, to be sure, but the one that will guide my inquiry in this paper—is the assemblage of tools (symbol systems such as notations and formalisms as well as other, non-symbolic tools, including models) that mathematicians avail themselves of on a daily basis. Such notational systems, it is argued, mediate—in the broadest sense—mathematical action and

[24] Ferreirós, *Mathematical Knowledge*, 28.

[25] Ibid., 14.

[26] Ibid., 42.

[27] Mancosu, *The Philosophy of Mathematical Practice*, 2.

[28] Sybille Krämer, "Writing, Notational Iconicity, Calculus: On Writing as a Cultural Technique," *Modern Language Notes* 118, no. 3 (2003): 518–37.

[29] Ferreirós, *Mathematical Knowledge*, 43.

activities, thereby shaping *mathematical practice*, both at the individual and collective level. Any approach that aims "to address epistemological issues having to do with fruitfulness, evidence, visualization, diagrammatic reasoning, understanding, explanation, and other aspects of mathematical epistemology which are orthogonal to the problem of access to 'abstract objects,'"[30] will be well-advised to also pay significant attention to the notations and formalisms that constitute the 'formats' and tools through which mathematical work unfolds.

Notations, Formalisms, Models

Anyone who begins to learn mathematics is immediately struck by its pervasive reliance on symbolic notations, starting from the simple '+'-sign designating addition—an operation that can still be expressed reasonably well in natural language—via the symbols for differentiation and integration (f', $\frac{df}{dx}$, $\int f(x)dx$), all the way to diagrammatic notations in, say, knot theory. The same pervasiveness of symbols, it could be argued, can be found in physics and chemistry, yet the fact that mathematical notations have proven useful also across the empirical sciences suggests that, by analyzing the more basic function of mathematical notations, we may be able to learn something fundamental about the role of symbolic notations in our epistemic practices.

At the most basic level, for the purposes of this paper, notations can be understood as symbol systems of sorts. While this is not the place for developing a full definition of what constitutes a symbol system, typically such a system would require that any well-formed arrangement of tokens, e.g. of physical marks on a piece of a paper, can be registered semantically as instances of a character, and that certain other features (such as rules for substituting, rearranging, and interpreting symbols) enable a competent user to manipulate and interpret any such expressions within that system. Yet, notations are more than just syntactically articulate symbol systems that allow for the precise specification of what any well-formed expression is designating: notations also function as *inferential resources*. In making this claim, we are no longer confining ourselves to features immanent to the notational system; instead, the claim is about the function of notations to human users. That is, there is no underlying assumption in what follows that a physical system that manipulates symbols, such as a digital computer, "has the necessary and sufficient means for intelligent action,"[31] nor do we need to take a stand on this (or any other similarly controversial) issue. Mathematical notations, and similar symbolic tools in the physical sciences (such as the operator formalism in quantum mechanics, Feynman diagrams in high-energy physics, structural formulas in chemistry, etc.) are, for present purposes, conceived of as thoroughly human constructs: they are what we teach students when they are being inducted into a

---

[30] Mancosu, *The Philosophy of Mathematical Practice*, 1–2.
[31] Allen Newell and Herbert Simon, "Computer Science as Empirical Inquiry: Symbols and Search," *Communications of the ACM* 19, no. 3 (1976): 113–26, here 116.

discipline, what scientists and mathematicians use on a daily basis in their own work and in communication with their peers, and they are subject to changes and modifications over time.

In science, mathematical notations—often referred to as 'mathematical formalisms' by scientific practitioners—have, of course, long been recognized as eminently useful. What makes mathematical formalisms such a valuable resource for those working in the empirical sciences, is nicely captured by Mary Hesse's remark that

> Mathematical formalisms, when used as hypotheses in the description of physical phenomena, may function like the mechanical model of an earlier stage in physics, without having in themselves any mechanical or other physical interpretations.[32]

Hesse is here referring to the 'mechanical analogies' of nineteenth-century physics, which accounted for physical phenomena "in terms of mechanical models whose behavior is known apart from the experimental facts to be explained." She further notes two desiderata of physical hypotheses that are specified in a certain way: first, that "it must be possible to deduce the data from the hypothesis when the symbols in the latter are suitably interpreted;" second, that "it is necessary that the hypothesis itself should be capable of being thought about, modified and generalized, without necessary reference to the experiments,"[33] so that derivations—of numerical predictions, or more theoretical results—can take place with a certain degree of autonomy from the empirical data. Mathematics, then, is not merely a convenient shorthand that allows for the expression of physical relationships, but it furnishes us with inferential resources that can guide subsequent inquiry. As Hesse puts it, it is in "the nature of the mathematical formalism itself" that

> any particular piece of mathematics has its own ways of suggesting modification and generalisation; it is not an isolated collection of equations having no relation to anything else, but is a recognisable part of the whole structure of abstract mathematics, and this is true whether the symbols employed have any concrete physical interpretation or not.[34]

Mathematical formulations of physical hypotheses do not merely record physical facts, but express them in a way that has an 'internal dynamic' built into them—*in virtue of* being so formulated. This "internal dynamic," as Richard I. G. Hughes notes, "is supplied, at least in part, by the deductive resources of the mathematics," which is "one reason why mathematical models are the norm in physics."[35] Hughes here relies on current scientific usage of the term 'mathematical model' which, for the most part, refers to a set of mathematical equations that are to be

---

[32] Mary Hesse, "Models in Physics," *The British Journal for the Philosophy of Science* 4, no. 15 (1953): 198–214, here 199–200.

[33] Ibid., 200.

[34] Ibid., 200.

[35] Richard I. G. Hughes, "Models and Representation," *Philosophy of Science* 64 (1997): 325–36, here 332.

interpreted as (partial) representations of, and in this sense 'stand in for,' certain aspects of empirical reality. Yet, a similar case could be made for material models in mathematics itself—that is, for models *of mathematical concepts and relationships* (as briefly mentioned above in the first section). Where symbolic derivations are made compelling by the logical force of deductive reasoning, material manipulations of 'hands-on' models derive their power from the vivacity and immediacy of direct sensory experience; both are constrained by the rules and affordances of their respective representational media. Ideally—this, at least, was the goal of many of those trafficking in material models of mathematics—logical derivation (of equations and theorems) and haptic manipulation (of material models) would coincide.[36]

When enriched with rules for the manipulation and physical interpretation of well-formed characters and formulas, symbolic notations that originate in mathematics may constitute what I have elsewhere called "mature mathematical formalisms" viz.

> a system of rules and conventions that deploys (and often adds to) the symbolic language of mathematics; it typically encompasses locally applicable rules for the manipulation of its notation, where these rules are derived from, or other systematically connected to, certain theoretical or methodological commitments.[37]

What makes a mathematical formalism 'mature' in this sense, is partly a matter of how entrenched it is in a given scientific (sub)discipline; partly, it depends on its fruitfulness—again, judged by the standards of the corresponding scientific field—in facilitating problem-solving and generating novel questions that may then guide future inquiry. As Hesse also hints at, not just any "particular piece of mathematics"[38] will automatically qualify as a (mature) mathematical formalism, since the versatility of mathematics as an inferential resource will typically underdetermine the specific rules and conventions to be adhered to—whose function, after all, it is to delimit and guide our scientific inferences. Mature mathematical formalisms in physics thus occupy an interesting middle ground between physical theory and

---

[36] A good example of this is Hermann Wiener, who specialized in the creation of wire models of geometrical shapes and who proclaimed that "[e]very step that we make in the process of computation can also be represented by the geometrical constructions, so that we will deal with an *intuitive method*" English translation cited after Michael Friedman, *A History of Folding in Mathematics: Mathematizing the Margins* (Cham: Birkhäuser, 2018), 178). "Jeder Schritt, den wir dabei rechnerisch machen, lässt sich auch an den geometrischen Gebilden anschaulich verfolgen, so dass wir es also mit einem *anschaulichen Verfahren* zu thun haben." (Hermann Wiener, "Über geometrische Analysen," *Berichte über die Verhandlungen der königlich-sächsischen Gesellschaft der Wissenschaften zu Leipzig, Mathematisch-physische Classe* 24 (Leipzig: S: Hirze 1890), 245–67, here 246.)

[37] Axel Gelfert, "Mathematical Formalisms in Scientific Practice: From Denotation to Model-based Representation," *Studies in History and Philosophy of Science* 42, no. 2 (2011): 272–86, here 272.

[38] Hesse, "Models in Physics," 200.

mathematics-at-large. Typically, the rules and conventions that govern their application—that is, the criteria that determine whether a given output of the formalism (a series of creation and annihilation operators in quantum many-body physics, say) is formally correct by the lights of the locally applicable rules and conventions (and thus can be interpreted as, for example, a series of changes in a many-body system)—ensures that certain theoretical constraints (say, preservation of particle number) are automatically satisfied. This may only work for a relevant subclass of theoretically permissible scenarios, beyond which the formalism would lose its legitimacy, but within its domain of applicability, the formalism may be said to both *enforce* and *constrain* the underlying general physical theory for the cases in question.[39] In this regard, mature mathematical formalisms may be considered as forming the basis for a specific form of "operative writing," which functions both as "a medium for representing a realm of cognitive phenomena" and as "a tool for operating hands-on with these phenomena in order to solve problems or to prove theories"[40] pertaining to the corresponding realm of phenomena.

Mature mathematical formalisms in the sciences have an important role to play in generating scientific representations.[41] This orientation towards representing empirically observable phenomena in the world might seem to detract from the relevance of what has been said so far to the realm of mathematics itself. After all, it is one thing to recognize the enabling and constraining role of mathematical formalisms in generating scientific representations, with which to make inferences about empirical phenomena; it is quite another to argue that similar considerations also extend to the role of notations in mathematics in general. Yet, the two cases may be less dissimilar than it might seem. Just as, in science, we are faced with decisions about which aspects of the world to represent—which is often not a free choice, but is heavily constrained by what is feasible, given our background knowledge—so, in mathematics, there is "theoretical reason for thinking that different notations would capture different truths," since any particular choice of a notation consisting of countably many basic elements entails that "many [mathematical] facts would have to go unrepresented in that notation"[42]—which is, of course, why mathematics as a whole operates with a great diversity of coexisting notations. While mathematics does not need to grapple with the vagaries of an imperfect physical reality, which has been known to harbor many empirical surprises and which resists attempts to fully understand it, this does not mean that mathematical relationships are readily transparent. The resistance we experience when working through mathematical problems is not due to hidden empirical

[39] On this point, see: Axel Gelfert, "Symbol Systems as Collective Representational Resources: Mary Hesse, Nelson Goodman, and the Problem of Scientific Representation," *Social Epistemology Review and Reply Collective* 4, no. 6 (2015): 52–61, here 58.

[40] Krämer, "Writing, Notational Iconicity," 522.

[41] On this point, see: Axel Gelfert, *How to Do Science with Models: A Philosophical Primer* (Cham: Springer 2016), Chapter 5.

[42] Brown, *Philosophy of Mathematics*, 95.

factors, but is a form of *cognitive resistance*—working through conceptual relationships in mathematics, as Husserl puts it in the passage quoted above, is "even with the most extensive practice, laborious" ("selbst bei grösster Uebung mühsam"). A well-chosen notation can overcome, or at least mitigate, this resistance. Again, the notion of "operative writing," or "operative systems of notation"[43] is helpful here. Just as the graphic-visual dimension of (ordinary) writing, with its various ways of structuring recorded speech (e.g., via punctuation), makes abstract relationships—in this case, grammatical structure—"accessible to the perceptual register of the 'aisthetic'" (ibid.), so mature mathematical formalisms, too, may well become entrenched as ways of depicting abstract structures and relations.

Practices, Agents, Actions

A good mathematical notation must be apt in two ways: at the individual level, it must be easily accessible by cognitively limited beings like us, for example by utilizing distinct and visually salient signs, along with straightforward rules for how to manipulate them. At the collective level, notations should not only be easily communicable, so as to facilitate the exchange of mathematical results and approaches, but should—ideally—aid the advancement of mathematics as a discipline. Both aspects are frequently acknowledged by those writing on the philosophical significance of mathematical notation, as in Husserl's emphasis on "sense-perceptible" signs that are "convenient to work with" (and which help to make "the conceptual method entirely superfluous," at least locally) and in Alfred North Whitehead's remark that "[b]y relieving the brain of all unnecessary work, a good notation sets it free to concentrate on more advanced problems," enabling us to "make transitions in reasoning almost mechanically by the eye, which otherwise would call into play the higher faculties of the brain."[44] Mathematical notation would hardly compel much interest, were it not for the fact that, as one mathematical practitioner puts it,

> notation can have a decisive influence on the development of mathematics: Differences of notation hindered or promoted mathematical advances, and seemingly small changes in notation or concepts led to new and profound mathematical results.[45]

Similar considerations also apply to diagrams, even where these may not meet the formal requirements for notations—which is why it has been argued that, more generally, "the introduction and development of systems of representation that

---

[43] Krämer, "Writing, Notational Iconicity," 522.

[44] Alfred North Whitehead, *An Introduction to Mathematics* (London: Williams & Northgate 1911), 58–61.

[45] Klaus Truemper, *The Construction of Mathematics: The Human Mind's Greatest Achievement* (Plano: Leibniz Company 2017), 38–39.

are indispensable for the practice, such as, symbols, notations and diagrams" is a "crucial target for the study of the practice of mathematics."[46]

Important though it is to recognize that mathematical practice merits at least as much philosophical attention as, say, the question of applicability, the notion of 'a practice'—in the singular—is of limited use when analyzing specific instances or historical episodes of mathematics. For one, mathematical practice—just like scientific practice—is really a plurality of interconnected practices (of conjecturing, demonstrating, and so forth). While such recognition of the plurality of practices goes some way toward correcting the traditional—overly abstract—view of mathematics as a single, unitary intellectual endeavor, in an important sense it is still too global, since it continues to maintain the illusion of a 'practice without practitioners.' Conjecturing, demonstrating, proving a theorem (and re-proving it, perhaps more elegantly!), refuting a conjecture: these are not impersonal, 'disembodied' steps in the advancement of mathematics as a discipline; instead, they are moves that are pursued and attempted—sometimes successfully, sometimes not—by individual agents (or, increasingly, groups of agents).

The need for an *agential* perspective in the philosophy of mathematics is perhaps most evident in discussions of mathematics education, where *learning* mathematics—as well as instructing others in various mathematical techniques— is, sometimes casually, equated with '*doing* mathematics.' Traditional philosophy of mathematics, with its emphasis on foundational issues and cutting-edge mathematical research, has been largely silent on the murky territory of mathematics instruction. Yet, it only takes a moment's reflection to realize that, without successful induction of every new cohort into the mathematical techniques of its time, mathematics as a discipline would disintegrate. This is not to say that mathematics is constituted by a fixed set of established techniques; rather—like any other scientific discipline and collective human endeavor—it needs to be underwritten by a set of 'live' practices. The basic idea is familiar from Kuhnian accounts of science: On the one hand, we have the global notion of a scientific 'paradigm' (or 'disciplinary matrix') that supposedly shapes a given discipline; on the other hand, the continued existence and reproduction of anything resembling a unified discipline would remain an utter mystery, were it not for the recurrence of specific instances of shared techniques and applications—Kuhn's *exemplars*—which correspondingly are used both in scientific training and as heuristic tools for tackling new problems. In mathematics, such exemplars will crucially "have to do with the manipulations and rules that an agent will apply while confronting a certain configuration of [symbols]."[47] Mathematicians themselves speak of "common tricks that you might see in proofs across a variety of [branches of mathematics],"[48]

[46] Silvia De Toffoli and Valeria Giardino, "Envisioning Transformations—The Practice of Topology," in *Mathematical Cultures*, ed. Brendan Larvor (Basel: Birkhäuser 2016), 25–50, here 30.

[47] Ferreirós, *Mathematical Knowledge*, 50.

[48] Laura Alcock, *How to Study as a Mathematics Major* (Oxford: Oxford University Press 2013), 103.

they exhort their students "to look for what is linked to what"[49] and, if one's initial search for a problem solution is unsuccessful, "to find consolation with some easier success, [and to] *try to solve first some related problems.*"[50] It is easy to see in these (and similar) pronouncements an "explicit recognition of the importance of exemplars in the practice of mathematics"[51] and, by extension, also a recognition of the centrality of agents who *master* a practice via acquiring the requisite know-how on the basis of working with exemplars.

Yet stable practices do not just come down to mere regularities in whatever it is that agents do in a certain context. For something to be recognizable as a distinct practice, we need standards of what is a correct or valid move within that practice. And for that, we need to look at the level of *individual actions.* This is why I wish to argue that we must move not only from a wholesale notion of 'practice' (in the singular) to a multiplicity of interconnected practices (in the plural), and from practices to those who engage with it—that is, human agents—but even further down, as it were, to the analysis of individual actions. The guiding assumption, of course, is that actions themselves display certain similarities, such that different agents can be said to perform, if not the same action, then at least the same *type of action.* Identifying distinctive patterns of action, thus, is a necessary complement to an agent-centered philosophy of mathematical practice.

Epistemic Actions and Their Limits

Before turning in the next section to distinctively *mathematical actions*—that is, actions which can reasonably be considered as being constitutive of, or at least as contributing to, mathematical practice—the present section will lay the necessary groundwork by distinguishing between two broader classes of actions. Following the work of David Kirsh and Paul Maglio (1994), I shall distinguish between *pragmatic actions*, which are performed in order to "bring one physically closer to a goal,"[52] and *epistemic actions* that "are performed to uncover information that is hidden or hard to compute mentally."[53] Arguably, the philosophy of action has traditionally focused on pragmatic actions which, broadly speaking, adhere to the standards of instrumental rationality, whereas epistemic actions would previously have been assimilated to the (more or less pre-theoretical) domain of heuristic behavior.

---

[49] Ibid., 139.

[50] George Polya, *How to Solve It: A New Aspect of Mathematical Method* (Princeton: Princeton University Press 1957), 114.

[51] Ferreirós, *Mathematical Knowledge*, 57.

[52] David Kirsh and Paul Maglio, "On Distinguishing Epistemic from Pragmatic Action," *Cognitive Science* 18, no. 4 (1994): 513–49, here 513.

[53] Ibid.

Writing from the perspective of a computational approach to problem solving, Kirsh and Maglio give the following qualitative characterization of what epistemic actions aim for:

> Epistemic actions – physical actions that make mental computation easier, faster, or more reliable – are *external* actions that an agent performs to change his or her own computational state.[54]

Why, one might ask, is there a need for the concept of 'epistemic action,' and for distinguishing it from 'pragmatic action,' which is itself hardly a simplistic concept and may well accommodate multiple functions and interpretations? Why introduce a new, distinct category in the first place? The novel point of the concept of 'epistemic action,' according to Kirsh and Maglio, is its ability to resolve a puzzle that would result if we had to evaluate actions solely in virtue of their contribution to a proximate *physical* goal:

> At times, an agent ignores a physically advantageous action and chooses instead an action that seems physically disadvantageous.[55]

This appears to be at odds with the traditional focus—common, for example, in early theories of AI—on how an intelligent agent 'chooses physically useful actions.' Only by allowing for epistemic actions—that is, actions performed in order to change the agent's own cognitive state—can one recognize such apparent physically disadvantageous moves as also being goal-oriented, not towards a proximate physical outcome, but to *cognitive goals*:

> When viewed from a perspective which includes epistemic goals – for instance, simplifying mental computation – such actions once again appear to be a cost-effective allocation of the agent's time and effort.[56]

It is important to be clear about what the introduction of the novel concept of 'epistemic action' is, and is not, meant to challenge. What is not at issue is the general idea that actions are goal-oriented at some structural level; rather, the existence of epistemic actions undermines the "assumption that the point of action is always pragmatic,"[57] along with what one might call a *linear model* of the relationship between cognition and action, according to which agents first identify goals, then plan accordingly, and finally carry out actions designed to achieve those goals. What epistemic actions demonstrate is that, on occasion, action can also "be undertaken *in order to alter the way that cognition proceeds*."[58]

As an application and illustration of what they have in mind, Kirsh and Maglio use the well-known computer game *Tetris*. The reasoning behind their choice of

---

[54] Ibid., 514.
[55] Ibid.
[56] Ibid.
[57] Ibid., 526.
[58] Ibid.

example is as follows: Tetris "is a fast, repetitive game requiring split-second decisions of a perceptual and cognitive sort," as variously shaped blocks move into view and need to be placed in such a way as to minimize any gaps between them. When a new block (or 'zoid,' as Kirsh and Maglio call it) appears, the 'clock starts ticking;' one could wait for the shape to become fully visible, but this might then not leave enough room for maneuvering the block to its optimal position. Rotating the block early can disclose additional information about its shape, by rotating into view features (e.g. 'hooks') that have not yet appeared as the block moves downward. More importantly, every action in the game inevitably contributes in measurable (positive or negative) ways to the desired final outcome, by bringing a given block closer to (or removing it further from) its final position in the playing field (also known as the 'matrix'). This means that one can clearly distinguish between moves that contribute to the pragmatic goal—to minimize gaps and eliminate completed rows of blocks—and those that fail to do so:[59]

> Thus, if epistemic actions are found in the time-limited context of Tetris, they are likely to be found almost everywhere.[60]

A further advantage of Kirsh and Maglio's choice of example, solely for the purposes of experimental study design, is the fact that it is easy to recruit participants for this game, which many people are already familiar with—even if they do not know some of the regularities (e.g. that certain types of blocks always emerge from the same columns of the matrix). As a result, they were able to carry out empirical-psychological experiments identifying the various actions that Tetris players (of different proficiencies) routinely engage in.

In particular, Kirsh and Maglio identify a number of specific moves that, they argue, are best interpreted as epistemic actions. For example, their data show that players are more likely to rotate blocks which are not yet fully visible and which are ambiguous in both shape and position, as compared with blocks that are merely ambiguous in shape. (Blocks of different shapes move gradually into view from the top of the matrix, so until a block has fully emerged—and depending on whether identifying markers of the block are still hidden from view—there remains some uncertainty as to which specific shape one will be encountering next; see Fig. 1.) Yet, early rotation can only effectively help with disambiguating shape, whereas the desired physical outcome—creating gap-free rows of blocks—depends predominantly on moving the (correctly oriented) block in its final position, once its shape is known. This suggests that, from an outcome-oriented perspective, the two cases should not be treated significantly differently, yet subjects do invest time and effort into (superfluous) early rotations. Thus, Kirsh and Maglio argue, "[e]arly rotation is a clear example of an epistemic action."[61] They are, however, aware that

---

[59] Ibid., 516.
[60] Ibid., 516.
[61] Ibid., 529.

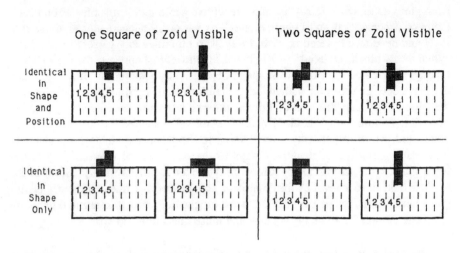

Fig. 1 'Zoids' moving into view; at the top, the visible portions are identical both in position and in shape; at the bottom, they are identical in shape alone. Republished with permission of Annual Reviews Inc. from: David Kirsh and Paul Maglio, "On Distinguishing Epistemic from Pragmatic Action," *Cognitive Science* 18, no. 4 (1994): 528, Fig. 8; permission conveyed through Copyright Clearance Center Inc., all rights reserved

one might try arguing against this view by suggesting that there is pragmatic value in orienting the zoid early, and so its epistemic function is not decisive. Such an explanation, however, fails to explain why partial displays that are ambiguous in shape and position are rotated more often than those that are not ambiguous in shape and position.[62]

The reasoning here is that non-ambiguous shapes need to be rotated no less often (and ambiguous ones no more often) in order to place them properly—that is, in order to achieve a satisfactory realization of the explicit goal, as defined by the rules of the game. If ambiguous shapes are, in fact, rotated more often, then this additional expenditure of time and effort, so the argument goes, is best explained by the status of such rotations as *epistemic actions* that aim at, for example, preparing individual decision-paths for the application of some learned strategy. This is not to say that such actions are somehow irrational, but rather that their performance cannot be solely justified as contributing directly to the determinate goal as explicitly defined within the game. Rather, such actions promote distinctly *epistemic goals*, regardless of their immediate contribution to the outcome demanded by the game.

It would be easy to misunderstand the term 'epistemic action' as referring to any action that advances, or otherwise affects, our epistemic state—which, of course, would be most actions. Turning my upper body in order to scratch my heel inevitably leads me to turn my head, which entails that I may perceive objects in

---

[62] Ibid., 529.

my environment—and gain perceptual knowledge about them—which I would otherwise have missed, had I continued to look straight ahead. Yet turning one's body to scratch one's heel is not an epistemic action. Neither are 'focusing attention' or 'directing one's gaze to a source of new information.' Including these and similar *sensor actions* under the umbrella term 'epistemic action,' simply because they result in new information, would rob the term of its specific meaning—which, according to Kirsh and Maglio, pertains to the issue of "control of activity."[63] Epistemic actions are ordinary actions that come at a cost, when judged in terms of their contribution to a (pre-defined) physical goal. Contrary to the view that actions are predominantly performed after intelligent planning—by searching for the most advantageous position among all the available physical states, and then attempting to reach that physical outcome—proponents of the idea of 'epistemic action' argue that informational states are an integral part of the state-space of options that a cognizer must consider. This point might seem trivial. Are there not many techniques—such a symbolic representation—that allow cognizers to bridge the gap between informational states and physical representations (which may then be manipulated through ordinary actions)? While there are indeed, Kirsh and Maglio insist that assimilating informational actions to the realm of symbolic representation risks obscuring the "less appreciated" fact "external actions can be [valuable] for simplifying the mental computation that takes place in tasks which are not clearly symbolic."[64]

It is this intended restriction of the term 'epistemic action' to non-symbolic actions (or at least to types of action that are not *clearly* symbolic) that poses a problem for any attempt to identify symbolic operations in mathematics with epistemic actions. Yet this is precisely how at least some recent philosophers of mathematics have viewed epistemic actions, and the use of symbolic notations, in mathematics. Thus, De Cruz and De Smedt (2013), argue:

> Mathematical symbols are epistemic actions, because they enable us to represent concepts that are literally unthinkable with our bare [unassisted] brains.[65]

One natural way to understand this statement is to read it as elliptical: while symbols themselves cannot be actions, *manipulations* of such symbols may well be. The claim then becomes one about the uses to which mathematical symbols can be put—through individual acts of manipulating token instances, designating mathematical entities with new symbols, enforcing certain notational rules, etc.—in order to enable inferences that would otherwise exceed our cognitive capacity.

Yet, the identification of symbolic manipulations with epistemic actions in mathematics remains problematic. Consider one of De Cruz and De Smedt's core examples: the introduction, by Leonhard Euler, of the imaginary unit i, which rendered calculations using complex numbers tractable—especially when combined

---

[63] Ibid., 515.
[64] Ibid., 514.
[65] De Cruz and De Smedt, "Mathematical Symbols," 3.

with Jean Robert Argand's innovation of the *complex plane*, in which complex numbers can be 'located,' and thereby uniquely represented, according to their real and imaginary parts:

> Once [*i* had been] introduced, mathematicians no longer needed to worry about square roots of negative numbers, because the symbolism effectively masks this cognitively intractable operation [...]. Denoting a cognitively intractable operation with a symbol makes it more manipulable, which effectively enables mathematicians to overcome human cognitive limitations.[66]

While this general characterization is eminently plausible, and also coheres with our earlier discussion (see section "Notations, formalisms, models" above), its designating the introduction of *i* as an epistemic action is problematic, at least if one applies the criteria spelled out by Kirsh and Maglio. For one, it is not clear which underlying *physical* action poses a puzzle. Recall that the concept of epistemic action was intended to explain why certain effortful physical actions that seem to be incompatible with goal-oriented behavior could nonetheless be vindicated as serving cognitive ends. Two options naturally present themselves. First, one could try to reinterpret the example as one that, while falling short of the status of 'epistemic action,' nonetheless exhibits similar surface features. For example, introducing *i* as a placeholder for $\sqrt{-1}$, thereby enabling us to perform calculations that we could not otherwise simultaneously survey in our mind, may be seen as just another form of *cognitive offloading*. This interpretation is supported by De Cruz and De Smedt's endorsement of Whitehead's remark that "a good notation sets [the brain] free to concentrate on more advanced problems, and in effect increases [its] mental power." While cognitive offloading is *also* one of the functions of epistemic actions explicitly mentioned by Kirsh and Maglio,[67] the two are not identical, and so, on this first interpretation, the introduction of the imaginary unit is an example of the former, yet not of the latter.

However, there is a second option—which may be implicitly intended by De Cruz and De Smedt—according to which we need to widen the scope of the term 'epistemic action' so as to pick out actions whose primary value to the cognizer consists in enabling cognitive processes that would not otherwise be available. Perhaps, then, Kirsh and Maglio's focus on the relationship between planning and intelligent behavior—though initially well-motivated for the purposes of critiquing certain types of AI approaches to the so-called "planning problem"[68]—is simply too limiting and stands in need of widening. After all, once it is acknowledged that this epistemic activity needs to be properly accounted for, AI practitioners may decide to expand the relevant state space of possible configurations to also include informational states along with physical ones;[69] yet, even then it would

---

[66] De Cruz and De Smedt, "Mathematical Symbols", 9.

[67] Kirsh and Maglio, "On Distinguishing," 514.

[68] For a survey, see: Robert Inder, "Planning and Problem Solving," in *Artificial Intelligence*, ed. Margaret A. Boden (San Diego: Academic Press 1996), 23–53.

[69] On this point see: Kirsh and Maglio, "On Distinguishing," 546.

be useful to distinguish those actions that *primarily* aim at the advancement of a cognizer's epistemic state from those that predominantly serve pragmatic goals. On this more 'liberal' view, the term 'epistemic action' would simply refer to the first kind of actions, yet with no restriction on how these actions can manifest themselves (e.g., symbolic or non-symbolic) and without insisting that a 'hard' trade-off be involved between the epistemic and pragmatic contributions of a specific action. Such an approach would also need to acknowledge the contribution of other factors, e.g. specific qualities of the requisite representational media: the introduction of the imaginary unit i as a *mere* placeholder for $\sqrt{-1}$ might not have been nearly as compelling, had it not been quickly supplemented by Argand's more easily visualizable representation in the complex plane.[70]

While I have broad sympathies for being more 'liberal' than Kirsh and Maglio about the extension of the term 'epistemic action,' their position is arguably both well motivated and possesses the virtue of specificity. Hence, in the remainder of this paper, I will be guided by the idea that, when it comes to giving a reconstruction of mathematical practice, it is worth adhering to Kirsh and Maglio's (restrictive) notion of 'epistemic action'—up to a point, that is. This, I believe, will shed light on what one might call different *scales of action*—some of which pertain to individual actions, others to broader disciplinary activities. As we shall see, there is no reason to restrict the term epistemic action to non-symbolic activities; however, the symbolic dimension of notations and formalisms is best introduced only once we have gained a better understanding of the richness—and the range—of relevant mathematical activities.

What 'Epistemic Actions' in Mathematics Might Be

Let us grant, for the purpose of argument, that mathematical notation and the introduction of new symbols are not epistemic actions in the narrow sense intended by Kirsh and Maglio. That is, they do not promote cognitive goals at the overt expense of proximate pragmatic goals. Or at least, they do not do so in any more than the unspecific sense, in which any pursuit of 'pure inquiry' may be said to be cognitively costly, inasmuch as it makes demands on our cognitive and attentional resources without offering any immediately identifiable pragmatic 'pay-off' in return. Yet, the cost of deploying a symbolic notation tends to be low and, since the symbolic realm offers less resistance to our manipulations than the physical world around us, it does not typically detract from our proximate physical goals. This diagnosis should not come as a surprise. For, the existence—and even the 'bringing into existence,' i.e. the coining—of a new term does not, in and of itself,

---

[70] See, for example: David Burton, *The History of Mathematics: An Introduction* (New York: McGrawHill, 2007), chapter 11.

do any epistemic work. If anything, it is specific acts of deploying a term or symbol that need to be assessed in terms of their contribution to our cognitive and/or pragmatic goals.

Does this mean that the notion of 'epistemic action' has no place in the analysis of mathematical practice? As already suggested in the previous section, the answer is no. Yet, in order to appreciate the contribution of epistemic actions, narrowly understood, to mathematical practice, one needs to adopt a wider perspective. Rather than offering a systematic account of the conditions under which epistemic actions can be expected to facilitate mathematical thinking (or to advance the discipline of mathematics in other ways), I shall proceed by way of example. In particular, I shall highlight four areas where one might fruitfully look for epistemic actions in the course of studying mathematical practice. The four examples in question are the use of gestures and other anticipatory bodily movements (in particular in instructional settings), the application of material models to mathematics, the initially puzzling disciplinary practice of re-proving theorems (which, after all, should be epistemically redundant), and—finally—the creation of collectively shared symbolic and notational resources. As I shall argue, it is the latter example—the emergence of mature notational formalisms—which leads to a partial vindication of symbol use as a type of epistemic action, provided we make a slight (though well-motivated) modification of the very notion of epistemic action in question.

The Use of Gestures and Symbolic Operations in Instructional Settings

Since the publication, in 1986, of Eric Livingston's book *The Ethnomethodological Foundations of Mathematics*, it has been a commonplace in the science studies of mathematical practice that "the heart of mathematical work is at and around the blackboard."[71] Sometimes, this is put the other way around, for example by Michael Barany and Donald MacKenzie (2014) when they emphasize that mathematicians work with—and prefer working with—*chalk*.

Coming out of such ethnographic studies are close-up observations of how mathematical instruction occurs in concrete settings. As an example, consider this excerpt from Christian Greiffenhagen's and Wes Sharrock's ethnovideographic study of the use of "Gestures in the blackboard work of mathematics instruction" (2005), which records how an instructor works through a particular derivation with his students and, in the course of doing so, introduces a specific mathematical notation (see Fig. 2):

---

[71] Christian Greiffenhagen and Wes Sharrock, "Gestures in the Blackboard Work of Mathematics Instruction," in *Interacting Bodies. Proceedings of 2nd Conference of the International Society for Gesture Studies* (June 2005), 1–24, here 1, online: http://gesture-lyon2005.ens-lyon.fr/article.php3?id_article=234 (accessed November 23, 2021).

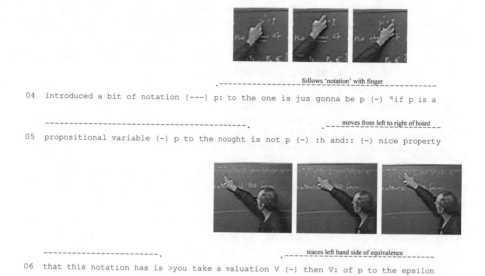

follows 'notation' with finger

04 introduced a bit of notation (---) p: to the one is jus gonna be p (-) °if p is a

moves from left to right of board

05 propositional variable (-) p to the nought is not p (-) :h and:: (-) nice property

traces left hand side of equivalence

06 that this notation has is >you take a valuation V (-) then V: of p to the epsilon

Fig. 2 Excerpt from the transcript of the instructor's classroom demonstration. From: Christian Greiffenhagen and Wes Sharrock, "Gestures in the blackboard work of mathematics instruction," in *Interacting Bodies. Proceedings of 2nd Conference of the International Society for Gesture Studies* (June 2005), 1–24, here 12. © Used with kind permission by Christian Greiffenhagen; all rights reserved

Going over this new notation prepares the way for commenting *upon* this new notation, i.e., commenting on a 'nice property that this notation has' (lines 5-6). This is spelled out by tracing the equivalence relation that has been written in the top right corner of the board (lines 6-8). Mathematically, the 'nice property' consists in the possibility of moving from the typically used *symbols* for propositions (namely 'p' or 'not p') to *numbers* (1 or 0) representing these two possibilities. In effect, it is putting in place a powerful mechanism to 'talk' about propositions of the formal language.[72]

This passage is significant for three reasons. First, it illustrates the importance of bodily actions—such as hand gestures, orienting one's body and directing one's gaze—to the effective instruction of learners, even in the context of a seemingly abstract operation such as introducing a new mathematical notation. Physical actions and verbal assertions nicely complement each other, as indicated by the instructor's verbal testimony, which highlights certain commendable features ('nice properties') of the notation in question. Second, the transcribed sequence documents actions that can properly be called 'epistemic actions' (in Kirsh and Maglio's sense), such as the extensive giving of 'meta-commentary' on the part of the instructor—including, among other things, a number of evaluative statements and endorsements which, at least on one level, detract from the proximate goal of proving the theorem being derived. Engaging in 'meta-talk,' after

---

[72] Ibid., 4.

all, delays the finalization of the proof—the proximate goal, even by the lights of the instructional aim of giving a sample demonstration—and so can only be understood as well-motivated if it serves other, specifically *cognitive* goals. Third, there is a clear *anticipatory* element in the way that gestures, verbalization, and the sequence of steps in the derivation link up:

> We can note the *anticipatory character* of the gestures, of the use of hands, and of the lecturer positioning himself with respect to the board. [...] these are coordinated with the oral exposition, often in such a way that the positioning, the gesture and the identification of the relevant point in the written out proof are designed to coincide.[73]

This echoes the case of early rotation of zoids in Kirsh and Maglio's *Tetris* example: some actions are being carried out simply in order to highlight *relevance*—again emphasizing the primarily *epistemic* character of such actions.

Of course, not all actions an instructor performs while demonstrating a proof on the blackboard, or while carrying out similar instructional work, are epistemic actions in the narrow sense. Apart from obviously irrelevant actions (e.g., interrupting the writing process in order to scratch an itch), there are also actions a presenter may carry out—including, again, gestures such as pointing—in order to direct his or her own attention, or that of the audience. Such *sensor actions*, for the reasons discussed earlier, would not be epistemic actions of the kind distinguished by Kirsh and Maglio. Other actions, though technically superfluous and even delaying the completion of the proximate task—such as completing a mathematical proof—aim at *organizing* or *structuring* a proof. Some may serve "to locate the license provided by an earlier entry for a current move"[74]; that is, they contribute directly to, or make explicit, the justificatory relationships that underwrite a given sequence. Depending on which among the "network of practices"[75] we focus on—the instructional goal of conveying mathematical relationships to the audience, or the goal of establishing the truth of a theorem—such actions may contribute only indirectly to the explicitly stated practical goal. The 'remainder,' as it were,—that is, the extent to which these actions constitute detours from the explicit goal—may well be best interpreted as due to *epistemic goals*.

Applying Material Models to Mathematics

The use of three-dimensional models in mathematics—which were variously made of plaster, paper, wire, and other materials—was by no means an oddity in late 19th-century mathematics. Indeed, as a number of historical studies in recent years have shown, demand for such models was high enough to sustain a significant level of commercial activity. Companies would offer series of dozens of models, mainly

---

[73] Ibid., 4.
[74] Ibid., 9.
[75] Ferreirós, *Mathematical Knowledge*, 42.

for use in instructional settings, and there was vigorous trade and export (e.g., from Germany to the United Kingdom). Felix Klein, in a lecture held at a mathematical congress that was part of the Chicago Columbian Exposition in 1893, describes the rationale for the use of models in mathematics instruction as follows: "Collections of mathematical models and courses in drawing are calculated to disarm, in part at least, the hostility directed against the excessive abstractness of the university education."[76] Others, such as Hermann Wiener,[77] emphasized that "the capacity to *form concepts* through abstraction" is trained "far more via visual aids [*Anschauungsmittel*] of all sorts" (and especially via models, because of their 'immediacy') than via calculating and even geometrical drawings.[78] Such pronouncements might make it seem that, just as in the earlier case of gestures and symbolic derivations using chalk on a blackboard, the use of material models served primarily pedagogical purposes. If the goal had been exclusively that of instilling mathematical knowledge in one's audience, rather than creating *new* knowledge, then it might be possible to sidestep the question of whether the actual deployment of models in mathematics had any genuinely epistemic features. Models in mathematics would simply be effective teaching tools, and the decision to use them would be an instrumentally rational choice to make—there would hardly be any reason to consider the effect that such models had on advancing the instructor's epistemic state.

Arguably, however, the function of material models in 19th-/early-twentieth century mathematics went well beyond that of serving as an effective teaching tool. Besides functioning as pedagogical tools, material models served as *heuristic devices* and *interpretative tools* for practicing mathematicians. Consider the development of non-Euclidean geometry in the second half of the nineteenth century. At a time when most mathematicians took the truth of Euclidean geometry to be self-evident and exclusive—even though indirect challenges and doubts had surfaced as early as in the *eighteenth* century (ironically, in the course of attempted *vindications* of Euclid)—the Italian mathematician Eugenio Beltrami (1835–1900) began to study mathematical objects that, while being represented as curved surfaces in three-dimensional Euclidean space, exhibited internal relations that could be considered as realizations of non-Euclidean geometries (of the type proposed by Nikolai Lobachevsky and János Bolyai). In particular, such structures included surfaces of constant negative curvature, which could be thought of as being generated

---

[76] Felix Klein, *The Evanston Colloquium Lectures on Mathematics* (London: Macmillan and Co. 1894), 99.

[77] See: footnote 36.

[78] Hermann Wiener, "Über den Wert der Anschauungsmittel für die mathematische Ausbildung," *Jahresbericht der Deutschen Mathematiker-Vereinigung* 22, no. 1–2 (1913), 294–96, here 296: "Die Möglichkeit, durch Abstraktion *Begriffe zu bilden*, hängt von der Fähigkeit ab, den Stoff mit sich herumzutragen, und diese wird im Unterricht bis zu einem gewissen Grade beim Lösen von Aufgaben durch Zeichnung und Rechnen erworben, aber weit mehr durch Anschauungsmittel aller Art, und wegen der vorhin geschilderten Unmittelbarkeit der Eindrücke sind die Modelle dazu besonders geeignet." (English translation by A.G.).

Fig. 3 One of Eugenio
Beltrami's models of the
pseudosphere, preserved at
the Department of
Mathematics of the
University of Pavia. © Used
with permission from the
Department of Mathematics
of the University of Pavia,
Fulvio Bisi and Maurizio
Cornalbaa; all rights reserved

by rotating a suitable curve around an axis in three-dimensional space. Material
models of such surfaces—or, more precisely, material models that *approximated*
portions of such surfaces—could be made from plaster or, in Beltrami's case, even
from paper (see Fig. 3 for one of the paper models of these surfaces, which Beltrami made in 1869). On the one hand, it is clear that any *actual* encounter, or
action, involving material models is merely an empirical occurrence that does not,
in and of itself, lend weight to any particular mathematical conclusion. On the
other hand, as props for the theoretical imagination, Beltrami's models opened
up ways of conceiving of concrete components (e.g. line segments marked on a
plaster model) as representing abstract elements that, in turn, realized a system of
geometrical principles.[79] The very co-existence of such models eased the way for
acknowledging that principles of geometry allowed for multiple 'interpretations'—
while, at the same time, demanding that a proper theoretical appreciation of the
new geometries would require transcending, in thought, the limitations associated
with the models' concreteness and materiality. As Moritz Epple puts it, Beltrami's
models constituted "one of the early epistemic objects of mathematics *beyond*
the sphere of ordinary spatial intuition that would become so characteristic of
mathematical modernity."[80] By extension, the concrete uses and manipulations of
material models in mathematics typically do not serve proximal, pragmatic goals,
but are instead best thought of as epistemic actions—even when, as in Beltrami's
case, they aim to challenge the accepted bounds of knowledge, as it were.

---

[79] On this point, see the contribution of Moritz Epple in this volume: "'Analogies,' 'Interpretations,' 'Images,' 'Systems,' and 'Models': Some Remarks on the History of Abstract Representation in the Sciences Since the Nineteenth Century."

[80] Ibid., 294.

Re-proving Theorems

One potential objection to the analysis of what one might call individual 'micro-level' actions—among others, anticipatory uses of gestures, pausing to highlight aspects of the derivation that do not themselves directly contribute to the outcome, and so forth—is that these are common to all contexts of instruction and, thus, are not specific to mathematics. They would, in other words, fail the specificity test for any prospective 'meso-level' philosophy of mathematical practice, where such an approach would not content itself with merely noting commonalities between mathematics and other practical endeavors, but would aim to pinpoint genuinely *mathematical* aspects of the various interlinked practices that give rise to mathematical knowledge.

As a more compelling case which is not limited to instructional contexts and which does appear to be specific to the practice of mathematics, consider the puzzling practice of *re-proving of mathematical theorems*. From an outcome-oriented perspective, which highlights the "discovery and proof of *new theorems* [as] the *sine qua non* of mathematical endeavor,"[81] re-proving appears to be redundant since, by definition, it does not contribute new theorems to the corpus of mathematical knowledge. After all, we do not stand to gain new mathematical knowledge from proving an already known theorem again. This raises legitimate questions:

> What reasons are there for re-proving known results? And how do mathematicians judge whether a proof is conceptually distinct from those that have been given before?[82]

A practice-oriented philosophy of mathematics—one that is less fixated on the final outcomes and abstract claims of mathematics—is better able to make sense of this phenomenon. For, as John Dawson notes, new proofs are often "presented as such, because they are deemed to be particularly *novel, ingenious*, or *elegant*, or (like the free climber who disdains mechanical aids) because they attain the goal through restricted means."[83] In other words, novel proofs are valued not because they generate new results, but because they promote values such as elegance or ingenuity. Sometimes a new proof may lead to the desired outcome more efficiently—by shortening the derivation—but even then it is technically redundant. Sometimes, a novel proof may not even shorten the derivation, but may lengthen it—just like the free climber who may take longer to reach the mountain top than his better-equipped competitors—yet may succeed in achieving the same, already known outcome under conditions of self-imposed constraints. Its value, then, consists not in contributing novel mathematical conclusions, but in furthering epistemic goals such as methodological knowledge or knowledge of how things 'hang together.' In much the same way as Beltrami's material models were suggestive of new

---

[81] John Dawson, "Why Do Mathematicians Re-prove Theorems?," *Philosophia Mathematica* 3, no. 14 (2006): 269–86, here 269.
[82] Ibid., 269.
[83] Ibid., 279.

interpretations of geometrical principles, novel proofs can also point to new intra-mathematical relationships and properties and suggest new lines of inquiry. Far from being redundant, then, the act of re-proving a theorem may be seen to be an *epistemic action* that serves (perhaps tacit) epistemic goals.

Notations as 'Institutionalized' (Long-Term) Epistemic Actions?

At the beginning of this section, I promised that a proper consideration of the emergence of collectively shared notational resources may yet vindicate the thought that certain deployments of notations should count as epistemic actions. In order for this to be possible, the initial notion of 'epistemic action' needs to be adjusted, but only moderately so. This avoids losing all specificity of the term—as would inevitably be the case if one were to call any action (symbolic or non-symbolic) with epistemic consequences an 'epistemic action.' A full discussion would need to look at the specific affordances of a range of specific formalisms and notations as well as the specific ways in which they enhance, or extend, human cognitive abilities.

In the context of the present paper, I can only refer to a select few examples and hint at some general reasons why it may be legitimate to consider the introduction and deployment of notations a type of epistemic action. An important motivation for entertaining this possibility is the oft-noted role of notations in cognitive offloading. Whitehead, in a passage already quoted earlier, gives vivid expression to this sentiment:

> By relieving the brain of all unnecessary work, a good notation sets it free to concentrate on more advanced problems, and in effect increases [our] mental power [...]. In mathematics, granted that we are giving any serious attention to mathematical ideas, the symbolism is invariably an immense simplification [...] [B]y the aid of symbolism, we can make transitions in reasoning almost mechanically by the eye, which otherwise would call into play the higher faculties of the brain.[84]

At a less global level, similar claims have been advanced for specific mathematical notations, formalisms, and diagrammatic methods.

One branch of mathematics that has received some attention in this regard is knot theory. Thus, Brown devotes half a chapter of his *Philosophy of Mathematics* (1999) to knot theory and the various properties and relationships that different notations—or, 'different forms of representation' (since, for Brown, mathematics seeks ways of representing an independently existing realm of mathematical entities and relations)—highlight. He is keen to point out—unsurprisingly, given his philosophical commitments—that different notations will typically capture different truths; that is, mathematical reality exceeds what any one notational system may be able to represent. Notations 'create' as much as 'reveal' the boundaries of

---

[84] Whitehead, *An Introduction*, 58–61.

what can be known about the properties of mathematical objects.[85] Taking their lead from Brown's discussion, Silvia De Toffoli and Valeria Giardino (2014) have analyzed the cognitive features and epistemic roles of diagrams in knot theory in greater detail. In particular, they propose an "*operational* account for knot diagrams, based on: (i) the *moves* allowed on them and (ii) the *space* they define."[86] Diagrammatic notations, such as those used in knot theory, are "effective in promoting inference" because they enable the cultivation of "a form of manipulative imagination" that is developed and enhanced through continued use of shared symbolic resources.[87] A properly cultivated manipulative imagination, according to De Toffoli and Giardino, prompts experts "to re-draw diagrams and calculate with them, performing *epistemic* actions."[88] It would be easy for a casual reader to misread this as a definition of what *makes* something an epistemic action in the narrow sense; thanks to our earlier discussion, we know better. Yet, De Toffoli and Giardino are entirely right to highlight the close affinity that exists between (certain types of) epistemic actions and routinized inferential strategies. What remains to be shown is how this affinity arises and how it can guide mathematical practice.

Extending our cognitive reach and facilitating inferences about the world is arguably one of the key motivations for introducing a novel formalism or notation. It takes only a moment's reflection, however, to realize that this desired effect cannot be a matter of a single act of deploying a particular notation. Instead, it requires a long-term commitment—both at the individual level and, if a formalism is to become a source of scientific progress, at the collective level—to deploy the same types of operations in order to denote the same types of inferential moves. The emergence—and often-deliberate invention—of notational formalisms aims at the creation of common symbolic and epistemic resources, which anyone who has been trained in their usage can then 'tap into.' As we have seen, the term 'epistemic action' has occasionally been co-opted—notably by De Cruz and De Smedt, as well as by De Toffoli and Giardino—in order to describe just such (epistemically oriented) activities of notational innovation. Yet, at the very least, what is required for such an interpretation to be defensible, is a *modified* type of epistemic action, namely *long-term epistemic action*, in contradistinction to *short-term epistemic action*.[89]

Whereas short-term epistemic actions "tune interactions between agent(s) and settings in such a way that a strong mutual fit and fluent problem-solving for single and concrete tasks in specifiable local conditions"[90]—as in specific instances of rotating a 'zoid' in a game of Tetris—*long-term epistemic actions* allow an agent

---

[85] See: Brown, *Philosophy of Mathematics*, 94–95.

[86] Silvia De Toffoli and Valeria Giardino, "Forms and Roles of Diagrams in Knot Theory," *Erkenntnis* 79, no. 4 (2014): 829–42, here 830.

[87] Ibid., 839.

[88] Ibid., 830.

[89] I am drawing here on work by Mark-Oliver Casper, "Long Term Epistemic Actions," *Avant* 8, no. 1 (2017): 119–130.

[90] Casper, "Long Term Epistemic Actions," 122.

"to plan, design and put [subsequent] short-term epistemic actions to use;"[91] this may also require robust commitments to certain ways of prioritizing problems—which, in turn, determine the way in which an agent would respond if things were different. All of this rings true of mathematical notations and formalisms, which often serve the purpose of preparing problem situations for systematic analysis. Once a problem has been 'formalized'—that is, has been rendered representable within the symbolic resources afforded by the notation—what we can do *with it* is heavily constrained: only certain symbolic manipulations will be deemed admissible by the lights of the formalism at hand. Formalization requires cognitive effort—and, as in the case of mathematics, often needs to be aided by physical actions (and, sometimes, material models) in order to 'work things out.' But, in the case of fruitful formalisms and notations, this effort—as well as the constraints that come with the choice of a given formalism—will be more than compensated for by the ease with which they allow us to take further steps of inquiry and, for example, derive meaningful and relevant results. We forego some leeway in the description of our problem situation, but in turn gain greater inferential depth—and, one hopes, greater insight into the problem as a whole. Whatever cognitive cost may be incurred in the form of first having to invent, and then having to master, a new notation or formalism, may simply be the price we have to pay for greater long-term pay-offs. It is this aspect of symbolic notations in mathematics—the fact that formalization does not always make *specific* derivation easier, because it requires considerable mastery, yet nonetheless promises greater epistemic returns 'further down'—rather than the mere fact of cognitive offloading that renders their development and mastery a case of epistemic actions, at least in the long haul.

---

[91] Ibid., 126.

Matrices—Compensating the Loss of *Anschauung*

Gabriele Gramelsberger

Introduction

In the nineteenth century, critical comments against the eighteenth century Kantian turn to *Anschauung* in mathematical science emerged.[1] Immanuel Kant's concept of intuition (*Anschauung*) guaranteed mathematics' ontological relevance when referring to the world. This was particularly the case as the basis for the experimental science *ex datis* of Galileo Galilei and Isaac Newton. Therefore, Kant interwove mathematics with the spatiotemporal ability of men to recognize objects.[2] However, such a concept of mathematics, as the captious critics would say, was naively based on the visuality of Euclidean geometry and its constructive nature of creating mathematical objects with tools such as compasses or inspired by counting with fingers, as Gottlob Frege sneered against Kant.[3] Most prominently Carl Friedrich Gauss argued in a letter to Friedrich Wilhelm Bessel on January 27th 1829 that the loss of *Anschauung* in his non-Euclidean geometry would cause hue and cry

---

[1] To look at something = *anschauen* in German. *Anschauen* can refer to visual objects but also to mental images and ideas.

[2] Immanuel Kant, *Critique of Pure Reason*, trans. Paul Guyer and Allen W. Wood (Cambridge: Cambridge University Press, 1998 [1781/1787]).

[3] Gottlob Frege, *Die Grundlagen der Arithmetik. Eine logisch mathematische Untersuchung über den Begriff der Zahl*, ed. Christian Thiel (Hamburg: Meiner, 1988 [1884]).

G. Gramelsberger (✉)
Theory of Science and Technology (TST), RWTH Aachen University, Theaterplatz 14,
D-52062 Aachen, Germany
e-mail: gramelsberger@humtec.rwth-aachen.de

© The Author(s) 2022
M. Friedman and K. Krauthausen (eds.), *Model and Mathematics:
From the 19th to the 21st Century*, Trends in the History of Science,
https://doi.org/10.1007/978-3-030-97833-4_12

of the "Boeoter," the Kantians of his time.[4] Gauss, afraid of being bashed by his philosopher colleagues for rethinking the "fifth postulate" of Euclidean geometry (parallel postulate), withheld the publication of his new geometry on which he had been working since the 1790s.[5] His fear was that the reputation of rethinking the fifth postulate might fall into common rank with the work on the quadrature of the circle and the perpetual motion machine, respectively.

This episode illustrates how contested the Kantian turn and how fierce the dispute on *Anschauung* was during the nineteenth century. It also reveals the struggle between Platonists and early constructivists in the Kantian notion later leading to the *Grundlagenkrise* of mathematics in the early twentieth century. A struggle that was transferred into physics in a different manner by Werner Heisenberg and Erwin Schrödinger fighting over *Anschauung* in early quantum theory.[6] Finally, the dispute on *Anschauung* and its loss, respectively, caused mathematician and phenomenologist Edmund Husserl to state that European science was in crisis in a lecture in 1935 on the eve of the collapse of European civilization.[7]

This spectrum of debates on *Anschauung* opens up the epistemological ground for asking about the power of mathematics to envision worlds beyond human *Anschauung*. A power that can lead to victory as well as crisis, but generally speaking to new physical sciences. However, based on the loss of *Anschauung* in geometry, models came into play in order to replace *Anschauung* (intuition) by *Anschaulichkeit* (palpable visuality).[8]

Immanuel Kant's Philosophy of Applied Mathematics

The concept of *Anschauung* in Kant's *Critique of Pure Reason* (*Kritik der reinen Vernunft*) is neither related to visuality nor to images.[9] *Anschauung* is the foundational ability of human beings to have sensory impressions. Sensory impressions, and this is the Kantian turn in epistemology, are not received passively but result actively from the ability of *Anschauung* to process sensory data and thus enable human beings to recognize objects. However, and this is the decisive point, the

---

[4] Carl Friedrich Gauß, "Brief an Bessel vom 27. Januar 1829," in *Die Theorie der Parallellinien von Euklid bis Gauß. Urkunden zur Geschichte der Nichteuklidischen Geometrie*, ed. Friedrich Engel and Paul Stäckel (Leipzig: Teubner, 1885 [1829]), 226.

[5] Gauß, "Brief an Bessel."

[6] Werner Heisenberg, "Über den anschaulichen Inhalt der quantentheoretischen Kinematik und Mechanik," *Zeitschrift für Physik* 43 (1927): 172–98; Henk de Regt, "Erwin Schrödinger, Anschaulichkeit, and Quantum Theory," *Studies in the History and Philosophy of Modern Physics* 28, no. 4 (1997): 461–81.

[7] Edmund Husserl, *The Crisis of European Science and Transcendental Philosophy*, trans. David Carr (Evaston, IL: Northwestern University Press, 1970 [1935]).

[8] Felix Klein, "Vergleichende Betrachtungen über neuere geometrische Forschungen," *Mathematische Annalen* 43 (1893 [1872]): 63–100, 94.

[9] Kant, *Critique*.

procedure of processing sensory data is not arbitrary but ordered: either simultaneously (space) or successively (time).[10] This formal procedure of ordering is called 'synthesis' in Kant's work.

Anschauung, however, is only able to process sensory data, not to interpret them conceptually. The latter is the task of the mind—a separate ability in Kant's concept. It is this separation of *Anschauung* and mind that provides the basis for criticizing pure reason for Kant. Pure reason is the empty—empty of sensory data—ability of producing ideas, sometimes illusive, and abstract concepts in the mind, as Kant argued. Only together, mind and *Anschauung* produce our experience of the world. It is *Anschauung* that gives ideas and concepts content, although it is arrived at only quantitatively and gradually that there is something (sensory data). *Anschauung*, in other words, indicates the existence of things but does not denote their meaning.

Although Kant introduced these two separate abilities of *Anschauung* and mind, he interconnected both by the procedure of transcendental schematism.[11] Transcendental schematism realizes the operation of interconnecting *Anschauung* and mind in time by synthesis; thus, creating synthetic judgments a priori. The various modes of synthesis define the categories of the mind like quantity, quality, relation, and modality, ordering and integrating sensory impressions in terms of various modes of temporality.[12] Thus, the term 'schema' does not refer to a visual schema or image. It is not true, as Philip Kitcher later accused Kant, that Kant had mental images or "cartoons" in mind in order to apprehend synthetic judgments a priori.[13] But what Kant had in mind were rule-based procedures for ordering sensory data simultaneously (space) and successively (time) and subordinating them to the temporal modes of the category. It was Edmund Husserl who advanced Kant's theory of temporal modes.[14]

However, this ordering of sensory data was conceived homogenously by Kant.[15] Homogenous and uniform ordering of time and space is a characteristic of Newtonian mechanics, which keeps the latter in the realm of uniform motion and linearity as well as in the realm of Euclidean geometry and its regular forms: lines, circles, spheres, cubes, triangles, etc. In the *Critique of Pure Reason*, Kant never argued,

---

[10] Kant, *Critique*, B 33.

[11] Kant, *Critique*, B 177.

[12] Kant, *Critique*, B 106.

[13] "It is hard to understand how a process of looking at mental cartoons [by Kant] give us knowledge, unless it were knowledge of a rather unexciting sort, concerned only with the particular figure before us. Hilbert and the intuitionists, who follow Kant in claiming that the fundamental mode of mathematical knowledge consists in apprehension of the properties of mentally presented entities, fail to explain how mathematics is anything more than a collection of trivial truths, concerned only with the properties of those mental entities which mathematicians chance to have discerned or those mental constructions which they happen to have effected" (Philip Kitcher, *The Nature of Mathematical Knowledge* (New York, Oxford: Oxford University Press, 1984), 50).

[14] Edmund Husserl, *Zur Phänomenologie des inneren Zeitbewusstseins*, vol. 10 of *Husserliana*, ed. Rudolf Boehm (Den Haag: Martinus Nijhoff, 1969 [1905]).

[15] Kant, *Critique*, B 751.

in any way, that spatial data must be three-dimensional. Although neither favoring three-dimensionality manifested in the constructible objects of Euclidean geometry nor limiting space to three dimensions, Kant's homogenous form of synthesis enabled him to succeed in providing a concise epistemological ground for the positivistic science of his time. In particular, he was able to provide a concise epistemological ground for Newtonian physics.[16]

By rooting mathematics in his formal concept of *Anschauung*—mathematical judgments are synthetic judgments a priori for Kant—he was overcoming the dichotomy of analytic (logical) and synthetic judgments (empirical ones). Furthermore, this root in *Anschauung* ensured the ontological relevance of mathematics when referring to the world resulting from the indexical ability of synthetic judgments a priori. In retrospect, Kant's *Critique of Pure Reason* can be called the first textbook for the philosophy of applied mathematics.

Loosing the root in *Anschauung* exiled mathematics into the realm of pure reason and of solely analytic judgments; it cut the ties to the world and encapsulated mathematics in a purely formal endeavor of dealing with symbols—only obeying the criteria of being not contradictory. However, self-consistency is the logical precondition for the existence of mathematical objects but does not imply the existence beyond mathematics. Without *Anschauung* mathematical objects were not linked to the world anymore. The emergence of pure mathematics during the nineteenth century was the result of the loss of the Kantian link (synthetic judgments a priori) between synthetic (empirical) and analytic (logical) judgments, or in other words: of the loss of *Anschauung*.

The Loss of *Anschauung* in the Nineteenth Century and the Declaration of *Anschaulichkeit* as a Model in Geometry

The loss of *Anschaulichkeit* in nineteenth century geometry is not necessarily linked to a loss of *Anschauung* in the formal sense presented by Kant. *Anschaulichkeit* and *Anschauung* are usually mixed up in the mathematical literature of the nineteenth century by wrongly assigning visuality to Kant's concept of *Anschauung*—as it is the case for *Anschaulichkeit*.[17] However, the loss of *Anschaulichkeit* in nineteenth century geometry resulted from an increasing process of abstracting and systemizing geometry which led to the development of mathematical group theory. By questioning Euclid's 'fifth postulate' (parallel postulate) new objects were introduced into geometry. Already in 1766, Johann Heinrich Lambert contested that the sum of the angles of a triangle has to be

---

[16] See, for instance: Michael Friedman, *Kant and the Exact Sciences* (Cambridge: Harvard University Press, 1992); Michael Friedman and Alfred Nordmann, eds., *The Kantian Legacy in Nineteenth-Century Science* (Cambridge: The MIT Press, 2006).

[17] See, for instance: Frege, *Die Grundlagen*, §5.

always 180° as in Euclidean geometry.[18] By introducing triangles with less than 180° (imaginary geometry) and more than 180° (spherical geometry) Lambert invented non-Euclidean geometry. Mathematicians like Adrien-Marie Legendre, Jean-Baptiste D'Alembert, Gauss and others followed. It was Gauss in 1829 who coined the term "Nicht-Euclidische Geometrie" in his letter to Bessel,[19] although Nikolai Lobatschewski und János Bolyai had already axiomatized these early versions of non-Euclidean (hyperbolic) geometry in the 1820s.[20]

The loss of *Anschaulichkeit* in nineteenth century mathematics resulted from various developments; questioning Euclid's parallel postulate was only one aspect.[21] Questioning three-dimensionality as well as geometrical properties of figures in the emerging projective geometry added an even more abstract view to geometry. Furthermore, replacing the synthetic-constructive method of geometry by the analytic-algebraic method—a paradigmatic change in mathematical media—opened up the door for describing *unanschauliche* geometries. '*Unanschaulich*' thereby refers to geometries which are not constructible with compasses anymore.[22] It was René Descartes in 1637 who developed analytic geometry replacing geometric constructions by algebraic descriptions.[23] However, Descartes rejected terms like a^4 and thus stuck to the translatability of Euclidean geometry into algebra and vice versa—a limitation which was rearticulated as the 'dictum of geometric construction' by Jean-Victor Poncelet in the beginning of the nineteenth century;[24] a dictum which dominates descriptive geometry for engineers and architects until today but was given up by projective geometers in the mid of the nineteenth century.

The development of projective geometry into abstract group theory has been retrospectively described by Felix Klein,[25] who successfully used group theory to reunite the scattered field of geometry.[26] It has been reconstructed by the historian

---

[18] Johann H. Lambert, "Theorie der Parallellinien," *Leipziger Magazin für reine und angewandte Mathematik* 1 (1786 [1766]): 137–64, 325–58.

[19] Gauß, "Brief an Bessel."

[20] Friedrich Engel and Paul Stäckel, *Die Theorie der Parallellinien von Euklid bis Gauß: Urkunden zur Geschichte der Nichteuklidischen Geometrie* (Leipzig: Teubner, 1895), 239ff.

[21] Klaus T. Volkert, *Die Krise der Anschauung: Eine Studie zu formalen und heuristischen Verfahren in der Mathematik seit 1850* (Göttingen: Vandenhoeck & Ruprecht, 1986).

[22] Sybille Krämer, *Symbolische Maschinen: Die Idee der Formalisierung in geschichtlichem Abriß* (Darmstadt: Wissenschaftliche Buchgesellschaft, 1988).

[23] René Descartes, "Géométrie," in René Descartes, *Discours de la méthode* (Leyden: Jan Mair, 1637), 295–413.

[24] Jean-Victor Poncelet, *Traité des propriétés projectives des figures* (Paris: Gauthier-Villars, 1822).

[25] Felix Klein, "Über die sogenannte Nicht-Euklidische Geometrie (erster Aufsatz)," in Felix Klein, *Gesammelte mathematische Abhandlungen*, vol. 1, ed. Robert Fricke und Alexander M. Ostrowski (Berlin: Springer, 1921 [1872]), 254–305; Felix Klein, "Über die sogenannte Nicht-Euklidische Geometrie (zweiter Aufsatz)," in Felix Klein, *Gesammelte mathematische Abhandlungen*, vol. 1, ed. Robert Fricke und Alexander M. Ostrowski (Berlin: Springer, 1921 [1873]), 311–43.

[26] Explicitly, via Klein's 'Erlanger Program' from 1872 (see: Klein, "Vergleichende Betrachtungen").

Hans Wussing.[27] From a group theoretic perspective, geometric figures become manifolds and 'constructing' turns into operating algebraically with symbol transformations on these manifolds. In group theory, geometric research is focused on attributes which remain invariant under transformations of space (e.g. geometric transformations such as translation, mirroring, rotation). Groups were defined "by means of the laws of combination of its symbols [...] in dealing purely with the theory of groups, no more concrete mode of representation should be used than is absolutely necessary."[28] Euclidean geometry, in this respect, became one of many groups, called the 'main group.' The main group describes the geometric attributes which remain invariant: dimension, orthogonality, parallelism, and indices. However, for the most general group, the projective group, only indices remain invariant under transformation.

The price for reuniting the scattered field of geometry by introducing abstract group theory was not only the irreversible loss of *Anschaulichkeit* but also of *Anschauung* for geometry. The translatability of geometry into algebra and vice versa had to be given up. Thus, as Klein argued, what was left over was *Anschaulichkeit* as a "model" for non-Euclidean geometry; for instance, the model of a constructible, plane model of elliptic non-Euclidean geometry.[29] Therefore, Klein differentiated two forms of geometry: an abstract form of geometry (projective group) for which *Anschaulichkeit* demonstrates some aspects by using visual models for pedagogical reasons[30]; and "eigentliche" geometry as the mathematical science of space and thus spatial *Anschauung*.[31] The increasing amount of engineering students in the second half of the nineteenth century, as Anja Sattelmacher has outlined in her study on material models in mathematics, led to a collection

---

[27] Hans Wussing, *Die Genesis des abstrakten Gruppenbegriffes: Ein Beitrag zur Entstehung der abstrakten Gruppentheorie* (Berlin: VEB Deutscher Verlag der Wissenschaften, 1969).

[28] Arthur Cayley quoted in William Burnside, *Theory of Groups of Finite Order* (Cambridge: Cambridge University Press, 1897), vi.

[29] Klein, "Vergleichende Betrachtungen," 94.

[30] "Die Anschauung hat [... hier] nur den Werth der Veranschaulichung" (ibid.). English translation by Mellen Woodman Haskell: Felix Klein, "A Comparative Review of Recent Researches in Geometry," *Bulletin New York Mathematical Society* 2 (1892—1893): 215–249, here 244: "Space-perception has then only the value of illustration, [...]."

[31] "Es gibt eine eigentliche Geometrie, die nicht, [...], nur eine veranschaulichte Form abstracterer Untersuchungen sein will. In ihr gilt es, die räumlichen Figuren nach ihrer vollen gestaltlichen Wirklichkeit aufzufassen und (was die mathematische Seite ist) die für sie geltenden Beziehungen als evidente Folgen der Grundsätze räumlicher Anschauung zu verstehen. Ein Modell—mag es nun ausgeführt und angeschaut oder nur lebhaft vorgestellt sein—ist für diese Geometrie nicht ein Mittel zum Zwecke, sondern die Sache selbst" (Klein, "Vergleichende Betrachtungen," 94). English Translation: Klein, "Comparative Review," 244: "There is a true geometry which is not, like the investigations discussed in the text, intended to be merely an illustrative form of more abstract investigations. Its problem is to grasp the full reality of the figures of space, and to interpret— and this is the mathematical side of the question—the relations holding for them as evident results of the axioms of space-perception. A model, whether constructed and observed or only vividly imagined, is for this geometry not a means to an end, but the subject itself."

of models for teaching engineers such as wire models of functions.[32] In his volume on *Anschauliche Geometrie*, David Hilbert provided several photographs of material models, for instance of a wire model of a hyperboloid and a wood model of ellipsoids.[33] However, as Hilbert outlined in his introduction, his *Anschauliche Geometrie* is more a popular science book than an academic, which would provide "Freude" (joy).[34] Luis Couturat in 1905 described this development in mathematics as a paradigmatic shift from *Anschauung* as the old medium of mathematical discovery to the new medium of logic and deduction in formal symbol systems.[35]

Matrices as New Tools for Compensating the Loss of *Anschauung* in Physics

The new medium of mathematical discovery solely belonged to the realm of analytic judgments, dismissing Kant's synthetic judgments a priori. However, dismissing Kant's synthetic judgments a priori released mathematics, and in particular geometry, from its ties to reality.[36] Thus, for physics, the "Unreasonable Effectiveness of Mathematics in the Natural Sciences" became an unsolvable miracle.[37] As Eugene Wigner stated in his essay: "The miracle of the appropriateness of the language of mathematics for the formulation of the laws of physics is a wonderful gift which we neither understand nor deserve."[38] However, it was not a miracle but the consequence of exiling mathematics into the realm of pure reason and, thus, of cutting its ties to the world.

What this cut would mean for physics can be learned from another fierce debate about *Anschauung*; this time for the emerging quantum theory in the 1920s

---

[32] Anja Sattelmacher, "Geordnete Verhältnisse. Mathematische Anschauungsmodelle im frühen 20. Jahrhundert," *Berichte zur Wissenschaftsgeschichte* 36 (2013): 294–312; Herbert Mehrtens, "Mathematical Models," in *Models: The Third Dimension of Science*, ed. Soraya de Chadarevian and Nick Hopwood (Stanford, CA: Stanford University Press, 2004), 277–306.

[33] David Hilbert and Stefan Cohn-Vossen, *Anschauliche Geometrie* (Berlin: Springer, 1932), 14, 17.

[34] Ibid., IV.

[35] Luis Couturat, *Les Principles des Mathématiques, avec un appendice sur la philosophie des Mathématiques de Kant* (Paris: Alcan, 1905).

[36] Ernst Cassirer as well as Alfred North Whitehead and Bertrand Russell tried to rehabilitate Kant's synthetic judgments a priori (Ernst Cassirer, "Kant und die moderne Mathematik. Mit Bezug auf Bertrand Russells und Luis Couturats Werke über die Prinzipien der Mathematik," *Kant-Studien* 12 (1907): 1–40; Alfred North Whitehead and Bertrand Russell, *Principia Mathematica*, 2 vols. (Cambridge: Cambridge University Press, 1910)).

[37] Eugene Wigner, "The Unreasonable Effectiveness of Mathematics in the Natural Sciences," in *Symmetries and Reflections: Scientific Essays of Eugene P. Wigner*, ed. Eugene Wigner (Bloomington: Indiana University Press, 1967 [1960]), 222–37; Albert Einstein, *Geometrie und Erfahrung* (Berlin: Springer, 1921).

[38] Wigner, "The Unreasonable Effectiveness," 237; See: Axel Gelfert, "Applicability, Indispensability, and Underdetermination: Puzzling Over Wigner's 'Unreasonable Effectiveness of Mathematics,'" *Science & Education* 23, no. 5 (2013): 997–1009.

Fig. 1 Square matrix.
Graphics: © Gabriele
Gramelsberger, all rights
reserved

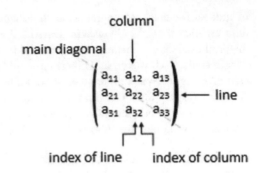

between Werner Heisenberg and Erwin Schrödinger.[39] Following Albert Einstein's dictum that every entity of a physical theory had to be observable, i.e. measurable,[40] Heisenberg pleaded for giving up the traditional idea of trajectories of electrons because such trajectories were not measurable. This un-observability resulted from the elimination of the relation between position and momentum (Heisenberg's uncertainty principle). Furthermore, instead of a continuous movement from one position to the other electrons jump between energy plateaus (quantization). However, such an uncertain and discontinuous behavior could not be conceptualized by using the old vocabulary of physics: differential equations, because movement was no longer conceivable as a function of time like in traditional physics. Thus, in Heisenberg's opinion, quantum physics had to develop a new tool in order to conceptualize this strange quantum behavior properly. His achievement was to describe the movement of electrons as the square schema of transitions resulting from the rule of multiplication of quantum mechanical quantities.[41]

Heisenberg's approach to conceptualize the strange quantum behavior and, thus, to compensate the loss of *Anschauung* consisted of introducing matrices as new tools for physics. However, it was not Heisenberg but his collaborators Max Born and Pascual Jordan who introduced matrices for conceptualizing the discontinuous behavior of electrons.[42] They identified Heisenberg's rule of multiplication of quantum mechanical quantities as the well-known rule of multiplication of matrices shown along the main diagonal of the matrix (Fig. 1).

---

[39] Mara Beller, "The Rhetoric of Antirealism and the Copenhagen Spirit," *Philosophy of Science* 63 (1996): 183–204; de Regt, "Erwin Schrödinger."

[40] Albert Einstein, "Zur Elektrodynamik bewegter Körper," *Annalen der Physik* 17 (1905): 891–921.

[41] Werner Heisenberg, "Über quantentheoretische Umdeutung kinematischer und mechanischer Beziehungen," in *Zur Begründung der Matrizenmechanik*, ed. Max Born, Werner Heisenberg, and Pascual Jordan (Stuttgart: Battenberg, 1962 [1925]), 31–45.

[42] Max Born and Pascual Jordan, "Zur Quantenmechanik," in *Zur Begründung der Matrizen-mechanik*, ed. Max Born, Werner Heisenberg, and Pascual Jordan (Stuttgart: Battenberg, 1962 [1925]), 46–76.

However, matrices have already been invented in the mid of the nineteenth century by Arthur Cayley. In his "A Memoir on the Theory of Matrices," Cayley defined a matrix as a set of quantities arranged in the form of square arrays.[43] He realized "that the square arrays themselves actually had algebraic properties."[44] In particular, square matrices combine properties of tables with diagrams. As tables they provide a definite order of quantities in lines and columns, as diagrams they have algebraic properties for calculation. Although "Hamilton must be regarded as the originator of the theory of matrices, as he was the first to show that the symbol of a linear transformation might be made the subject-matter of a calculus,"[45] it was Cayley who had developed the characteristic form of matrices and who had articulated the rules for using matrices as algebraic tools. Related to Euclidean space, square matrices describe the geometric transformations of the main group like translation, rotation, and mirroring.

Based on these tools, Heisenberg, Born, and Jordan were able to articulate the states of electrons organized in lines (p position) and columns (q momentum). Along the main diagonal one can follow the probability of phase transitions of the electrons. Based on this new matrix analysis, "the laws of the new mechanics have been completely articulated. Every other law of quantum mechanics must be derived from these basic laws."[46] As outlined above, it was Heisenberg's aim to introduce only such elements to his matrix analysis which were observable, i.e. measurable. p and q are conjugated operators which cannot be measured at the same time. However, comparing the measurements of both states is possible but only statistically as it is displayed in the diagonal of the matrix. It is this unique experimental circumstance, called the uncertainty principle, which led to the development of new tools as Dirac highlighted in accordance with Heisenberg:

> Heisenberg puts forward a new theory, which suggests that it is not the equations of classical mechanics that are in any way at fault, but that the mathematical operations by which physical results are deduced from them require modification.[47]

---

[43] Arthur Cayley, "A Memoir in the Theory of Matrices," in Arthur Cayley, *The Collected Mathematical Papers*, vol. 2 (Cambridge: Cambridge University Press, 1889 [1855]), 475–96, here 475; see also: Arthur Cayley, "Recherches sur les matrices dont les termes sont des fonctions linèaires d'une seule indéterminée," in *The Collected Mathematical Papers*, vol. 2, ed. Arthur Cayley, (Cambridge: Cambridge University Press, 1889 [1855]), 217–20.

[44] Cayley quoted in Henry Taber, "On the Theory of Matrices," *American Journal of Mathematics* 12, no. 4 (1890): 337–96, here 337.

[45] Taber, "On the Theory of Matrices," 337.

[46] Born and Jordan, "Zur Quantenmechanik," 58: Damit "sind die Grundgesetze der neuen Mechanik vollständig gegeben. Alle sonstigen Gesetze der Quantenmechanik, denen Allgemeingültigkeit zugesprochen werden soll, müssen aus ihnen heraus zu beweisen sein." (English translation by G.G.).

[47] Paul Dirac, "The Fundamental Equations of Quantum Mechanics," *Proceedings of the Royal Society* 109 (1925): 642–53, here 642.

Early Twentieth Century Debate on *Anschauung* and *Anschaulichkeit* in Physics

Opposed to Heisenberg's matrix mechanics, Schrödinger stated that it would be much more helpful to use the traditional idea of oscillation modes merging into each other and, thus, conceptualizing phase transitions instead of jumping electrons.[48] Schrödinger's version of wave mechanics draws on the old vocabulary of physics: differential equations. His reason was the desire for *Anschaulichkeit*, as Henk W. de Regt has pointed out aptly:

> The association of visualisability with understanding rather than with realism may be elucidated by considering the German word *Anschaulichkeit*, which is the term Schrödinger used in his writings. This word does not only mean 'visualisability' but also 'intelligibility.'[49]

However, the debate between Heisenberg and Schrödinger can be reconstructed as a mismatching debate on *Anschauung* (Heisenberg) and *Anschaulichkeit* (Schrödinger). While Heisenberg put the emphasis on the loss of spatiotemporal *Anschauung* in the Kantian notion, thus asking for new concepts of *Anschaulichkeit*, Schrödinger ignored the loss of *Anschauung* in favor of the traditional spatiotemporal concept of (Newtonian) *Anschaulichkeit*. Of course, Schrödinger was aware that his vote for a familiar style of *Anschaulichkeit* did not describe quantum reality but was a helpful 'image' or 'model.'[50] Nevertheless, he rejected Heisenberg's approach because of its ugly *Unanschaulichkeit*. Schrödinger introduced his 1935 essay on the current situation of quantum mechanics[51] with a section on entitled "The physics of models." In this section, he argued for the imagination of models which are more accurate than empirical knowledge can ever be. Mathematics and geometry are the languages for articulating such models. However, as Schrödinger pointed out, while geometry is static these models must include temporal behavior. Thus, if enough elements and relations between these elements are known, then these models can be used for predicting future states of temporal behavior. If these predictions can be validated empirically by observation, the underlying hypotheses were valid. If not, it leads to progress of knowledge by adapting the underlying hypotheses and models. "This is the way how slowly a better adaptation of the image, i.e. of our thoughts, to the facts can be achieved."[52]

---

[48] Erwin Schrödinger, "Quantisierung als Eigenwertproblem (I)," *Annalen der Physik* 79 (1926): 361–76.

[49] de Regt, "Erwin Schrödinger," 461.

[50] Erwin Schrödinger, "Die gegenwärtige Situation in der Quantenmechanik," *Die Naturwissenschaften* 23, no. 48 (1935): 807–12.

[51] Schrödinger, "Die gegenwärtige Situation."

[52] Schrödinger, "Die gegenwärtige Situation," 52: "Denn im Grunde ist das die Art, wie allmählich eine immer bessere Anpassung des Bildes, das heißt unserer Gedanken, an die Tatsachen gelingen kann." (English translation by G.G.).

This 'model style of physics' was new and was later called the 'hypothetic-deductive research style' by philosophers of science.[53] Different than the inductive-deductive research style of Newtonian physics, hypothetic-deductive thinking was firstly introduced by Einstein's unique theory of relativity in 1905.[54] As the term 'hypothetic-deductive' reveals, scientific thinking starts with a hypothesis that has to be evaluated afterwards. In case of the predictions of Einstein's general theory of relativity,

> three 'crucial tests' are usually cited as experimental verifications [...]: the red shift of spectral lines emitted by atoms in a region of strong gravitational potential, the deflection of light rays that pass close to the sun, and the precession of the perihelion of the orbit of the planet Mars.[55]

From a philosophical point of view, the transformation from the inductive-deductive into the hypothetic-deductive research style is the pivotal point where theories have to turn into models and were explanations have to turn into predictions. From now on, a 'good science' is a model-based predictive one which derives computable models of its theories. In turn, a model is only scientifically valuable if it can compute some predictions which can be evaluated empirically afterwards. However, models usually predict spatiotemporal behavior based on partial differential equations. For instance, Einstein's general theory of relativity has to be re-articulated as a mathematical model involving ten coupled partial differential equations in order to compute any predictions. However, such a complex model is neither exactly computable nor does it give exact predictions; it can only be numerically simulated, i.e. approximated.[56]

Surreality of the New Physics

The loss of *Anschauung* also caused major disputes in philosophy at this time. The new physics was seen as a problem by Edmund Husserl but embraced by Gaston Bachelard. Husserl, in his lecture on "The Crisis of European Sciences and Transcendental Phenomenology," argued that the loss of *Anschauung* also means the loss of the subject for science and this, in turn, is accompanied by losing reference to everyday lifeworld.[57] Science, and first of all physics, has turned into an abstract endeavor unrelated to humankind and its concerns. For Husserl, rationality

---

[53] Karl Popper, *The Logic of Scientific Discovery* (London: Routledge, 1959 [1935]).
[54] Einstein, "Zur Elektrodynamik."
[55] Leonard I. Schiff, "On Experimental Tests of the General Theory of Relativity," *American Journal of Physics* 28, no. 4 (1960): 340–43, here 340.
[56] Gabriele Gramelsberger, *From Science to Computational Sciences: Studies in the History of Computing and its Influence on Today's Society* (Chicago: The University of Chicago Press, 2015).
[57] Husserl, *The Crisis*.

has failed because the link to humans as rational beings has been given up. Science has turned into a remote endeavor ignoring most aspects of rationality such as values and aesthetics by highlighting only instrumental and logical rationality.

Bachelard, however, celebrated the new physics and exactly this highlighting of instrumental and logical rationality in his book *Le nouvel esprit scientifique*.[58] Einstein's and Heisenberg's new physics is turning the naïve imaginations of everyday experiences—an epistemological obstacle in Bachelard's opinion[59]—into mathematical activities constructing the world anew and enabling new (observational) experiences. Mathematics' operational lucidity is replacing the Cartesian model of lucidity. For this paradigmatic shift in scientific thinking Bachelard gives the example of conceiving the scattering of light. In classical physics, scattering was conceived as deflection of particles by a mirror based on Newton's laws concerning moving bodies and their interactions. In 1871, John Strutt (Lord Rayleigh) developed a first equation of scattering of light[60] and this equation, as Bachelard stated, mirrored the naïve imagination of deflection geometrically-*anschaulich*. However, the new equation of Hendrik Kramers and Werner Heisenberg from 1924[61] overcame these naïve epistemological obstacles. For Bachelard, the equation of Kramers and Heisenberg marks a paradigmatic shift in scientific thinking, although Heisenberg, Born, and Jordan had developed this equation in 1925 just before the introduction of matrix mechanics.[62] Nevertheless, it involved the new quantum mechanical concept of interaction between matter and radiation. Although for Strutt the path of a beam of light is deflected differently depending on the wavelengths of the light, the deflection was conceived deterministically; while for the equation of Kramers and Heisenberg radiation was modified by the influence of matter. Thus, not the *Anschauung* of elastically vibrating electrons but the abstract idea of an inelastic scattering of light with certain frequencies, where the frequencies correspond to possible transitions within the atom, characterized the new approach. In order to express these possible transitions Kramers and Heisenberg used vector analysis as a mathematical notation system for expressing scattering as sums of all contributing absorption and emission amplitudes. For Bachelard, not *Anschauung* but the use of vector notation was guiding the new scientific thinking

---

[58] Gaston Bachelard, *Le nouvel esprit scientifique* (Paris: Presses Universitaires de France, 1934).

[59] "La science [...] s'oppose absolument à l'opinion. S'il lui arrive, sur un point particulier, de légitimer l'opinion, c'est pour d'autres raisons que celles qui fondent l'opinion; de sorte que l'opinion a, en droit, toujours tort. L'opinion pense mal; elle ne pense pas: elle traduit des besoins en connaissances" (Gaston Bachelard, *Essai sur la connaissance approchée* (Paris: Vrin, 1986), 169).

[60] John (Lord Rayleigh) Strutt, "On the scattering of light by small particle," in *Scientific Papers (1869–1881)*, vol. 1, ed. John Strutt (Cambridge: Cambridge University Press, 1899 [1871]), 104–11.

[61] Hendrik Kramers and Werner Heisenberg, "Über die Streuung von Strahlung durch Atome," *Zeitschrift für Physik* 31, no. 1 (1924): 681–708.

[62] Max Born, Werner Heisenberg, and Pascual Jordan, *Zur Begründung der Matrizenmechanik* (Stuttgart: Battenberg, 1962 [1925]).

as a complex interplay of notational and numerical conventions.[63] This new scientific thinking provides a more general and, thus, realistic view of the world. Today, the modern equation of scattering of light, considering all possible transitions, is using operator analysis.[64] Jun John Sakurai's transition operator T probabilistically describes all possible transitions between the initial and final states of a scattering atom. Reality, in Bachelard's interpretation, turns into a special case of possibility.

This development from naïve, geometrically-*anschaulich* inspired imaginations to reality as a special case of possibility is owed to mathematics and its increasing complex notations: from vectors over matrices to tensors. Or, put differently: Reality becomes the model! Reality turns into the hypothetical-deductive program of realizing notational predictions empirically. Bachelard called this new scientific thinking 'surrational,' that means it goes beyond the rational view of classical determinism.[65] Just as the microscope has opened up access to new worlds, the new mathematical notations opened up new worlds because they allowed for new generalizations. Rationality tries to understand given reality, while surrationality tries to understand all possibilities of reality. Of course, the prerequisite of surrationality is the overcoming of *Anschauung*.

Conclusion

Celebrating new physics means celebrating matrices. Matrices, like vectors and tensors, are powerful epistemic tools for mathematics as well as other sciences.[66] They can be modified by basic algebraic operations like addition, scalar multiplications, and transposition. Matrices can be used to express vectors and tensors as well as to articulate and work with linear equations which are widely applied in physics and engineering, for instance to approximate the behavior of non-linear systems. Furthermore, matrices are used for carrying out linear transformations in group-theoretic geometry, for representing complex numbers by particular real 2×2 matrices, and for providing normal-form representations of games in game theory, for instance the payoff matrix of the prisoner's dilemma.[67]

Usually, mathematics, science, as well as philosophy do not pay much attention to their semiotic media for expressing thoughts like notational systems. But when these media become the basis for achieving new insights into new worlds—turning induction into hypothetical-deduction—then some thoughts should be invested in

---

[63] Bachelard, *Le nouvel esprit*, 77.

[64] Jun John Sakurai, *Modern Quantum Mechanics* (Boston: Addison-Wesley, 1993).

[65] Gaston Bachelard, *Philosophie du non. Essai d'une philosophie du nouvel esprit scientifique* (Paris: Presses Universitaires de France, 1940).

[66] Gabriele Gramelsberger, "Matrizen – Schriftbilder diskontinuierlicher Zeitmodelle," in *Visuelle Zeitgestaltungen*, series *Bildwelten des Wissens* 15, ed. Claudia Blümle, Claudia Mareis, and Christof Windgätter (Berlin: De Gruyter, 2020), 61–69.

[67] John von Neumann and Oskar Morgenstern, *Theory of Games and Economic Behavior* (Princeton: Princeton University Press, 1944).

the impact of these semiotic media on science. For instance, Mark Steiner has asked in his study *The Applicability of Mathematics*: "How, then, did scientists arrive at the atomic and subatomic laws of nature?" His answer: "By mathematical analogy."[68] If there is no *Anschauung* only mathematical analogy based on symbol manipulations remains as a tool for scientific imagination, research becoming an investigation of "representational systems [… rather] than nature."[69] In Steiner's reading, Schrödinger also turned away from *Anschaulichkeit*:

> In sum, Schroedinger began with a sine wave of fixed frequency, based on an analogy to an optical wave, where the frequency is given by a fixed energy field. In writing down the 'wave' equation by taking derivatives, Schroedinger completely abstracted away from this intuition, ending with an equation having no direct physical meaning; one with superposed solutions; one with solutions having no 'wavelike' qualities at all.[70]

When Heisenberg in 1932 postulated the existence of positrons and neutrons by analyzing the local *SU(2)* rotation symmetry with a 2×2 matrix,[71] he employed a mathematical analogy based on a fictitious isospin space. In this fictitious isospin space, the rotation of 180° obtains a neutron from a proton and a full 720° rotation returns a neutron or a proton to its initial isospin state. "It seems clear," as Steiner stated, "that the mathematics is doing all the work in this analogy."[72] Finally, Steiner concluded:

> It is the formalism itself (and not what it means) that is the fundamental subject of physics today. […] discoveries made this way relied on symbolic manipulations that border on the magical. I say 'magical' because the object of study of physics became more and more the formalism of physics itself, as though the symbols were reality.[73]

However, as long as the hypothetic-deductive research style requires experimental verification, 'magic' can turn into reality.[74] But if this is not possible as in current cosmology of multiverse theories and string theory, respectively, science can either criticize the empirical necessity as an "overzealous Popperism"[75]—signing away its spirit on the miracle of a purely mathematical isomorphism[76]—or, it can pose the question anew about the role of *Anschauung* for applying mathematics in science.

---

[68] Mark Steiner, *The Applicability of Mathematics as a Philosophical Problem* (Cambridge: Harvard University Press, 1998), 3.

[69] Steiner, *The Applicability*, 7.

[70] Steiner, *The Applicability*, 81.

[71] Werner Heisenberg, "Über den Bau der Atomkerne," *Zeitschrift für Physik* 77 (1932): 1–11.

[72] Steiner, *The Applicability*, 87.

[73] Steiner, *The Applicability*, 176, 136.

[74] Popper, *The Logic*.

[75] Helge Kragh, "The Most Philosophically of All the Sciences: Karl Popper and Physical Cosmology," *Preprint PhilSci-Archive* (2012): 41. URL: http://philsci-archive.pitt.edu/9062/ (accessed November 23, 2020).

[76] James R. Brown, "The Miracle of Science," *Philosophical Quarterly* 32 (1982): 232–44.

Part III

From Production Processes to Exhibition Practices

Interview with Anja Sattelmacher: Between Viewing and Touching—Models and Their Materiality

Michael Friedman, Karin Krauthausen, and Anja Sattelmacher

KK: You recently published your book *Anschauen, Anfassen, Auffassen: Eine Wissensgeschichte mathematischer Modelle*,[1] where you discuss material mathematical models from the nineteenth and early twentieth century, concentrating mainly on Germany and particularly on the processes of manufacturing, collecting, and distributing these models. Before discussing these issues in more detail, we would like to begin with your understanding of 'model.' Against the background of your research in the history of science and cultural studies, what would you say a mathematical model is?

---

[1] Anja Sattelmacher, *Anschauen, Anfassen, Auffassen: Eine Wissensgeschichte Mathematischer Modelle* (Wiesbaden: Springer Spektrum, 2021). The book is based on the dissertation with the same title which was defended in 2017 at the Department of History, Humboldt-Universität zu Berlin. I am deeply indebted to my supervisor Anke te Heesen, who not only gave the first impulse to this project but also guided and accompanied it through all the years with infinite patience and care.

M. Friedman (✉)
The Cohn Institute for the History and Philosophy of Science and Ideas, The Lester and Sally Entin Faculty of Humanities, Tel Aviv University, Ramat Aviv, Tel Aviv, Israel
e-mail: friedmanm@tauex.tau.ac.il

K. Krauthausen
Cluster of Excellence "Matters of Activity. Image Space Material," Humboldt-Universität zu Berlin, Unter den Linden 6, 10099 Berlin, Germany
e-mail: Karin.Krauthausen@hu-berlin.de

A. Sattelmacher
Institut für Musik- und Medienwissenschaft, Humboldt-Universität zu Berlin, Unter den Linden 6, 10099 Berlin, Germany
e-mail: anja.sattelmacher@hu-berlin.de

M. Friedman and K. Krauthausen (eds.), *Model and Mathematics: From the 19th to the 21st Century*, Trends in the History of Science, https://doi.org/10.1007/978-3-030-97833-4_13

AS: In my research the term 'model' is closely tied to the concept of *Anschauung*,[2] with an emphasis on the material context. Both of these tend to be overlooked when considering the conceptual superstructure of mathematics. If one focuses on materiality, then the mathematical model is something that is subjected to a process of production, which then leads to a three-dimensional, haptic object. A proper account of the cultural and epistemic techniques and practices at work in this production process takes us beyond a history of the discipline of mathematics. These practices extend into a number of very different fields of study, ranging from the applied arts to industrial engineering to the fine arts.

KK: And mathematical education is another of these fields in which the model is firmly rooted.

AS: Exactly: educational practices, mathematical knowledge, and, of course, artisanal know-how. In my book I show that many model builders have backgrounds with close ties to the fields of craft or engineering or architecture. It is not by chance that it was precisely these mathematicians who so strongly embraced and promoted the making of models.[3]

MF: The beginnings of the tradition of making mathematical models lie in the nineteenth century in France, Germany, and England (see Fig. 1a for models made by the French mathematician Théodore Olivier and Fig. 1b for models made in Germany by Ludwig and Alexander Brill). In what way do these traditions differ, and how are they anchored in the respective cultures of science and knowledge?

AS: The way science was studied and taught in France or Germany was of course very different, particularly since France had a centralist and Germany a more 'federalist' system. Furthermore, the culture of model building in France began earlier and initially arose in the military sector (that is, in the École Polytechnique, which had its roots in the military) before being used for civilian purposes.[4] That then also gave rise to the concept of the civil engineer.[5] This development started around 1830 with models by the mathematician Olivier (see Fig. 1a), a graduate of the École polytechnique. That is the period in which the modern understanding of the engineer emerged, and it was then that these model-building practices started.

---

[2] On the concept of *Anschauung* (intuition), see the discussion in the interview below.

[3] See also: Ernst Seidl, Edgar Bierende, and Frank Loose, ed., *Mathematik mit Modellen: Alexander von Brill und die Tübinger Modellsammlung* (Tübingen: Museum der Universität Tübingen, 2018).

[4] See: Sattelmacher, *Anschauen, Anfassen, Auffassen*, chapter 3. See also the article by Frédéric Brechenmacher in this volume.

[5] Sattelmacher, *Anschauen, Anfassen, Auffassen*, 77: "Die Gründung der ECAM [École centrale des arts et manufactures, 1829] erfolgte unter dem Eindruck, dass es einen neuen Berufstand, den '*ingénieur civile*' benötigte, um die (aufgrund der industriellen Revolution eintretenden) neuen wirtschaftlichen und technischen Herausforderungen des Landes besser bewältigen zu können." ["ECAM [École centrale des arts et manufactures] was founded [in 1829] with the belief that a new profession, the '*ingénieur civile*,' was needed to respond to the new economic and technical challenges faced by the country in the wake of the Industrial Revolution."].

a   b

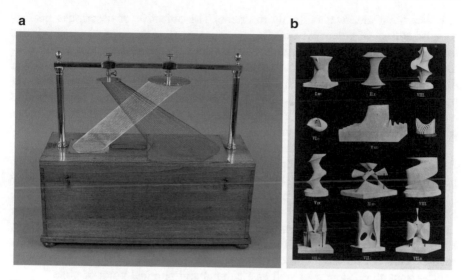

Fig. 1 a A string model made by Théodore Olivier of an intersection of an oblique cylinder and an oblique conoid (yellow and blue), ca. 1830–1845, fruitwood, brass, thread, lead weights (Union College Permanent Collection, 1868.28 UCPC). Courtesy of the Union College Permanent Collection, all rights reserved. **b** Various mathematical models from plaster at the inner cover of Ludwig Brill, ed., *Catalog mathematischer Modelle für den höheren mathematischen Unterricht* (Darmstadt: L. Brill, 1881)

This situation cannot be directly transferred to Germany, although at that time there was of course academic research and teaching on descriptive geometry in Germany too. But in Germany model building remained a local matter—that is, it only took place in universities where there were professors of mathematics who were interested in model building and actively promoted it. Of course, German mathematicians around 1870 also travelled to France and looked at the mathematical models at the Musée des Arts et Métiers. There is not a continuous line, but certain techniques and conceptions were adopted from France. That can be seen, for example, in the founding of a Polytechnikum in Karlsruhe, one of the oldest technical colleges in Germany, which was based on the idea of the École polytechnique. The first professors taught in Karlsruhe on the basis of drawings by the mathematician Gaspard Monge, who in 1794, along with the engineer Lazare Nicolas Marguerite Carnot, founded the École polytechnique. Christian Wiener, a professor of descriptive geometry at Karlsruhe Polytechnikum, was also aware of Monge's works and probably used them as a basis for his own work. Already in the 1860s he constructed models for the teaching of higher mathematics. It is with Wiener that the German model-building tradition begins—a tradition continued by his son, Hermann Wiener, who was very active in model building.

MF: Did French and German model building have the same aim: roughly speaking, to visualize an abstract mathematical formula?

AS: The basic idea was certainly the same. The objective of descriptive geometry—or, at first, of geometric drawing—was to visualize the abstract formula.[6] But in France, at the end of the eighteenth and the beginning of the nineteenth century, there was a greater focus on architectural design,[7] that is, on the construction of buildings. In descriptive geometry there was also an interest in how the line is materialized—that is, the course the line takes when, for example, it rotates around a pole, the forms this gives rise to (for instance, a cylinder), and what an engineer can do with this in his or her practice.

In Germany, on the other hand, the focus was on *Anschauung*, and that gave rise to the idea of using descriptive geometry and the mathematical model for pedagogical purposes, as a teaching aid—for example, in schools. Which is to say that in Germany the epistemic objectives were broader. It should be borne in mind, however, that there is a gap of more than seventy years between the beginning of the model tradition in France and the dissemination of model building in Germany, and that much changed in this time. This relates to the different ways in which the models were constructed, but, of course, also to the development of mathematics itself—from an epistemic as well as a cultural and political point of view.

But ultimately you are right: model building is concerned with visualizing a mathematical formula. For me, however, the important question is another one: what is actually happening in such a process of model building? What do the different steps that are necessary look like concretely? How is the drawing made, and how on the basis of the drawing does one decide which materials should be used? What tools and what practical knowledge is linked to the processing of the materials? So there is a chain of multiple translations, from the drawing to the model. In my book I did not only want to look at the beginning and the end of the

---

[6] In the program for his lessons at the École normale de l'an III at the beginning of 1795, Gaspard Monge emphasized the geometric visualization of mathematical and physical problems and defined the aim and objectives as following: "The first [objective of descriptive geometry] is to accurately represent, with drawings that have only two dimensions, objects that have three dimensions and are likely to be rigorously defined. From this point of view, it [descriptive geometry] is a necessary language for the man of genius who designs a project, for those who must direct its execution, and finally for the artists who must themselves execute its various parts. The second objective of descriptive geometry is to deduce from the exact description of the bodies everything that necessarily follows from their respective shapes and positions. In this sense, it is a way of seeking the truth." (in: Jean Dhombres, ed., *L'Ecole normale de l'an III. Vol. 1, Leçons de mathématiques. Laplace–Lagrange–Monge* (Paris: Éditions Rue d'Ulm, 1992), 305–6) ["Le premier est de représenter avec exactitude, sur des dessins qui n'ont que deux dimensions, les objets qui en ont trois, et qui sont susceptibles de définition rigoureuse. Sous ce point de vue, c'est une langue nécessaire à l'homme de génie qui conçoit un projet, à ceux qui doivent en diriger l'exécution, et enfin aux artistes qui doivent eux-mêmes en exécuter les différentes parties. Le second objet de la géométrie descriptive est de déduire de la description exacte des corps tout ce qui suit nécessairement de leurs formes et de leurs positions respectives. Dans ce sens, c'est un moyen de rechercher la vérité […]."].

[7] See: Joël Sakarovitch, *Epures d'architecture: De la coupe des pierres à la géométrie descriptive XVIe-XIXe siècles* (Basel/Boston: Birkhäuser, 1998).

process, I wanted to consider the whole procedure of model building—that is, to describe the practices that generate knowledge. These are epistemic practices.

KK: This leads to two far-reaching questions: First, what role does '*Anschauung*' play in the discourses and practices of mathematical model building? And, second, what exactly do the material and practical steps in the translation of an abstraction (the mathematical formula) into a concrete model look like? Let us perhaps first proceed with the meaning of *Anschauung*. The concept does not exist in the same way in other languages, and particularly in English it has either to be reformulated or to be reduced by speaking, for instance, of 'intuition.' Could you describe what the concept of *Anschauung* reveals in the German discussion about mathematics and model building?

AS: Without the concept of *Anschauung*, the mathematical model cannot be thought at all, at least not in the sense that the mathematical model had in the nineteenth century. *Anschauung* does not only mean that I look at something (*etwas anschauen*), or that I observe something. It goes beyond visual perception, because, first, it sets a thought process in motion, and, second, this process can be learned, practiced, taught, and, if you like, conditioned. *Anschauung* generates knowledge, and *Anschauung* instigates a learning process. Therefore, in nineteenth-century sources in the field of education, the concept of *Anschauung* is ubiquitous. It is not only used by the leading educational reformers of the early nineteenth century—such as Friedrich Fröbel and Johann Heinrich Pestalozzi—but basically by anyone involved with students or with formal education, that is, with pedagogical work.

In this context, one needs to refer to the discussions about formal education taking place in Germany at the time. In the nineteenth century there were two conflicting ideologies: on the one hand, the new humanism, which put its hopes in classical education with Greek and Latin; and, on the other, an educational reform movement that had a more practical orientation. The latter espoused the concept of *Anschauung* in order to argue for the acquisition of newer languages and for the teaching of an application-oriented mathematics. The mathematics professor Felix Klein was one of the most prominent figures to embrace the concept of *Anschauung* and to opt for model building among his students. He was well acquainted with these discourses and was also aware of the political situation of education in Germany.

The German-speaking mathematicians working with models in the nineteenth century had internalized what *Anschauung* meant—that is, that it also aims at a learning process and at a process of 'making graspable.' Indeed, *Anschauung* does not only mean that I *show* the object to my students, and that in this way they grasp something; rather, *Anschauung* also means that the students construct the model themselves, and this constitutes the actual learning process. Hence, *Anschauung* is also related to the touching of materials—that is why my book is called *Anschauen, Anfassen, Auffassen* (Viewing, touching, grasping). The model is not only observed with the eyes; the other senses are addressed too, in particular the sense of touch, since it is with the hands that one builds the model.

Fig. 2 **a** Glass model of a heliozoan (1885), Blaschka workshop, Dresden, now in the zoological collection of Universität Tübingen. Photo: Peter Neumann. © Stadtmuseum Tübingen, all rights reserved. **b** Wax model of the vascular system of a human liver (1912–1914), Adolf Ziegler workshop, Freiburg im Breisgau (Institut für Anatomie und Zellbiologie, Universität Heidelberg). © Universität Heidelberg, all rights reserved

KK: Over the last thirty to forty years the history of science has been interested in the concrete practices and procedures that accompany, promote, or even simply make possible the production of knowledge.[8] Here, it is no longer a matter of insight or truth but of 'knowledge in the making' or of artisanal knowledge. Could one say that in the mathematics of the nineteenth century there was a greater awareness of the pedagogical and, further, of the epistemological meaning of the work with material and of the concrete process of production—the "motorische Empfindung" (motor sensation) as Klein calls it?[9]

AS: I would certainly agree with that, and it is not only limited to mathematics. If we look at the models made by the Blaschkas (Fig. 2a),[10] with these works it was certainly a matter of acquiring knowledge through model building. Or if one

---

[8] See, for example: Bruno Latour and Steve Woolgar, *Laboratory Life: The Construction of Scientific Facts* (Beverly Hills: Sage, 1979); Karin Knorr-Cetina, *The Manufacture of Knowledge: An Essay on the Constructivist and Contextual Nature of Science* (Oxford: Pergamon Press, 1981); Steven Shapin and Simon Schaffer, *Leviathan and the Air-Pump: Hobbes, Boyle, and the Experimental Life* (Princeton: Princeton University Press, 1985); Hans-Jörg Rheinberger, *Toward a History of Epistemic Things: Synthesizing Proteins in the Test Tube* (Stanford: Stanford University Press, 1997).

[9] Felix Klein, "Über Arithmetisierung der Mathematik. Vortrag gehalten in der öffentlichen Sitzung der K. Gesellschaft der Akademie der Wissenschaften zu Göttingen am 2. November 1895," in Felix Klein, *Gesammelte mathematische Abhandlungen*, ed., Robert Fricke and Alexander Ostrowski (Berlin and Heidelberg: Springer, 1922), 2nd vol., 232–40, here 237.

[10] Leopold Blaschka and his son Rudolph Blaschka were famous for producing hundreds of glass models of marine animals and plants. See also: Lorraine Daston, "The Glass Flowers," in *Things*

thinks of Adolf Ziegler's wax models (Fig. 2b),[11] in this case too the intention was to generate and pass on knowledge, whereby knowledge here arose both in the making of the object itself and through the interaction with the model. I would say that that was a particular characteristic of the time.

MF: I would like to return to the role of *Anschauung* at the end of the nineteenth century, since it was then that the 'crisis of *Anschauung*' erupted, in the sense that mathematicians discovered several mathematical objects and domains that were sometimes regarded as 'monsters' and considered as objects which cannot be visualized, one of the famous examples of these objects being the Weierstrass function, a continuous but nowhere differentiable function.[12] Another example is mentioned in an observation by Christian Wiener concerning a surface described by 'moving' one Weierstrass function along another one. The described surface, which exactly like the Weierstrass curve is nowhere differentiable, "cannot be represented by means of drawing or a model."[13] How is this crisis of *Anschauung* to be reconciled with the tradition of model building? How did the mathematicians that championed model building—such as Klein, Alexander Brill, or Hermann Wiener—react to this crisis and the problem of the mathematical objects that elude *Anschauung*?

AS: One could say that for Klein this did not present a problem because in his opinion not every conceivable mathematical phenomenon should be expressed visually. Models were a means to understand one aspect of mathematics, not the whole of mathematics. Apart from that, however, the crisis of *Anschauung* was fundamentally important. This crisis did not only last one or two years but from the end of the nineteenth century to the 1920s. And yet this crisis and mathematical model building did not have so much to do with one another, because the mathematicians working in a purely theoretical way were concerned with entirely different questions and issues. Of course, the model builders were aware that non-Euclidean hyperbolic geometry reduced the possibility of visualization and also of

---

that Talk: Object Lessons from Art and Science, ed. Lorraine Daston (New York: Zone Book, 2014), 223–56.

[11] Adolf Ziegler was a nineteenth-century wax modeler in developmental physiology and anatomy. He and his son Friedrich crafted a series of wax models, many of which were sold to medical schools around the world. See also: Nick Hopwood, "Plastic Publishing in Embryology," in *Models: The Third Dimension of Science*, ed. Soraya de Chadarevian and Nick Hopwood (Stanford: Stanford University Press, 2004), 170–206.

[12] See: Klaus Volkert, *Die Krise der Anschauung: Eine Studie zu formalen und heuristischen Verfahren in der Mathematik seit 1850* (Göttingen: Vandenhoeck & Ruprecht, 1986); Klaus Volkert, "Die Geschichte der pathologischen Funktionen: Ein Beitrag zur Entstehung der mathematischen Methodologie," *Archive for History of Exact Sciences*, 37, no. 3 (1987): 193–232; Marek Jarnicki and Peter Pflug, *Continuous Nowhere Differentiable Functions: The Monsters of Analysis* (Cham: Springer, 2015).

[13] Christian Wiener, *Lehrbuch der darstellenden Geometrie*, vol. 2 (Leipzig: Teubner, 1887), 35: The surface "[kann] nicht durch Zeichnung oder ein Modell dargestellt werden [...]."

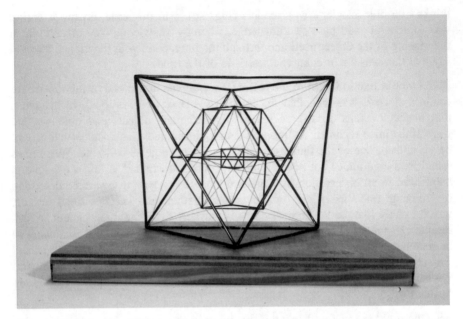

Fig. 3 Victor Schlegel's model of a projection of the regular four-dimensional 24-cell into three-dimensional space (Göttingen Collection of Mathematical Models and Instruments, model 352). © Mathematical Institute, Georg-August-University Göttingen, all rights reserved. See also: Victor Schlegel, *Ueber den sogenannten vierdimensionalen Raum* (Berlin: H. Riemann, 1888)

Anschauung.[14] However, there were certainly attempts to build models for objects of non-Euclidean geometry. There was no clear opposition in the sense of their being, on the one hand, a Euclidean mathematics that was an object of *Anschauung* and could be visualized and, on the other, a non-Euclidean mathematics that was purely abstract and could not be visualized. Another example of the attempt to go beyond the possible is the modeling of four-dimensional space (see Fig. 3 for Victor Schlegel's model of a convex regular four-dimensional polytope).[15]

However, one has to admit that after the peak of the crisis of *Anschauung* models became less important within the class curriculum at technical colleges. In the 1930s they gradually disappeared from the classroom and were relegated to university collections. The production of models completely stopped. That had already begun of course when set theory became more important, and, for this, other modes of academic teaching were used. At that point the mathematical models migrated to the glass cabinets of the collections and disappeared from production.

---

[14] For a discussion on Eugenio Beltrami's paper models of the hyperbolic plane from the end of the 1860s, see the contribution by Moritz Epple in this volume.

[15] See: Michael Friedman, "Coloring the Fourth Dimension? Coloring Polytopes and Complex Curves at the End of the Nineteenth Century," in *Meaningful Color: Epistemology of Color in the Sciences*, ed. Bettina Bock von Wülfingen (Berlin and New York: De Gruyter, 2019), 81–98.

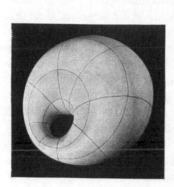 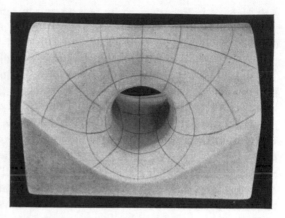

Fig. 4 Photos of plaster models of Dupin cyclide. From David Hilbert and Stefan Cohn-Vossen, *Anschauliche Geometrie* (Berlin: Springer, 1932), 193. © Springer, all rights reserved

MF: The loss in the importance of models can be gathered from the famous lecture "Die Krise der Anschauung" given by Hans Hahn in 1932.[16] Although Hahn, as he states himself, used drawings to illustrate the mathematical objects that triggered this crisis, he does not mention any models. On the other hand, also in 1932, the book *Anschauliche Geometrie* by David Hilbert and Stefan Cohn-Vossen was published,[17] and this work contains numerous photos of models (see, for example, Fig. 4). So, to put the question concretely, what role do models play in the 1930s?

---

[16] Hans Hahn, "Die Krise der Anschauung," in Herman F. Mark, Hans Thirring, Hans Hahn, Karl Menger, and Georg Nöbeling, *Krise und Neuaufbau in den exakten Wissenschaften: Fünf Wiener Vorträge* (Leipzig and Vienna: Deuticke, 1933), 41–64. The lecture was held as part of a series of five lectures in Vienna titled "Krise und Neuaufbau in den exakten Wissenschaften." In another lecture from this series, "Die vierte Dimension und der krumme Raum," by Georg Nöbeling, Nöbeling's usage of the term 'model' points to a shift in the way this term was used: "Klein succeeded in specifying a system of things [*System von Dingen*] within the three-dimensional Euclidean space, whereby, if one terms these things 'points,' 'straight lines,' and 'planes,' then all of the non-Euclidean axioms hold. In order to obtain this model, one considers the interior of a sphere of a three-dimensional Euclidean space [...]," (Georg Nöbeling, "Die vierte Dimension und der krumme Raum," in Herman F. Mark, Hans Thirring, Hans Hahn, Karl Menger, and Georg Nöbeling, *Krise und Neuaufbau in den exakten Wissenschaften* (Leipzig and Vienna: Franz Deuticke, 1933), 65–92, here 77–78 ["Es ist nämlich Klein gelungen, innerhalb des dreidimensionalen euklidischen Raumes ein System von Dingen anzugeben, welches, wenn man die Dinge als 'Punkte,' 'Geraden' und 'Ebenen' bezeichnet, sämtlichen nicht-euklidischen Axiome genügt. Man betrachtet, um dieses Modell zu erhalten, das Innere einer Kugel eines dreidimensionalen euklidischen Raumes [...]."]). In addition, Nöbeling points out that the material or illustrative aspects of 'point,' 'line,' or 'plane' are irrelevant and that all that matters are the relations between the concepts (ibid., 79).

[17] David Hilbert and Stephan Cohn-Vossen, *Anschauliche Geometrie* (Berlin: Springer, 1932). The book was translated into English as: David Hilbert and Stephan Cohn-Vossen, *Geometry and the Imagination* (New York: Chelsea Publishing, 1952).

AS: Here, it is important to be precise and to remember that Hilbert's book did not appear suddenly; it was based on a series of lectures titled "Anschauliche Geometrie" that Hilbert gave between 1920 and 1921 in Göttingen, and at that time he was involved in an intense study of the mathematical models found there.[18] These lectures were then published in 1932, which means that another ten years lie between the propositions and their publication. Moreover, these lectures were intended for teaching, and the book too should be understood as a textbook—that is not the same thing as a research finding that is addressed to other researchers. It is important to differentiate. How do mathematicians work at their desks when they want to solve mathematical problems? And how do they then teach what they have discovered at their desks during this research? One can gather from Hilbert's book that he valued the pedagogical and epistemic function of models. From a historical perspective, I would not want to make such a clear distinction between mathematical model builders or representatives of *Anschauung* on one side and formalists on the other,[19] at least not for the 1920s.

KK: The conflict between descriptive geometry and abstract or formal mathematics cannot be decided on the basis of the significance of the built models. Could one say, however, that another tension becomes noticeable at this point, one that would continue to grow in the twentieth century: the difference between the applied sciences and a 'pure' mathematics?

AS: Yes, I think that it is precisely here that one finds the division. That can clearly be seen if we look at the history of the mathematical institute in Göttingen.[20] Basically, descriptive geometry in Germany around 1900 became increasingly disconnected from mathematics and migrated to the architecture faculty. In Göttingen Klein wanted to build a science campus with a wind tunnel for applied physics etc. This was a place for the applied sciences so to speak, and it was precisely there that mathematics should also be located. The building for mathematics was opened in 1929—hence, only after Klein's death—but with this location mathematics was indeed brought much closer to the applied sciences. In purely institutional terms, however, mathematics in Göttingen still remained tied to the faculty of philosophy—and these contradictory alliances brings to light a conflict in which Klein also found himself. Klein had a humanist education and explicitly stated in his biography: "mathematics is essentially a pure *Geisteswissenschaft* [humanities discipline]."[21] But he also wanted it to serve the applied sciences. Of course,

---

[18] Sattelmacher, *Anschauen, Anfassen, Auffassen*, 42–43.

[19] See: Herbert Mehrtens, "Mathematical Models" in *Models: The Third Dimension of Science*, ed. Soraya de Chadarevian and Nick Hopwood (Stanford: Stanford University Press, 2004), 276–306.

[20] See: Anja Sattelmacher, "Zwischen Ästhetisierung und Historisierung: Die Sammlung geometrischer Modelle des Göttinger mathematischen Instituts," *Mathematische Semesterberichte* 61 (2014): 131–43.

[21] Felix Klein, *Vorträge über den mathematischen Unterricht an den höheren Schulen. 1. Teil: Von der Organisation des mathematischen Unterrichts*, ed. Rudolf Schimmack (Leipzig: B.G. Teubner, 1907), 137: "[…] die Mathematik ist an sich eine reine Geisteswissenschaft."

the ambivalent position of mathematics also related to mathematical models. And this ambivalence was partly a result of university politics: the chemists had their laboratory, the biologists had their formaldehyde specimens, and because the mathematicians did not only have chalk and board, pencil and paper, but also models, they now also had something to show, and that entitled them to demand more space. If one looks at Klein's correspondence,[22] then what one sees is indeed this question: how big do the spaces of the mathematical institute need to be in order to house all the models? Hence, the mathematical models in Göttingen played an important role in the founding of the institute.

The history of this Göttingen building gives us a good picture of the turf war that had a formative role in the mathematics of the 1920s. On the one hand, this building was built in the heyday of mathematical formalism and, on the other, the building was literally built around the mathematical models. Every mathematics student that enters the Auditorium Maximum in Göttingen had to pass these models. And that was at a time when these models actually no longer played a role at all.

MF: I would like to return to the second topic, which was mentioned before, namely the material techniques. The models were made from different materials: plaster, threads, card, etc. Can one assign different mathematical meanings, or rather interpretations, to the various materials the models are made from? Or do the materials refer in each case to different epistemic practices in or beyond mathematics?

AS: That is an interesting question, and one that was actually at the beginning of my work on mathematical models. I travelled throughout Germany visiting the collections of mathematical models, and in doing so—at least within descriptive geometry—I frequently came across the same forms (for example, a cylinder), but often in different materials, that is, in plaster or wood or brass and thread. I asked myself why these various materials were used and what kind of epistemic meaning was linked with this. I think that there were entirely pragmatic reasons for this, because the materials are of course also expressions of their times. So the models that work with brass and thread were made at a completely different time from the plaster models, and these in turn were made at a different time from the wire models. There is a chronological order of the materials that points beyond mathematics. The use of brass and silk thread is derived from the electromagnetism of the eighteenth century, and that was ultimately borrowed by the model-building mathematicians from the physicists of the École polytechnique. The early French models are reminiscent of certain instruments in the experiments of Michael Faraday or André-Marie Ampère.[23]

If we look at the model of a rotating cylinder by Olivier (Fig. 5a), then the

---

[22] Sattelmacher, *Anschauen, Anfassen, Auffassen*, 139ff.

[23] Ibid., 108–9. See also: Friedrich Steinle, *Explorative Experimente: Ampère, Faraday und die Ursprünge der Elektrodynamik* (Stuttgart: Steiner, 2005).

a   b

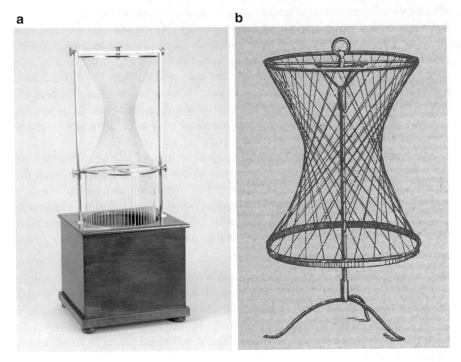

Fig. 5 **a** One of Théodore Olivier's models from the 1830s: a rotating cylinder with a wooden box (Collection of Historical Scientific Instruments, Harvard University). © Collection of Historical Scientific Instruments at Harvard University, all rights reserved. **b** A string model of a rotating cyliner. From Hermann Wiener and Peter Treutlein, *Verzeichnis von H. Wieners und P. Treutleins Sammlungen Mathematischer Modelle. Für Hochschulen, Höhere Lehranstalten und technische Fachschulen* (Leipzig: Teubner, 1912), 12

mathematics on this cylinder is limited—the 'pure' model as such is of course only this cylinder, but this in turn is attached to a wooden box and a brass frame—a little like in the physics instruments or the physics models of the time. The wooden box has a function, but this function seems to have more of an aesthetic origin, as with the physics models too. In the more recent mathematical models, these additional constructions mostly disappear or are reduced. When one looks at a later example—for instance, by Hermann Wiener from 1912 (see Fig. 5b)—then one sees that the cylinder now stands alone. One can rotate the cylinder without having to turn some screw on the side of the frame.

KK: From a mathematical point of view it is to some extent the same model, but from an aesthetic point of view it has a different material finish, and above all it does not have the massive 'frame' represented by the wooden box.

AS: Exactly. In 1910 it is still the same model of a cylinder that can transform, but now without the frame.

One has to remember that the mathematicians did not actually have a tradition of experiment, and therefore for their modeling practice they drew inspiration from the natural sciences. The company Pixii père et fils, which built the early models in France and in particular the models for Olivier, also constructed experimental assemblages—which are basically also models—for Ampère and Faraday, so it is not surprising, therefore, that these models resemble each other.

Of course, there are also biographical contingencies. Olivier was from Lyon and therefore from a center of the silk manufacturing industry, which partially explains the use of silk thread. However, it was not uncommon in France to use silk thread in scientific models, especially in the physics of magnetism. In Germany, on the other hand, initially other materials were used. Alexander Brill produced his mathematical models around 1870/80 using plaster. Brill at that time was a professor at the Technische Universität in Munich, and plaster was also the material used by the art students to make casts. The plaster cast collection in Munich was certainly another influence. Joseph Kreittmayr, who made the plaster models for Brill also cast models for the Bayerisches Nationalmuseum.[24] In order to produce the mathematical models, Brill first had to find the craftspeople and organize the production, and he therefore drew on techniques and infrastructures that were already established. Moreover, plaster was cheap, and once the negative form had been made it was simple to reproduce the mathematical model.

Hermann Wiener's models belong to yet another tradition. Although Wiener had studied with Brill and both were also related (they were cousins), Wiener made his models at a time when plaster no longer appeared adequate to the requirements of mathematical model making. Instead he used wire, which was flexible and could be easily shaped. Wiener criticized plaster models for not being able to show the essential: the movement of the line in space.[25] He wanted to show this movement in the model, but could not do this with plaster. Although one could score the lines and then use the model to examine their course, one could not manipulate the model. With Olivier's models, that was still possible: there were screws that could transform the cylinder into a paraboloid. In Wiener's case it is the hinge that enables the transformation from the planar to the spatial (and vice versa). This can be seen clearly in his model of an ellipsoid (see Fig. 6b).

The model has to be flexible so that one can show the generation and the course of a curve. But models should also be robust and capable of being produced in large numbers. For this reason they need to be standardized. Wiener's model designs meet these different requirements. In Fig. 6a one can see his construction drawing for the ellipsoid. With such a drawing, any precision mechanic

---

[24] Ludwig Brill, ed., *Catalog mathematischer Modelle für den höheren mathematischen Unterricht* (Darmstadt: L. Brill, 1881), 2.

[25] Hermann Wiener, "Entwicklung geometrischer Formen," in *Verhandlungen des 3. Internationalen Mathematiker-Kongresses in Heidelberg vom 8. bis 13. August 1904*, ed. Adolf Krazer (Leipzig: Teubner, 1905), 739–50, here 739–40.

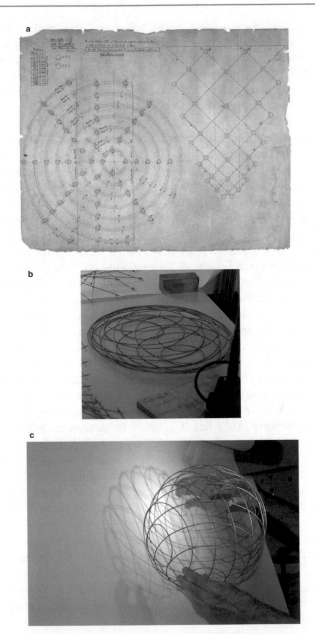

Fig.6 **a** Construction drawing by Hermann Wiener—in this case, for an elliptic paraboloid. Wiener draws the wire sizes at a ratio of 1:1 to the model so that the points at which the wire should be cut and soldered (or connected by a hinge) can be transferred directly from the drawing to the model. Private collection Friedhelm Kürpig, undated. © Friedhelm Kürpig, all rights reserved. **b** Model of an ellipsoid in folded form, built following the construction drawing by Hermann Wiener. Model by Friedhelm Kürpig. Photo: © Anja Sattelmacher, all rights reserved. **c** The same model of the ellipsoid (as in (**b**)), now unfolded. Photo: © Anja Sattelmacher, all rights reserved

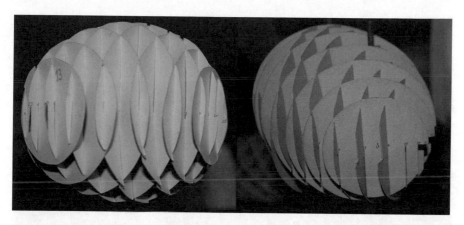

Fig. 7 Markings on sections of a sphere. The numbers refer to the points at which the sections should be slotted together (Göttingen Collection of Mathematical Models and Instruments, models 13 and 19). Photo: © Anja Sattelmacher, all rights reserved

schooled in model building could and can reproduce this model. That was not so simple with the plaster models because they were less standardized.[26]

KK: Can you describe in more detail the work process necessary for the design and the making of the plaster models?

AS: To produce a prototype of a model—for instance, of an ellipsoid or a hyperbolic paraboloid—the individual sections for the solid were first calculated and then cut out from a zinc sheet. This procedure was identical to the production of the flexible card models (Figs. 7 and 8), only this time the sections slotted together are made from zinc sheet. Once assembled as a model, the structure has to be held together with card, bits of wood, modeling clay, or scraps of metal in such a way that the form—for instance, a sphere—was preserved and no longer moved. It was subsequently filled with modeling clay.

The negative of the surface produced in this way was hollow on the inside and could now be filled with plaster—assuming the models were produced as hollow casts. That means that an approximately three-centimeter layer of plaster was poured into the body.

---

[26] After Wiener ceased his model production at the beginning of the twentieth century, his model making practices were taken up by a range of model makers, including Friedhelm Kürpig. Until recently, Kürpig rebuilt models from Wiener's collection using special tools and machines. For a photographic documentation of Kürpig's practice, see: Sattelmacher, *Anschauen, Anfassen, Auffassen*, 217–30, 237–38, and 261–63. Shortly after this book was published, the floods that devastated Germany in the summer of 2021 also damaged Kürpig's workshop. All the historic tools and machines are gone for now; the models, however, remain intact. Kürpig's work can be seen at: https://kürpig.de/ (accessed November 7, 2021).

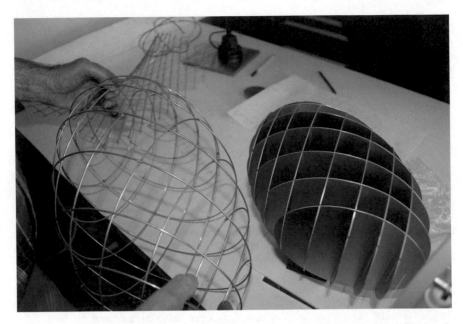

Fig. 8 Two models of an ellipsoid: on the right a version made of cut steel plate following the method used in Brill's card models; on the left a version made of bent wire following Hermann Wiener. Models by Friedhelm Kürpig. Photo: © Anja Sattelmacher, all rights reserved

KK: So the work process involves a chain of translations and probably also a number of different actors. It begins with the mathematical formula, which is translated into a drawing, which is then in turn translated into card or steel and slotted together to form a three-dimensional structure. From this one obtains a casting mold, if I have understood you correctly. But alternatively a wooden mold can be lathed by a professional carpenter, which can likewise be used for the creation of a mold for the plaster.

AS: The intermediate stage with the wooden mold does not always take place. When the sections are made of a tougher material—for example, thicker card—one can pour the modeling clay directly into the form, and then one already basically has the master mold for the plaster models.

MF: The people in the model-building workshops were not necessarily mathematicians; there were also craftspeople—carpenters, for example. How much mathematical knowledge did the craftspeople need to have? Mathematicians of course were not present at all stages of the work.

AS: To answer this question, one has to take a more detailed look at when which people were involved in the production process of the models. If one considers the process from the formula to the finished model, then at the beginning, for the calculation, rather a lot of mathematicians and mathematics students tended to be involved, and at the end fewer. If we take the card model already mentioned as an example, this consists of cut sections, and both the calculation and the drawing of the curves would have been carried out by mathematicians, because that has to be very precise. The prototypes of the sections were therefore prepared and also cut out by mathematicians. But, for the plaster models, the actual work of reproduction, such as the production of the master mold and the pouring of the plaster, was outsourced to the workshops. Once the prototype was ready, this was followed by purely artisanal techniques, and these were always the same regardless of whether one was casting a Greek Minerva or a model based on a mathematical formula. This being said, right at the end of the modeling process, the mathematicians were brought back to score the curves (compare the photos of Hilbert's plaster models in Fig. 4, on which the scored curves are clearly visible). Although, of course, it is difficult to say in retrospect who actually did the scoring.

MF: But the scoring would have had to have been guided in some way, and such templates could only have been created by a mathematician, right?

AS: Exactly, such templates were made by mathematicians. Who actually did the scoring? I can imagine that it took place in the workshops—hence, one can assume the scoring was carried out, for instance, by those doing the casting, on the basis of templates, since these models were produced and distributed serially. One could not score the curves in the master mold, because plaster expands during the casting and would cover over the lines.

A few sources, such as the recorded recollections of a female descendant of the model publisher Martin Schilling, suggest that this manual, often very precise and delicate work was carried out for the most part in a domestic setting by the wives of the mathematicians. However, this is not mentioned or documented by the mathematicians themselves.[27]

MF: If one looks at the process of manufacturing the models (as you have just described it) in the framework of seminars, then this may have been suitable when one taught a class of ten to fifteen students, since there probably would have been a need for assistance and close contact between the persons involved. But, if I understand correctly, the number of students at the end of the nineteenth century in Germany did not remain fixed over the years. Did the change in student numbers affect the way models were used?

AS: Here, we come to an important point, since one can in fact ascertain that in the 1910s projective media became more important for teaching purposes. That was certainly linked to the growing number of students at the technical colleges. Wiener in Darmstadt, for example, had considerably more students than Brill and

---

[27] See, for example: Sattelmacher, *Anschauen, Anfassen, Auffassen*, 173–242; see also: Nils Güttler, "Unsichtbare Hände," *Archiv für die Geschichte des Buchwesens* 68 (2013): 133–53; Donald L. Opitz, Staffan Bergwik, and Brigitte Van Tiggelen, eds., *Domesticity in the Making of Modern Science* (Basingstoke: Palgrave Macmillan, 2016); Carla Bittel, Elaine Leong, Christine von Oertzen, eds., *Working with Paper: Gendered Practices in the History of Knowledge* (Pittsburg: University of Pittsburg Press, 2019).

Klein in Munich. In their modeling seminar, Klein and Brill had six to fifteen students, whereas Wiener had around fifty. So naturally he needed to consider how it would still be possible to work with models with such a large number of students—that is, what these models would need to look like to be useable in a large lecture hall; he was not satisfied by the idea of simply passing the models around. Therefore, he constructed the models in such a way that they could also be projected, quite simply with the help of a light source.[28] An example of these new projection techniques can be seen in Fig. 9, showing a demonstration of Erwin Papperitz's kinodiaphragmatic (*kinodiaphragmatische*) projection apparatus.

MF: Wiener projected the models in his lectures?

AS: Yes. That was also his argument—if one can move the model and demonstrate how the form changes, then one can *show* the course of a curve, how the sections change. And wire for him was the ideal medium because, with the help of light, the wire models allowed spatial projections that still functioned as a material of *Anschauung* even in front of a larger group of students. But what this also shows is that Wiener lived at a time in which film and projective media were beginning to appear. At the beginning of the twentieth century, Wiener was aiming at effects that were clearly different from those of Brill in the 1870s—despite the fact that they were both mathematicians, came from a similar school, and that their models visualized the same mathematical objects.

MF: With these new media techniques, it is also a matter of another kind of knowledge transfer. In the German model seminars of the 1870s and 1880s, one of the tasks of more advanced students was to produce models themselves. So the models were not always intended to be looked at. Could one say that, with the new projection techniques, the haptic character of the models gets lost?

AS: One can certainly see it that way, since the production of models by the students themselves is either reduced or no longer takes place. That is also related to the fact that descriptive geometry had become less important in mathematics and likewise in teaching. Also the concept of *Anschauung* changes at that time.

MF: In what way?

AS: Between 1870 and 1910 one sees a gradual shift in the way one thought about mathematical models, and that also had consequences for the understanding of *Anschauung*. That can be seen clearly in Wiener's teaching practice, which is based on the idea that projection allows the eye to see the model and that the rest arises via the process of thought. Hence, the eye is the organ that senses the model…

---

[28] Sattelmacher, *Anschauen, Anfassen, Auffassen*, 251ff. See also: Anja Sattelmacher, "Modelle in den Schatten gestellt: Erwin Papperitz und die Entwicklung einer räumlichen Bewegungsäs-thetik um 1910," in *Long Lost Friends: Wechselbeziehungen zwischen Design-, Medien- und Wissenschaftsforschung*, ed. Claudia Mareis and Christof Windgätter (Zürich: Diaphanes, 2012), 39–60.

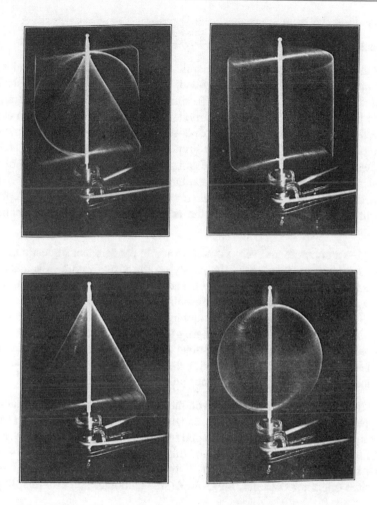

Die drei wichtigsten Rotationskörper, erzeugt mit der Schwungmaschine
durch Drahtfiguren.

Anordnung nach Leppin & Masche, Berlin.

Fig. 9 Four photos of a demonstration of Erwin Papperitz's kinodiaphragmatic projection apparatus showing various light and shadow projections. The first photo (top left) shows the conic
sections of a paraboloid resulting from the shadow projection. The second shows an intersection
curve on a sphere likewise produced by means of shadow. From Erwin Papperitz, *Methodik des
Mathematischen Unterrichts* (Leipzig: Quelle und Meyer, 1916), Plate 2

MF: …and not the hands.

AS: Exactly. A transfer from the hand to the eye takes place.

MF: With this disappearance of the haptic, the era of three-dimensional mathematical models essentially comes to an end. From the 1930s onwards basically no more model series were produced.[29] To finish we would like to talk about what seems to be a revival of mathematical models at the end of the twentieth century. How do you view the new digital models that can be produced and shown with the help of virtual reality software, 3D printers, touch screens, and other technical tools? Can one understand these new visualization techniques as a revival of mathematical models? Especially since these new techniques can now also visualize mathematical objects that in the nineteenth century were thought to be impossible to visualize. What in your opinion is the new epistemic and cultural meaning of these techniques?

KK: To give a concrete and very popular example for Michael's question, IMAGINARY is a platform that works rigorously on the visualization of mathematical formulae (e.g., of curves and surfaces), and it does this among other things by using 'swarm intelligence,' since the visualization program is put at the disposal of the website's users.[30] The visual modeling of the mathematical objects happens interactively and in real time: by changing the mathematical parameters, the user simultaneously changes the visualized object. This gives rise to a 'haptic moment,' simply because the objects change with the interventions of the user—as when Olivier turns the screws or when Wiener pulls on the hinges.

AS: The question is: what possibilities do these technologies offer and what limits? What types of models could I not yet realize ten years ago, and what can I realize today because the technology has changed in a certain respect? That does not have to be tied to a progress-oriented way of thinking but can be based on the simple fact that the technology has changed. If one wants to think the model process further, then one must consider, for example, the computer and also the 3D printer

---

[29] This is not to imply that material mathematical models were not used at all in research contexts after the 1930s; however, they were certainly marginalized. Examples of the continued use of these models are the material models of triply periodic minimal surfaces built by Alan Schoen during the late 1960s (see the interview with Myfanwy Evans in this volume), or the Costa surface, being a minimal surface of genus 1, discovered by Celso Costa in 1982 and visualized by David Hoffman and William Meeks in the same decade. With the latter's computer-based visualization one could observe properties which were only later proven (see: David Hoffmann, "The Computer-aided Discovery of New Embedded Minimal Surfaces" *The Mathematical Intelligencer* 9 (1987): 8–21). With the rise of these new methods of visualization, one could say that the tradition of material models was revived, although the technology and media used are different. Recently, Jacob Gaboury has shown that mathematical modeling played an important role for the development of computer graphics in the 1960s and 1970s. See: Jacob Gaboury, *Image Objects* (Cambridge and London: MIT Press, 2021).

[30] For an extensive discussion of IMAGINARY and its history, see the interview with Andreas Daniel Matt in this volume.

as an object that already contains various technologies itself, and not as something into which one enters something—as into a black box—and in the end something comes out. We historians of science will not get anywhere if we only consider the beginning and the end. We have to look at what kinds of 3D printers and touch screens existed ten years ago. What could they do, what kinds of technologies did they operate with? Why are we doing things differently now, and what haptic qualities does this give rise to? What does that generate in the model itself? As with the models of the nineteenth century, with today's visualizations too, one has to look very carefully at the practices and technologies that are involved.

Translated by Benjamin Carter.

Anja Sattelmacher is a historian working at the intersection of the history of science and media. She studied Media and Cultural Studies in Weimar and Lyon and holds a master's degree in Museum Studies from Macquarie University in Sydney. For her PhD on the history of knowledge of mathematical models at the history department of Humboldt-Universität zu Berlin, she studied numerous collections of material mathematical models throughout Germany, France, and beyond. What most struck her was the willingness to experiment with different materials for making the models. Sattelmacher is now an academic adjunct at Humboldt-Universität zu Berlin in the media studies department. For her current project, she examines collections of historic film documents to write about the history of political education in 1950s and 1960s (West) Germany. Together with Sarine Waltenspül and Mario Schulze, she recently published a Focus section in the journal *Isis* on research film.

Acknowledgements Michael Friedman and Karin Krauthausen acknowledge the support of the Cluster of Excellence "Matters of Activity. Image Space Material" funded by the Deutsche Forschungsgemeinschaft (DFG, German Research Foundation) under Germany's Excellence Strategy—EXC 2025—390648296.

Interview with Ulf Hashagen: Exhibitions and Mathematical Models in the Nineteenth and Twentieth Centuries

Michael Friedman, Ulf Hashagen, and Karin Krauthausen

KK: You will probably know this famous quote by Walther von Dyck from 1929: "To show the essence, content, and aims of mathematical research in its entirety cannot be the task of a museum."[1] Nevertheless, we would like to speak with you today about mathematics exhibitions in the long nineteenth and the early twentieth century. If you think specifically about the small exhibition of mathematical models at a mathematics convention in Göttingen in 1873, or about the exhibition of scientific instruments and apparatuses in the South Kensington Museum in London in 1876, or the important 1893 exhibition in Munich commissioned by the German Mathematical Society, or finally the opening of the mathematics

---

[1] "Wesen, Inhalt und Ziele der mathematischen Forschung in ihrer Gesamtheit vor Augen zu führen, kann nicht Aufgabe eines Museums sein." Walther von Dyck, "Mathematik," in *Das Deutsche Museum: Geschichte/Aufgaben/Ziele*, 2nd ed., ed. Verein deutscher Ingenieure/Conrad Matschoss (Berlin and Munich: VDI Verlag and R. Oldenbourg, 1929), 169–78, here 169.

M. Friedman (✉)
The Cohn Institute for the History and Philosophy of Science and Ideas, The Lester and Sally Entin Faculty of Humanities, Tel Aviv University, Ramat Aviv, 6997801 Tel Aviv, Israel
e-mail: friedmanm@tauex.tau.ac.il

U. Hashagen
Forschungsinstitut für Technik- und Wissenschaftsgeschichte, Deutsches Museum, Museumsinsel 1, 80538 München, Germany
e-mail: u.hashagen@deutsches-museum.de

K. Krauthausen
Cluster of Excellence "Matters of Activity. Image Space Material," Humboldt-Universität zu Berlin, Unter den Linden 6, 10099 Berlin, Germany
e-mail: Karin.Krauthausen@hu-berlin.de

section at the Deutsches Museum in 1925,[2] how would you describe the relation between mathematics and exhibition from a historical perspective? And how do mathematical models fit into this history of mathematics exhibitions?

UH: The first thing that needs to be said is that the nineteenth and the early twentieth century was an era of exhibitions, in which regional and national trade and industry exhibitions as well as world exhibitions played a major role in economic, cultural, and public life. Yet with the emergence of engineering colleges, the nineteenth century was also characterized by collections of 'disciplinary artifacts,' that is, by scholarly collections that were specific to a particular discipline. These two contexts had a profound influence on the development of mathematics exhibitions, and thus also on the development of the interrelationship between mathematics, exhibitions, and mathematical artifacts.

The 1873 exhibition on the occasion of the mathematics convention in Göttingen you mentioned was certainly an initial event for German mathematics. It was probably the first time that mathematical models—some of which came from Paris—as well as mathematical instruments and calculating machines were exhibited as part of a mathematics convention.[3] From today's point of view, this group of mathematicians showed an astonishing eagerness to engage with the 'mathematical artifacts,' that is, with objects that, in a tangible form, represented mathematics as a discipline. In the late nineteenth century these mathematical artifacts were primarily mathematical models, such as string or plaster models, that were produced in the context of contemporary geometrical research and constructed by mathematicians. These models and apparatuses come from a tradition that can be traced back above all to the string models used by Gaspard Monge and Théodore Olivier in their geometry classes and to the subsequent interest in mathematical models in France (see Fig. 1).[4]

The next large scientific exhibition in which mathematics played a role took place at the South Kensington Museum in 1876 (see Fig. 2). This was in fact a very

---

[2] On the London exhibition, see: South Kensington Museum, ed., *Handbook of the Special Loan Collection of Scientific Apparatus 1876* (London: Chapman and Hall, 1876) (on the geometric models, esp. 34–53). On the 1893 Munich exhibition, see: Walther Dyck, ed., *Katalog mathematischer und mathematisch-physikalischer Modelle, Apparate und Instrumente* (Munich: Wolf und Sohn, 1892); Walther Dyck, ed., *Nachtrag zum Katalog mathematischer und mathematisch-physikalischer Modelle, Apparate und Instrumente* (Munich: C. Wolf und Sohn, 1893); Walther Dyck, "Einleitender Bericht über die mathematische Ausstellung in München," *Jahresbericht der Deutschen Mathematiker-Vereinigung* 3 (1894): 39–56. On the genesis of the 1893 exhibition, see: Ulf Hashagen, *Walther von Dyck (1856–1934): Mathematik, Technik und Wissenschaftsorganisation an der TH München* (Stuttgart: Franz Steiner Verlag, 2003), 419–35.

[3] See: August Gutzmer, *Geschichte der Deutschen Mathematiker-Vereinigung von ihrer Begründung bis zur Gegenwart* (Leipzig: Teubner, 1904) 19–24.

[4] See: Felix Klein, *Vorlesungen über die Entwicklung der Mathematik im 19. Jahrhundert*, vol. 1 (Berlin: Springer, 1926), 78; Dyck, ed., *Nachtrag*, 51–52.

Fig. 1 A conoid string surface model by Fabre de Lagrange, France, 1872. Two equal circles in parallel planes divided equidistantly and connected by threads form a cone, a cylinder, and two conoids. This model was influenced by a series of models designed in 1830 by Theodore Olivier that could be distorted and rotated to provide a variety of geometric configurations. From "Collection," Science Museum Group, https://collection.sciencemuseumgroup.org.uk/objects/co5 9728/conoid-surface-model-surface-model-string, (accessed September 16, 2021). CC BY-NC-SA 4.0 (https://creativecommons.org/licenses/by-nc-sa/4.0/)

large, successful exhibition on apparatuses and instruments in science and technology.[5] Here, mathematics only formed a small subsection of the exhibition, and only two branches were exhibited: arithmetic and geometry. The section on arithmetic chiefly displayed slide rules and calculating machines, including contemporary models and historical machines such as the Pascaline from the seventeenth century. The section on geometry, on the other hand, exhibited instruments that were used for geometric drawing or to produce specific mathematical curves—but also three-dimensional mathematical models. This seems to have made a great impression on a number of German mathematicians, since in Germany the interest in geometric models and their production was only just emerging, and models from Germany could also be seen in the London exhibition. With this exhibition one can see how mathematical artifacts (i.e., mathematical models and mathematical

---

[5] See: Robert Bud, "Responding to Stories: The 1876 Loan Exhibition of Scientific Apparatus," *Science Museum Group Journal* 1 (Spring 2014), 10.15180; 140104, online journal, http://journal.sciencemuseum.ac.uk/browse/2014/responding-to-stories/ (accessed November 23, 2021).

Fig. 2 Room 2 of the loan collection, South Kensington Museum, London, 1876. From Robert Bud, "Responding to stories: The 1876 Loan Collection of Scientific Apparatus and the Science Museum," *Science Museum Group Journal* 1 (2014): Fig. 3, http://dx.doi.org/10.15180/140 104/003.

instruments) were used within a public science exhibition to 'construct' mathematics as a discipline. One can also see the context—also the national context—in which mathematical artifacts acquired their respective meaning. Many of the German mathematicians were extremely enthusiastic to discover such an extensive exhibition of mathematical models from France, Britain, and Germany—nothing like it had been seen before.

This enthusiasm also provided the incentive for the next large exhibition on mathematical models, apparatuses, and instruments, planned for the annual conference of the Gesellschaft Deutscher Naturforscher und Ärzte (Society of German Natural Scientists and Physicians) in 1892 in Nuremberg. Due to the 1892 cholera epidemic, the exhibition had to be cancelled and was eventually shown independently of the convention of natural scientists as part of the annual conference of the German Mathematical Society in Munich in 1893 (see Fig. 3). The Munich exhibition showed roughly the same group of mathematical artifacts—namely, the calculating machines and slide rules for arithmetic, and the drawing instruments

Fig. 3 Mathematics exhibition at the Technische Hochschule München, 1893, geometry section. Technical University of Munich | TUM Archive, Fotob. "Mathematische Ausstellung 1893," all rights reserved

and geometric models for geometry—but now in a different context than in 1876 in South Kensington. It is worth mentioning here that this led to a different interpretation not only of the artifacts but also of mathematics as a whole.[6] The exhibition consisted of four large sections—analysis, geometry, mechanics, and mathematical physics—each of which was presented in a space of its own and named after an eminent scientist. The exhibition was organized and curated by Walther Dyck,[7] who in 1884 was made a professor of higher mathematics and analytical mechanics at the Technische Hochschule München (Munich Engineering College). With this exhibition, Dyck not only set out to present mathematics as an academic discipline but also to show mathematics' potential for application and its utility for technology. Thus, the exhibition was addressed not only to mathematicians but also to natural scientists and engineers. In addition, Dyck wanted to show how mathematical devices and artifacts could be used to practice mathematics and to visualize objects in teaching and research. It is reasonable to assume that in the

---

[6] The four rooms were named after Gottfried Wilhelm Leibniz, René Descartes, Galileo Galilei, and Isaac Newton, and also showed their busts. See: Dyck, ed., *Katalog*; see, in particular: the introduction of Dyck ("Einleitung"), in ibid., iii–vi.

[7] On this exhibition, see: Hashagen, *Walther von Dyck*.

concept and implementation of this exhibition, the mathematical instruments and models played an entirely different role than in England in the mid-1870s.

KK: This orientation to applied mathematics and, conversely, to the mathematization of engineering teaching—was that not already part of the French tradition earlier on, that is, following the founding of the École polytechnique? And was this mathematization of engineering practice one of the reasons for the strong interest in mathematical models and the exhibitions containing these models?

UH: First of all, it has to be said that mathematical education at the German engineering colleges founded in the course of the nineteenth century ultimately tended to be oriented to the model of the École polytechnique in Paris. However, before 1850, the mathematical level of engineering courses at the polytechnic schools in Germany (the predecessor institutions of the engineering colleges) was comparatively low and the academic level of the Paris model remained unrivaled. It was only in the 1860s, with attempts made by German engineers to turn the polytechnic schools into engineering colleges, and the higher expectations connected with this, that there was a real improvement in the standard of mathematics teaching.[8] Here, however, mathematics had a dual purpose: first, to secure the scholarly standard of the engineering colleges and the formal education of the students; and, second—in analogy with its relation to the exact sciences—to serve as the scientific basis for the technical sciences.[9] However, in what measure and in what manner the tradition developed by Gaspard Monge of using geometric models in the classroom was transferred to or taken up by the German polytechnic schools and later by the engineering colleges has still not been systematically investigated by historians of mathematics. The first larger model collection at a polytechnic school or engineering college in Germany was almost certainly the one assembled in Karlsruhe by the mathematician Christian Wiener, who taught descriptive geometry there. It is probably safe to say that in the 1870s one still looked at mathematical models from France, but that in the 1890s that was no longer the case.

Your second question on the relation between the mathematization of engineering practice and the strong interest in mathematical models and exhibitions of models merits closer examination. I will attempt to give an answer with the example of the Technische Hochschule München and a mathematician who has already been mentioned, Walther von Dyck. Von Dyck received his scientific education at the Technische Hochschule München (the Königlich Bayerische Technische Hochschule München, today's TUM or Technical University of Munich), where he studied mathematics from 1875 to 1879. His teachers at the Technische

---

[8] For the general development of engineering education in the nineteenth century, see also: Jonathan Harwood, "Engineering Education between Science and Practice: Rethinking the Historiography," *History and Technology* 22, no. 1 (2006): 53–79.

[9] See: Susann Hensel, "Die Auseinandersetzungen um die mathematische Ausbildung der Ingenieure an den Technischen Hochschulen Deutschlands Ende des 19. Jahrhunderts," in *Mathematik und Technik im 19. Jahrhundert in Deutschland*, ed. Susann Hensel, Karl-Norbert Ihmig, and Michael Otte (Göttingen: Vandenhoeck & Ruprecht, 1989), 1–111.

Hochschule München were the mathematicians Alexander Brill and Felix Klein, the main founders of the mathematical model tradition in Germany. Brill and Klein encouraged mathematics students as well as engineering students, including Rudolf Diesel,[10] to build these models in their classes. It is worth mentioning here that they both saw a great need to convey an increasingly abstract 'pure' mathematics to the engineering students, who showed little appetite for this. Therefore, Brill and Klein developed the didactic principle of generating an interest in and of conveying complex and increasingly abstract mathematical concepts by means of geometric intuition [*Anschauung*]—not only in geometry but also in analysis.

However, the teaching conditions in mathematics at the Technische Hochschule München in the 1870s under Brill and Klein differed fundamentally from the situation in the 1890s when Dyck was teaching there and planning his first mathematics exhibition. Already during the early 1890s, Dyck felt a growing pressure to show the application potential and the real utility of the higher mathematics and analytical mechanics he was teaching. Whereas the model making continued by Dyck following the example of Brill and Klein was motivated at least in part by the teaching of mathematics students at the Technische Hochschule München, one has to see the 1893 exhibition conceived and curated by Dyck in the context of these general developments.

In the years that followed, a number of lucky coincidences came together that led Dyck to continue to occupy himself intensively with mathematics exhibitions and eventually to become the most important organizer of mathematics exhibitions of the nineteenth and early twentieth century. Thus, in 1903, Dyck's former schoolmate Oskar von Miller founded the Deutsches Museum in Munich, and Dyck as the rector of the Technische Hochschule München played a crucial role in its founding. Together with the engineer Carl von Linde, Oskar von Miller and the recently ennobled Walther von Dyck made up the first board of directors of the Deutsches Museum.[11] In the newly founded museum, mathematics was then presented in another way, in the form of a permanent exhibition. The mathematics section of the Deutsches Museum was shown for the first time in 1905 in a provisional building, and then permanently following the completion of its current home in 1925.

If one compares the photos of the 1893 exhibition with the photos of the 1925 exhibition (Figs. 3, 4 and 5), one can see, on the one hand, that Dyck learned a lot about how exhibitions can be used as a medium to convey mathematical concepts and, on the other, the extent to which the didactic ambition had grown. Of course, one of the tasks of the Deutsches Museum was to make mathematics accessible to the general public. For this reason, the museum attempted to present

---

[10] See: Hashagen, *Walther von Dyck*, 58–61; Stephan Finsterbusch, "Die ersten Diesel-Modelle," *Frankfurter Allgemeine Magazin* 52 (August 12, 2017), 42–43. See Fig. 11 for a series of models (of the real part of the Weierstrass $\wp$-function) constructed by Dyck's assistant Heinrich Burkhardt and the teacher trainee Adolf Wildbrett.

[11] See: Ulf Hashagen, "Ein unbekannter Mitbegründer des Deutschen Museums: Zum 150. Geburtstag des Mathematikers Walther von Dyck," *Kultur und Technik* 30 (2006), 43–45.

Fig. 4 A corner of the mathematics section of the exhibition at the Deutsches Museum, 1925 (Deutsches Museum, Munich, Archive BA BN 53303, all rights reserved). See also: Walther von Dyck, "Mathematik," in *Das Deutsche Museum: Geschichte, Aufgaben, Ziele*, 2nd ed., ed. Verein Deutscher Ingenieure/Conrad Matschoss (Berlin and Munich: VDI Verlag and R. Oldenbourg, 1929), 169–78, here 174

mathematics in different ways—for example, by using examples taken from art (see Fig. 5). This meant, however, that the mathematical models started to play a less prominent role. In photos of the period, one can see how these models were exhibited at a height of about two meters and moved into the corner. This can be compared—certainly a little exaggeratedly—to the way products are displayed in supermarkets: the important items are placed at the front. In our case, what happens is that, alongside the mathematical instruments and calculating machines, it is the didactic aspect (i.e., whatever should be conveyed with the images and with the help of art) that comes to the fore. The mathematical models, on the other hand, tend to be seen further back.

Of course, that is also related to the fact that, toward the end of the nineteenth century, the production of mathematical models became less and less important. Models became less important in research, and in mathematical institutes they increasingly acquired the status of a merely decorative accessory. In Germany around 1870, however, models still had a significant role to play: at German universities and engineering colleges, these mathematical artifacts were used for the institutionalization of mathematics, that is, to found mathematical institutes. In a physics institute with it measuring instruments and diverse apparatuses this was

Fig. 5 The section "The Beginnings of Perspective Drawing" in the mathematics exhibition at the Deutsches Museum, ca. 1925 (Deutsches Museum, Munich, Archive BA BN L 1663/11, all rights reserved). Most of these images have a strong connection with art; they include works by Albrecht Dürer, images showing the development of perspective in painting, and a photo of a relief perspective model of a theater stage by Ludwig Burmester. See also: Walther von Dyck, "Mathematik," in *Das Deutsche Museum: Geschichte, Aufgaben, Ziele,* 2nd ed., ed. Verein Deutscher Ingenieure/Conrad Matschoss (Berlin and Munich: VDI Verlag and R. Oldenbourg, 1929), 169–78, here 170

relatively straightforward. But to find a way to convey mathematics in a tangible manner using concrete objects and instruments was more problematic. For a certain period in the second half of the nineteenth century, this constitutive role was played by mathematical models. This can be seen, for example, in the founding of the mathematics department by Klein at the Universität Leipzig in 1881. Here, Klein first requested the acquisition of a collection of geometric models and only then the founding of his own mathematics institute. Later, at the Universität Göttingen, Klein continued the model tradition in a similar way. In exhibitions and mathematics congresses after 1893, however, including the International Congresses of Mathematicians, which occurred regularly after 1897, mathematical models began to play a less prominent role. In Zürich in 1897, in Paris in 1900, as well as in later International Congresses of Mathematicians, such as in Cambridge in 1912, one sees fewer mathematical models, which were increasingly replaced by mathematics books.[12] Hence, mathematics started to be presented differently, and that reflects a shift within the discipline. In a broader context, the mathematical artifacts and the mathematics exhibitions in Munich in 1893 and in the Deutsches Museum in 1905 can probably be interpreted as part of a general fin de siècle scientific culture.[13]

MF: I would like to come back to von Dyck's remark from 1929 that we quoted at the beginning. Von Dyck claims that mathematics—or, more precisely, "the essence" of mathematics—cannot be exhibited. On the other hand, more than forty years earlier, in 1886, Brill stated: "Often […] the model prompted subsequent investigations."[14] Hence, in Brill's time, mathematical models were still considered as epistemological objects. As you have already pointed out, at the beginning of the twentieth century, if not earlier, models are either exhibited as

---

[12] For an extensive overview on the International Congress of Mathematicians from the end of the nineteenth century to the present, see: Guillermo P. Curbera, *Mathematicians of the World, Unite!* (Wellesley: A K Peters, Ltd., 2009). Curbera notes that, at the International Congress of Mathematicians in Heidelberg in 1904, "the congress organized an 'Exhibition of Literature and Models,' with mathematical books, models, and apparatus that concentrated on recent materials, less than ten years old. The bibliographic exhibition consisted of scientific literature. More than 900 publications were exhibited […] [and] [m]ore then 300 mathematical models were shown […]." (Ibid., 32). At the 1912 congress in Cambridge, an "exhibition of books, models and machines (chiefly calculating machines) [was organized], which was arranged in two rooms […]." (Ibid., 51).

[13] See: Ulf Hashagen, "Mathematics on Display: Mathematical Models in Fin de Siècle Scientific Culture," *Oberwolfach Report* 12 (2015), 2838–41.

[14] "Dem Verfertiger eines Modells stand es frei, eine Abhandlung zu demselben zu schreiben, deren Veröffentlichung […] nicht wenig dazu anreizte, die oft mühsamen Rechnungen und Zeichnungen, welche der praktischen Ausführung zu Grunde lagen, durchzuführen. Öfter veranlaßte umgekehrt das Modell nachträgliche Untersuchungen über Besonderheiten des dargestellten Gebildes." ["The maker of a model was free to write a paper on this, whose publication […] played no small part in encouraging one to carry out the often-arduous calculations and drawings at the basis of the practical execution. Often, conversely, the model prompted subsequent investigations."] Alexander Brill, "Über die Modellsammlung des mathematischen Seminars der Universität Tübingen," *Mathematisch-naturwissenschaftliche Mitteilungen*, vol. 2 (1887), 69–80, here 77.

museal objects or disappear from view. How is the model's change of status (from scientific artifact to museal object) to be reconciled with the transformation of mathematics, or of the image of mathematics?

UH: To answer this question, let us perhaps first stay in the Munich context. Munich was a center of mathematics—here, the discipline was supported by two academic institutions, by the Ludwig-Maximilians-Universität and the Technische Hochschule München, although the latter only had the right to award doctorates after 1901. However, already in these two institutions, mathematical models were used in different ways. The model tradition developed principally at the Technische Hochschule, where it was introduced by Klein and Brill, who designed and constructed many models with their students and doctoral candidates. These models were then sold by Alexander Brill's brother Ludwig Brill through his publishing house—and later also by the Verlag Martin Schilling.[15] The model tradition had a heyday in Munich from 1875 to 1880 because Klein and Brill were working there. When Brill was offered a position at the Universität Tübingen in 1880, he took this tradition with him, and again established a model collection. In Munich the model tradition was principally continued by Dyck, who was made a professor at the Technische Hochschule München in 1884. Dyck's efforts were supported and continued by Sebastian Finsterwalder when the latter became a professor of mathematics at the Technische Hochschule at the beginning of the 1890s. In this way, the model tradition in Munich survived almost into the 1930s. Models continued to be actively built—on the one hand, by Finsterwalder and, on the other, by his student Robert Sauer, who was a privatdozent at the Technische Hochschule München.

Hence, Dyck and Finsterwalder, who remained professors in Munich into the 1930s, practiced a mathematics that did not submit to the modernization of mathematics represented by Bartel Leendert van der Waerden, Emmy Noether, David Hilbert, or Nicolas Bourbaki;[16] rather, they practiced what might be described as a traditional mathematics, a mathematics of the nineteenth century. Their geometric objects were less abstract and formal than, above all, concrete and intuitive. Intuition (*Anschauung*) played an important role both in the research process and in the publications. Whereas Dyck made his mark principally with mathematics (museum) exhibitions and carried out research on mathematical instruments (integraphs), for Finsterwalder, who was concerned with concrete and application-based mathematical problems—and who besides Carl Runge was probably the

---

[15] The first three editions of the Brill Verlag's catalogue of models were published in 1881, 1882, and 1885: Ludwig Brill, ed., *Catalog mathematischer Modelle für den höheren mathematischen Unterricht* (Darmstadt: L. Brill, 1881); Ludwig Brill, ed., *Catalog mathematischer Modelle für den höheren mathematischen Unterricht*, 2nd ed. (Darmstadt: L. Brill, 1882); Ludwig Brill, ed., *Catalog mathematischer Modelle für den höheren mathematischen Unterricht*, 3rd ed. (Darmstadt: L. Brill, 1885). Between the second and the third edition (between 1882 and 1885) hundreds of models were added to the catalogue, as well as elaborate theoretical explanations.

[16] On the rise of modern algebra during this period, see: Leo Corry, *Modern Algebra and the Rise of Mathematical Structures* (Basel: Birkhäuser, 2004).

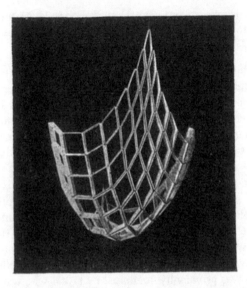

Fig. 6 Sebastian Finsterwalder illustrated transformations of surfaces with models of material nets. His article: Sebastian Finsterwalder, "Mechanische Beziehungen bei der Flächendeformation," *Jahresbericht der Deutschen Mathematiker-Vereinigung* 6 (1897): 43–90, contains numerous illustrations and photos of models. For the illustration above, see: ibid., 66, Fig. 14. In this article Finsterwalder notes: "Das abgebildete Netz ist von Herrn stud. math. Fr. Thiersch für das mathematische Institut der technischen Hochschule München gestellt worden" ["The illustrated net was made by Herr stud. math. Fr. Thiersch for the mathematics institute of the Technische Hochschule München"] (ibid., 67)

most important applied mathematician of his time—the models remained above all a research instrument with which he attempted to visualize and interpret mathematical problems and concepts (see Fig. 6).[17]

At the Ludwig-Maximilians-Universität, however, the situation was different. Here, there was a much smaller model collection compared to the Technische Hochschule München, and far from all the mathematicians showed an interest in models: whereas the mathematicians Gustav Bauer and Ferdinand Lindemann created a collection of models, and Karl Doehlemann had his students make numerous card and string models, which were shown in the cabinets of the seminar rooms, Alfred Pringsheim (an epigone of Weierstrassian analysis) firmly rejected the use of intuition [*Anschauung*] in the teaching of mathematics in colleges.[18] Moreover, these models initiated by the—in scholarly terms—probably least important

---

[17] See, for example: Sebastian Finsterwalder, "Mechanische Beziehungen bei der Flächendeformation," *Jahresbericht der Deutschen Mathematiker-Vereinigung* 6 (1899): 45–90.

[18] Alfred Pringsheim, "Über den Zahl- und Grenzwertbegriff im Unterricht," Jahresbericht der Deutschen Mathematiker-Vereinigung 6 (1899), 73–83; Micheal Toepell, *Mathematiker und Mathematik an der Universität München: 500 Jahre Lehre und Forschung* (Munich: Institut für Geschichte der Naturwissenschaften, 1996), 251–55.

mathematician at the university, largely lacked the research context of the models by Brill, Klein, and Finsterwalder at the Technische Hochschule München. They were also not as elaborately produced as the models at the Technische Hochschule München and nor were they sold as copies in large quantities by the publishers Ludwig Brill and Martin Schilling, but remained for the most part unique pieces produced by prospective schoolteachers.[19]

In other places and at other colleges, the way in which model collections were developed and deployed in the classroom depended likewise both on the institutional conditions and on the mathematicians teaching there—we still know relatively little about this. If one looks at the surviving model collections, then, besides the one found at todays Technical University of Munich, one finds larger collections above all in Dresden,[20] Karlsruhe, Tübingen, and Göttingen. At the engineering college in Karlsruhe it was the mathematician Christian Wiener who built the models himself and whose example was followed by his son Hermann Wiener, who later taught at the engineering college in Darmstadt. In Tübingen it was Alexander Brill who established and developed the collection, having switched to the university there from Munich.[21] The most remarkable collection, however, was probably the one at the Georg-August-Universität Göttingen.[22] This is principally due to the fact that Klein brought his concept of collections of mathematical models to Göttingen from Munich via Leipzig in 1886, and in Göttingen (as already in Munich and Leipzig) he again built up a large model collection. This was presented in glass cabinets in the mathematics institute (see Fig. 7) and included mathematical instruments and calculating machines. In Göttingen, however, it is questionable whether the models were still used for research and teaching purposes as had been the case for Klein at the beginning of his career in the 1870s.[23] It is reasonable to assume that from a certain point, perhaps at the turn of the century, that was only the case to a very limited extent.

MF: In this connection, one should perhaps also mention that Hilbert and Stefan Cohn-Vossen's book *Anschauliche Geometrie*, published in 1932, contains numerous photos of mathematical models.[24] The book was based on a lecture given by

---

[19] Martin Schilling, ed., *Catalog mathematischer Modelle für den höheren mathematischen Unterricht* (Leipzig: Martin Schilling, 1911).

[20] See: https://www.math.tu-dresden.de/modellsammlung/ (accessed August 18, 2021).

[21] See: https://www.unimuseum.uni-tuebingen.de/de/sammlungen/mathematische-modellsammlung. html (accessed August 18, 2021); Ernst Seidl, Frank Loose, and Edgar Bierende, eds., *Mathematik mit Modellen: Alexander von Brill und die Tübinger Modellsammlung* (Tübingen: Museum der Universität Tübingen—MUT, 2018).

[22] See also: Anja Sattelmacher, "Zwischen Ästhetisierung und Historisierung: Die Sammlung geometrischer Modelle des Göttinger mathematischen Instituts," *Mathematische Semesterberichte* 61 (2014): 131–43.

[23] David E. Rowe, "Mathematical Models as Artefacts for Research: Felix Klein and the Case of Kummer Surfaces," *Mathematische Semesterberichte* 60 (2013), 1–24.

[24] See, for example: David Hilbert and Stefan Cohn-Vossen, *Anschauliche Geometrie* (Berlin: Springer-Verlag, 1932), 14, 17, 175, 191, 193, 194, and passim.

Fig. 7 View of the model collection in Göttingen. The cabinets are still positioned as they were at the time of Richard Courant and Otto Neugebauer in 1929. © Collection of Mathematical Models and Instruments, Georg-August-University Göttingen, all rights reserved

Hilbert. For Hilbert, however, the models reproduced in the book were not research objects but objects whose purpose was above all to trigger enthusiasm, and in this way they lose their heuristic, research-oriented status.

UH: I can basically only agree with this assessment of Hilbert's book *Anschauliche Geometrie*. While Hilbert grants an important role to geometric intuition [*Anschauung*] in the research process, the vast number of drawings and photographed mathematical models from the Göttingen collection were principally intended to promote mathematics as an academic discipline.

If I may add an autobiographical observation: Beginning in the mid-1980s I studied mathematics in Göttingen myself. The cabinets with the models were found in the mathematics institute on the second floor, in the central hall. However, I never perceived these models as mathematical objects, at least not before I became a historian of mathematics. They were decoration in the best sense of the word. The most interesting for me was one in which Klein had mapped and traced the curves on a bust of the Apollo Belvedere. He wanted to prove that the classical Greek beauty ideal corresponded to the shape of certain mathematical curves (see Fig. 8). That has stayed with me, but I have almost no recollection of the other objects or models, since at the time I hardly paid attention to them.

Of course, what you said in connection with Hilbert's book about the function of the models as a trigger for enthusiasm could already be applied to the nineteenth

Fig. 8 A plaster bust of the Apollo Belvedere with parabolic curves (Göttingen Collection of Mathematical Models and Instruments, model 211). © Collection of Mathematical Models and Instruments, Georg-August-University Göttingen, all rights reserved. See also: David Hilbert and Stefan Cohn-Vossen, *Anschauliche Geometrie* (Berlin: Springer, 1932), 174. "F. Klein used the parabolic curves for a peculiar investigation. To test his hypothesis that the artistic beauty of a face was based on certain mathematical relations, he had all the parabolic curves marked out on the Apollo Belvidere, a statue renowned for the high degree of classical beauty portrayed in its features. But the curves did not possess a particularly simple form, nor did they follow any general law that could be discerned." (David Hilbert and Stefan Cohn-Vossen, *Geometry and the Imagination* (New York: Chelsea, 1952), 198, footnote 2)

century. The mathematicians teaching at German universities and engineering colleges at that time had to find a way to promote their discipline, and indeed not only to the students; they also had to 'sell' mathematics with respect to the other disciplines. Moreover, in the German Empire, the physics and chemistry professors as well as the professors of engineering increasingly received large new institute buildings with a relatively extensive assistant personnel. The size of the institute as well as the collection of teaching materials and the number of assistants that were employed to look after them were an expression of the importance the college attributed to the respective discipline.

In connection with your question, it seems remarkable that at the International Congresses of Mathematicians it was no longer primarily models that were exhibited, but books, even though the models were more striking and were able to

Fig. 9 The mathematical section of the German university exhibit at the World's Columbian Exposition, 1893. Reprinted by permission from Springer Nature: Karen V. H. Parshall and David E. Rowe, "Embedded in the Culture: Mathematics at the World's Columbian Exposition of 1893," *Mathematical Intelligencer* 15, no. 2 (March 1993): 40–45, here 45. © 2009, Springer Nature, all rights reserved

'enthuse.' That may be due in part to the increasing orientation of mathematics to another branch of scholarship, namely the humanities, as well as the increasing orientation to pure mathematics.

MF: The issue of the books leads me to a question that moves away a little from mathematics as an academic discipline. How did the publishers benefit from having their books shown? And what was their role in relation to the changes in the model tradition? Here, I am not only referring to the publishing houses of Brill and Schilling, which sold mathematical models, but in particular to the publishers whose books were shown at the world congresses.

UH: There is of course an important exhibition that we have not yet talked about, namely the mathematics exhibition that was shown in 1893 as part of the World's Columbian Exposition in Chicago (see Fig. 9).[25] This presentation came about because the Prussian government wanted to present German science at the world exhibition in Chicago. The Prussian Ministry of Education found a partner in Dyck, since at the time he was busy preparing the exhibition that was intended

---

[25] Karen H. Parshall and David E. Rowe, "Embedded in the Culture: Mathematics at the World's Columbian Exposition of 1893," *Mathematical Intelligencer* 15 (1993): 40–45; Hashagen, *Walther von Dyck*, 425–36.

for Nuremberg in 1892, but which was eventually shown in Munich in 1893. The mathematics exhibition in Chicago showed mathematical models just as they could also be seen in the 1893 exhibition in Munich. In addition, however, the Chicago exhibition also presented the production of German mathematics publishers in order to highlight the research achievements of German mathematics.[26] This was done, on the one hand, by displaying the collected works of famous German mathematicians and, on the other, by presenting the famous German mathematics journals, such as August Leopold Crelle's *Journal für die reine und angewandte Mathematik*, and *Mathematische Annalen*, while also showing portraits of eminent German mathematicians—and a bust of Carl Friedrich Gauß. So, in the case of this mathematics exhibition, one worked not only with models but also with important publications and publication series as well as with major names. The self-display at this large public exhibition in Chicago was therefore different from the self-display of mathematics in the 1876 exhibition at the South Kensington Museum, which was distinct in turn from the mathematics exhibitions accompanying the congresses of mathematicians, for which there was also a different public.

Remarkably, in 1925 in the Deutsches Museum, one then finds a presentation that in part has a strong connection to the conception of world exhibitions and also displays the great mathematicians as statues, but does not incorporate the books. As I mentioned earlier, the concept of the great mathematician was also found in Dyck's 1893 exhibition in Munich, where each exhibition space bore the name of a famous scientist, and each of the names stood for a different branch of mathematics. The personification of science through the names and busts of great scientists corresponds in a way to the focus on the great artists in art exhibitions. The books and scientific objects, on the other hand, were shown in order to present the respective discipline.

To come to your actual question: did the exhibitions also bring benefits to the publishers? They certainly brought major benefits to the publishing houses. The congresses attracted a large specialist public that saw what was being published or the publisher with which as mathematicians they could be published themselves. At least until World War I, publishing mathematics books was quite a profitable business. That was true, for example, for the Teubner Verlag, which in the late nineteenth and early twentieth century not only published numerous mathematics textbooks but also the *Encyklopädie der mathematischen Wissenschaften mit*

---

[26] See: Walther Dyck, ed., *Special-Katalog der Mathematischen Ausstellung* (Munich: Wolf, 1893). See also the series of lectures delivered by Klein at the 1893 exposition: Felix Klein, *Lectures on Mathematics* (New York: Macmillan, 1894); and especially his appendix to the lectures, Felix Klein, "The Development of Mathematics at the German Universities," 99–109, esp.: 108–9: "In conclusion a few words should here be said concerning the modern development of university instruction. The principal effort has been to reduce the difficulty of mathematical study by improving the seminary arrangements and equipments. [...] Collections of mathematical models and courses in drawing are calculated to disarm, in part at least, the hostility directed against the excessive abstractness of the university instruction." See also: Parshall and Rowe, "Embedded in the Culture," 40–45.

Einschluß ihrer Anwendungen. In addition, in 1912, the Teubner Verlag endowed a prize for the promotion of mathematics, which certainly also served to attract attention to the house and helped it to extend its network. After the war and with the onset of inflation, the business of the Verlag was thrown into a crisis. However, the Springer Verlag, which had already been founded in the nineteenth century, then followed in the footsteps of Teubner and developed into the next major mathematics publisher in Germany with an international standing.[27]

KK: I would like to come back to the question of the overall concept and guiding image of mathematics that was created in these exhibitions and who was actually responsible for this conception. In the case of the 1873 exhibition in Göttingen and the 1893 exhibition in Munich, it was the mathematicians who assembled and presented the objects and who therefore assumed the role of curators. But when the mathematical models end up in the museum context, they leave the academic context of teaching and research. In France it was even the case that a few models were made at the suggestion of the Musée des Arts et Métiers. How would you describe the influences and responsibilities during the conception of the first mathematics section in the Deutsches Museum?

UH: For the Deutsches Museum, which would become one of the world's largest museums of science and technology, one can say that its conception in the founding phase was shaped above all by the world exhibitions, and that it was also influenced by the engineering college in Munich. The influence of the Technische Hochschule München can be seen in the—in a few major points—similar basic conception of these two Munich institutions. The museum should serve to show the interaction between science and technology, and that corresponds to the function of mediating between research and application that in Dyck's view the engineering colleges should have at that time.

In this connection, it is perhaps interesting to note that, for Dyck's 1893 exhibition, this relation is presented somewhat differently again. Think of the scientists who met at the annual meetings of the Gesellschaft Deutscher Naturforscher und Ärzte (Society of German Natural Scientists and Physicians). What did they find there? Exhibitions. In that case, these were exhibitions of medical and scientific instruments, but one can immediately see why Dyck and others might have thought, in 1892 or 1893, of planning a mathematics exhibition.

In the founding phase of the Deutsches Museum at the end of the century, it was certainly also the case that scientists at that time did not only want to present the results of their research to a general bourgeois or aristocratic public; they also wanted to use the history and the heroic figures of their respective science for the sake of self-display. For the most part, the first exhibition curators and consultants at the Deutsches Museum were professors, who were brought into the museum to conceive exhibitions relating to particular disciplines and to select the artifacts. The

---

[27] Volker R. Remmert and Ute Schneider, *Eine Disziplin und ihre Verleger: Disziplinenkultur und Publikationswesen der Mathematik in Deutschland, 1871–1949* (Bielefeld: transcript, 2010).

museum continued this tradition for decades, as can be seen with the exhibitions on computer science and mathematics curated by the Munich computer science professor Friedrich L. Bauer in the 1980s and 1990s.

MF: This leads, for me, to a number of fundamental questions: How, at that time, did one understand the act of exhibiting? What was the task of exhibitions, in particular in museums? Were exhibitions intended to instruct and educate or to trigger enthusiasm? What role did 'visualization' play in these exhibitions?

UH: The short answer to your question is that to exhibit meant to 'construct' mathematics. If we look again at Dyck's conception from 1893, with this exhibition, he wanted to show the relation between mathematics and its application, and thus not only pure science but also applied mathematics. However, this occurred in a period of the arithmetization of mathematics and thus parallel to a process in which mathematical concepts that were introduced in a purely formal way began to appear whose practical application was initially rather unclear. When, beginning in the mid-1890s in the so-called anti-mathematics movement at the German engineering colleges, the engineers protested vehemently against this arithmetization and a mathematics that had become incomprehensible to them (such as Weierstrassian analysis), the Technische Hochschule München played a special role in these debates. Together with other professor colleagues, Dyck developed the concept of a strongly 'science-oriented' engineering college that was committed to a strong interaction between mathematics, physics, and the theoretical disciplines of engineering on one side and technical applications on the other.[28] Dyck was one of the first to recognize this crisis and he endeavored to counteract it with his exhibition on an applied and material mathematics. In this sense, he was, among other things, exhibiting his own conception of mathematics.

MF: Dyck's catalogue to the 1893 exhibition is called *Katalog mathematischer und mathematisch-physikalischer Modelle, Apparate und Instrumente* (Catalogue of Mathematical and Mathematical-Physical Models, Apparatuses, and Instruments). How do models, instruments, and apparatuses relate to one another in the different exhibition contexts (congress, world exhibition, museum)?

UH: First of all it is quite amazing that one sees these very different artifacts next to one another. Since what, for instance, does a calculating machine have to do with a geometric plaster model? They were made in completely different contexts, and they also operate in completely different contexts.

Included among the mathematical instruments, for example, are simple, familiar artifacts such as a compass, and less familiar devices such as the planimeter, a device used to measure a surface area by tracing a curve around the boundary

---

[28] Ulf Hashagen, "Der Mathematiker Walther von Dyck und die 'wissenschaftliche' Technische Hochschule," in *Oszillationen: Naturwissenschaftler und Ingenieure zwischen Forschung und Markt*, ed. Ivo Schneider, Helmut Trischler, and Ulrich Wengenroth (Munich: Oldenbourg, 2000), 267–96.

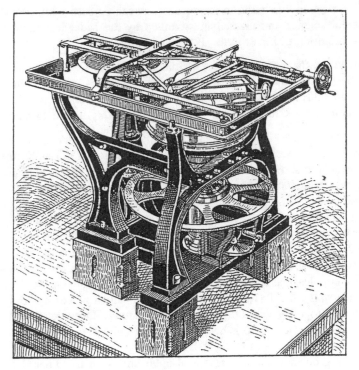

Fig. 10 Arnold Sommerfeld and Emil Wiechert's harmonic analyzer. From Walther Dyck, ed., *Katalog mathematischer und mathematisch-physikalischer Modelle, Apparate und Instrumente* (Munich: C. Wolf und Sohn, 1892), 214

of the shape to be measured. Examples of these are illustrated and thus visualized in Dyck's catalogue. One illustration shows a planimeter that geodesists use to calculate surface areas, another shows the harmonic analyzer that Arnold Sommerfeld constructed with Emil Wiechert (Fig. 10). In the catalogue one also finds entries on and illustrations of stepped drum calculating machines, like those used by the engineer Carl von Linde.[29] These instruments and apparatuses thus refer to mathematical, natural-scientific, and technical practitioners.

The models on the other hand stand for a kind of *Versinnlichung* [sensible rendering], or rather for the materialization of a geometric, or physical, intuition [*Anschauung*], as was the case with the mechanistic models for the visualization of

---

[29] On the planimeter, harmonic analyzer, and arithmometer, see: Dyck, ed., *Katalog*, 100ff., 214ff., and 150; Joachim Fischer, "Instrumente zur Mechanischen Integration: Ein Zwischenbericht," in *Brückenschläge: 25 Jahre Lehrstuhl für Geschichte der exakten Wissenschaften und der Technik an der Technischen Universität Berlin 1969–1994*, ed. Hans-Werner Schütt and Burghard Weiss (Berlin: Engel, 1995), 111–56.

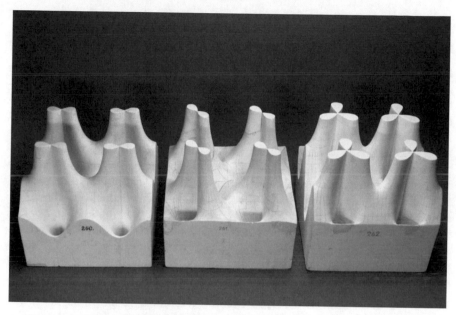

Fig. 11 The real part of the Weierstrass ℘-function (Göttingen Collection of Mathematical Models and Instruments). © Collection of Mathematical Models and Instruments, Georg-August-University Göttingen, all rights reserved. This series of models was designed under the direction of Walther Dyck and constructed by Dyck's assistant Heinrich Burkhardt and the teacher trainee Adolf Wildbrett

electrical processes exhibited in 1893 by Ludwig Boltzmann.[30] The models were part of a teaching practice, whereas the instruments and apparatuses referred to the mathematician or physicist who used them in his or her scientific practice. All of that was encompassed by mathematics—that is how I would understand this exhibition context around 1890.

If one looks at the later literature on instruments in (applied) mathematics, then one sees that the models eventually disappear. Thus, Friedrich Willer's 1926 book is called *Mathematische Instrumente* (Mathematical Instruments).[31] This now deals only with the application context and with building devices that can then be used by natural scientists, engineers, and, possibly, mathematicians. In the *Zeitschrift für Instrumentenkunde* (Journal for the Study of Instruments), the major journal for scientific instruments in Germany published from 1881 onward, these mathematical instruments and calculating machines are placed in another larger context, where they are found next to astronomical, geodetic, meteorological, nautical,

---

[30] Dyck, ed., *Katalog*, 405–8. Here, Boltzmann describes an "Apparat zur mechanischen Versinnlichung des Verhaltens zweier elektrischer Ströme (Bicycle)" ["Apparatus for the mechanical *Versinnlichung* of the behavior of two electrical currents (bicyclic)"] (ibid., 405).
[31] Friedrich Adolf Willers, *Mathematische Instrumente* (Berlin: de Gruyter, 1926).

and other instruments and apparatuses. Although one also finds didactic devices such as demonstration apparatuses, mathematical models are missing here to my knowledge.

KK: Mathematics drew inspiration for model making from physics among other places. But can one say, conversely, that physics also profited from geometric graphic representations and that this in turn elevated the scientific status of representations as a whole? Here, I am thinking, for instance, of James Clerk Maxwell, who utilized two-dimensional representations in magnetism as the basis for further analyses and calculations. In this connection, Maxwell speaks around 1850 of 'mechanical analogies' in physics and attributes a special value to these analogies.[32] Later, Ernst Mach also draws on Maxwell in an article on analogy and similarity as a 'leitmotif of research' and thus as a heuristic method. Can Maxwell's valorization of analogical representation and description be transferred to mathematical models?[33] Or, to put it another way, how did physics and mathematics behave toward each other in the late nineteenth century, when one thinks of the scientific deployment of three-dimensional representations, and thus also of models?

UH: If we look again at the 1893 Munich exhibition, at that time the physicists were completely aware that they were building models for the description of physical processes, and thus for the visualization of wave propagation and for the mechanical *Versinnlichung* [sensible rendering] of electrical processes. In this respect, Boltzmann's model for the mechanical representation of electricity is a very interesting exponent of this history. It seems to me, however, that what was happening at the same time in mathematics was much closer to mathematics itself. The mathematicians visualized or materialized the objects of their geometric or function-theoretical research. Dyck, for example, had models of the Weierstrass ℘-function built (see Fig. 11), which is to say he visualized this function in three-dimensions. Boltzmann, on the other hand, attempted to visualize electrical processes via something mechanistic, and thus via something with which the people of the time were already familiar and therefore better able to imagine. In the

---

[32] James Clerk Maxwell, "On Faraday's Lines of Force," [1855/6] in James Clerk Maxwell, *The Scientific Papers of James Clerk Maxwell*, vol. I, ed. William Davidson Niven (Cambridge: Cambridge University Press, 1890), 155–59.

[33] Ernst Mach, "Die Ähnlichkeit und die Analogie als Leitmotiv der Forschung," in *Annalen der Naturphilosophie*, vol. 1, ed. Wilhelm Ostwald (Leipzig: Veit & Comp., 1902), 5–14, here 5: "Die Analogie ist jedoch ein besonderer Fall der Aehnlichkeit. Nicht ein einziges unmittelbar wahrnehmbares Merkmal des einen Objectes braucht mit einem Merkmal des anderen Objectes übereinzustimmen, und doch können zwischen den Merkmalen des einen Objectes Beziehungen bestehen, welche zwischen den Merkmalen des anderen Objectes in übereinstimmender, identischer Weise wiedergefunden werden. [...] [M]an könnte dieselbe auch eine abstrakte Ähnlichkeit nennen." ["Analogy is, however, a special case of similarity. Not a single immediately perceptible feature of one object need coincide with a feature of another object, and yet relations can exist between the features of one object in exactly the same way as those between features of the other object. [...] [O]ne might also call this an abstract similarity"].

case of Boltzmann's model, a kind of detour is implied, whereas the mathematicians used their mathematical models to directly explain geometric problems or to interrogate these. While similarity also plays a role in the mathematical models, there is an epistemic difference between the way in which Boltzmann deployed his model and the way mathematicians deployed their models. This becomes clearer with the later works of Dyck's colleague Finsterwalder, since Finsterwalder questioned, for example, how certain geometric properties of surfaces could be realized through deformable, mechanical models, and in this way he established that these models could lead to new discoveries in geometric research.[34] Although both Boltzmann and Finsterwalder took the path of a mechanical *Versinnlichung* (of geometric structures or of electrical processes), the value of Boltzmann's model for the process of acquiring scientific knowledge is probably much more limited and only plays a role, if at all, at a certain stage.

KK: What you have just said reflects what Boltzmann writes in 1892 in his contribution to Dyck's catalogue: "At first, of course, all these mechanical models [by Maxwell] existed only in thought; they were dynamic illustrations in the imagination, and they could not be executed with this generality in practice."[35] And in 1902, in his article "Models" in the *Encyclopædia Britannica*, Boltzmann characterizes mathematical models in relation to their tangibility, and thus their concrete three-dimensional execution.[36]

---

[34] Finsterwalder, "Mechanische Beziehungen bei der Flächendeformation," 46: "Man kann sich versuchsweise auf den Standpunkt stellen, die geometrischen Eigenschaften der Flächen nach der Möglichkeit, sie auf mechanischem Wege herzustellen, zu beurteilen, und es ist interessant zu sehen, wie eine Reihe geometrisch wichtiger Eigenschaften auch einer einfachen Realisirung auf mechanischem Wege fähig sind. Andererseits aber wird man durch die Möglichkeit der mechanischen Realisirung auf die Betrachtung von geometrischen Eigenschaften aufmerksam, welche sich sonst der Untersuchung leicht entziehen würden, so dass die hier eingeschlagene Methode auch einigen heuristischen Wert beanspruchen darf." ["By way of experiment, one can attempt to judge the geometric properties of the surfaces according to the possibility of producing them by mechanical means, and it is interesting to see how a number of geometrically important properties are also capable of a simple realization by mechanical means. On the other hand, the possibility of mechanical realization makes one attentive to geometric properties that would otherwise easily escape from investigation, so that the method adopted here may also claim some heuristic value"].

[35] "Alle diese mechanischen Modelle [von Maxwell] bestanden vorerst freilich nur in Gedanken, es waren dynamische Illustrationen in der Phantasie, und sie konnten in dieser Allgemeinheit nicht praktisch ausgeführt werden." Ludwig Boltzmann, "Über die Methoden der theoretischen Physik," in *Katalog mathematischer und mathematisch-physikalischer Modelle, Apparate und Instrumente*, ed. Walther Dyck (Munich: C. Wolf und Sohn, 1892), 89–98, here 97.

[36] Ludwig Boltzmann, "Models," in: *Encyclopædia Britannica*, 10th ed., vol. 30 (Edinburgh, 1902), 788–91, here 788–89: "The term model denotes a tangible representation, whether the size be equal, or greater, or smaller, of an object which is either in actual existence, or has to be constructed in fact or in thought. [...] In pure mathematics, especially geometry, models constructed of *papier-mâché* and plaster are chiefly employed to present to the senses the precise form of geometrical figures, surfaces, and curves."

UH: When we talk about the relation between mathematics and physics, one has to bear in mind that Dyck's 1893 exhibition was the only exhibition in which he cast the net so widely and integrated physics, and thus also brought Boltzmann into the exhibition. That was unusual. The physicists were also included in the 1876 exhibition in the South Kensington Museum, but they were represented by their own section, because all sections could be seen there. In this respect Dyck's exhibition was special, which is partly explained by the fact that Dyck, at that time, was sitting in on Boltzmann's lectures, where he was grappling with mathematical physics, and also of course saw the engineers and their teaching models. To some extent, Dyck created his own universe, which can be seen in his work at the Technische Hochschule München and in the emerging engineering movement of the 1890s. But, in general, physics and mathematics do not seem to me to have had much to do with one another with regard to their models, and nor does this change later on. The mathematical instruments are then mostly used and described by so-called applied mathematicians such as Runge or Friedrich Adolf Willers or later Alwin Walther. One builds these instruments or machines—that is, planimeters, integraphs, or calculating machines—but one no longer sees physical models.

MF: You just mentioned applied mathematics and its special importance for the 1892/1893 exhibition. What was the relation between theoretical mathematics and applied mathematics in the other exhibitions we have spoken about?

UH: If we look at the 1876 exhibition at the South Kensington Museum, this included two branches of mathematics: arithmetic and geometry. Arithmetic was elementary mathematics in the best sense, and the machines exhibited were actually elementary mathematical machines. In addition, geometry was also to be seen there, but the two together did not make up the whole world of mathematics in 1876. Analysis, for example, was completely missing. Nor did one speak explicitly of applied mathematics there, but one simply assigned to the objects—that is, to the geometric models and the arithmetical machines—certain mathematical subdisciplines. In Dyck's 1893 exhibition, on the other hand, it was already a matter very clearly of applied mathematics, and this is also found as an explicit category in the accompanying catalogue.[37] That was due to the fact that, at that

---

[37] Indeed, the closing passage of the introduction of the 1892 catalogue, written by Dyck, emphasizes both pure and applied mathematics: "Möge Interesse und Mitwirkung Aller, die in diesem Jahre unser Vorhaben gefördert haben, uns dabei nicht fehlen, mögen insbesondere die Fachgenossen mit Rat und That den Zweck des ganzen Unternehmens durchführen helfen, ein vollständiges Bild zu geben, all' der mannigfachen Hilfsmittel, wie sie heute in Gestalt von Modellen, Apparaten und Instrumenten dem Unterricht und der Forschung in der reinen und angewandten Mathematik dienen!" ["May the interest and participation of all those who in this year have supported our undertaking not fail, may in particular our colleagues help to accomplish in word and deed the purpose of the whole undertaking, to provide a complete picture of all the manifold aids as they serve today in the form of models, apparatuses, and instruments for teaching and research in pure and applied mathematics!"] (Dyck, ed., *Katalog*, vi). The third section of the catalogue is

date, Dyck was already aware of the tension between the mathematics teaching at the Technische Hochschule and what the engineers could actually utilize from mathematics, but also from analytical mechanics. Thus, Dyck's exhibition anticipated certain aspects of a development that manifested itself later on—for instance, with Runge's appointment to the chair of applied mathematics at the Georg-August-Universität Göttingen in 1904. Through Runge's appointment, applied mathematics was institutionalized at the Universität Göttingen. At the same time, however, Runge should not only be seen as a precursor of modern 'numerical mathematics,' as this has predominantly been the case in the historiography of mathematics; he should also be interpreted as a precursor of 'practical mathematics' (*praktische Mathematik*), since, in Göttingen, besides numerical and graphic methods, he also integrated instrumental mathematical methods into his teaching and research. Beginning around 1910, the expression 'practical mathematics,' which is now no longer commonly used, stood for the triad of graphic, numerical, and instrumental methods for solving mathematical problems and has become known especially through the Institut für Praktische Mathematik at the Technische Hochschule Darmstadt, founded in the 1930s by Alwin Walther.[38]

Between 1893 and 1914, and thus in the period leading up to World War I, the development of applied mathematics was strongly promoted by the fact that the prevailing pure mathematics increasingly split off from the application contexts. In the 'agenda' of mathematics as a discipline, analysis was much more important than the application of analysis. Here, I am deliberately using the concept of agenda developed by the science historian Michael S. Mahoney[39] to emphasize what a discipline focuses on and thus what it pits itself against or the central tasks it has to solve. And these tasks were increasingly defined within mathematics as inner-mathematical problems, which ultimately led to mathematicians no longer expressing such a keen interest in application. This split and the counter movement was also apparent around 1900 in Klein's reorientation of the *Zeitschrift für Mathematik und Physik* (Journal of Mathematics and Physics), which was now called *Zeitschrift für angewandte Mathematik* (Journal of Applied Mathematics),

titled "Angewandte Mathematik" (ibid., 307–420). See also: Dyck, "Einleitender Bericht über die Mathematische Ausstellung in München," (1894); Hashagen, *Walther von Dyck*, 254–64.

[38] For a range of interpretations, cf.: Gottfried Richenhagen, *Carl Runge (1856–1927): Von der reinen Mathematik zur Numerik* (Göttingen: Vandenhoeck & Ruprecht, 1985); Ulf Hashagen, "Rechner für die Wissenschaft: 'Scientific Computing' und Informatik im deutschen Wissenschaftssystem 1870–1970," in *Rechnende Maschinen im Wandel*, ed. Ulf Hashagen and Hans Dieter Hellige (Munich: Deutsches Museum, 2011), 111–52, here 116; Ulf Hashagen, "Computers for Science-Scientific Computing and Computer Science in the German Scientific System 1870–1970," in *The German Research Foundation 1920–1970*, ed.: Mark Walker, Karin Orth, Ulrich Herbert, and Rüdiger vom Bruch (Stuttgart: Steiner, 2013), 135–50.

[39] Michael S. Mahoney, "Computer Science: The Search for a Mathematical Theory," in *Science in the 20th Century*, ed. John Krige and Dominique Pestre (Amsterdam: Harwood Academic Publishers, 1997), 617–34; Michael S. Mahoney, "Software as Science – Science as Software," in *History of Computing: Software Issues*, ed. Ulf Hashagen, Reinhard Keil-Slawik, and Arthur Norberg (Berlin: Springer Verlag, 2002), 25–48.

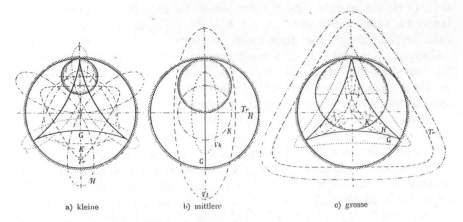

a) kleine   b) mittlere   c) grosse

Fig. 12 Drawing of hypocycloids in Walther Dyck's catalogue. From Walther Dyck, ed., *Katalog mathematischer und mathematisch-physikalischer Modelle, Apparate und Instrumente* (Munich: C. Wolf und Sohn, 1892), 336

and it led in effect also to the establishment of the first professorship of applied mathematics at the Georg-August-Universität Göttingen as well as to the establishment of further professorships for applied mathematics—for example, in Jena. Starting in 1895 mathematics faced the growing problem that its position at the engineering colleges was becoming endangered by the anti-mathematics movement I mentioned earlier. It had to prove that it was still important to engineers. In Munich one was aware of this problem, and the artifacts in Dyck's 1893 exhibition can all be seen in this context.

KK: To finish we would like to speak about visualization techniques, precisely because to exhibit mathematics also means to visualize it. If one looks at the photo of the 1925 exhibition in the Deutsches Museum (Fig. 4), then one sees that mathematical panels and graphic displays are very present. For the catalogue of the earlier exhibition, one sees the same thing: graphic displays and panels occupy a large space. What is the significance of the two-dimensional visualization methods in relation to the three-dimensional models?

UH: One of course has to consider what and how much one can visualize in a book or in an exhibition in 1850, 1890, 1910, or 1930. That is connected to the technical constraints of printing—for example, the possibility of reproducing photos. If you look at Dyck's catalogue from 1892/1893, you will see that it is relatively rich in illustrations, but it does not include any photos. That changes completely with Hilbert's book *Anschauliche Geometrie* almost forty years later. In Dyck's catalogue one finds many drawings, such as the hypocycloids (Fig. 12),[40] and in this

---

[40] A hypocycloid is generated by the passage of a point P on a circle k with radius r when rolling inside a circle K with radius R.

sense the catalogue makes full use of the potential to visualize objects, processes, or instruments. Furthermore, Dyck carries the model tradition forward in his own way by, as already mentioned, having function-theoretical models produced. But when he has the Weierstrass $\wp$-function modeled, this certainly does not take place within a research context. It is a purely didactic model, whose purpose is to show what this function looks like.

On the other hand, drawing on Henri Poincaré's research of the 1880s, he also worked on the qualitative theory of differential equations.[41] There too he attempted to visualize differential equation curves and thus, for example, to show the shape of the solution curves. In a letter to Dyck, Klein wrote that the latter should make use of his "artistic imagination" ["künstlerische Phantasie"] in his mathematical works.[42] Here, one sees a strong tradition of understanding mathematics visually and of concretely visualizing it, whether through artifacts (i.e., via tangible objects) or through drawings. And this occurs at a time in which modern mathematics was beginning to move away from *Anschauung* (intuition). There was therefore also the opposite movement that attempted to extract anything geometric from the theory of functions and to proceed in a purely arithmetical way, as with Karl Weierstraß.

Dyck and his colleague Finsterwalder were in the truest sense of the word counter-modern, if with 'modern' one thinks of Herbert Mehrtens's *Moderne—Sprache—Mathematik*.[43] But one can also consider this from another angle, since one of the greatest innovations of the German science system was that it managed between 1900 and 1925 to create a link between mathematics and engineering. That not only led in 1921 to the founding of the *Zeitschrift für angewandte Mathematik und Mechanik*, but also established a link between engineering, mathematics, and physics. What is classified as counter-modern from the perspective of the modernization of mathematics now becomes modern.

Translated by Benjamin Carter.

Ulf Hashagen is the head of the Research Institute for the History of Science and Technology at the Deutsches Museum and teaches at the Ludwig-Maximilians-Universität München (LMU). Following the study of mathematics in Marburg and Göttingen he obtained his doctorate in the history of science at the Ludwig-Maximilians-Universität München. From 1993 to 2000 he helped as a curator to set up the Heinz Nixdorf MuseumsForum in Paderborn and was jointly responsible for the scientific conception and realization of the permanent exhibition on the history of information technology. Since 2001 he works and carries out research at the Research Institute for the History of Science and Technology at the Deutsches Museum principally on the history of computing and the history of mathematical sciences in the nineteenth and twentieth centuries. A further area of research is the history of scientific and mathematical artifacts.

---

[41] Hashagen, *Walther von Dyck*, 265–97.

[42] Ibid., 280.

[43] Herbert Mehrtens, *Moderne—Sprache—Mathematik: Eine Geschichte des Streits um die Grundlagen der Disziplin und des Subjekts formaler Systeme* (Frankfurt am Main: Suhrkamp, 1990).

Acknowledgements Michael Friedman and Karin Krauthausen acknowledge the support of the Cluster of Excellence "Matters of Activity. Image Space Material" funded by the Deutsche Forschungsgemeinschaft (DFG, German Research Foundation) under Germany's Excellence Strategy—EXC 2025—390648296.

Interview with Andreas Daniel Matt: Real-Time Mathematics

Michael Friedman, Karin Krauthausen, and Andreas Daniel Matt

"Mathematically, the program [SURFER] visualizes real algebraic geometry in real-time. The surfaces shown are given by the zero set of a polynomial equation in the variables x, y, and z. All points in space that solve the equation are displayed and form the surface. [For] example[,] [by entering] $x^2 + y^2 + z^2 - 1 = 0$, [one obtains] a sphere. [...]

The great thing about SURFER is that you don't have to understand the underlying mathematics (algebraic geometry) a priori, you can experiment, try, follow your intuition and creativity and this way learn math[ematics] and create unique artwork like pictures or animations.

SURFER is the new, Java-based version of the program SURFER2008 that was developed for the IMAGINARY exhibition in the year of Mathematics 2008 in Germany. The program is platform-independent and runs on a Windows, Linux or Mac operating system."[1]

MF: We would like to start with the history of the project IMAGINARY and the exhibition concept behind it. As can be seen from the quotation, taken from

---

[1] This description of the software module SURFER can be found on the website of IMAGINARY. See: "program. SURFER," online: https://www.imaginary.org/program/surfer (accessed November 23, 2020).

M. Friedman (✉)
The Cohn Institute for the History and Philosophy of Science and Ideas, The Lester and Sally Entin Faculty of Humanities, Tel Aviv University, Ramat Aviv, 6997801 Tel Aviv, Israel
e-mail: friedmanm@tauex.tau.ac.il

K. Krauthausen
Cluster of Excellence "Matters of Activity. Image Space Material," Humboldt-Universität zu Berlin, Unter den Linden 6, 10099 Berlin, Germany
e-mail: Karin.Krauthausen@hu-berlin.de

A. D. Matt
IMAGINARY gGmbH, Mittenwalder Str. 48, 10961 Berlin, Germany
e-mail: andreas.matt@imaginary.org

© The Author(s) 2022
M. Friedman and K. Krauthausen (eds.), *Model and Mathematics: From the 19th to the 21st Century*, Trends in the History of Science,
https://doi.org/10.1007/978-3-030-97833-4_15

Fig. 1 "The equation $x^2 + z^2 = y^3(1 - y)^3$ of Citric [*Zitrus*] appears as simple as the figure itself. Two cusps mirror-symmetrically arranged rotate around the traversing axis."[2] Graphic by Herwig Hauser. CC BY-NC-SA-3.0 (https://creativecommons.org/licenses/by-nc-sa/3.0/)

the IMAGINARY website, the first exhibition used a computer program called SURFER, a program which, given their equation, enables the visualization of, among other things, surfaces in the real Euclidean space $\mathbb{R}^3$ (see Fig. 1 for an example of such a surface). Could you tell us how IMAGINARY and SURFER were developed? How did you come to the idea of visualizing mathematical objects?

ADM: It all started in 2007 when it was already clear that 2008 was going to be the German Year of Mathematics. The Mathematisches Forschungsinstitut Oberwolfach (MFO, Oberwolfach Research Institute for Mathematics) wanted to contribute to that year. At first one had the idea of establishing a mathematics museum in Oberwolfach. This idea was floating around, and the institute was looking for people to help realize it. At that time I was in touch with the Austrian mathematician Herwig Hauser, an algebraic geometer, who was experimenting with algebraic surfaces. What he did was to develop images of algebraic surfaces that he created using POV-Ray, an open-source ray-tracing program—though it took several hours to render a single image. He prepared these images and added, one may say, an aesthetic or artistic component to them. He began this project even before 2000, I believe, and had exhibited some of his pictures—for example, at the ICM (the International Congress of Mathematicians) in Madrid in 2006. So there was already a gallery of his pictures available.

In 2007, with Hauser, I had large-sized pictures of algebraic surfaces printed. He told me that one of his colleagues, Gert-Martin Greuel, an algebraic geometer who was the director of MFO at that time (between 2002 and 2013), was looking for somebody to carry out a public outreach project in mathematics, similar to a museum. So I had my job interview at MFO, and they hired me right

---

[2] "Zitrus (Citric)," website of IMAGINARY, https://www.imaginary.org/gallery/herwig-hauser-classic (accessed November 23, 2021).

away. Though the idea of a mathematics museum was a bit too physical and too big, monetarily speaking, it was clear that algebraic geometry nevertheless had to be involved. We had Hauser's pictures after all, and Gert-Martin Greuel agreed to present surfaces—it was also his topic of research—and the pictures looked attractive as well. The idea was to use touch screens. I had already experimented with big touch screens a few years before—recall that this is the pre-smartphone era; touch screens were not yet common; you really had to explain to people that they could touch the screen using their fingers. In a math and artificial intelligence exhibition I did with touch screens in 2002,[3] people could indeed interact in and with the exhibition using their fingers. It was then very clear for me that this real-time interactivity and feedback between the visitors and the object is a must.

My goal was to see whether we can just change the equation in real time and immediately present the image of the resulting surface. At that time it was considered impossible, as it took five hours of calculation time for one image. Hence, my first task was to find people who could implement this. At the Technische Universität Kaiserslautern there was a software called Surf, developed by Stephan Endrass and others.[4] It was open-source software, though without an impressive user interface—but it was able to do quite fast ray tracing of images. At the same time Henning Meyer and Christian Stussak joined the team—Christian was then doing his master thesis on real-time ray tracing using graphics cards, which was very new at that time as well.[5] Nowadays everybody uses graphics cards all the time to do faster operations for geometry—and not only for video games; that is, today one constantly uses the parallel processing power of CPUs. In 2007 we had another key person joining: Oliver Labs. He had just finished his PhD and was also working on algebraic geometry. He found a few very nice singular surfaces, on which I'll elaborate later. In addition, he had developed another software called Surfex, which had added a new interface to Surf.

---

[3] See: "Interaktive Roboter-Tage im Einkaufspark Sillpark Innsbruck, vom 13.–15. Juni 2002," http://fabula.tv/roboter/ (no longer accessible).

[4] See: "What is *surf*?," http://surf.sourceforge.net/ (accessed September 6, 2021). The program SURFER is based on Surf. In 1997 Stephan Endrass also found what is now known as the Endrass surface: a nodal surface (i.e., a surface whose singularities are only nodes) of degree eight with 168 real nodes. Note that the maximum number of nodes that a surface of degree eight may have is 174. See: Stephan Endrass, "A projective surface of degree eight with 168 nodes," *Journal of Algebraic Geometry* 6, no. 2 (1997): 325–34.

[5] In 2013 Christian Stussak obtained his PhD with the dissertation "On reliable visualization algorithms for real algebraic curves and surfaces" at Martin-Luther-Universität Halle-Wittenberg. In 2010 Stussak, together with Peter Schenzel, published a paper on their program RealSurf, which also discusses the IMAGINARY project. See: Christian Stussak and Peter Schenzel, "RealSurf— A Tool for the Interactive Visualization of Mathematical Models," in *Arts and Technology: ArtsIT 2009—Lecture Notes of the Institute for Computer Sciences, Social Informatics and Telecommunications Engineering*, vol. 30, ed. Fay Huang, and Reen-Cheng Wang (Berlin, Heidelberg: Springer, 2010), 173–80.

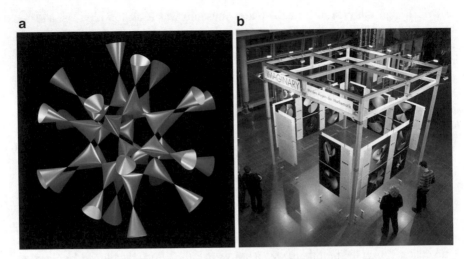

Fig. 2 **a** The Barth sextic: "This surface of degree 6 (sextic) was constructed by Wolf Barth in 1996. Altogether, it has 65 singularities when also counting the 15 [singular points at infinity] [...]. 65 is the maximum possible number of singularities on a sextic as shown in 1997 by Jaffe and Ruberman." From: Oliver Labs, "Barth Sextic," Website of IMAGINARY: https://www.imaginary.org/gallery/oliver-labs (accessed November 23, 2020). See also: David B. Jaffe and Daniel Ruberman, "A sextic surface cannot have 66 nodes," *Journal of Algebraic Geometry* 6, no. 1(1997): 151–68. Graphic by Oliver Labs. CC BY-NC-SA-3.0 (https://creativecommons.org/licenses/by-nc-sa/3.0/). **b** One of IMAGINARY's exhibitions (2007–2008), displaying digital prints of algebraic surfaces. From *Endbericht IMAGINARY: Wanderausstellung des Mathematischen Forschungsinstituts Oberwolfach, Dezember 2007–Dezember 2008*, 2. Photo: Andreas Daniel Matt. CC BY-NC-SA-3.0 (https://creativecommons.org/licenses/by-nc-sa/3.0/)

The key issue was, having all of that research on hand, to develop an accessible user interface on Surf. In the end we didn't change too much. We did the Windows port, which was complicated because Surf was pure Linux software. This was how the first exhibition functioned. The user interface was comprised of just some parameters, some buttons on the touch screen—one could change parameters. We also presented a gallery of surfaces (see Fig. 2b), including some surfaces Labs found, and also surfaces with nodal singularities, surfaces which have the maximum number of singularities of this kind (see for example Fig. 2a for the Barth surface).

From that point onward we continued adding interactivity. At the very first exhibition it was simply that. At the second, however, we added a printer, giving the visitors the opportunity to take with them a print of the surface they created. One could also see the formula on the printout and leave it in a new user gallery in the exhibition. Other visitors could then retype the formulas and play with surfaces done by other visitors. After this we allowed downloading the program, adding some instructions online. We organized several competitions in which people sent us images they created with SURFER. They were online competitions—so we had the very first online exhibition competition, which was called "Kunst aus Formeln" by *ZEIT online*.[6]

The online competition was an open one—all of the participants could see the other entries as well before the competition ended. This created a community of people playing with SURFER, ranging from professional mathematicians to hobby mathematicians, artists, designers, architects, etc. I sometimes compare SURFER to a digital camera—in a sense, we created a digital camera with which you can take pictures of abstract mathematical objects. Changing the lens or the camera settings changes the equation. You could really see how people interacted and how they also came to discover the flaws of the camera by going into extreme settings. Of course, it's still impossible to have a hundred percent correct visualization of surfaces, even for curves.

KK: This leads me to the following question. What were the challenges that you encountered when developing the program? Since, eventually, what is presented on the screen are pixels and not mathematical points.

ADM: Yes, we're talking about surfaces, but these are eventually just pixels on the screen, and one has to deal with infinitely small lines or points. Hence, we constantly improved the software and even completely rewrote the core for a new software version—this was a new ray-tracing program with different algorithms which were faster and more exact. Coming back to Surf and SURFER, to this interactive component, the whole trick was to psychologically make it feel as if it was being calculated and presented in real time. The problem was that it took a lot of time. You really have to solve a high degree polynomial for each point, for each pixel—and this isn't trivial at all. What you need is high processing power, and then you have to calculate it in parallel. Lastly, you need to make it feel as if it's in real time. If one, for example, rotates the surface presented—I mean with a finger—for rotation one needs a certain number of frames per second in order to make it feel as if it's happening in real time, something like 20 or 30 pictures per second on the screen, as in film. In the end I think we used 10 frames a second

---

[6] N.N., "Kunst aus Formeln," *ZEIT online* (May 30, 2008), https://www.zeit.de/online/2008/23/bg-matheskulpturen (accessed November 6, 2021); *Endbericht IMAGINARY: Wanderausstellung des Mathematischen Forschungsinstituts Oberwolfach, Dezember 2007–Dezember 2008*, 3, website of IMAGINARY, http://data.imaginary2008.de/Endbericht-IMAGINARY-2008.pdf (accessed November 6, 2021): See also: Björn Schwentker, "Mathematik Skulpturenprojekt: Anleitungen", *ZEIT online* (March 11, 2008), https://www.zeit.de/online/2008/04/mathe-wettbewerb-anleitungen (accessed, November 6, 2021). "From February to December 2008, mathematics-art competitions using the program SURFER were organized together with media partners (*Die Zeit, Spektrum der Wissenschaften*). In total, more than 8000 pictures together with their formulas were created for the competitions." ["Gemeinsam mit Medienpartnern (Die Zeit, Spektrum der Wissenschaften) wurden von Februar bis Dezember 2008 Mathematik-Kunst-Wettbewerbe mit dem Programm SURFER veranstaltet. Insgesamt wurden über 8000 Bilder zusammen mit ihren Formeln für die Wettbewerbe erstellt." *Endbericht IMAGINARY: Wanderausstellung des Mathematischen Forschungsinstituts Oberwolfach, Dezember 2007–Dezember 2008*, 3, website of IMAGINARY, http://data.imaginary2008.de/Endbericht-IMAGINARY-2008.pdf (accessed November 6, 2021)] (Trans. by M.F./K.K.).

and we lowered the resolution. Even now, if you look at the program, it adapts, there's a process of adaptation. Now, it actually adapts automatically, depending on your CPU. This adaptive rendering is extremely important, so it feels interactive. I can rotate it, I can change the equation, and the results are immediate. It then becomes sharper while you simply look at it, and you don't even notice it—it happens automatically.

If I return to the technological aspect, we had those big touch screens that nobody had. This technology, which was fascinating at that time, was completely new. Nobody could imagine that we had SURFER. Everyone was shocked how quickly we could do everything. It was the algorithms and also the design—it was such a beautifully designed mathematics exhibition.

MF: From your description, it seems as if the very concept and meaning of 'exhibition' has changed. Did this happen in other ways too?

ADM: Indeed, one important example is worth mentioning—it happened at the very first exhibition in Munich. During the exhibition, we received an email from a teacher of a local school in Bavaria. She told me that she loved the exhibition and asked if she could have it at her school. I immediately agreed. I asked the other organizers and contributors whether I could send the teacher the digital files of the images, so she could just go ahead and print them. And indeed she printed them and organized them in a similar structure. All the contributors were perfectly fine with this—I mean everyone had a 'sharing philosophy.' This was the first case of what we have been doing now for more than 10 years in many, many countries. We can call it a kind of open-source remote exhibition. The different museums, institutes, universities simply copy and add their own flavor, presenting it as their exhibition, while they base their exhibition on the content that's already there.

KK: How did the project and the exhibition continue?

ADM: The IMAGINARY project, in the first years, was always kind of jumping from one year to the next. It was planned as a one-year travelling exhibition in Germany, 12 cities,[7] and that was it. But then it turned out that it was kind of longer. There were requests from other institutions. There was a very big exhibition by the Newton Institute in Cambridge, and then in Austria in various places. It started expanding in the European context, and at the same time we had our first remote open-source exhibition in Kyiv, Ukraine. Remote means that we were not even present in the preparation of the exhibition. Then we received some more funding—it was interesting because IMAGINARY always raised funds. We had the initial start-up capital from the Ministry of Education and Research. The exhibition travelled quite a lot. We still have an agreement with the Spanish Royal Mathematics Society, and I think that, since 2010, they've done more than 20

---

[7] Among the cities were Munich, Berlin, Stuttgart, Potsdam, Leipzig, Cologne, and Saarbrücken. There were also many special exhibitions during this year (2008). See: *Endbericht IMAGINARY*, 2.

exhibitions, more than we've done in Germany. This was based on the original materials, but of course they renewed the computers, they bought some new touch screens (the technology changed a lot, clearly), although they still have the same images.

These days we have a few types of exhibitions. There are images, films (and the hands-on aspects as well), which could be data for 3D printing, but also anything you build physically, and there are also the interactive programs, which are software-related exhibitions, where the software plays a major part. Of course, in this case, one would have a touch screen or some hardware attached. There are also the accompanying texts, pedagogical aspects, and the instructions to consider, but this structure generally proved to be very good and is still in place. In 2013 we launched a platform with support by the Klaus Tschira Foundation, which made it easier for us to distribute and deploy the content, but also to collect content as well. That is, people can just upload things themselves; we can curate it online; it's kind of automatic.

MF: I want to go back to the program itself. In which program language was it written? You also mentioned that the algorithms changed during the years. How did that happen? The last question, which is also very much related, concerns the role of the mathematician: does one actually need mathematical knowledge of algebraic geometry for these algorithms?

ADM: The original SURFER was written in C++. The original user interface used the GTK library. The new core was written in Java, which was also a kind of decision taken at that time, because we hoped—something that eventually proved to be wrong—that Java could be used in a browser. So there was this kind of Java application, which worked for a couple of years. We therefore rewrote everything in Java, including the user interface. Of course, software development is a very tricky part, because you have to rely on new technologies, but at the same time the technologies change. Thus, the languages changed, the libraries changed, and they eventually stop at some point. Hence, the new user interface library, which used JavaFX 1, was discontinued. A noncompatible new version, JavaFX 2, appeared, so we had to completely redo the whole user interface in order to be able to continue working with it. At that time—and that was indeed a good decision, which was also a professional one in terms of architecture—we separated the rendering core from the user interface.

One of the latest trends at that time was also doing everything in the browser using graphics cards. There were a lot of technological decisions to be made. For example, there was the question of whether we could really use graphics cards. If we used them, we could speed up SURFER by a thousand times. But there was a problem. The code required to render certain algebraic surfaces was too complex for certain GPUs, and you would never know in advance which GPU/driver combination would fail for which surface. So it was just not practical to deploy this to the general public in 2011 when the WebGL specification had been released.

This leads me to the role the mathematician plays in this process and the algorithm behind it. Our developer is a computer scientist with a PhD in computer

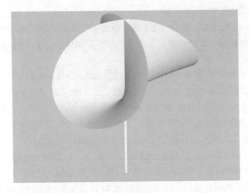

Fig. 3 The Whitney umbrella is a self-intersecting surface (along the z axis), having the equation $x^2 = y^2 z$. The negative part of the z axis is also considered as a part of this surface, being a single line continuing from below the surface. Republished with permission of American Mathematical Society from Eleonore Faber and Herwig Hauser, "Today's Menu: Geometry and Resolution of Singular Algebraic Surfaces," *Bulletin (new series) of the American Mathematical Society* 47 (2010): 373–417, here 382, Fig. 7; permission conveyed through Copyright Clearance Center Inc., all rights reserved

science, but he works with mathematicians. He has a very strong interest in and knowledge of algebraic geometry and mathematics. I think that this is a necessary condition to work on SURFER—you must also understand the mathematics behind it.

There are a lot of known algorithms—for example, you have different algorithms to solve polynomials in higher degrees. It's as if you have good ones, exact ones, or more or less exact, faster ones—but of course we need the fastest. Then you compromise on quality or combine algorithms, and in the end the programmer uses one algorithm. The original Surf had an option where you could switch between algorithms, so there were a lot that were implemented; you could switch and then also see the results—you could then see the errors in the visualization. This is where you need mathematicians to present test cases. Oliver Labs gave us twenty equations that can't be visualized well—then you can test them and subsequently see what's obtained. One of them was the famous Whitney umbrella (see Fig. 3), with a singular line continuing it below.

Explicitly, the problem with the visualization of the Whitney umbrella and this 'singular line' continuing below it is that one isn't able to ray-trace this line, because you can't 'catch' a line. Ray-tracing works as follows: you send rays to different pixels and you have to find the pixel; but, if it's just one point you would like to catch, you won't be able to do that because, even if you're a little bit next to it, it's still not there. This is the difficulty also with singularities, i.e., with singular points. You have to apply other tricks in this case.

Labs gave us these test cases; this showed the limits or limitations of our visualization power. As mathematicians, we want this to be as accurate as possible, but there's always a trade-off between accuracy and speed. If the computer power increases, you have more cores, so you can do more parallel processing. Such programming wouldn't be possible without mathematicians. It would also not be possible without computer scientists who know the mathematics.

KK: As you noted, there are mathematical questions that evolve from trying to visualize certain surfaces. I wonder where and how these visualizations can influence in turn the realm of pure mathematics?

ADM: There is one example which comes to my mind of how IMAGINARY and SURFER can stimulate research in pure mathematics. This concerns what I was talking about in terms of the maximum number of singularities for a given degree of a surface. Let's say that you have an equation of a surface with the highest exponent being six. It has been proven that the maximum number of nodal singularities of such a surface is 65. The Barth surface or the Barth sextic, having 65 nodes, which I mentioned before (see Fig. 2a), is an example of such a surface.[8]

MF: Just to be sure—obviously, the singularities could also be complex;[9] so how do you visualize these complex parts?

ADM: We don't visualize them. We show only the real points. And then it's a question of whether you count the complex singularities or not (i.e., in real algebraic geometry, only the real singular points are counted, as one works only over the real numbers). If one works over the complex numbers, as is usually done, you also count complex points, and in that case you need another tool, such as the software Singular, which you can use to count also such complex singularities.

To return to our discussion on surfaces, if we now consider degree seven surfaces, Oliver Labs found an equation of such a surface with 99 singularities (that is, nodes), but it's been proven that the maximum number of such singular points which a degree seven surface can have is between 99 and 104.[10] So the question arises: can one construct a surface, or find its explicit equation, which has between 100 and 104 nodes? Constructing such equations is highly complex. It's not at all trivial. But we had thousands of users—mathematicians and nonmathematicians— who were playing all of the time with the equations, trying to find such surfaces. My hope was that one would find an equation of a degree seven surface with 100, but so far that still hasn't happened.

MF: You have just mentioned that a community was forming around the attempts to find such a surface, but before that you also mentioned another way of communicating this knowledge visually to people—a three-dimensional printer could also be used to print these surfaces. Can one indeed just print these surfaces so easily?

---

[8] See: Jaffe and Ruberman, "A sextic surface cannot have 66 nodes."

[9] If (x, y, z) is a point on the surface $z = z(x, y)$, then 'complex' here means that at least one of the coordinates is of the form $a + bi$, when $b \neq 0$ and $i = -1$.

[10] In 2006 Labs found a degree seven surface with 99 real nodes. See: Oliver Labs, "A septic with 99 real nodes," *Rendiconti del Seminario Matematico della Università di Padova* 116 (2006), 299– 313.

Fig. 4 A smoothed Togliatti quintic, 30 cm, 2012, by MO-Labs (https://mo-labs. com/en/sculpture/). A Togliatti surface is a nodal surface of degree five with 31 nodes. In 1980 Arnaud Beauville proved that the maximum number of nodes for a surface of this degree is 31. Photo, creation of 3d data, and equation adapted for 3d printing from Wolf Barth's original equation by Oliver Labs. © Oliver Labs, all rights reserved

ADM: For example, one can't print the Barth sextic as it looks in (real) space, since it looks as if it's composed of several pieces (as seen in Fig. 2a). These components look as if they're connected at only one point. If one prints them like that, they would just break. Thus, you really have to find a printer that's so strong that one single point would hold a structure—and, in any case, you would have to smooth it in some way. So you can print, but what one obtains isn't exact. What I mean is that you can't see the singular points. The other problem is the thickness. These surfaces are mathematically 'infinitely thin.' So how can you print that? You have to make them thick, but how do you do that? You always have two layers and you fill the area between. You need to print the volume. Such problems are not trivial. Oliver Labs now has a company called MO-Labs. He does precisely this, he has his own tools, he does a lot of manual adjustments, and he has three-dimensional triangulations of most of the surfaces now, with which he is able to print these surfaces (see Fig. 4).[11]

KK: When one compares the visualized surfaces produced by SURFER with the material models obtained using a 3D printer, I wonder what kind of models we

---

[11] In 2019 Oliver Labs developed another method of three-dimensionally visualizing these surfaces—that is, by lasering them in a glass cube, which enables a more accurate presentation of the singular points. See: website of *Mathe-in-Glas*, https://www.mathe-in-glas.de/ (accessed November 6, 2021).

have with SURFER. The models produced by SURFER are interactive, they can be transformed and changed in real time, they're embedded in a sort of virtual reality as much as they're realistic and as quasi-haptic as possible. Hence, how exactly are these objects considered as models? What is their epistemic value as models?

ADM: IMAGINARY's main motivation is to communicate, to enable a transfer of ideas, of knowledge. The central idea is to always spark something with the person who is confronted with that given model. It's really about producing different types of motivations in the user. And then, having opened that door, these incentives can lead one to develop something, to make a new program, to create something, to play with it. Since this motivation, or what sparks such motivation, is very important, we can't separate it from when we look at the model. Hence, it's a key ingredient of the models we use. That's the reason why I always look at the design, the aesthetic component. Even if it were the best mathematical model, if it doesn't appear pleasing, I wouldn't use it. There are these sorts of criteria. Also, if it's not participatory enough, I couldn't use it to catch people's attention.

So what are the models? Epistemically, the models are definitely surrounded by these communication criteria, which are essential. Maybe the latter are even the key issue. A technology which allows and prompts aesthetical and participatory elements should be used, even if this means that the model would subsequently change. Were a new real-time three-dimensional printer to appear which is perfect, we would use it right away. Three-dimensional printing is a good example, because we don't really use it, since it's simply not up to the standards of real time that we require. Each criterion has a minimum. To wait for five hours for something to be printed is just boring. Being interactive, aesthetic, participatory—these criteria really define the desired aspects of the model.

Maybe one should also look at the model in a more abstract way. What the ideal tool or object is for us, for example, is a very interesting question. What's the ideal object? SURFER as an exhibition was somehow ideal, because it's attractive, it's interactive, and it has a lot of mathematics behind it—it's there right in front of your eyes; the mathematics isn't hidden somewhere. Because of its aesthetics, it also attracts all types of audience. You can three-dimensionally print it, you can show it—there are quite simply a lot of possibilities, which is what makes it ideal. We're still considering this when creating other exhibitions. It's not as though we're only working with algebraic geometry, and not even only with geometry; we also try to have applied mathematical topics in many types of other exhibitions, but so far we still believe that SURFER is somehow unique. The disadvantage of it is that it starts on a higher level. For example, in museums, as a stand-alone exhibition, it doesn't work that well, since people would just see these equations, and if they don't know what the signs by which exponents are shown mean, they won't understand. Thus, there are some small details...

MF: ...which can be understood only by mathematicians.

ADM: Yes. Thus, there's this one problem with SURFER, which makes it unsuitable for complete stand-alone exhibitions. That said, as soon as you have somebody there to introduce it, then the audience will stay for hours. You have other exhibitions where you start right away, but after five minutes it becomes boring. SURFER doesn't get boring. It really does have this infinite number of possibilities to present.

MF: You were talking about the problematics of 3D printing the virtual models: the incapability of the material to be thin enough, or that there are singularities. The problem was how to present them. Are there other problems that SURFER should tackle? This relates to a question we discussed earlier: which object is presented? If we use virtual reality on the computer screen, we can visualize points or singularities in a very precise way. The question is how you actually transfer such things to the three-dimensional world without causing a visual distortion.

ADM: One possible way is to use AR [augmented reality], which is a simpler approach. You have the digital projection in three dimensions in a real space. With that you have the option of seeing real singularities in space. One may try this with HoloLens, which we're using. The object is then floating—you can look at it and walk around it. Of course, I would love to have something like what I previously mentioned, that is, real-time three-dimensional printing.

In terms of challenges, even if you can improve the algorithm, using more GPU [graphics-processing-unit] power, making it super fast, at some point SURFER will reach its limit. It's similar to the situation with the visualization errors of algebraic surfaces: you give me an equation and I can show it to you, I can print it in two- or three-dimensions, I can have different displays, etc., but there will also be problems; at some point a limit is reached.

KK: Making—or, more precisely, creating—is an essential part of the exhibition, and of the entire procedure. The exhibition has something that drives curiosity, which also propels research forward and motivates a certain playful aspect; it's an open process and it's not functionalized in only one direction. With these visualization techniques, is there a threshold crossed or a new exchange established between mathematics and 'play' or even playfulness through which this creative process unfolds?

ADM: The philosophy is that you, as the user, should be able to create something that surprises me, that I would respond to with astonishment. Even if I'm the author, the programmer, the visitor has the possibility of doing something that creates this kind of effect of amazement on me. That should be part of every object. To give a short example: Many people make polyhedra out of paper[12]— you can even buy such things preprinted now; it's merely a question of cutting and

---

[12] The making (or, more precisely, folding) of polyhedra out of paper is a tradition that started in the sixteenth century. See: Michael Friedman, *A History of Folding in Mathematics: Mathematizing the Margins* (Cham: Birkhäuser, 2018), 29–83.

folding. Personally, I find this boring, I look at it and ask myself why I should even bother. A lot of people have fun with it, however. For those people, it's as if they construct it themselves. For me, this isn't the case at all—it's simply reproducing. You have no option to surprise the author. You may make it perfectly, but it still doesn't have this aspect of creativity. What I do is to give people a tool, a software with which polyhedra can be designed. You can then create your own cardboard and print it out, so you have something that's your own creation. You can even change the tool and come up with new ideas.

Another important aspect of the making/creation process is that it occurs within a community, of which I'm then part. Here, one can do things together. This is the part that has to do with making. I think it's extremely important to have the process with the object. The how-to should be there, but it should not be simply a question of me giving you the print and then you just putting it together. The how-to should go back to the source as much as possible, so that every option remains open to create something new.

KK: To continue with this issue of community: this also points toward a possibly new pedagogical approach. Is it being taken up in universities? For example, are mathematics departments using your approach?

ADM: No, generally they're not. I've given some courses at universities, but not in Germany. I taught an entire three semester course in Argentina. It was called "Interactions between science and art for mathematicians," but we also had other scientists participating as well. We had a large IMAGINARY conference where we tried to create new products for mathematics education. One project dealt with creating a university curriculum for mathematics, "Interactive mathematics communication." Our idea is to have an open IMAGINARY approach, an open curriculum that people could build on.

MF: If one examines historically how material models and mathematical education were and are interwoven,[13] then, at the end of the nineteenth century in Europe, and especially in Germany, there was an influential culture of production of material mathematical models, of surfaces, of curves, which were also an integral part of courses given at universities. For example, the students had to produce their own models; they had to develop the skills to produce the material models from wood, string, metal, plaster, etc. in order to get a grade.

---

[13] See, for example: Anja Sattelmacher, *Anschauen, Anfassen, Auffassen: Eine Wissensgeschichte Mathematischer Modelle* (Wiesbaden: Springer Spektrum, 2021); Anja Sattelmacher, "Geordnete Verhältnisse: Mathematische Anschauungsmodelle im frühen 20. Jahrhundert," *Berichte zur Wissenschaftsgeschichte* 36, no. 4, 2013, 294–312; Anja Sattelmacher, "Zwischen Ästhetisierung und Historisierung: Die Sammlung geometrischer Modelle des Göttinger mathematischen Instituts," *Mathematische Semesterberichte* 61, no. 2, 2014, 131–43; Gerd Fischer, ed., *Mathematische Modelle aus den Sammlungen von Universitäten und Museen*, 2 vols. (Braunschweig: Vieweg, 1986).

Given IMAGINARY's and SURFER's interaction and interactivity aspects, which bring the visitors to be actively involved, do you see this as a new mathematical culture? Or can one consider it as a renewed continuation of older mathematical cultures?

ADM: in March 2017 I was at a workshop in Wuppertal called "Mathematik und ihre Öffentlichkeiten,"[14] where I was invited to present IMAGINARY. The talk by Ulf Hashagen, who gave his talk before me, was about the mathematical exhibitions in Europe around 1900. At the end of my own talk there was a comment pointing out that what I was doing with the IMAGINARY exhibition was the same as what had been done previously, that IMAGINARY was nothing different, nothing new. However, I responded to this by saying that IMAGINARY not only has a much more international scope but it's also far more interactive. Nevertheless, one is probably always part of a culture. Perhaps it's only logical that IMAGINARY was created in Germany.

I think that what's really new is where IMAGINARY breaks with tradition. What we can do right now—and it's a trend in many fields—is to attempt to break hierarchies. It's not a question of experts coming to teach and to present on one side and a passive audience on the other side. You may be an expert in mathematical equations, for example, someone else is an expert in visual intuition, someone in creative ideas, algorithms, the list goes on. We're not dealing any longer with this kind of strict segmentation of fields or disciplines. I think IMAGINARY can break the hierarchy. I think that this is what's new; there's a different and much vaster audience involved.

This isn't the only new aspect here; also the aspect of interactive geometry should be stressed. If the visual technology is fast, then you may arrive at new results by just playing around, playing in a good sense, experimenting, because you can test much more. Here again, we're dealing with the question of real time. If I can change parameters in real time, I might discover things that I couldn't discover otherwise, since I can't do such computations in my mind (or with a single material model). I could have intuitions, but I couldn't test them. When I don't have to wait days for the images, for the testing, then this real-time mathematics could have a major impact.

Andreas Daniel Matt is the director of IMAGINARY, a Berlin-based nonprofit organization for the communication of current research in mathematics. He has a PhD in Mathematics in the field of machine learning, and worked from 2007 until 2016 for the Mathematisches Forschungsinstitut Oberwolfach, a Leibniz Institute, where he co-initiated IMAGINARY. In 2013 he received the Media Award of the German Society of Mathematics for his contribution to the communication of mathematics. In 2020 IMAGINARY was awarded the Mariano Gago Ecsite Award for Sustainable Success in Science Engagement.

---

[14] The workshop took place on March 23 and 24, 2017, at the IZWT, Bergische Universität Wuppertal, and was organized by Maria Remenyi.

IMAGINARY is currently involved in a major international training and teaching program in AI and mathematics, and has developed a new digital and physical open-source AI exhibition (https://www.i-am.ai/) and a new exhibition on the connection between mathematics and music (https://www.imaginary.org/exhibition/la-la-lab-the-mathematics-of-music). Other current projects are an interactive picture book which combines storytelling and mathematics (https://www.mathina.eu/) and a mobile mini museum on mathematical modeling for the climate crisis (https://10mm.imaginary.org/). IMAGINARY is also coordinating the website and communication for the UNESCO International Day of Mathematics, a project by the International Mathematical Union (https://www.idm314.org/). And yes, SURFER is still in our minds and hearts: we recently presented a web ray tracer of surfaces, developed by Aaron Montag at TU Munich, see https://love.imaginary.org.[15]

Acknowledgements Michael Friedman and Karin Krauthausen acknowledge the support of the Cluster of Excellence "Matters of Activity. Image Space Material" funded by the Deutsche Forschungsgemeinschaft (DFG, German Research Foundation) under Germany's Excellence Strategy—EXC 2025—390648296.

---

[15] For the mathematical form of a heart, see: Siobhan Roberts, "The Perfect Valentine? A Math Formula," *The New York Times*, https://www.nytimes.com/2019/02/14/science/math-algorithm-valentine.html (accessed November 6, 2021).

Printed in the United States
by Baker & Taylor Publisher Services